Evolution, Development and Ecology of Anemonefishes

Anemonefishes, one of the most popular and recognizable fishes in the world, are much more than film characters; they are also emerging model organisms for studying the biology, ecology, and evolution of coral reef fishes. They are a group of 28 species often employed to study patterns and processes of social organization, intra- and inter-specific competition, sex change, mutualism, dispersal and connectivity of fish populations, habitat selection, pigment pattern formation, lifespan, and predator–prey interactions. This multi-authored book covers all these areas and provides an update on the research done with this model and the perspective it opens for the future.

Key Features

- Contains basic and up-to-date information on an emerging fish model
- Allows non-specialist readers to grasp the relevance of a wide research area
- Provides accurate and easy-to-access information on each of the 28 species
- Includes guidance for establishing a breeding colony
- Documents that anemonefishes are useful model organisms for ecological, developmental, and climate research

Evolution, Development and Ecology of Anemonefishes

Model Organisms for Marine Science

Edited by

Vincent Laudet and Timothy Ravasi

CRC Press
Taylor & Francis Group
Boca Raton London New York

CRC Press is an imprint of the
Taylor & Francis Group, an **informa** business

Cover photo: OIST/Kina Hayashi 林希奈
Licensed under CCBY4.0 (attribution)

First edition published 2023
by CRC Press
6000 Broken Sound Parkway NW, Suite 300, Boca Raton, FL 33487-2742

and by CRC Press
4 Park Square, Milton Park, Abingdon, Oxon, OX14 4RN

CRC Press is an imprint of Taylor & Francis Group, LLC

Library of Congress Cataloging-in-Publication Data

Names: Laudet, Vincent, editor. | Ravasi, Timothy, editor.
Title: Evolution, development and ecology of anemonefishes : model organisms for marine science / edited by Vincent Laudet and Timothy Ravasi.
Description: First edition. | Boca Raton : CRC Press, 2022. | Includes bibliographical references and index.
Identifiers: LCCN 2022020451 (print) | LCCN 2022020452 (ebook) | ISBN 9780367645816 (hardback) | ISBN 9780367645779 (paperback) | ISBN 9781003125365 (ebook)
Subjects: LCSH: Anemonefishes. | Coral reef fishes--Ecology. | Coral reef fishes--Evolution.
Classification: LCC QL638.P77 E96 2022 (print) | LCC QL638.P77 (ebook) | DDC 597/.72--dc23/eng/20220721
LC record available at https://lccn.loc.gov/2022020451
LC ebook record available at https://lccn.loc.gov/2022020452

ISBN: 9780367645816 (hbk)
ISBN: 9780367645779 (pbk)
ISBN: 9781003125365 (ebk)

DOI: 10.1201/9781003125365

Typeset in Times
by Deanta Global Publishing Services, Chennai, India

We would like to dedicate this book to Gerald R. Allen.

No ichthyologist had a larger contribution on the topic of anemonefishes than he. He described no less than nine species and wrote several books on these fish and their sea anemone hosts that still shape our knowledge of anemonefish.

Contents

PART III Reproduction and Social Behavior

PART IV Ecology

PART V Human Impact and Conservation

Preface

Anemonefishes, which were immortalized in the film *Finding Nemo*, are arguably one of the most recognized fishes on Earth. But these iconic fish are much more than film characters; they are also a most important species for studying the ecology and evolution of coral reef fishes. They have been central to innovative research into the scale of larval dispersal and population connectivity in marine fishes and how this influences the efficacy of marine protected areas. They are also used to study the ecological effects of environmental disturbances in marine ecosystems, including climate change and ocean acidification. As a model species, they allow studying patterns and processes of social organization, sex change, mutualism, habitat selection, lifespan, and predator–prey interactions. Perhaps more than any other species, anemonefishes have become a mainstay of research into the chemical, molecular, behavioral, population, conservation, genomics, and climate-change ecology of marine fishes.

Anemonefishes have a mutualistic relationship with sea anemones. Wild adults and juveniles live exclusively in association with a sea anemone, where they gain shelter from predators. This close association of clownfishes and other anemonefishes with sea anemones makes them excellent species for studying aspects of marine mutualisms and habitat selection.

The growing importance of anemonefishes in ecological, developmental, evolutionary, and climate research can be also found in the recent development of several genomic resources for clownfishes such as the genomes and transcriptomes assemblies of the anemonefish *A. percula*, *A. ocellaris*, *A. melanopus*, *A. frenatus*, and *A. clarkii* (see Chapter 2).

More recently anemonefishes are becoming a model for evo/devo research (Roux et al. 2020). Indeed, the genomic and developmental basis of their pigmentation patterns, the mechanisms allowing the social control of growth and sex change, and the way by which the young juveniles detect the sea anemone and establish a symbiotic relationship can also be studied using the last technological developments offered by functional genomics.

Here, we gathered the world's experts in different research fields on these fascinating fishes. Each on their speciality, they have provided an excellent collection of manuscripts highlighting the past and current implications of anemonefishes as model organisms in several aspects of marine science research.

We have organized these 25 chapters into five distinct parts:

1. The first part, "Evolution, Biogeography, and Genetics", presents anemonefish in their context within damselfish (Chapter 1) and states our current knowledge of the phylogeny and phylogeography of this group (Chapter 3), including the recent analyses that suggest the importance of hybridization in anemonefish radiation (Chapter 4). This part also discusses the genomes of anemonefish (Chapter 2) and how these resources facilitate the use of modern tools for functional genomics such as CRISPR/Cas9 genome editing (Chapter 5).
2. The second part, "Life History and Development", focuses on the biphasic life stage of these fishes: that is, their post-embryonic development and metamorphosis (Chapter 6), their unusual longevity (up to 30 years or more) (Chapter 8), and their neuroendocrinology (Chapter 11). This part also highlights the peculiar features that make them relevant biological model organisms such as pigment pattern formation (Chapter 7), visual ecology (Chapter 9), and sound communication (Chapter 10).
3. The third part, "Reproduction and Social Behavior", explores the rich social life of these fish through their ability to change sex (Chapter 12), their reproduction (Chapter 13), the parental care they provide to their eggs (Chapter 15), and, more generally, their elaborated hierarchical organization (Chapter 14).
4. The fourth part, "Ecology", dives into the ecology of these fish, an area that has been under intense scrutiny over the years. It discusses their habitat selection strategies (Chapter 16), how 3D analysis of reefs allows a better understanding of their habitats (Chapter 17), and the symbiosis with the giant sea anemone (Chapter 19). This part also presents updates on cohabitation and competition (Chapter 18), as well as larval dispersal (Chapter 20) and its influence on anemonefish population dynamics.
5. Lastly, the fifth part, "Human Impact and Conservation", presents the effect that humans have on anemonefish populations worldwide by discussing the impact of aquarium trade (Chapter 21), climate change (Chapter 23), and pollution (Chapter 24). Furthermore, this part also presents anemonefish husbandry and how to use them as model organisms to tackle a number of different biological questions (Chapter 22), and lastly, their conservation status and effective management strategies (Chapter 25).

Before letting our readers enjoy the various contributions by all our co-authors, we must explain three choices we have made: each of these points can be discussed and thus we think it is important that we provide some explanations.

1. The first is about the English name of these fishes: should we call them "anemonefish" (which refers to their unique symbiosis with giant sea anemones) or "clownfish" (which refers to their brilliant color and their bold behavior)? It seems to us that the main distinctive characteristic of these fish which explains much of their biology is their symbiosis, and this is why, following most authors in this field, we refer to them as anemonefish. We use the term clownfish for *Amphiprion percula* and *Amphiprion ocellaris*, two sister species forming a natural subgroup within anemonefish.
2. The second choice is about generic names. In the past, it has been considered that anemonefish correspond to two genera: *Amphiprion*, which includes most of the species, and the monospecific genus *Premnas*, with the spinecheek anemonefish *Premnas biaculeatus*. However, recent phylogenetic analysis (Litsios and Salamin 2014) and a thorough systematic analysis of damselfish (Tang et al. 2021) have suggested that *Premnas* is, in fact, related to the clownfishes *A. ocellaris* and *A. percula* and should therefore be called *Amphiprion biaculeatus*. Since this issue is very

TABLE P.1
Anemonefish and Giant Symbiotic Sea Anemone Species

List of the 28 anemonefish species and their common names:
Amphiprion akallopisos, Bleeker, 1853: Skunk anemonefish
Amphiprion akindynos, Allen, 1972: Barrier reef anemonefish
Amphiprion allardi, Klausewitz, 1970: Two-bar anemonefish
Amphiprion barberi, Allen, Drew, and Kaufman, 2008: Barber's anemonefish
Amphiprion biaculeatus, Cuvier, 1816: Spinecheek anemonefish or maroon clownfish
Amphiprion bicinctus, Rüppell, 1830: Two-band anemonefish
Amphiprion chagosensis, Allen, 1972: Chagos anemonefish
Amphiprion chrysogaster, Cuvier, 1830: Mauritian anemonefish
Amphiprion chrysopterus, Cuvier, 1830: Orange-fin anemonefish
Amphiprion clarkii, J. W. Bennett, 1830: Yellow-tail anemonefish
Amphiprion ephippium, Bloch, 1790: Saddle anemonefish
Amphiprion frenatus, Brevoort, 1856: Tomato anemonefish
Amphiprion fuscocaudatus, Allen, 1972: Seychelles anemonefish
Amphiprion latezonatus, Waite, 1900: Wide-band anemonefish
Amphiprion latifasciatus, Allen, 1972: Madagascar anemonefish
Amphiprion mccullochi, Whitley, 1929: White-snout anemonefish
Amphiprion melanopus, Bleeker, 1852: Fire anemonefish
Amphiprion nigripes, Regan, 1908: Maldive anemonefish
Amphiprion ocellaris, Cuvier, 1830: False clownfish
Amphiprion omanensis, Allen and Mee, 1991: Oman anemonefish
Amphiprion pacificus, Allen, Drew, and Fenner, 2010: Pacific anemonefish
Amphiprion percula, Lacepède, 1802: Orange clownfish
Amphiprion perideraion, Bleeker, 1855: Pink skunk anemonefish
Amphiprion polymnus, Linnaeus, 1758: Saddleback anemonefish
Amphiprion rubrocinctus, Richardson, 1842: Red anemonefish
Amphiprion sandaracinos, Allen, 1972: Yellow skunk anemonefish
Amphiprion sebae, Bleeker, 1853: Sebae anemonefish
Amphiprion tricinctus, Schultz and Welander, 1953: Three-band anemonefish

Two species considered hybrids:
Amphiprion leucokranos, Allen, 1973: White-bonnet anemonefish
Amphiprion thiellei, Burgess, 1981: Thielle's anemonefish

List of the ten giant sea anemone species in which anemonefish live and their common names:
Cryptodendrum adhaesivum, Klunzinger, 1877: Pizza anemone
Entacmaea quadricolor, Rüppell and Leuckart, 1828: Bubble-tip anemone
Macrodactyla doreensis, Quoy and Gaimard, 1833: Corkscrew tentacle sea anemone
Heteractis aurora, Quoy and Gaimard, 1833: Beaded sea anemone
Heteractis crispa, Hemprich and Ehrenberg, 1834: Sebae anemone
Heteractis magnifica, Quoy and Gaimard, 1833: Magnificent sea anemone
Heteractis malu, Haddon and Shackleton, 1893: Malu anemone
Stichodactyla gigantea, Forsskål, 1775: Giant carpet anemone
Stichodactyla haddoni, Saville-Kent, 1893: Haddon's carpet sea anemone
Stichodactyla mertensii, Brandt, 1835: Mertens' carpet sea anemone

FIGURE P.1 The cover page (left) of volume II of the scientific report of the Commodore Perry expedition which contains, on page 356, the description of *Amphiprion frenatus*, illustrated (right) in a plate by Patterson.

recent, we let each author use the name of their choice, though we favour this new vision and think that all anemonefish species should be included in the *Amphiprion* genus.

3. There is a discussion about the number of anemonefish species. Numerous studies, including the chapters of this book, have shown that there are between 28 and 30 described species of anemonefish (see Table P.1). This discrepancy comes from two species, *A. leucokranos* and *A. thiellei*, which are suspected to be hybrids. There is strong evidence for *A. leucokranos*, which is a hybrid between *A. chrysopterus* and *A. sandaracinos*, as recently demonstrated by Geoff Jones (Gainsford et al. 2015). The data are less clear for *A. thielli*, a species for which there is no definitive genomic proof that its hybrid condition exists. However, here we advocate the use of 28 species for consistency and clarity. As progress is made in this field, particularly in genomic studies of species with vast distribution areas such as *A. larkia* or *A. chrysopterus*, new species may be described in the years to come.

This book would have not been possible without the support of the Okinawa Institute of Science and Technology (OIST) which has recently invested substantial resources in marine sciences and in particular in coral reef and anemonefish research. We hope that this book will put Okinawa back on the map in anemonefish research. While preparing this book, we discovered that one of the iconic species of anemonefish, the tomato anemonefish *Amphiprion frenatus*, was described by James Carson Breevort in 1856 from specimens captured during the Commodore Perry expeditions in Japanese waters in 1852–1854. In Breevort's report entitled "Notes on Some Figures of Japanese Fish", it is indicated that the specimens, described from drawings made by Patterson, were captured in "Lew Chew", the name of the Ryukyu Islands at that time (though the precise place where the fish were collected is unknown). Thus, Okinawa is, indeed, a historical place for anemonefish research! (See Figure P.1 and Table P.1.)

Vincent Laudet and Timothy Ravasi

REFERENCES

Gainsford, A., L. Van Herwerden, and G. P. Jones. 2015. Hierarchical behaviour, habitat use and species size differences shape evolutionary outcomes of hybridization in a coral reef fish. *Journal of Evolutionary Biology* 28: 205–222.

Litsios, G., and N. Salamin. 2014. Hybridisation and diversification in the adaptive radiation of clownfishes. *BMC Evolutionary Biology* 14: 245.

Roux, N., P. Salis, S. H. Lee, L. Besseu, and V. Laudet. 2020. Anemonefish, a model for Eco-Evo-Devo. *EvoDevo* 11: 20.

Tang, K. L., M. L. J. Stiassny, R. L. Mayden and R. DeSalle. 2021. Systematics of damselfish. *Ichthyology and Herpetology* 109: 258–318.

Acknowledgements

We would like to thank our colleague Geoff Jones for his help and advice during the early phase of the preparation of this book. We are also greatly thankful for the expert editorial work and dedication of Yoko Shintani who has been instrumental in assembling the final version of the manuscript. We gratefully acknowledge Marcela Herrera Sarrias and Roger Huerlimann who read all the chapters for a thorough consistency check. Yoko Fujitomi, Natacha Roux, and Manon Mercader have also been helpful for the preparation of the book. We also thank all the independent reviewers for their time and dedication as they greatly improved this book. Finally, we also deeply thank our colleagues, the authors of the 25 chapters that have given a vivid and exhaustive account of anemonefish research. We hope that this book will be helpful for this community and beyond.

Editors

Vincent Laudet has studied the role of nuclear hormone receptors for 20 years, in particular the thyroid hormone receptor in evolution and development using several model organisms such as amphioxus or zebrafish. Between 2015 and 2020 he was director of the marine station of Banyuls-sur-Mer in France. There, his group focused on the evolution of life history strategies, in particular the recruitment of coral reef fish larvae to the reef and the role played by thyroid hormones in triggering and coordinating this process. Since March 2020 he has been Professor at OIST (Okinawa Institute of Science and Technology Graduate University) and Research Fellow at Academia Sinica (Taiwan) where his group develops the clownfish *Amphiprion ocellaris* as a model organism for Eco/Evo/Devo studies. He focuses in particular on the function and the plasticity of the brilliant pigment patterns of these iconic fishes. His team actively collaborate with Tim Ravasi on the genomics and ecology of anemonefishes from the Ryukyu archipelago. Vincent Laudet's laboratory has contributed to more than 250 scientific papers and 40 reviews (h-index: 78) including two books in molecular endocrinology, molecular evolution and developmental biology.

Timothy Ravasi showed for the first time that climate change stressors such as ocean warming and acidification are able to induce genomics and epigenomics changes in tropical fish; specifically his team was able to: (i) demonstrate for the first time that tropical fish are able to restore their energy metabolism if parents are reared at high water temperature (Transgenerational Acclimation); (ii) identify those molecular pathways that underline this Transgenerational Acclimation; (iii) provide the first evidence that selective DNA methylation of specific loci is one of the epigenetics mechanisms used by fish to transfer the information of a new environment to the next generation (iv) unveil for the first time, the molecular mechanisms that underline sex change in clownfishes. Furthermore, his team sequenced the genomes of the orange clownfish *Amphiprion percula*, the false clownfish *Amphiprion ocellaris*, Clark's anemonefish *Amphiprion clarkii*, and the cinnamon clownfish *Amphiprion melanopus*, which today are among the most complete fish genomes that have ever been sequenced. Between 2009 and 2019 he was a Tenured Professor at the King Abdullah University of Science and Technology (KAUST) in Saudi Arabia. There, his group focused on developing genomics approaches and protocols to study coral reef fish, their ecology and their responses to climate stressors. Since August 2019 he has been a Tenured Professor at Okinawa Institute of Science and Technology Graduate University (OIST) and Adjunct Professor at the Australian Research Council Center of Excellence for Coral Reef Studies, James Cook University, where his group develops the clownfish as a model organism for climate change studies. Timothy Ravasi's laboratory has contributed to more than 140 scientific papers and 4 book chapters that have been cited more than 25,000 times with an h-index of 58.

Contributors

Catherine A. Abbott
Flinders University
Adelaide, Australia

Tina A. Barbasch
Boston University
Boston, Massachusetts USA

Ricardo Beldade
Pontificia Universidad Católica de Chile
Santiago, Chile

Giacomo Bernardi
University of California Santa Cruz
Santa Cruz, California USA

Michael L. Berumen
King Abdullah University of Science and Technology
Thuwal, Saudi Arabia

Laurence Besseau
SU CNRS UMR Biologie Intégrative des Organismes Marins-BIOM
Banyuls-sur-Mer, France

Marc Besson
University of Bristol
Bristol, UK

Rebecca Branconi
University of Exeter
Exeter, UK

Karen Burke da Silva
Flinders University
Adelaide, Australia

Peter M. Buston
Department of Biology
Boston University
Boston, Massachusetts USA

Laura Casas
Institute of Marine Research (IIM-CSIC)
Vigo, Spain

Alessandro Cellerino
Scuola Normale Superiore
Pisa, Italy

Fabio Cortesi
The University of Queensland
Brisbane, Australia

Carmen R. B. da Silva
Monash University
Melbourne, Australia

Fanny de Busserolles
The University of Queensland
Brisbane, Australia

Ross DeAngelis
The University of Texas at Austin
Austin, Texas USA

Jennifer M. Donelson
James Cook University
Townsville, Australia

Mélanie Dussenne
SU CNRS UMR Biologie Intégrative des Organismes Marins-BIOM
Banyuls-sur-Mer, France

Emily K. Fobert
Flinders University
Adelaide, Australia

Bruno Frédérich
University of Liège
Liège, Belgium

Gemma F. Galbraith
James Cook University
Townsville, Australia

Benjamin Geffroy
MARBEC Univ Montpellier
Montpellier, France

Alexander Goikoetxea
MARBEC Univ Montpellier
Montpellier, France

Hugo B. Harrison
James Cook University
Townsville, Australia

Akihisa Hattori
Shiga University
Hikone, Japan

Kina Hayashi
University of the Ryukyus
Nishihara, Japan

Song He
King Abdullah University of Science and Technology
Thuwal, Saudi Arabia

Cassie M. Hoepner
Flinders University
Adelaide, Australia

Geoffrey P. Jones
James Cook University
Townsville, Australia

Masato Kinoshita
Kyoto University
Kyoto, Japan

Marleen Klann
Okinawa Institute of Science and Technology
Onna, Japan

David Lecchini
PSL Research University
French Polynesia

Anna Marcionetti
University of Lausanne
Écublens, Switzerland

Manon Mercader
Okinawa Institute of Science and Technology
Onna, Japan

Marc Metian
International Atomic Energy Agency (IAEA)
Monaco

Suzanne C. Mills
CRIOBE, EPHE, Université PSL, UPVD, CNRS, UAR CRIOBE
Moorea, French Polynesia

Laurie John Lee Mitchell
The University of Queensland
Brisbane, Australia

Alison Monroe
Texas A&M University Corpus Christi
Corpus Christi, Texas USA

Philip L. Munday
James Cook University
Townsville, Australia

Mirko Mutalipassi
Stazione Zoologica Anton Dohrn
Naples, Italy

Coltan Gable Parker
University of Illinois
Urbana-Champaign, Illinois USA

Eric Parmentier
University of Liège
Liège, Belgium

Serge Planes
Laboratoire d'Excellence CORAIL
France

Simon Pouil
Oak Ridge National Laboratory
Oak Ridge, Tennessee USA

Benoit Pujol
PSL Université Paris
Paris, France

James Davis Reimer
University of the Ryukyus
Nishihara, Japan

Justin S. Rhodes
University of Illinois
Urbana-Champaign, Illinois USA

Pascal Romans
Sorbonne Université
Paris, France

Natacha Roux
Sorbonne Université
Paris, France

Theresa Rueger
Newcastle University
Newcastle, UK

Taewoo Ryu
Okinawa Institute of Science and Technology Graduate University (OIST)
Onna, Japan

Nicolas Salamin
University of Lausanne
Écublens, Switzerland

Pauline Salis
Sorbonne Université
Paris, France

Sarah Schmid
University of Lausanne
Écublens, Switzerland

Celia Schunter
The University of Hong Kong
Hong Kong

Maya Srinivasan
James Cook University
Townsville, Australia

Eva Terzibasi Tozzini
Stazione Zoologica Anton Dohrn
Naples, Italy

Valerio Tettamanti
The University of Queensland
Brisbane, Australia

Simon R. Thorrold
Woods Hole Oceanographic Institution
Falmouth, Massachusetts USA

Sakuto Yamanaka
Kyoto University
Kyoto, Japan

Part I

Evolution, Biogeography, and Genetics

1 A Phylogenetic Context
The Diversification of Damselfishes (Pomacentridae)

Bruno Frédérich

CONTENTS

1.1 INTRODUCTION

Damselfishes (Teleostei: Pomacentridae) represent a species-rich group of primarily marine fishes (rare in brackish water). They are distributed worldwide, inhabiting coastal habitats of tropical and temperate seas, and they represent a major component of reef communities (rocky and coral reefs) (Allen 1991; Frédérich and Parmentier 2016). In August 2021, Eschmeyer's catalog of fishes referred to 424 valid damselfish species distributed in four subfamilies and 29 genera (Table 1.1) (Eschmeyer et al. 2021; Tang et al. 2021). New species are still regularly described, as demonstrated by the description of approximately three species per year during the last ten years (Eschmeyer et al. 2021). The diagnosis of damselfish is provided in Parmentier and Frédérich (2016). In this chapter devoted to the diversity of damselfishes (Figure 1.1), I will first highlight two ecological specializations only present in Pomacentridae. Then, I will aim (a) to discuss the phylogenetic position of damselfishes within the ray-finned fishes (Actinopterygii), (b) to summarize the recent advances in the systematics of damselfishes, and (c) to describe the pattern of ecological diversification observed in damselfishes.

1.2 FARMERS AND CLOWNFISHES: UNIQUENESS OF DAMSELFISHES AMONG RAY-FINNED FISHES

The Pomacentridae are a highly diverse group of reef fishes, especially regarding their diversity of ecology, morphology, and behavior (Allen 1991; Frédérich and Parmentier 2016). The Pomacentridae include solitary and gregarious species (Fishelson 1998). The majority of solitary species are highly territorial, defending a small area against intruders such as fishes and mobile invertebrates. Some of these territorial damselfishes are considered algal farmers, cultivating and protecting distinct crop assemblages (Hata and Kato 2004; Ceccarelli et al. 2005). As reviewed by Hata and Ceccarelli (2016), damselfishes manage their farms in at least three ways. Firstly, territorial defence alters herbivory that occurs inside territories, leading to an algal assemblage which differs from that found outside territory boundaries. Secondly, damselfishes prepare substratum for their farm by, for example, killing corals, and they also invest in farm maintenance by "cleaning up" debris from their territory. Finally, fish selectively remove unpalatable algae to promote the growth of their preferred algae (i.e., weeding behavior). The types of algal farms vary among territorial damselfishes: there is a continuum between some species producing intensive monoculture of palatable algal species on a small territory (e.g., *Stegastes nigricans*) and others managing extensive mixed-culture of algae on a larger territory (e.g., *Stegastes obreptus*). Cultivation of algae was also reported for limpets (Branch 1981) and some herbivorous cichlids (Hata et al. 2014) but the farming behavior of damselfishes appear to be the most advanced in aquatic environments, even leading to obligate plant-herbivore cultivation mutualism (Hata and Kato 2006). Farming behavior evolved multiple times during the evolutionary history of damselfishes (Frédérich et al. 2013; McCord et al. 2021) and the research on damselfish adaptations related to farming is currently ongoing (Olivier et al. 2014, 2016a, 2021).

The gregarious damselfishes show diverse social structures, including species with either monogamous or polygamous adults (Fishelson 1998). The best examples of monogamous pomacentrid species are the brightly colored clownfishes (*Amphiprion* spp.), which form permanent reproductive pairs with a high level of fidelity. Beyond this, the clownfishes, a monophyletic lineage within Pomacentridae (Litsios et al. 2012b), are well known for

DOI: 10.1201/9781003125365-2

TABLE 1.1
List of Subfamilies, Tribes, and Genera Following the New Taxonomic Classification of Pomacentridae

Subfamily	Tribes	Genus	N
Microspathodontinae			70
		Hypsypos	
		Lepidozygus	
		Mecaenichthys	
		Microspathodon	
		Nexilosus	
		Parma	
		Plectroglyphidodon	
		Similiparma	
		Stegastes	
Glyphisodontinae			21
		Abudfeduf	
Chrominae			122
		Azurina	
		Chromis	
		Dascyllus	
		Pycnochromis	
Pomacentrinae			211
	Amphiprionini	*Amphiprion*	
	Cheiloprionini	*Cheiloprion*	
		Chrysiptera sensu stricto	
		"Chrysiptera"	
		Dischistodus	
		Pomachromis	
	Hemiglyphidodontini	*Acanthochromis*	
		Altrichthys	
		Amblyglyphidodon	
		Hemiglyphidodon	
		Neoglyphidodon	
	Pomacentrini	*Amblypomacentrus*	
		Neopomacentrus	
		Pomacentrus	
		Pristotis	
		Teixeirichthys	

Source: Tang et al. (2021).

their symbiosis with tropical sea anemones that was first reported in 1868 (Collingwood 1868). This intimate relationship, unique in the animal kingdom, has become a textbook example of mutualistic interactions (Fautin 1986, 1991) and a great deal of attention has been given to this symbiosis. As witnessed by the present book, research about the biology of clownfishes is ongoing in many areas.

1.3 PHYLOGENETIC POSITION OF DAMSELFISHES WITHIN RAY-FINNED FISHES (ACTINOPTERYGII)

With more than 30,000 species, the ray-finned fishes (actinopterygians) represent one of the most successful radiations in the history of vertebrates. The great majority of actinopterygians (99.8%) are teleost fishes (Nelson et al. 2016) and among this species-rich clade, Müller (1843) distinguished a taxon called Pharyngognathi acanthopterygii with the following traits: (1) the lower pharyngeal bones are coalesced forming jaws; (2) part of the rays of the dorsal, anal, and ventral fins are not articulated forming spines; and (3) the swim bladder is deprived of a pneumatic duct. As synthesized by Parmentier and Frédérich (2016), the Pomacentridae (Labroidei ctenoidei *sensu* Müller) was one of the first families constituting this order made by Müller (1843, 1844) with cichlids (Cichlidae) and labrids (Labridae).

The possession of pharyngeal jaw apparatus, or "pharyngognathy", and associated morphological characteristics were studied and used as systematic characters (Müller 1843; Stiassny 1981; Stiassny and Jensen 1987). As currently recognized, pharyngognathy involves three prominent modifications to the typical pharyngeal jaw apparatus of percomorphs: (1) left and right lower jaw elements (fifth ceratobranchials) that are united to make a single lower pharyngeal jaw; (2) a muscular sling that directly connects the underside of the neurocranium with the lower pharyngeal jaw; and (3) a mobile diarthrotic articulation of the upper pharyngeal jaws with the neurocranium (Stiassny 1981; Stiassny and Jensen 1987; Wainwright et al. 2012). These morphological and functional features were proposed as synapomorphies uniting the Labroidei clade that included Cichlidae, Embiotocidae, Labridae, Odacidae, Scaridae, and Pomacentridae (Liem and Greenwood 1981; Kaufman and Liem 1982). In parallel to an interest in systematics, this fascinating system of pharyngeal jaws was deeply studied from a functional point of view (Liem 1973; Liem and Osse 1975). Liem argued that the acquisition of an extra set of jaws in cichlids and other Labroidei has enabled a diversification of food preparation techniques and therefore feeding habits (Liem 1973; Liem and Osse 1975; Liem and Greenwood 1981). First, the flexibility of this highly integrated pharyngeal jaw apparatus would be a major factor that has enabled this diversity of feeding habits. Then, such a duplication of jaws (oral and pharyngeal jaws) was viewed as an evolutionary novelty leading to the functional decoupling between prey capture and prey processing (Liem 1973; Hulsey 2006), ultimately allowing the independent morphological diversification of both jaws systems. The functional morphology of pharyngeal jaws apparatus varies among Labroidei lineages (Stiassny and Jensen 1987; Wainwright et al. 2012). In damselfishes, the characteristics of the lower pharyngeal jaw (LPJ) are a Y-shape (and width is greater than the length), no trace of a central sutural union, a well-developed median keel on the ventral face of the bone, and tooth rows arranged radially across the LPJ with teeth located over the median region of the jaw (Kaufman and Liem 1982; Stiassny and Jensen 1987). Although it is not found in all the pomacentrids (e.g., *Microspathodon* and *Chromis*), pharyngo-cleithral articulations can join the expanded lateral horns of the LPJ to the cleithrum (Liem 1973; Liem and Greenwood 1981; Stiassny and Jensen 1987). It is expected that the support

FIGURE 1.1 Sample of eight species illustrating damselfish diversity. Photos were gratefully provided by Florent Charpin (*Stegastes*, *Abudefduf*, *Chromis*, and *Dascyllus* – reefguide.org) and Mark Erdmann (*Amphiprion*, *Amblyglyphidodon*, *Pomacentrus*, and *Chrysiptera*).

provided by the articulation of the LPJ with the shoulder girdle increases the total biting force that can be exerted on prey (Galis and Snelderwaard 1997).

Our knowledge of the phylogeny of ray-finned fishes has significantly advanced in recent years thanks to the phylogenetic analyses including many genes, morphological characters and fossil data (e.g., Near et al. 2012; Broughton et al. 2013; Faircloth et al. 2013; Chen et al. 2014; Davesne et al. 2016; Betancur et al. 2017; Alfaro et al. 2018; Hughes et al. 2018). Among other advances in our understanding of the evolution of ray-finned fishes, molecular phylogenetic studies revealed polyphyly of the traditional clade of Labroidei and thus showed that pharyngognathy evolved multiple times during the radiation of actinopterygians (Mabuchi et al. 2007; Wainwright et al. 2012). Labridae (including odacines and scarines which are nested within Labridae (Bellwood 1994; Clements et al. 2004; Westneat and Alfaro 2005) are separated from the remainder of the traditional labroid lineages (Cichlidae, Embiotocidae, and Pomacentridae). These three families are now included in a clade of 40 families and more than 4,800 species which were named Ovalentaria (taxonomic level: series) for their characteristic demersal, adhesive eggs with chorionic filaments (Wainwright et al. 2012; Betancur et al. 2017). In addition to cichlids, damselfishes, and surfperches, Ovalentaria includes familiar lineages of fishes such as blennies, silversides, dottybacks, and mullets. If the monophyly of Ovalentaria is strongly supported, interrelationships among the major lineages of Ovalentaria are still not well-resolved (Wainwright et al. 2012; Eytan et al. 2015). To date, there is no consensus about which lineage is the most closely related to Pomacentridae.

One major anatomical characteristic of damselfish is the cerato-mandibular ligament (c-md) that joins the ceratohyal of the hyoid bar to the lower jaw, at the level of the coronoid process (Stiassny 1981; Olivier et al. 2016a) (Figure 1.2). Although secondarily lost in some species (Frédérich et al. 2014), this ligament appears to be a synapomorphic trait within Pomacentridae (Stiassny 1981). Recent works have highlighted that the c-md is involved in at least two major functions of damselfish behavior: sound production (Parmentier et al. 2007 Colleye et al. 2012) and feeding (Olivier et al. 2015, 2016b). Both behaviors are based on the same principle: the c-md allows rapidly closing the lower jaws in a few milliseconds, without the help of the *adductor mandibulae* muscles. The slam of the oral jaws causes teeth collision creating a vibrational wave and the resulting sounds (Colleye et al. 2012). Olivier et al. (2021) recently demonstrated that the possession of two mouth-closing systems enabled grazing damselfishes to have a forceful and extremely fast bite, challenging thus the functional trade-off between force and velocity. Currently, it is hypothesized that the cmd would have operated as a fundamental key to the process of diversification in damselfishes (Frédérich et al. 2014; Olivier et al. 2016a, 2021). According to the expected importance of the cmd, Olivier et al. (2016a) checked the presence of such a ligament in eight groups of Ovalentaria. Only Pomacentridae and Pseudochrominae share the presence of a cmd ligament but its insertion differs between the two groups, suggesting a difference in its function (Olivier et al. 2016a). Conversely to a potential morphological link between Pomacentridae and Pseudochrominae, Tang et al. (2021) provided molecular data supporting the hypothesis that the Embiotocidae would be the sister group of Pomacentridae. Regarding these conflicting results (Eytan et al. 2015; Olivier et al. 2016a; Tang et al. 2021), additional works concerning the identity of the damselfish sister group are clearly needed.

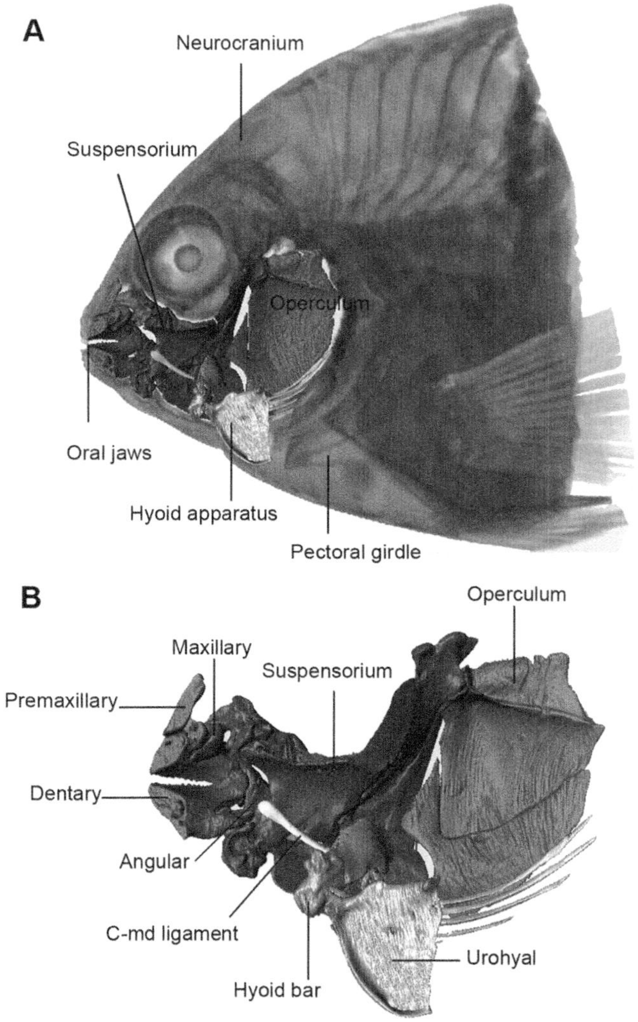

FIGURE 1.2 Illustration of the cerato-mandibular (c-md) ligament in damselfishes. (A) Left lateral view of *Stegastes rectifraenum*. The left oral jaw, suspensorium, opercle, and hyoid bar have been removed allowing view of the right part of the hyoid apparatus in the buccal cavity. The cerato-mandibular ligament is highlighted in green on this 3-D reconstruction. (B) Zoom on the 3-D reconstruction.

1.4 SYSTEMATICS OF DAMSELFISHES

The family Pomacentridae is monophyletic. Recent phylogenies of damselfishes, mostly based on DNA sequence data, agree with the presence of four major clades reflecting the current taxonomic classification with four subfamilies: Microspathodontinae, Chrominae, Glyphosodontinae, and Pomacentrinae (Cooper et al. 2009; Cooper and Santini 2016; McCord et al. 2021; Tang et al. 2021). The systematics of damselfishes was reviewed by Cooper and Santini (2016) and even more recently by Tang et al. (2021) and McCord et al. (2021). Here, I briefly summarize these two most exhaustive studies providing phylogenetic hypotheses with the largest taxon sampling (322 and 345 species) (Figure 1.3). Discordances between these two studies built on different numbers of traditional Sanger loci (5 mtDNA & 3 nuDNA in Tang et al. [2021]; 7 mtDNA & 5 nuDNA in

McCord et al. [2021]) will be pointed out but further details can be found within.

The subfamily Microspathodontinae, including nine genera (Table 1.1), is the sister group of all other pomacentrids (Figure 1.3). This clade includes the largest damselfishes with *Parma* species living around Australia and New Zealand (>200 mm of standard length) and the tribe Microspathodontini, the so-well named "giant damselfishes" (i.e., *Hypsipops*, *Nexilosus*, *Similiparma*, and *Microspathodon* [>300 mm SL]) (Cooper and Santini 2016) which are confined to the Atlantic and Eastern Pacific. Both *Plectroglyphidodon* and *Stegastes* are not monophyletic. A clade including the great majority of *Plectroglyphidodon* species and some *Stegastes* appears to be the sister lineage to Microspathodontini. Accordingly, Tang et al. (2021) suggested classifying this first group of "*Stegastes*" as *Plectroglyphidodon*. The monotypic *Lepidozygus* is the sister lineage of a clade made by the rest of *Stegastes* species and *Plectroglyphidodon lacrymatus*. Then Tang et al. (2021) referred now to *Stegastes lacrymatus*.

The subfamily Glyphisodontinae is made of the genus *Abudefduf*, which includes 21 species. On one hand, Tang

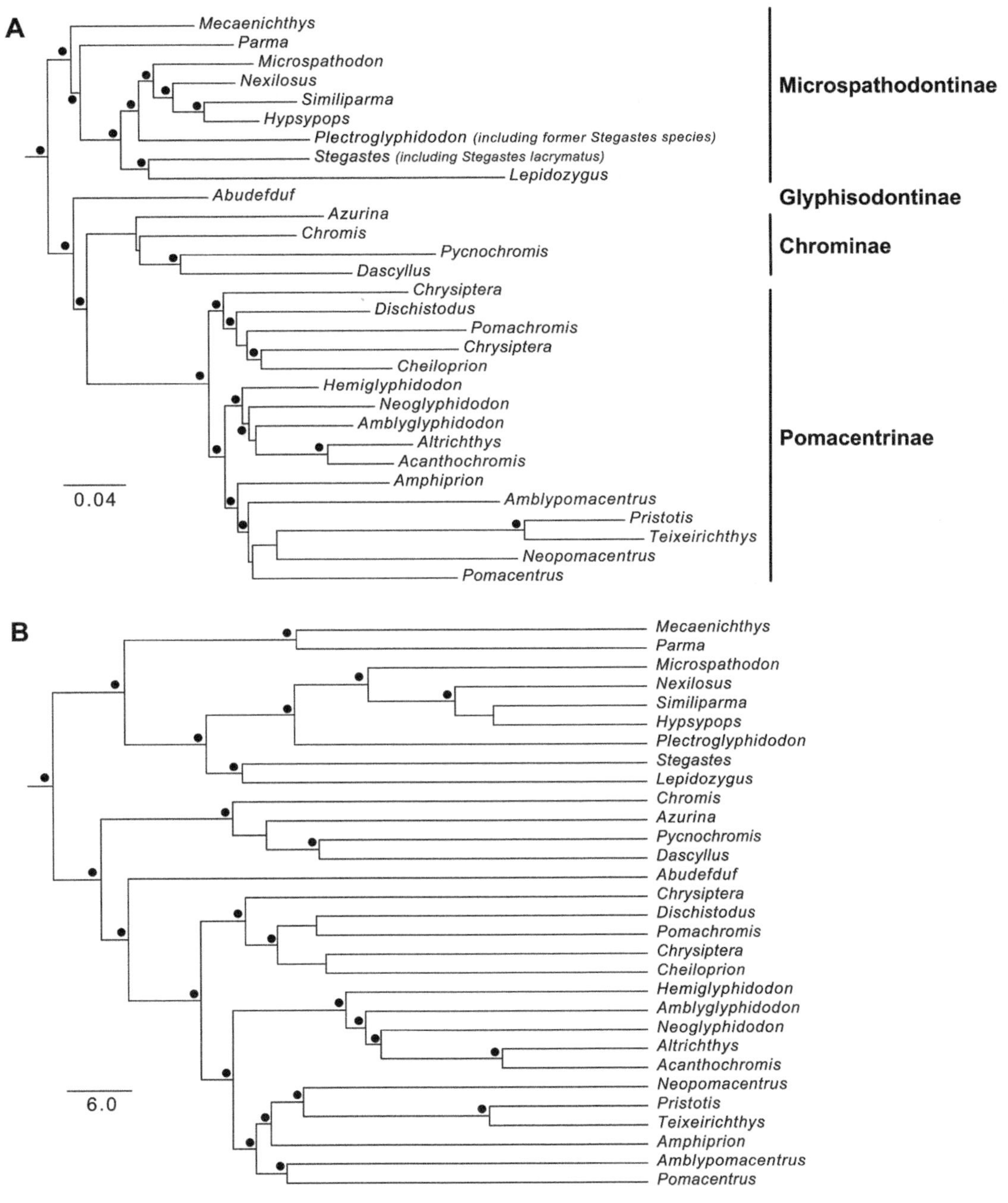

FIGURE 1.3 (A) Maximum likelihood topology of the molecular phylogeny from Tang et al. (2021) and (B) time-calibrated topology of the phylogeny from McCord et al. (2021) illustrating the relationships among the recognized genera of damselfishes. Bootstrap support values greater than 90% and nodal values with Bayesian posterior support levels above 0.9 are indicated with black dots on (A) and (B), respectively.

et al. (2021) retrieved this monophyletic group as a sister to a clade formed by the subfamilies Chrominae and Pomacentrinae. On the other hand, the analyses of McCord et al. (2021) support Glyphisodontinae as the sister group of Pomacentrinae. Such disagreement is not new (see discussion in Tang et al. 2021) and the exact position of Glyphisodontinae is still unresolved (Figure 1.3).

The subfamily Chrominae, including four genera (Table 1.1), appears as the sister group to the subfamily Pomcentrinae in the phylogeny of Tang et al. (2021). Conversely, in McCord et al. (2021)'s phylogeny, Chrominae are sisters to a clade formed by the subfamilies Glyphisodontinae and Pomacentrinae. The Chrominae is dominated by representatives of the polyphyletic genus *Chromis,* which are currently distributed in three disjunct clades. Tang et al. (2021) solved this polyphyly by breaking up the putative *Chromis* species into three different genera: species most closely related to *Azurina hirundo* are now referred to as *Azurina* (e.g., *Chromis cyanea* becomes *Azurina cyanea*), species forming the sister clade of *Dascyllus* are now referred to as *Pycnochromis*, and the other species fall into the clade of *Chromis sensu stricto.* The tree topology of McCord et al. (2021) agrees with this except for a small group of *Chromis* that are outside the main group.

The Pomacentrinae is the largest subfamily, grouping 15 of the 29 genera and holding half of all currently recognized species. This subfamily is divided into four tribes by Tang et al. (2021): Amphiprionini, Cheiloprionini, Hemiglyphidodontini, and Pomacentrini. The Cheiloprionini are the sister tribe to the remainder of the subfamily. The Amphiprionini is the tribe grouping all the clownfishes, constituted of the genera *Premnas* and *Amphiprion*. Based on their robust phylogenetic data and earlier studies, Tang et al. (2021) treat *Premnas* as a junior synonym of *Amphiprion.* Thus, *Amphiprion biaculeatus* should be recognized as the valid species. Among other genera (Table 1.1), the Pomacentrini includes the genus *Pomacentrus* which becomes the largest genus in the family (81 species) after the restructuration of the polyphyletic *Chromis* (108 species) (Eschmeyer et al. 2021). The phylogenetic analyses of McCord et al. (2021) identified five clades in Pomacentrinae, which are partially concordant with the ones of Tang et al. (2021), but some nodes were only weakly supported in both analyses.

Fossil records of damselfish are scarce, which is limited to six described taxa unquestionably assigned to Pomacentridae (Cooper and Santini 2016; Cantalice et al. 2020). The earliest record of the family dated from the Paleocene (*Chaychanus gonzalezorum*, 63 million years ago: mya) (Cantalice et al. 2020). Three fossil taxa are from the deposits of Monte Bolca in Italy (Middle Eocene, 50 mya) and two others dated from the Miocene (Cooper and Santini 2016). Currently, no fossil clownfish was found and described. Further details about the damselfish fossil records are available in Bellwood and Sorbini (1996), Carnevale and Landini (2000), Cooper and Santini (2016), and Cantalice et al. (2020). Fossil data combined with the estimation of the tempo of lineage diversification provided by time-calibrated phylogenies (Litsios et al. 2012a; Frédérich et al. 2013; McCord et al. 2021) suggest that the early diversification of damselfishes occurred just after the Cretaceous-Paleogene boundary (66 mya). The lineages leading to the four extant subfamilies originated during the Eocene and then observed major diversification events during the Miocene-Oligocene (Figure 1.4). Dates of origin and estimates of divergence times for damselfish lineages are detailed in McCord et al. (2021). Briefly, the Microspathodontinae diverged from other pomacentrids at 55 mya and the subclade of giant damselfishes (Microspathodontini) began to diversify ~26 mya. The Chrominae originated 51 mya and extant lineages diverged ~38 mya. Within Chrominae, *Dascyllus*, *Chromis*, and *Pycnochromis* radiated mainly during the Miocene. The Glyphisodontinae diverged from the Pomcantrinae ~49 mya and living *Abudefduf* began to diversify ~31 mya. Major subclades of Pomacentrinae diverge from each other between ~42 and ~34 mya. The diversification of the species-rich group of *Pomacentrus* occurred during the last ~27 million years. The tribe Amphiprionini is relatively young (Litsios et al. 2012b; Frédérich et al. 2013), and according to McCord et al. (2021), it diverged from a common ancestor ~18 mya. Most clownfish species arise only 3–5 mya (Figure 1.4).

1.5 ECOLOGICAL RADIATION OF DAMSELFISHES

With 424 species, the Pomacentridae are an example of a highly successful adaptive radiation. The pomacentrids present a large diversity of habitat use, feeding, morphology, behavior, and color pattern (Frédérich and Parmentier 2016), and resource partitioning is certainly one of the key factors of the process of diversification in damselfishes.

Meekan et al. (1995), Ormond et al. (1996), Pratchett et al. (2016), and Komyakova et al. (2019) are a few examples of detailed comparative analyses of habitat uses in damselfishes. Most of the tropical species live amongst living or dead coral formations on the barrier reef (outer reef slope, reef flat) and in the lagoon (micro-atolls, coral heads, fringing reef). The habitat of numerous species can be restricted to one zone. For example, *Chrysiptera annulata* (*Amblypomacentrus annulatus sensu* [Tang et al. 2021]) lives only on the reef flat of the Great Reef of Toliara in Madagascar (Lepoint et al. 2016). *Stegastes nigricans* and *Dascyllus aruanus* occur strictly in the lagoon (Meekan et al. 1995; Lecchini and Galzin 2005; Gajdzik et al. 2016). On the other hand, some species can be encountered both on the barrier reef and in the lagoon: *Pomacentrus wardi* at Heron Island (Robertson and Lassig 1980) and *Dascyllus flavicaudus* at Moorea Island (Gajdzik et al. 2016). The distribution of the species at small spatial scales is mainly related to the depth, the presence/absence of conspecific,

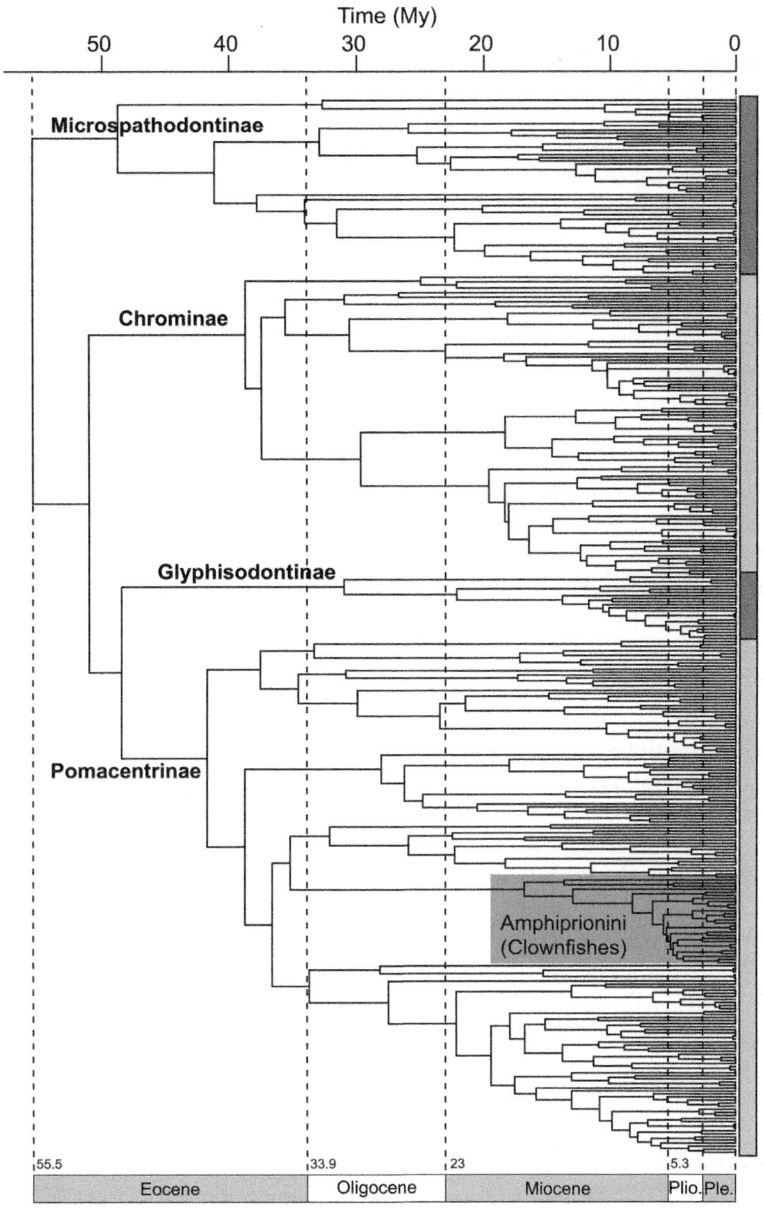

FIGURE 1.4 Consensus time tree from McCord et al. (2021) showing lineage diversification in damselfishes. The four subfamilies and the tribe of clownfishes are identified.

the presence/absence of predators, the coral cover, and/or the kind of substrates (Pratchett et al. 2016; Komyakova et al. 2019). It is worth noting that, among the diversity of habitats available in the tropical coastal environment, relatively few pomacentrids are encountered around mangroves and seagrass beds at the adult stage. For example, *C. annulata* lives in subtidal seagrass meadows found on the reef flat of Toliara Reef (Lepoint et al. 2016) and the presence of some species of the genus *Dischistodus* was reported in a seagrass bed of a fringing reef at Iriomote Island, Southern Japan (Nakamura et al. 2003). In temperate seas, the damselfishes mainly occur in rocky areas (e.g., *Chromis chromis* in the Mediterranean Sea) but some live closely associated with kelp forests (e.g., *Hypsypops rubicundus* along the coast of California) (Allen 1991). Ontogenetic habitat shifts are present in damselfishes but it is relatively uncommon (Komyakova et al. 2019).

The trophic diversity of damselfishes was extensively studied by *in situ* observations of feeding events, the analyses of stomach contents, and the use of trophic markers such as stable isotopes and fatty acids (reviewed in Frédérich et al. 2016b). Damselfishes may be grouped into three main trophic guilds, established on functional demands, and referring to what and where the prey is caught (Frédérich et al. 2009): (1) the pelagic feeders that feed mainly on planktonic copepods, (2) the benthic feeders that mainly graze on filamentous algae, and (3) an intermediate group including species that forage for their prey in the pelagic and benthic environments in variable proportions (e.g., planktonic and benthic copepods, small vagile invertebrates, and filamentous algae). At least three damselfishes are known to be corallivorous species: *Cheiloprion labiatus*, *Plectroglyphidodon johnstonianus*, and *Plectroglyphidodon dickii* (Allen 1991; Kuo and Shao 1991; Ho et al. 2009), and may be grouped within the benthic feeders. The division among these three main trophic guilds is not strict. Indeed, a continuum exists between exclusive zooplankton feeders and algivorous species, and it can be difficult to precisely assign some species to one of the three categories due to feeding plasticity (Frédérich et al. 2016b).

Interestingly, dietary specializations, habitat uses, and social behaviors are tightly associated in damselfishes. As observed in the damselfish assemblage of the reef at Moorea Island (Gajdzik et al. 2016), most of the pelagic feeders form large aggregations (either shoals or schools) up to 20 m depth, live in areas bathed by the open ocean or just behind the barrier reef crest, and are associated with live or dead corals. Benthic feeders mostly display territorial, solitary behavior, and they can establish their shelter on various types of substrates. Species from the intermediate group are generally gregarious and forage in protected areas at usually shallower depths than pelagic feeders.

Beyond the study of niche partitioning among sympatric damselfish species from different regions, the evolution of their ecological diversity was also studied at the genus and the family levels (Cooper and Westneat 2009; Aguilar-Medrano 2013, 2017; Frédérich et al. 2013, 2016a; Aguilar-Medrano and Barber 2016; Gajdzik et al. 2019; McCord et al. 2021). Since the 1970s, numerous ecomorphological studies revealed interspecific variation in various morphological functional traits and discussed their adaptive significance (reviewed in Frédérich et al. 2016b). Head shape, oral jaws, pharyngeal jaws, oral and pharyngeal teeth, gill rakers, and intestine length all appeared to be ecologically relevant traits allowing the discrimination of the three damselfish feeding guilds. The study of body form and pectoral fins also allowed the discrimination of functional groups related to habitat partitioning and swimming mode (Frédérich et al. 2016a).

The combination of ecological and morphological data with various phylogenetically informed comparative analyses has demonstrated high rates of evolutionary change in the trophic ecology of damselfishes (Cooper and Westneat 2009; Frédérich et al. 2013; Gajdzik et al. 2019; McCord et al. 2021). What is unusual about this radiation is that instead of invading a large diversity of ecological niches, it has progressed by rapidly and repeatedly converging on similar ecomorphological states (ecotypes). Cooper and Westneat (2009) refer to this pattern as "reticulate adaptive radiation". Frédérich et al. (2013) who confirmed this evolutionary pattern with additional species and using other phylogenetic comparative methods, preferred the term "iterative evolution". Both refer to the repetitive occurrence of similar morphologies, ecologies, or behaviors during the evolutionary progression of a lineage. This pattern of repeated convergence was already described for overall skull shape (Cooper and Westneat 2009; Aguilar-Medrano et al. 2011), bite mechanics (Cooper and Westneat 2009), oral jaws (Frédérich et al. 2013), farming behavior, and trophic ecology (Cooper and Westneat 2009; Frédérich et al. 2013), but it is highly expected that other phenotypic traits evolve along the same pattern. One of the ecological outcomes of this evolutionary pattern is the production of highly similar damselfish assemblages in different geographic regions. Even if the number of species varies among regions, Gajdzik et al. (2018) showed consistent levels of eco-functional diversity in coral reef damselfish assemblages in Toliara reef (Madagascar), Dongsha atoll (Taiwan), and Moorea Island (French Polynesia). Every damselfish assemblage, mainly driven by niche-related processes, hosted species whose niches were highly differentiated and evenly distributed in eco-functional spaces (Gajdzik et al. 2018).

Beyond the picture of an iterative ecomorphological radiation, it appears that the pattern of transitions between ecotypes (i.e., the three trophic states: benthic feeder, intermediate omnivore, and pelagic feeder) is not random and the frequency at which these transitions occurred is relatively unbalanced (Gajdzik et al. 2019; McCord et al. 2021). Analyses of the evolution of the three ecotypes revealed that direct transition between the "specialist" benthic and pelagic feeders does not happen when the frequency of intermediate omnivore transitioning to the two "specialist" ecotypes is high. Accordingly, Gajdzik et al. (2019) suggested that the intermediate trophic guild may operate as a stepping-stone state towards specialized strategies in damselfishes.

Globally, the tempo of lineage diversification for the Pomacentridae is quite constant through time (Frédérich et al. 2013). However, recent works revealed that diversification rates are dependent on fish body size and trophic ecology (Gajdzik et al. 2019; McCord et al. 2021). Speciation rates were the highest among medium-sized damselfishes in comparison with small and large species (McCord et al. 2021). Concerning variation in diversification rates among ecotypes, the results from Gajdzik et al. (2019) and McCord et al. (2021) slightly differ but the differences could certainly be explained by the taxon sampling in phylogenies and the used comparative analyses. Both studies estimated that the benthic feeders are characterized by the lowest rate of diversification in comparison with the pelagic planktivores and the intermediate omnivores.

Recent advances in damselfish phylogenetics with large taxon sampling and associated chronograms provide the tools to study the successful radiation of Pomacentridae in coral and rocky reef environments. Additional works are certainly still needed to decipher all the factors explaining their success, even if their morphology and their versatility are probably key components allowing the observed easy shifts among a limited set of trophic ecotypes. By developing their symbiotic relationships with giant sea anemones, clownfishes represent a peculiar tribe within Pomacentridae. Our understanding of clownfish evolution characterized by their singular adaptations is challenging and the progress of clownfish research must pass through the identification of their sister lineage within Pomacentrinae. Unfortunately, recent exhaustive phylogenies are not yet congruent about sister groups of clownfishes but the best candidates are certainly the genera *Pomacentrus*, *Neopomacentrus*, and *Amblypomacentrus*.

ACKNOWLEDGEMENTS

I gratefully acknowledge the editor, Vincent Laudet, Mark Westneat and one anonymous reviewer for their insightful comments and helpful criticism of the original manuscript. I thank M. Erdmann and F. Charpin, who kindly shared their photos of damselfishes allowing the making of Figure 1.1.

REFERENCES

Aguilar-Medrano, R. 2013. Body shape evolution of *Chromis* and *Azurina* species (Percifomes, Pomacentridae) of the eastern Pacific. *Animal Biology* 63: 217–232.

Aguilar-Medrano, R. 2017. Ecomorphology and evolution of the pharyngeal apparatus of benthic damselfishes (Pomacentridae, subfamily Stegastinae). *Marine Biology* 164: 21.

Aguilar-Medrano, R. and P. H. Barber. 2016. Ecomorphological diversification in reef fish of the genus Abudefduf (Percifomes, Pomacentridae). *Zoomorphology* 135: 103–114.

Aguilar-Medrano, R., B. Frédérich, E. De Luna, and E. F. Balart. 2011. Patterns of morphological evolution of the cephalic region in damselfishes (Perciformes: Pomacentridae) of the Eastern Pacific. *Biological Journal of the Linnean Society* 102(3): 593–613.

Alfaro, M. E., B. C. Faircloth, R. C. Harrington, L. Sorenson, M. Friedman, C. E. Thacker, C. H. Oliveros, et al. 2018. Explosive diversification of marine fishes at the Cretaceous-Palaeogene boundary. *Nature Ecology and Evolution* 2: 688–696.

Allen, G. R. 1991. *Damselfishes of the World*. Melle: Mergus.

Bellwood, D. R. 1994. A phylogenetic study of the parrotfishes family Scaridae (Pisces: Labroidei), with a revision of genera. *Records of the Australian Museum, Supplement* 20: 1–86.

Bellwood, D. R. and L. Sorbini. 1996. A review of the fossil record of the Pomacentridae (Teleostei: Labroidei) with a description of a new genus and species from the Eocene of Monte Bolca, Italy. *Zoological Journal of the Linnean Society* 117: 159–174.

Betancur, R. R., E. O. Wiley, G. Arratia, A. Acero, N. Bailly, M. Miya, G. Lecointre, et al. 2017. Phylogenetic classification of bony fishes. *BMC Evolutionary Biology* 17: 162.

Branch, G. M. 1981. The biology of limpets: Physical factors, energy flow, and ecological interactions. *Oceanography and Marine Biology – An Annual Review* 19: 235–380.

Broughton, R. E., R. Betancur-R., C. Li, G. Arratia, and G. Ortí. 2013. Multi-locus phylogenetic analysis reveals the pattern and tempo of bony fish evolution. *PloS Currents* 1(16): 5.

Cantalice, K. M., J. Alvarado-Ortega, and D. R. Bellwood. 2020. †*Chaychanus gonzalezorum* gen. et sp. Nov.: A damselfish fossil (Percomorphaceae; Pomacentridae), from the Early Paleocene outcrop of Chiapas, Southeastern Mexico. *Journal of South American Earth Sciences* 98: 102322.

Carnevale, G. and W. Landini. 2000. A fossil damselfish (Pisces, Pomacentridae) from the late Miocene of central Italy. Biological and biogeographical considerations. *Palaeontographia Italica* 87: 65–70.

Ceccarelli, D. M., G. P. Jones, and L. J. McCook. 2005. Effects of territorial damselfish on an algal-dominated coastal coral reef. *Coral Reefs* 24: 606–620.

Chen, W.-J., F. Santini, G. Carnevale, J.-N. Chen, S.-H. Liu, S. Lavoué, and R. L. Mayden. 2014. New insights on early evolution of spiny-rayed fishes (Teleostei: Acanthomorpha). *Frontiers in Marine Science* 1: 53.

Clements, K. D., M. E. Alfaro, J. L. Fessler, and M. W. Westneat. 2004. Relationships of the temperate Australasian labrid fish tribe Odacini (Perciformes; Teleostei). *Molecular Phylogenetics and Evolution* 32: 575–587.

Colleye, O., M. Nakamura, B. Frédérich, and E. Parmentier. 2012. Further insight into the sound-producing mechanism of clownfishes: What structure is involved in sound radiation? *Journal of Experimental Biology* 215: 2192–2202.

Collingwood, C. 1868. *Rambles of a Naturalist on the Shores and Waters of the China Sea*. London: John Murray.

Cooper, W. J. and F. Santini. 2016. A revised damselfish taxonomy with a description of the tribe microspathodontini (giant damselfishes). In *Biology of Damselfishes*, eds B. Frédérich and E. Parmentier, 13–30. Boca Raton: CRC Press/Taylor & Francis.

Cooper, W. J., L. L. Smith, and M. W. Westneat. 2009. Exploring the radiation of a diverse reef fish family: Phylogenetics of the damselfishes (Pomacentridae), with new classifications based on molecular analyses of all genera. *Molecular Phylogenetics and Evolution* 52: 1–16.

Cooper, W. J. and M. W. Westneat. 2009. Form and function of damselfish skulls: Rapid and repeated evolution into a limited number of trophic niches. *BMC Evolutionary Biology* 9: 24.

Davesne, D., C. Gallut, V. Barriel, P. Janvier, G. Lecointre, and O. Otero. 2016. The phylogenetic intrarelationships of spiny-rayed fishes (Acanthomorpha, Teleostei, Actinopterygii): Fossil taxa increase the congruence of morphology with molecular data. *Frontiers in Ecology and Evolution* 4: 129.

Eschmeyer, W. N., R. Fricke, and R. van der Laan. 2021. *Eschmeyer's Catalog of Fishes: Genera, Species, References*. Electronic version accessed 15 August 2021.

Eytan, R. I., B. R. Evans, A. Dornburg, A. R. Lemmon, E. M. Lemmon, P. C. Wainwright, et al. 2015. Are 100 enough? Inferring acanthomorph teleost phylogeny using anchored hybrid enrichment. *BMC Evolutionary Biology* 15: 113.

Faircloth, B. C., L. Sorenson, F. Santini, and M. E. Alfaro. 2013. A phylogenomic perspective on the radiation of ray-finned fishes based upon targeted sequencing of ultraconserved elements (UCEs). *PloS ONE* 8(6): e65923.

Fautin, D. G. 1986. Why do anemonefishes inhabit only some host actinians? *Environmental Biology of Fishes* 15: 171–180.

Fautin, D. G. 1991. The anemonefish symbiosis: What is known and what is not. *Symbiosis* 10: 23–46.

Fishelson, L. 1998. Behaviour, socio-ecology and sexuality in damselfishes (Pomacentridae). *Italian Journal of Zoology* 65: 387–398.

Frédérich, B., W. J. Cooper, and R. Aguilar-Medrano. 2016a. Ecomorphology and iterative ecological radiation of damselfishes. In *Biology of Damselfishes*, eds B. Frédérich and E. Parmentier, 183–203. Boca Raton: CRC Press/Taylor & Francis.

Frédérich, B., G. Fabri, G. Lepoint, P. Vandewalle, and E. Parmentier. 2009. Trophic niches of thirteen damselfishes (Pomacentridae) at the Grand Récif of Toliara, Madagascar. *Ichthyological Research* 56: 10–17.

Frédérich, B., D. Olivier, L. Gajdzik, and E. Parmentier. 2016b. Trophic ecology of damselfishes. In *Biology of Damselfishes*, eds B. Frédérich and E. Parmentier, 153–167. Boca Raton: CRC Press/Taylor & Francis.

Frédérich, B., D. Olivier, G. Litsios, M. E. Alfaro, and E. Parmentier. 2014. Trait decoupling promotes evolutionary diversification of the trophic and acoustic system of damselfishes. *Proceedings of the Royal Society B: Biological Sciences* 281: 20141047.

Frédérich, B. and E. Parmentier. 2016. *Biology of Damselfishes*. Boca Raton: CRC Press/Taylor & Francis.

Frédérich, B., L. Sorenson, F. Santini, G. J. Slater, and M. E. Alfaro. 2013. Iterative ecological radiation and convergence during the evolutionary history of damselfishes (Pomacentridae). *American Naturalist* 181: 94–113.

Gajdzik, L., R. Aguilar-Medrano, and B. Frédérich. 2019. Diversification and functional evolution of reef fish feeding guilds. *Ecology Letters* 22: 572–582.

Gajdzik, L., E. Parmentier, L. N. Michel, N. Sturaro, K. Soong, G. Lepoint, and B. Frédérich. 2018. Similar levels of trophic and functional diversity within damselfish assemblages across Indo-Pacific coral reefs. *Functional Ecology* 32(5): 1358–1369.

Gajdzik, L., E. Parmentier, N. Sturaro, and B. Frédérich. 2016. Trophic specializations of damselfishes are tightly associated with reef habitats and social behaviours. *Marine Biology* 163: 249.

Galis, F. and P. Snelderwaard. 1997. A novel biting mechanism in damselfishes (Pomacentridae): The pushing up of the lower pharyngeal jaw by the pectoral girdle. *Netherlands Journal of Zoology* 47: 405–410.

Hata, H. and D. Ceccarelli. 2016. Farming behaviour of territorial damselfishes. In *Biology of Damselfishes*, eds Frédérich, B. and E. Parmentier, 122–152. Boca Raton: CRC Press/Taylor & Francis.

Hata, H. and M. Kato. 2004. Monoculture and mixed-species algal farms on a coral reef are maintained through intensive and extensive management by damselfishes. *Journal of Experimental Marine Biology and Ecology* 313: 285–296.

Hata, H. and M. Kato. 2006. A novel obligate cultivation mutualism between damselfish and Polysiphonia algae. *Biology Letters* 2: 593–596.

Hata, H., A. S. Tanabe, S. Yamamoto, H. Toju, M. Kohda, and M. Hori. 2014. Diet disparity among sympatric herbivorous cichlids in the same ecomorphs in Lake Tanganyika: Amplicon pyrosequences on algal farms and stomach contents. *BMC Biology* 12: 90.

Ho, C. T., Y. C. Fu, C. L. Sun, S. J. Kao, and R. Q. Jan. 2009. Plasticity of feeding habits of two *Plectroglyphidodon* damselfishes on coral reefs in southern Taiwan: Evidence from stomach content and stable isotope analyses. *Zoological Studies* 48: 649–656.

Hughes, L. C., G. Ortí, Y. Huang, Y. Sun, C. C. Baldwin, A. W. Thompson, D. Arcila, et al. 2018. Comprehensive phylogeny of ray-finned fishes (Actinopterygii) based on transcriptomic and genomic data. *Proceedings of the National Academy of Sciences of the United States of America* 115: 6249–6254.

Hulsey, C. D. 2006. Function of a key morphological innovation: Fusion of the cichlid pharyngeal jaw. *Proceedings of the Royal Society B-Biological Sciences* 273: 669–675.

Kaufman, L. S. and K. F. Liem. 1982. Fishes of the suborder Labroidei (Pisces: Perciformes): Phylogeny, ecology, and evolutionary significance. *Breviora* 472: 1–19.

Komyakova, V., P. L. Munday, and G. P. Jones. 2019. Comparative analysis of habitat use and ontogenetic habitat-shifts among coral reef damselfishes. *Environmental Biology of Fishes* 102: 1201–1218.

Kuo, S.-R. and K.-T. Shao. 1991. Feeding habits of damselfishes (Pomacentridae) from the southern part of Taiwan. *Journal of the Fisheries Society of Taiwan* 18: 165–176.

Lecchini, D. and R. Galzin. 2005. Spatial repartition and ontogenetic shifts in habitat use by coral reef fishes (Moorea, French Polynesia). *Marine Biology* 147: 47–58.

Lepoint, G., L. N. Michel, E. Parmentier, and B. Frédérich. 2016. Trophic ecology of the seagrass-inhabiting footballer demoiselle *Chrysiptera annulata* (Peters, 1855); comparison with three other reef-associated damselfishes. *Belgian Journal of Zoology* 146: 21–32.

Liem, K. F. 1973. Evolutionary strategies and morphological innovations: Cichlid pharyngeal jaws. *Systematic Zoology* 22: 425–441.

Liem, K. F. and P. H. Greenwood. 1981. A functional approach to the phylogeny of the pharyngognath teleosts. *Integrative and Comparative Biology* 21: 83–101.

Liem, K. F. and J. W. M. Osse. 1975. Biological versatility, evolution, and food resource exploitation in African cichlid fishes. *American Zoologist* 15: 427–454.

Litsios, G., L. Pellissier, F. Forest, C. Lexer, P. B. Pearman, N. E. Zimmermann, and N. Salamin. 2012a. Trophic specialization influences the rate of environmental niche evolution in damselfishes (Pomacentridae). *Proceedings of the Royal Society B: Biological Sciences* 279: 3662–3669.

Litsios, G., C. A. Sims, R. O. Wueest, P. B. Pearman, N. E. Zimmermann, and N. Salamin. 2012b. Mutualism with sea anemones triggered the adaptive radiation of clownfishes. *BMC Evolutionary Biology* 12: 212.

Mabuchi, K., M. Miya, Y. Azuma, and M. Nishida. 2007. Independent evolution of the specialized pharyngeal jaw apparatus in cichlid and labrid fishes. *BMC Evolutionary Biology* 7: 10.

McCord, C. L., C. M. Nash, W. J. Cooper, and M. W. Westneat. 2021. Phylogeny of the damselfishes (Pomacentridae) and patterns of asymmetrical diversification in body size and feeding ecology. *PloS ONE* 16(10): e0258889.

Meekan, M. G., A. D. L. Steven, and M. J. Fortin. 1995. Spatial patterns in the distribution of damselfishes on a fringing coral-reef. *Coral Reefs* 14: 151–161.

Müller, J. 1843. Beiträge zur Kenntnis der natürlichen Familien der Fische. *Archiv für Naturgeschichte*, 9: 292–381.

Müller, J. 1844. Uber den bau und die grenzen der Ganoiden und über das natürliche system der fische. *Abhandlungen der Königlichen Akademie der Wissenschaften* 117–216.

Nakamura, Y., M. Horinouchi, T. Nakai, and M. Sano. 2003. Food habits of fishes in a seagrass bed on a fringing coral reef at Iriomote Island, southern Japan. *Ichthyological Research* 50: 15–22.

Near, T. J., R. I. Eytan, A. Dornburg, K. L. Kuhn, J. A. Moore, M. P. Davis, P. C. Wainwright, et al. 2012. Resolution of ray-finned fish phylogeny and timing of diversification. *Proceedings of the National Academy of Sciences of the United States of America* 109: 13698–13703.

Nelson, J. S., T. C. Grande, and M. V. H. Wilson. 2016. *Fishes of the World*. 5th Edition. Hoboken: John Wiley & Sons, Inc.

Olivier, D., B. Frédérich, A. Herrel, and E. Parmentier. 2015. A morphological novelty for feeding and sound production in the yellowtail clownfish. *Journal of Experimental Zoology Part A: Ecological Genetics and Physiology* 323(4): 227–238.

Olivier, D., B. Frédérich, and E. Parmentier. 2016a. Cerato-mandibular ligament: A key trait in damselfishes? In *Biology of Damselfishes*, eds B. Frédérich and E. Parmentier, 295–313. Boca Raton: CRC Press/Taylor & Francis.

Olivier, D., B. Frédérich, M. Spanopoulos-Zarco, E. F. Balart, and E. Parmentier. 2014. The cerato-mandibular ligament: A key functional trait for grazing in damselfishes (Pomacentridae). *Frontiers in Zoology* 11: 63.

Olivier, D., E. Parmentier, and B. Frédérich. 2016b. Insight into biting diversity to capture benthic prey in damselfishes (Pomacentridae). *Zoologischer Anzeiger* 264: 47–55.

Olivier, D., S. van Wassenbergh, E. Parmentier, and B. Frédérich. 2021. Unprecedented biting performance in herbivorous fish: How the complex biting system of pomacentridae circumvents performance trade-offs. *American Naturalist* 197: E156–E172.

Ormond, R. F. G., J. M. Roberts, and R. Q. Jan. 1996. Behavioural differences in microhabitat use by damselfishes (Pomacentridae): Implications for reef fish biodiversity. *Journal of Experimental Marine Biology and Ecology* 202: 85–95.

Parmentier, E., O. Colleye, M. L. Fine, B. Frédérich, P. Vandewalle, and A. Herrel. 2007. Sound production in the clownfish *Amphiprion clarkii*. *Science* 316: 1006.

Parmentier, E. and B. Frédérich. 2016. Meet the damselfishes. In *Biology of Damselfishes*, eds B. Frédérich and E. Parmentier, 1–12. Boca Raton: CRC Press/Taylor & Francis.

Pratchett, M. S., A. S. Hoey, S. K. Wilson, J.-P. A. Hobbs, and G. R. Allen. 2016. Habitat-use and specialisation among coral reef damselfishes. In *Biology of Damselfishes*, eds B. Frédérich and E. Parmentier, 84–121. Boca Raton: CRC Press/Taylor & Francis.

Robertson, D. R. and B. Lassig. 1980. Spatial distribution patterns and coexistence of a group of territorial damselfishes from the Great Barrier Reef. *Bulletin of Marine Science* 30: 187–203.

Stiassny, M. L. J. 1981. The phyletic status of the family Cichlidae (Pisces, Perciformes): A comparative anatomical investigation. *Netherlands Journal of Zoology* 31: 275–314.

Stiassny, M. L. J. and J. S. Jensen. 1987. Labroid intrarelationships revisited: Morphological complexity, key innovations, and the study of comparative diversity. *Bulletin of the Museum of Comparative Zoology* 151: 269–319.

Tang, K. L., M. L. J. Stiassny, R. L. Mayden, and R. DeSalle. 2021. Systematics of damselfishes. *Ichthyology & Herpetology* 109: 258–318.

Wainwright, P. C., W. L. Smith, S. A. Price, K. L. Tang, J. S. Sparks, L. A. Ferry, K. L. Kuhn, et al. 2012. The evolution of pharyngognathy: A phylogenetic and functional appraisal of the pharyngeal jaw key innovation in labroid fishes and beyond. *Systematic Biology* 61: 1001–1027.

Westneat, M. W. and M. E. Alfaro. 2005. Phylogenetic relationships and evolutionary history of the reef fish family Labridae. *Molecular Phylogenetics and Evolution* 36: 370–390.

2 Anemonefish Genomics

Nicolas Salamin, Celia Schunter, Alison Monroe, Taewoo Ryu, and Timothy Ravasi

CONTENTS

2.1 INTRODUCTION

The evolution of the anemonefish lineage has been the focus of numerous phylogenetic studies to decipher its evolutionary history (Litsios et al. 2012; Litsios and Salamin 2014; Tang et al. 2021). Most analyses have focused on a small number of genes, either nuclear or mitochondrial, but the logical next step has been to reach a better understanding of the genomic architecture of the lineage. The availability of high-quality and complete genomic data provides valuable information to identify the mechanisms responsible for mutualistic interactions, the particular social structure seen in anemonefish, and to characterize the genes involved in the phenotypic differences between species. This will lead to further studies that improve our understanding of adaptation and evolution in this fascinating group of fishes.

The first genomes of anemonefish (Lehmann et al. 2019; Marcionetti et al. 2018) were an important step in our understanding of the genetic mechanisms behind the evolution of this group. It gave access to resources for three species (*Amphiprion frenatus*, Marcionetti et al. 2018; *Amphiprion percula*, Lehmann et al. 2019; *Amphiprion ocellaris*, Ryu et al. 2022) that cover the main divergence in the group. Different approaches were used to build the genome assemblies. The former obtained high coverage via short Illumina reads, which led to an assembly containing all the essential genes but with a high number of scaffolds. The latter adopted a thorough data collection combining short and long reads with coverage that enabled the reconstruction of a chromosome level assembly. However, the main summary statistics obtained by the two studies were congruent (Marcionetti et al. 2019), which suggests that the genomic architecture within the genus is conserved.

The *A. percula* genome (Lehmann et al. 2019) has a total assembly size of 908.9 Mb, which represents almost 95% of the predicted genome size. It recovered 26,597 genes, 85% of which were functionally annotated into proteins. The high quality of the assembly enabled the 365 scaffolds to be assembled into 24 chromosomes, with only 2.1% of the assembled sequences unassigned. The gene density across the chromosomes was fairly even, with an average of 29.7 ± 3.46 genes per Mb on each chromosome (288 genes were not placed into the chromosomes). The short-read sequencing of Marcionetti et al. (2018) led to a lower-quality assembly (17,801 scaffolds with a total assembly size of 791 Mb), but the functional content was similar, with 26,917 genes found and 94.9% of them functionally annotated.

2.2 ANEMONEFISH PHYLOGENOMICS

The genomic resources were further expanded by the sequencing of nine other anemonefish species (*A. akallopisos*, *A. perideraion*, *A. melanopus*, *A. polymnus*, *A. sebae*, *A. ocellaris*, *A. nigripes*, *A. bicinctus* and *Premnas biaculeatus*; Marcionetti et al. 2019) as well as recently a chromosome-level genome of *A. ocellaris* from Okinawa island (Ryu et al., 2022). The assembly quality was similar to the *A. frenatus* genome (total assembly size: 798.9 ± 3.2 Mb; number of genes: $28{,}696 \pm 788$; percentage of annotation: 93.2 ± 0.6). The analyses of all the orthologous genes between the ten anemonefish species and other fish genomes further showed that the rate of gene duplication within anemonefish is not different from what is observed in damselfish or cichlids (Figure 2.1A). The availability of these new genomes further clarified the phylogenetic relationships between anemonefish (Figure 2.1B). For instance, as already suggested by Tang et al. 2021, the genus *Premnas* should not be separated from the genus *Amphiprion* because the level of divergence is within the range of what is observed between *Amphiprion* species (Figure 2.1B). This was further reinforced by the fact that across the genome, gene trees estimated from 100 Kb windows display an ambiguous placement for *Premnas,* either as the basal species of anemonefish or as sister to *A. ocellaris* and *A. percula* (Figure 2.1C). It has been proposed that the key genomic characteristic that drives rapid diversification is

DOI: 10.1201/9781003125365-3

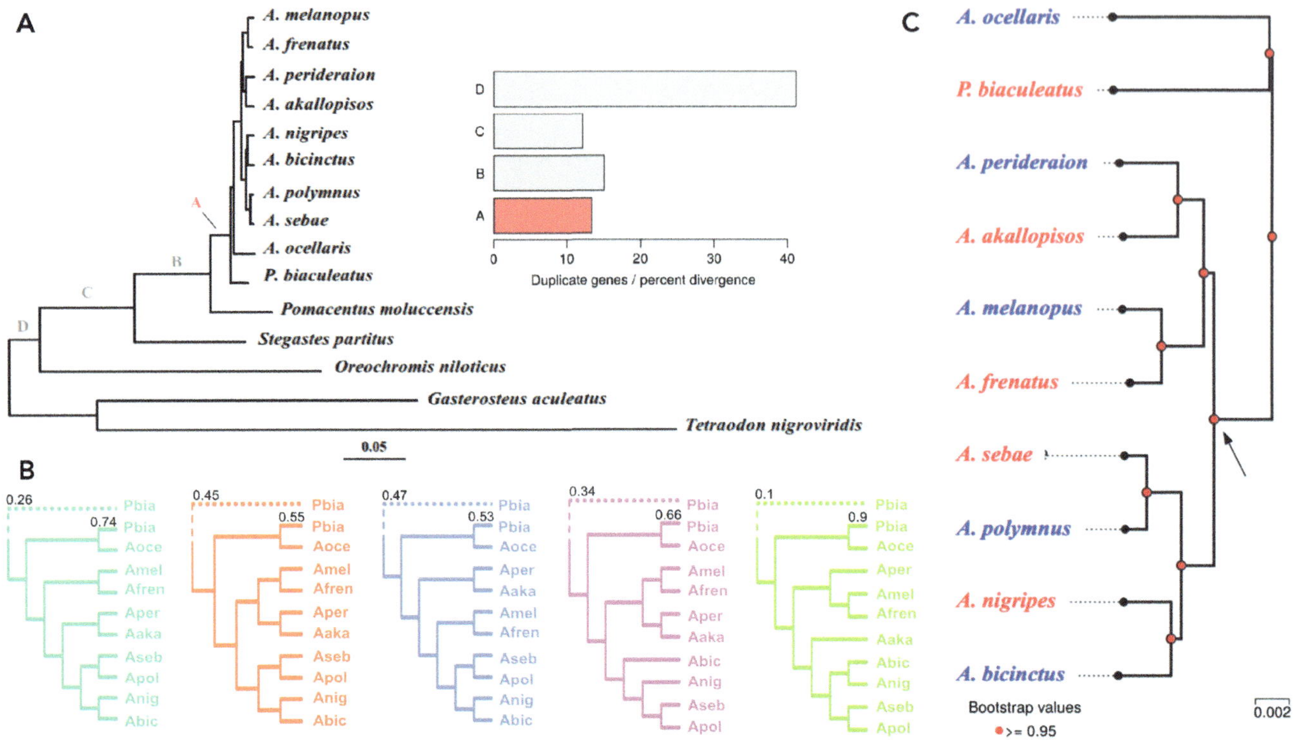

FIGURE 2.1 A. Phylogenetic tree based on the analyses of all the orthologous genes between the ten anemonefish species and other fish genomes. B. Phylogenetic relationships between anemonefish based on the alignment of fully sequenced genomes. C. Phylogenetic placement of the genus *Premnas* with respect to the genus *Amphiprion* with a level of divergence within the range of what is observed between *Amphiprion* species (adapted from Marcionetti and Salamin 2022).

the access to ancient genetic variation through gene flow (Berner and Salzburger 2015). There are clear signs in the genome that hybridization has played a role in anemonefish evolution (Litsios and Salamin 2014) and the several known hybrid species (e.g., Gainsford et al. 2020) show that this process is still ongoing. Further genomic studies should better characterize the level of hybridization and the role played by this genomic reshuffling in the evolution and diversification of the group.

In addition to the access to ancient genetic variation, other genomic features often observed in adaptive radiation are structural variants, changes in regulatory sequences (Berner and Salzburger 2015; Brawand et al. 2014; Dasmahapatra et al. 2012; Jones et al. 2012; Lamichhaney et al. 2015) and, more recently, high levels of heterozygosity (Ronco et al. 2021). This has not yet been fully characterized in anemonefishes and there is a need to evaluate the role of these elements to better understand their diversification and the functional relevance of these genomic features. Chromosome-level assemblies, like the one available for *A. percula*, will facilitate the analysis of structural variants and changes in regulatory sequences which modify gene expression and play a key role in the evolution of phenotypes such as morphology, colouration, and behavior, especially in closely related taxa (reviewed in Stern and Orgogozo 2008; Wray 2007).

The emergence of adaptive phenotypic traits may also be promoted by few alterations in both coding and non-coding DNA sequences. Within the ten available genomes, a set of 13 genes were identified as playing a key role in the onset of the mutualism acquisition (Marcionetti et al. 2019). Two of these (Versican core protein and Protein O-GlcNAcase) show particularly interesting functions associated with *N*-acetylated sugars, which are known to be involved in sea anemone discharge of toxins. Similar bioinformatic analyses are currently ongoing to understand the molecular footprint during the anemonefish diversification, but these analyses focus only on the protein-coding genes. We are still missing an understanding of the role played by non-coding elements of the genome. Preliminary work on anemonefish identified conserved non-coding regions, likely containing regulatory sequences such as transcription factor binding sites, and evaluated their evolution using the approach of Brawand et al. (2014). However, the small level of divergence within the anemonefishes and the difficulty in identifying the structure of these non-coding elements means that for now the results are still inconclusive and further work is needed.

The genomic characterization of anemonefish has seen an impressive advance over the last few years. This has provided interesting new insights into their evolution, but more work is necessary to fully understand the fine-scale differences existing between the species as well as the role played by genomic features in the evolution of the group. New next-generation sequencing techniques (long reads, Hi-C, ATAC-seq) could bring valuable resources to push anemonefish forward as a genetic model system.

2.3 ANEMONEFISH TRANSCRIPTOMICS

The development of RNA-seq in the past decade has provided the tools to map and quantify the transcriptome in a wide variety of organisms (Wang et al. 2009). This relatively low-cost method provides high-resolution data without the need for extensive genomic resources (Qian et al. 2014). Using RNA-seq, researchers can identify the molecular pathways involved in biological processes such as development, adaptation, immunology, and response to environmental stress (Figure 2.2; Connon et al. 2018; Qian et al. 2014). The integration of gene expression measurements with physiological and population-level measurements has driven ecological research forward while providing key information on adaptive phenotypes. This has been mainly helped by recent advancements in bioinformatic techniques (Connon et al. 2018). In fish, RNA-seq has expanded transcriptomic studies to include research on many commercially and ecologically important species, including anemonefish (Casas et al. 2016; Salis et al. 2019; Schunter et al. 2021).

The transcriptome is dynamic compared to the genome, and it is useful when measuring the changing cellular processes in developmental biology (Martin and Wang 2011). These developmental changes in gene expression can help link the genotype of an individual with its phenotype (Xu et al. 2017). In fish, the embryonic to larval stages are especially important and persisted stress during this process can impact the long-term survival of adult fish (Fu et al. 2019). Early research was focused on zebrafish, but the increasing affordability of RNA-seq has led to the examination of other species including common sole, bighead carp, channel catfish, and Mahi Mahi (Ferraresso et al. 2013; Fu et al. 2019; Ma et al. 2020; Vesterlund et al. 2011; Xu et al. 2017).

In anemonefish, RNA-seq studies have looked at developmental gene expression related to sex change in *A. bicinctus*, pigment cells that determine color patterns in *A. ocellaris and A. percula*, as well as opsin expression in 11 different species to analyze their visual ecology and behaviors (Casas et al. 2016; Maytin et al. 2018; Mitchell et al. 2021; Salis et al. 2019; Steib et al. 2019). Studying sex differentiation in anemonefish can provide key insights into the cellular processes behind functional hermaphroditism, a strategy widely used in coral reef fishes (Casas et al. 2016; Kobayashi et al. 2013). Recent research has produced detailed descriptions of the embryonic life stages of *A. ocellaris* (Salis et al. 2021) which will be an important resource for future studies examining developmental transcriptomic changes in anemonefish. Understanding these molecular mechanisms will help determine survival rates throughout various life stages and serve as important baselines for further research examining environmental changes.

Transcriptomics has been used to identify gene expression changes due to environmental factors, such as temperature, salinity, pH, and pollution, in a large number of marine fishes (Oomen and Hutchings 2017). Results vary depending on species, length of exposure, magnitude of change, and especially life stage of the fish when these stressors occur. However, there are some consistently impacted pathways independent of the aforementioned variables including, metabolic performance when exposed to increased temperatures (Bernal et al. 2018; Narum and Campbell 2015; Veilleux et al. 2015), neurotransmitter signalling under changes in pH (Porteus et al. 2018; Schunter et al. 2018), and the cellular stress response in those exposed to various stressors (Huth and Place 2016). Several studies have integrated these molecular pathways with observed physiological and behavioral measurements, creating a whole-organism view of responses to environmental changes (Bernal et al. 2018; Porteus et al. 2018; Shama et al. 2014). This data can help inform about acclimation and adaptive potential, especially when predicting the effects of future ocean climate scenarios.

New research has examined the impacts of elevated pCO_2 on the brain transcriptome of the orange clownfish, *Amphiprion percula* (Schunter et al. 2021). Overall, this research found small gene expression changes between pCO_2 conditions, especially compared to research from other damselfishes. Within these differentially expressed genes, this study identified changes in circadian rhythm regulators and those controlling hormone changes, similar to pathways found in other studies on coral reef fish under elevated pCO_2 levels (Schunter et al. 2016, 2018, 2021). This is the first study researching the impacts of environmental changes in anemonefish and the field is wide open to continue examining other impacts.

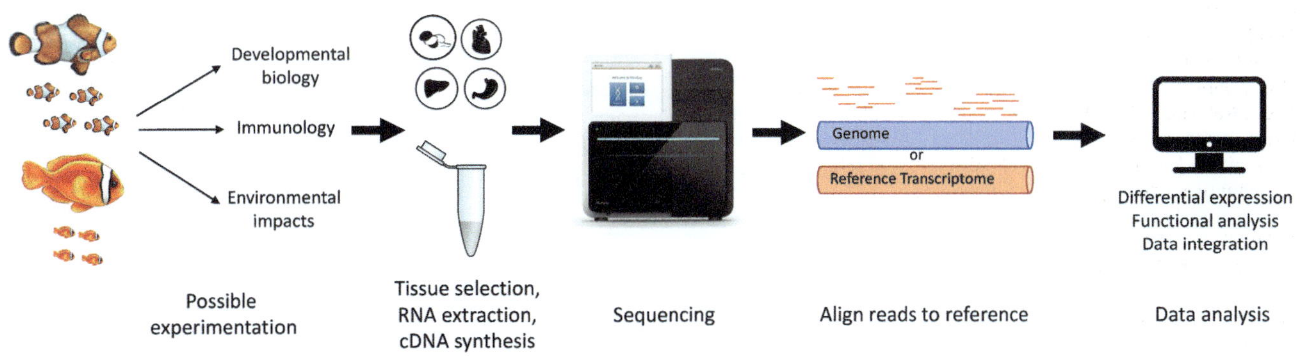

FIGURE 2.2 Schematic of possible applications and transcriptomics techniques to be used for anemonefish research.

The advancements of RNA-seq technology combined with the growing genomic resources for anemonefish (e.g., Lehmann et al. 2019; Marcionetti et al. 2018, 2019) make this group an excellent candidate for integrative studies. Also, their relationship with host anemones offers a unique opportunity to examine the molecular processes behind symbiosis between two taxonomic groups. Combining molecular processes with physiological changes under environmental changes or between various life stages will provide powerful insight into anemonefish ecology.

2.4 ANEMONEFISH PROTEOMICS

Proteomics is the quantification of all proteins present in an organism, tissue or cell at a point in time and is complementary to other omics techniques, such as transcriptomics (Aslam et al. 2017) (Figure 2.3). The proteome can provide greater insight into cellular phenotypes by measuring the abundance of proteins and identifying their functional information (Aebersold and Mann 2016; Tang et al. 2015) (Figure 2.3). Variation over time and across cells as well as post-translational modifications create a dynamic and complex research field that has lagged behind other -omics research (Aebersold and Mann 2016; Liu et al. 2016). However, proteomics often has a stronger correlation to observed phenotypes than transcriptomics or genomics, making it an important tool in identifying molecular pathways behind biological characterizations (Liu et al. 2016; Tang et al. 2015).

Conventional methods in proteomics focus on using established biochemistry methods to isolate specific proteins to study their structure and function (Aebersold and Mann 2016). Research has been concentrated on disease and drug development in humans and model organisms (i.e., mice), using targeted methods where the proteins in question were already known, and measurement assays were already developed (Edwards et al. 2011). This has led to a specific set of intensely studied proteins over the past decades, despite increases in genetic knowledge. However, recent technological advancements in mass spectrometry have provided the tools to accurately and reliably quantify amino acids at a proteome-wide scale (Aebersold and Mann 2016).

One popular method, which started to gain traction due to possible use in non-model organisms, is iTRAQ (isobaric tags for relative and absolute quantification, Figure 2.3). Through this approach, different biological samples are labelled and processed together on a mass spectrometer. Then, the measured relative abundance of the peptides or proteins is compared. It has recently been used in a wide array of studies in non-model organisms and in ecological contexts such as behavior or responses to environmental change (Effertz et al. 2014; Xu et al. 2016). A study on one Pomacentridae fish species identified protein changes in the brain under elevated ocean acidification conditions (Tsang et al. 2020). The biggest limitation to this method is the number of possible relative comparisons. With iTRAQ labelling, the number of samples that can be compared directly is limited to the number of labels, which are generally either four or eight. Hence, pooling samples within one label is commonly used to increase the number of individuals measured and therefore results cannot be compared across experiments (Evans et al. 2012).

A newer mass spectrometry method, named sequential window acquisition of all theoretical spectra (SWATH-MS), is able to identify and quantify thousands of proteins in one measurement (Gillet et al. 2012; Figure 2.3). It is label-free, making it relatively cheap, and it has been shown to have high reproducibility across different labs (Collins et al. 2017). This method uses data-dependent acquisition (DDA) on the mass spectrometer to create a spectral library against which samples quantified with data-independent acquisition (DIA) can be mapped (Gillet et al. 2012; Huang et al. 2015; Figure 2.3). Once a spectral library has been created, it can theoretically be used in different labs to identify proteome level changes across individuals (Rosenberger et al. 2017). Currently, this

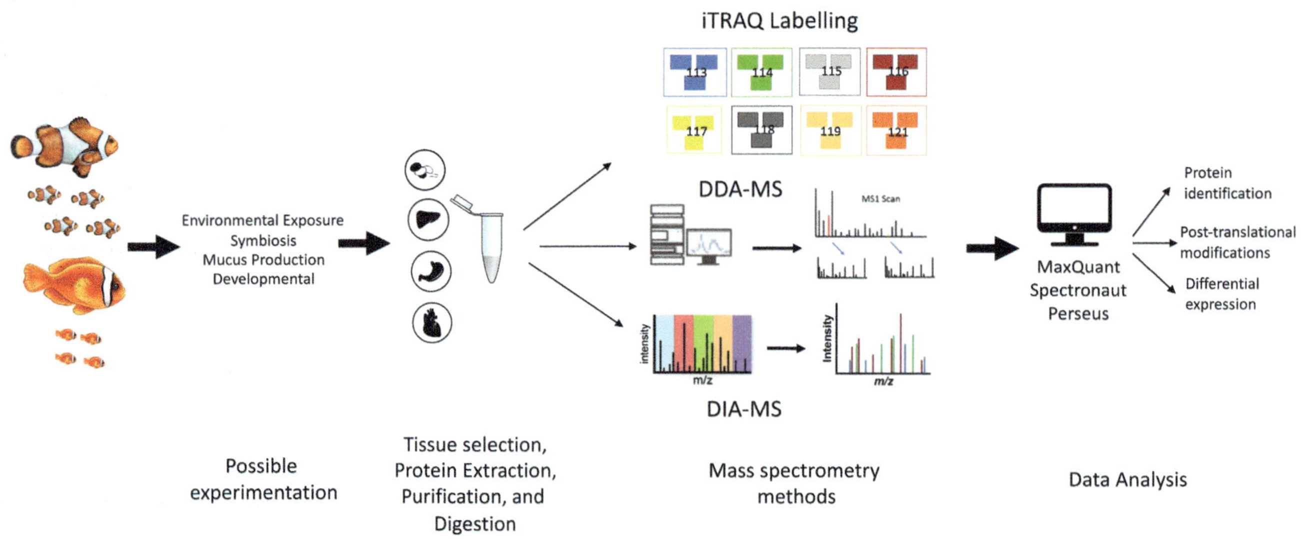

FIGURE 2.3 Schematic of possible applications and proteomics techniques to be used for anemonefish research.

method has been used to understand molecular mechanisms defining complex physiological phenotypes in several model organisms, including humans, mice, Arabidopsis, and zebrafish (Blattmann et al. 2019; Braccia et al. 2018; Bruderer et al. 2015; Collins et al. 2017; Krasny et al. 2018; Rosenberger et al. 2014; Zhang et al. 2019). A recent study provided the first step to applying this method to a wide range of non-model organisms and wild individuals with intrinsic individual variation (Monroe et al. 2020). The research evaluated the effectiveness of SWATH-MS in identifying proteomic expression differences in a closely related coral reef associated species to anemonefish, *Acanthochromis polyacanthus* (Monroe et al. 2020). This method provides the ability to detect significant differentially expressed proteins from ecologically relevant pathways across individuals exposed to variable environmental conditions.

The advancement of new techniques and the strong ties of the proteome to observed phenotypes, makes proteomics a powerful analytical tool in molecular ecology. Rapid developments in quantitative methods in the past decade, increasing reproducibility and data density, have turned quantitative proteomics into a reality (Gillet et al. 2012; Rosenberger et al. 2017; Tang et al. 2015). Powerful mass spectrometry analyses and bioinformatic advancements have created a mainstream way to examine ecologically relevant, proteome level changes in non-model fish species (Forné et al. 2010). This allows for wide-ranging use of proteomics to study many aspects concerning anemonefishes. Despite this usefulness, proteomics has only been employed to study the host anemone in the context of toxicity and drug development (Domínguez-Pérez et al. 2018). We encourage more studies to focus on the protein level with powerful mass spectrometry analyses to better understand ecological and molecular processes such as development, responses to environmental change (e.g., Monroe et al. 2020; Tsang et al. 2020), and optimization of aquacultural and husbandry conditions (e.g. Díaz-Jiménez et al. 2020). Proteomics can also be used to evaluate processes driving symbiosis with the host anemone, behavior, reproduction, and parental care in *Amphiprion* species.

2.5 CONCLUSIONS

In this chapter, we described several advances in genomics technologies that substantially transformed the role of anemonefish as a group in the understanding of evolution, ecology, physiology, and genetics of coral reef fishes. For example, as described in the "Anemonefish Phylogenomics" section, the availability of several chromosomes-scale genomes for anemonefish species allowed researchers, for the first time, to resolve an accurate phylogeny of this group of fishes and in the process highlighted interesting aspects of their mutualistic lifestyle with host anemones, their unique color patterns, and their development. The "Transcriptomics and Proteomics" sections demonstrated how these genome-wide technologies have been recently applied to non-model organisms, and how they can improve our understanding of the molecular mechanisms underlying anemonefishes' responses to predicted future climate conditions, sex change, social structure, and development. To conclude, the rapid development of genomic technologies has driven the availability of high-quality genome-wide datasets for anemonefish species. These datasets will have a transformative impact on anemonefish coral reef fish research in general, and will further establish these fishes as important model organisms for ecology, genetics, and developmental biology.

ACKNOWLEDGEMENTS

We would like to thank Anna Marcionetti and Sarah Schmid for their discussions and feedback on the genomic part of this chapter. Thank you to the Okinawa Institute of Science and Technology & Graduate University (OIST) and the Australian Research Council Centre of Excellence for Coral Reef Studies, James Cook University, for the support to T. R. and T. W. R.

REFERENCES

Aebersold, R., and M. Mann. 2016. Mass-spectrometric exploration of proteome structure and function. *Nature 537*: 347–355.

Aslam, B., M. Basit, M. A. Nisar, M. Khurshid, and M. H. Rasool. 2017. Proteomics: Technologies and their applications. *Journal of Chromatographic Science 55*(2): 182–196.

Bernal, M. A., J. M. Donelson, H. D. Veilleux, T. Ryu, P. L. Munday, and T. Ravasi. 2018. Phenotypic and molecular consequences of stepwise temperature increase across generations in a coral reef fish. *Molecular Ecology 27*(22): 4516–4528.

Berner, D., and W. Salzburger. 2015. The genomics of organismal diversification illuminated by adaptive radiations. *Trends in Genetics 31*(9): 491–499.

Blattmann, P., V. Stutz, G. Lizzo, J. Richard, P. Gut, and R. Aebersold. 2019. Data descriptor: Generation of a zebrafish SWATH-MS spectral library to quantify 10,000 proteins. *Scientific Data 6*: 1–11.

Braccia, C., M. P. Espinal, M. Pini, D. De Pietri, and A. Armirotti. 2018. A new SWATH ion library for mouse adult hippocampal neural stem cells. *Data in Brief 18*: 1–8.

Brawand, D., C. E. Wagner, Y. I. Li, M. Malinsky, I. Keller, S. Fan, O. Simakov, et al. 2014. The genomic substrate for adaptive radiation in African cichlid fish. *Nature 513*(7518): 375–381.

Bruderer, R., O. M. Bernhardt, T. Gandhi, S. M. Miladinović, L-Y. Cheng, S. Messner, T. Ehrenberger, et al. 2015. Extending the limits of quantitative proteome profiling with data-independent acquisition and application to acetaminophen-treated three-dimensional liver microtissues. *Molecular & Cellular Proteomics 14*(5): 1400–1410.

Casas, L., F. Saborido-Rey, T. Ryu, C. Michell, T. Ravasi, and X. Irigoien. 2016. Sex change in clownfish: Molecular insights from transcriptome analysis. *Scientific Reports 6*(1): 1–19.

Collins, B. C., C. L. Hunter, Y. Liu, B. Schilling, G. Rosenberger, S. L. Bader, D. W. Chan, et al. 2017. Multi-laboratory assessment of reproducibility, qualitative and quantitative performance of SWATH-mass spectrometry. *Nature Communications 8*(1): 1–11.

Connon, R. E., K. M. Jeffries, L. M. Komoroske, A. E. Todgham, and N. A. Fangue. 2018. The utility of transcriptomics in fish conservation. *Journal of Experimental Biology 221*(2): jeb148833.

Dasmahapatra, K. K., J. R. Walters, A. D. Briscoe, J. W. Davey, A. Whibley, N. J. Nadeau, A. V. Zimin. 2012. Butterfly genome reveals promiscuous exchange of mimicry adaptations among species. *Nature 487*(7405): 94–98.

Díaz-Jiménez, L., M. P. Hernández-Vergara, and C. I. Pérez-Rostro. 2020. Protein/lipid ratio for the growth of juvenile clownfish, *Amphiprion ocellaris. Journal of the World Aquaculture Society 51*(3): 666–678.

Domínguez-Pérez, D., A. Campos, A. A. Rodríguez, M. V. Turkina, T. Ribeiro, H. Osorio, V. Vasconcelos, et al. 2018. Proteomic analyses of the unexplored sea anemone *Bunodactis verrucosa. Marine Drugs 16*(2): 42.

Edwards, A. M., R. Isserlin, G. D. Bader, S. V. Frye, T. M. Willson, and F. H. Yu. 2011. Too many roads not taken. *Nature 470*(7333): 163–165.

Effertz, C., S. Müller, and E. von Elert. 2014. Differential peptide labeling (iTRAQ) in LC–MS/MS based proteomics in daphnia reveal mechanisms of an antipredator response. *Journal of Proteome Research 14*(2): 888–896.

Evans, C., J. Noirel, S. Y. Ow, M. Salim, A. G. Pereira-Medrano, N. Couto, J. Pandhal, et al. 2012. An insight into iTRAQ: Where do we stand now? *Analytical and Bioanalytical Chemistry 404*(4): 1011–1027.

Ferraresso, S., A. Bonaldo, L. Parma, S. Cinotti, P. Massi, L. Bargelloni, and P. P. Gatta. 2013. Exploring the larval transcriptome of the common sole (*Solea solea L.*). *BMC Genomics 14*(1): 1–22.

Forné, I., J. Abián, and J. Cerdà. 2010. Fish proteome analysis: Model organisms and non-sequenced species. *Proteomics 10*(4): 858–872.

Fu, J., W. Zhu, L. Wang, M. Luo, F. Song, and Z. Dong. 2019. Dynamic transcriptome sequencing and analysis during early development in the bighead carp (*Hypophthalmichthys nobilis*). *BMC Genomics 20*(1): 1–14.

Gainsford, A., G. P. Jones, J. A. Hobbs, F. M. Heindler, and L. Herwerden. 2020. Species integrity, introgression, and genetic variation across a coral reef fish hybrid zone. *Ecology and Evolution 10*(21): 11998–12014.

Gillet, L. C., P. Navarro, S. Tate, H. Röst, N. Selevsek, L. Reiter, R. Bonner, et al. 2012. Targeted data extraction of the MS/MS spectra generated by data-independent acquisition: A new concept for consistent and accurate proteome analysis. *Molecular & Cellular Proteomics 11*(6): O111.016717.

Huang, Q., L. Yang, J. Luo, L. Guo, Z. Wang, X. Yang, W. Jin, et al. 2015. SWATH enables precise label-free quantification on proteome scale. *Proteomics 15*(7): 1215–1223.

Huth, T. J., and S. P. Place. 2016. RNA-seq reveals a diminished acclimation response to the combined effects of ocean acidification and elevated seawater temperature in *Pagothenia borchgrevinki. Marine Genomics 28*: 87–97.

Jones, F. C., M. G. Grabherr, Y. Chan, P. Russell, E. Mauceli, J. Johnson, R. Swofford, et al. 2012. The genomic basis of adaptive evolution in threespine sticklebacks. *Nature 484*(7392): 55–61.

Kobayashi, Y., Y. Nagahama, and M. Nakamura. 2013. Diversity and plasticity of sex determination and differentiation in fishes. *Sexual Development: Genetics, Molecular Biology, Evolution, Endocrinology, Embryology, and Pathology of Sex Determination and Differentiation 7*(1–3): 115–125.

Krasny, L., P. Bland, N. Kogata, P. Wai, B. A. Howard, R. C. Natrajan, and P. H. Huang. 2018. SWATH mass spectrometry as a tool for quantitative profiling of the matrisome. *Journal of Proteomics 189*(January): 11–22.

Lamichhaney, S., J. Berglund, M. S. Almén, K. Maqbool, M. Grabherr, A. Martinez-Barrio, M. Promerová, et al. 2015. Evolution of Darwin's finches and their beaks revealed by genome sequencing. *Nature 518*(7539): 371–375.

Lehmann, R., D. J. Lightfoot, C. Schunter, C. T. Michell, H. Ohyanagi, K. Mineta, S. Foret, et al. 2019. Finding Nemo's genes: A chromosome-scale reference assembly of the genome of the orange clownfish Amphiprion percula. *Molecular Ecology Resources 19*(3): 570–585.

Litsios, G., and N. Salamin. 2014. Hybridisation and diversification in the adaptive radiation of clownfishes. *BMC Evolutionary Biology 14*(1): 245.

Litsios, G., C. A. Sims, R. O. Wüest, P. B. Pearman, N. E. Zimmermann, and N. Salamin. 2012. Mutualism with sea anemones triggered the adaptive radiation of clownfishes. *BMC Evolutionary Biology 12*(1): 212.

Liu, Y., A. Beyer, and R. Aebersold. 2016. On the dependency of cellular protein levels on mRNA abundance. *Cell 165*(3): 535–550.

Ma, X., B. Su, Y. Tian, N. J. C. Backenstose, Z. Ye, A. Moss, T. Y. Duong, et al. 2020. Deep transcriptomic analysis reveals the dynamic developmental progression during early development of channel catfish (*Ictalurus punctatus*). *International Journal of Molecular Sciences 21*(15): 1–22.

Marcionetti, A., and N. Salamin. 2022. Insights into the genomics of clownfish adaptive radiation: The genomic substrate of the diversification. *bioRxiv* 2022.05.12.491701, doi: https://10.1101/2022.05.12.491701.

Marcionetti, A., V. Rossier, J. A. M. Bertrand, G. Litsios, and N. Salamin. 2018. First draft genome of an iconic clownfish species (*Amphiprion frenatus*). *Molecular Ecology Resources 18*(5): 1092–1101.

Marcionetti, A., V. Rossier, N. Roux, P. Salis, V. Laudet, and N. Salamin. 2019. Insights into the genomics of clownfish adaptive radiation: Genetic basis of the mutualism with sea anemones. *Genome Biology and Evolution 11*(3): evz042.

Martin, J. A., and Z. Wang. 2011. Next-generation transcriptome assembly. *Nature Reviews Genetics 12*(10): 671–682.

Maytin, A. K., S. W. Davies, G. E. Smith, S. P. Mullen, and P. M. Buston. 2018. De novo transcriptome assembly of the clown anemonefish (*Amphiprion percula*): A new resource to study the evolution of fish color. *Frontiers in Marine Science 5*(8): 284.

Mitchell, L. J., K. L. Cheney, M. Lührmann, J. Marshall, K. Michie, and F. Cortesi. 2021. Molecular evolution of ultraviolet visual opsins and spectral tuning of photoreceptors in anemonefishes (Amphiprioninae). *Genome Biology and Evolution 13*(10): evab184.

Monroe, A. A., H. Zhang, C. Schunter, and T. Ravasi. 2020. Probing SWATH-MS as a tool for proteome level quantification in a nonmodel fish. *Molecular Ecology Resources 20*(6): 1647–1657.

Narum, S. R., and N. R. Campbell. 2015. Transcriptomic response to heat stress among ecologically divergent populations of redband trout. *BMC Genomics16*(1): 1–12.

Oomen, R. A., and J. A. Hutchings. 2017. Transcriptomic responses to environmental change in fishes: Insights from RNA sequencing. *FACETS 2*(2): 610–641.

Porteus, C. S., P. C. Hubbard, T. M. Uren Webster, R. van Aerle, A. V. M. Canário, E. M. Santos, and R. W. Wilson. 2018. Near-future CO2 levels impair the olfactory system of a marine fish. *Nature Climate Change 8*: 737–743.
Qian, X., Y. Ba, Q. Zhuang, and G. Zhong. 2014. RNA-Seq technology and its application in fish transcriptomics. *OMICS: A Journal of Integrative Biology 18*(2): 98.
Ronco, F., M. Matschiner, A. Böhne, A. Boila, H. H. Büscher, A. E. Taher, A. Indermaur, et al. 2021. Drivers and dynamics of a massive adaptive radiation in cichlid fishes. *Nature 589*(7840): 76–81.
Rosenberger, G., I. Bludau, U. Schmitt, M. Heusel, C. L. Hunter, Y. Liu, M. J. Maccoss, et al. 2017. Statistical control of peptide and protein error rates in large-scale targeted data-independent acquisition analyses. *Nature Methods 14*(9): 921–927.
Rosenberger, G., C. C. Koh, T. Guo, H. L. Röst, P. Kouvonen, B. C. Collins, M. Heusel, et al. 2014. A repository of assays to quantify 10,000 human proteins by SWATH-MS. *Scientific Data 1*: 1–15.
Ryu, T., M. Herrera, B. Moore, M. Izumiyama, E. Kawai, V. Laudet, and T. Ravasi. 2022. A chromosome-scale genome assembly of the false clownfish, *Amphiprion ocellaris. G3 Genes|Genomes|Genetics* 12(5): jkac074.
Salis, P., S. Lee, N. Roux, D. Lecchini, and V. Laudet. 2021. The real Nemo movie: Description of embryonic development in *Amphiprion ocellaris* from first division to hatching. *Developmental Dynamics 250*(11): 1651–1667.
Salis, P., T. Lorin, V. Lewis, C. Rey, A. Marcionetti, M-L. Escande, N. Roux, et al. 2019. Developmental and comparative transcriptomic identification of iridophore contribution to white barring in clownfish. *Pigment Cell & Melanoma Research 32*(3): 391–402.
Schunter, C., M. D. Jarrold, P. L. Munday, T. Ravasi. 2021. Diel CO_2 fluctuations alter the molecular response of coral reef fishes to ocean acidification conditions. *Molecular Ecology 30*(20): 5105–5118.
Schunter, C., M. J. Welch, G. E. Nilsson, J. L. Rummer, P. L. Munday, and T. Ravasi. 2018. An interplay between plasticity and parental phenotype determines impacts of ocean acidification on a reef fish. *Nature Ecology and Evolution 2*(2): 334–342.
Schunter, C., M. J. Welch, T. Ryu, H. Zhang, M. L. Berumen, G. E. Nilsson, P. L. Munday, et al. 2016. Molecular signatures of transgenerational response to ocean acidification in a species of reef fish. *Nature Climate Change 6*(11): 1014–1018.
Shama, L. N. S., A. Strobel, F. C. Mark, and K. M. Wegner. 2014. Transgenerational plasticity in marine sticklebacks: Maternal effects mediate impacts of a warming ocean. *Functional Ecology 28*(6): 1482–1493.
Stieb, S. M., F. de Busserolles, K. L. Carleton, F. Cortesi, W. S. Chung, B. E. Dalton, A. Hammond, et al. 2019. A detailed investigation of the visual system and visual ecology of the Barrier Reef anemonefish, *Amphiprion akindynos. Scientific Reports 9*(1): 1–14.
Stern, D. L., and V. Orgogozo. 2008. The loci of evolution: How predictable is genetic evolution? *Evolution; International Journal of Organic Evolution 62*(9): 2155–2177.
Tang, K. L., M. L. J. Stiassny, R. L. Mayden, and R. DeSalle. 2021. Systematics of damselfishes. *Ichthyology & Herpetology 109*(1): 258–318.
Tang, X., Q. Meng, J. Gao, S. Zhang, H. Zhang, and M. Zhang. 2015. Label-free quantitative analysis of changes in broiler liver proteins under heat stress using SWATH-MS technology. *Scientific Reports 5*: 1–15.
Tsang, H. H., M. J. Welch, P. L. Munday, T. Ravasi, and C. Schunter. 2020. Proteomic responses to ocean acidification in the brain of juvenile coral reef fish. *Frontiers in Marine Science 7*(7): 605.
Veilleux, H. D., T. Ryu, J. M. Donelson, L. Van Herwerden, L. Seridi, Y. Ghosheh, M. L. Berumen, et al. 2015. Molecular processes of transgenerational acclimation to a warming ocean. *Nature Climate Change 5*(12): 1074–1078.
Vesterlund, L., H. Jiao, P. Unneberg, O. Hovatta, and J. Kere. 2011. The zebrafish transcriptome during early development. *BMC Developmental Biology 11*: 30.
Wang, Z., M. Gerstein, M. Snyder. 2009. RNA-Seq: A revolutionary tool for transcriptomics. *Nature Reviews Genetics 10*(1): 57–63.
Wray, G. A. 2007. The evolutionary significance of cis-regulatory mutations. *Nature Reviews Genetics 8*(3): 206–216.
Xu, D., L. Sun, S. Liu, L. Zhang, H. Yang. 2016. Understanding the heat shock response in the sea cucumber, *Apostichopus japonicus*, using iTRAQ-Based proteomics. *International Journal of Molecular Sciences 17*(2): 150.
Xu, E. G., E. M. Mager, M. Grosell, J. D. Stieglitz, E. S. Hazard, G. Hardiman, and D. Schlenk. 2017. Developmental transcriptomic analyses for mechanistic insights into critical pathways involved in embryogenesis of pelagic mahi-mahi (*Coryphaena hippurus*). *PLOS ONE 12*(7): e0180454.
Zhang, H., P. Liu, T. Guo, H. Zhao, D. Bensaddek, R. Aebersold, and L. Xiong. 2019. Arabidopsis proteome and the mass spectral assay library. *Scientific Data 6*: 278.

3 Biogeography and Genetic Barriers in *Amphiprion* Anemonefishes

Song He, Benoit Pujol, Serge Planes, and Michael L. Berumen

CONTENTS

3.1 INTRODUCTION

Anemonefish are a focal species for much research on connectivity, particularly concerning larval dispersal studies. Certain anemonefish species have become model organisms for empirical studies of larval dispersal using parentage analysis techniques (Chapter 20). As the species are typically extremely site-attached, they are useful for understanding larval connectivity because of the very limited potential for adult movements. The study of biogeography patterns can thus provide insights into the underlying connectivity processes in an evolutionary and larger geographic context.

As discussed in the previous chapter (Chapter 2), 28 species of anemonefish have been recognized and accepted. They are distributed throughout the tropical Indo-Pacific Ocean basins. Litsios et al. (2014) assessed genetic data for 27 species (*Amphiprion* and *Premnas*) in a biogeographical context. While most species are distributed within the Indo-Australian archipelago, one clade colonized the eastern shores of Africa. No difference in diversification rate between the main radiation and the African clade was detected (Litsios et al. 2014). Hence, the biogeographic history of the anemonefish is characterized as a replicated adaptive radiation that occurred in different regions. The anemonefish first originated and diversified in the Indo-Australian archipelago and geographically independent radiation occurred in the Indian Ocean off East Africa (Litsios et al. 2014).

3.2 BIOGEOGRAPHY AND GENE FLOW

Six biogeographic regions were assessed by Litsios et al. (2014) for anemonefishes (Figure 3.1). Not every anemonefish species has a distribution range limited to a single region; in fact, more than one-third of species occur in multiple regions. To investigate the species boundaries status of those widely distributed species, we compared various genetic distances within and between species (Figure 3.2). As one of the most common markers archived in GenBank data for anemonefish species, sequences from the cytochrome b (Cytb) fragment were selected for generating the specific genetic comparison database.

All the accessible Cytb sequences for 13 anemonefish species that are distributed in more than one of the biogeographic regions, or for which we can make a comparison with sister species (i.e., *Amphiprion bicinctus*, *A. latifasciatus*, *A. chagosensis*, *A. nigripes*, *A. akindynos*, *A. polymnus*, *A. sebae*, *A. perideraion*, *A. akallopisos*, *A. chrysopterus*, *A. melanopus*, *A. clarkii*, and *A. percula*) were downloaded (accession numbers are available in Appendix 3.1) and aligned using Geneious Prime 2022.1.1 (www.geneious.com). For each species, the whole mitochondrial sequence was used to help create the alignment. Records with short aligned lengths or with low sequence qualities were discarded from the next step in the comparison of genetic distance. Pairwise genetic distance comparisons were conducted in MEGA6 (Tamura et al. 2013) by calculating the *p*-distance for each sequence pair. Codon positions included were 1st+2nd+3rd+Noncoding. All positions containing gaps and missing data were eliminated. A very low *p*-distance (i.e., close to 0) indicates high genetic similarity whereas higher *p*-distances indicate less gene flow (i.e., more genetic difference). A *p*-distance of 0.1 could generally be regarded to indicate a strong species boundary and a *p*-distance below 0.05 could indicate intraspecific variation. However, the application of specific thresholds in *p*-distance evaluation is somewhat subjective and the implications may vary among taxonomic groups. For example, hybridization has been documented to occur in species with *p*-distances as high as 0.17 in groupers (Chen et al. 2017) and 0.13 in *Dascyllus* spp. (He et al. 2017).

For intraspecific comparisons, the maximum values of *p*-distance from samples collected in different

DOI: 10.1201/9781003125365-4

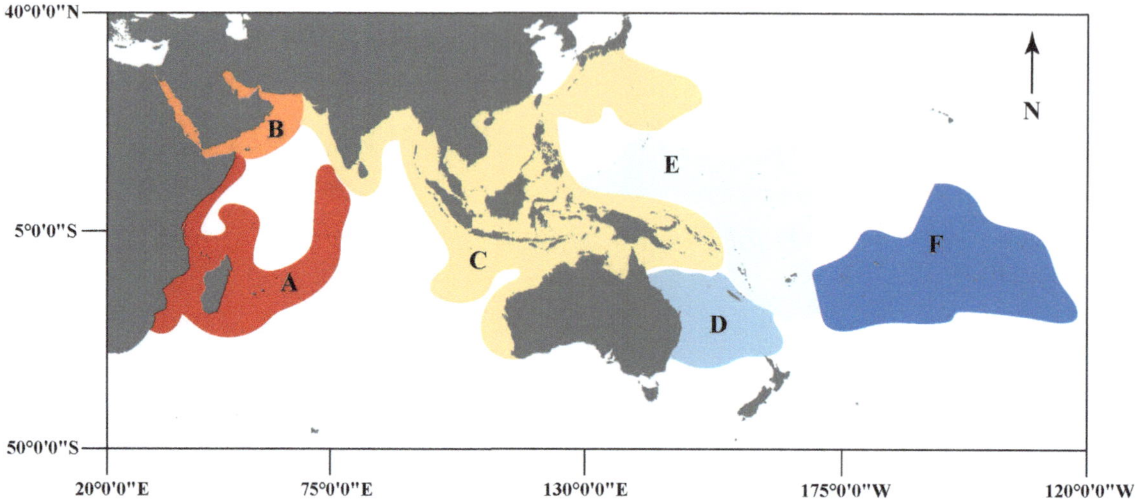

FIGURE 3.1 Biogeographic regions for anemonefishes (redrawn approximating Litsios et al. 2014). A. Western Indian region; B. north-western Indian region; C. central Indo-Pacific region; D. south-western Pacific region; E. central Pacific region; F. Polynesian region.

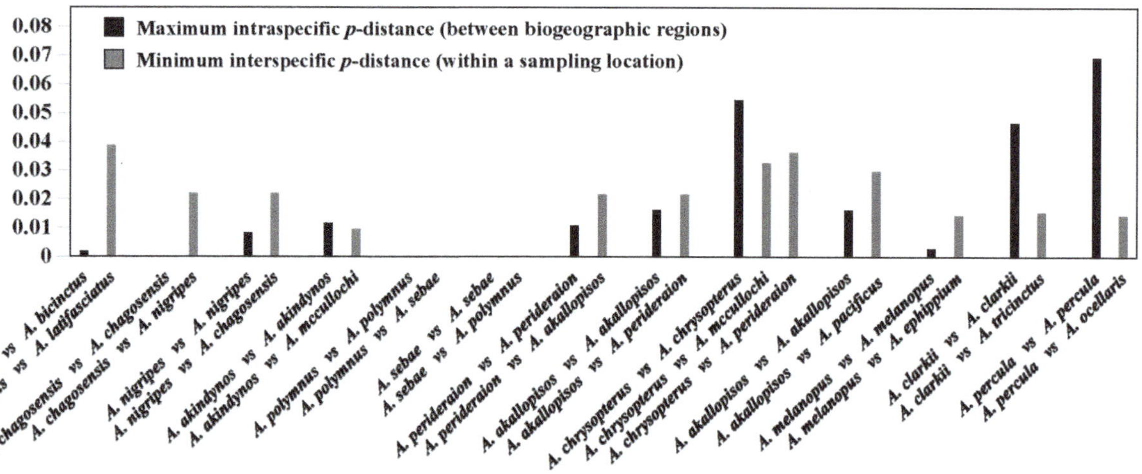

FIGURE 3.2 Genetic distance comparison between inter- and intraspecific pairs of anemonefish with cross-regional biogeographic distributions. Vertical bars indicate the *p*-distance (genetic dissimilarity) at the Cytb mitochondrial gene for each comparison (indicated on the x-axis). As indicated in the legend, the genetic variations within a species are shown in black vertical bars and represent the maximum *p*-distance value for sequences analyzed for samples of the same species but from different biogeographic regions. Interspecific comparisons are shown in grey vertical bars and represent the minimum *p*-distance between species pairs sampled from the same location.

biogeographic regions were selected to indicate the upper limit of the intraspecific variation. These values are indicated in black bars in Figure 3.2. For comparison of closely related or sister species (i.e., interspecific comparisons), sequences were carefully selected to represent individuals from within the same sampling location or from the nearest location possible within the region. These values are indicated in gray bars in Figure 3.2. For interspecific comparisons, the minimum *p*-distance values for each comparison were selected, indicative of the species boundaries and the potential for genetic exchange between the species pairs. Overall, this analysis allows us to assess whether there is a greater genetic distance within individuals of a species compared to the genetic distance between species. In principle, interspecific genetic distance should be greater than intraspecific genetic distance. Exceptions to this expectation could identify opportunities for future investigations of other mechanisms maintaining species boundaries among closely related taxa. For species with distributions spanning multiple biogeographic regions, identifying larger-than-expected intraspecific genetic distance may also be the first step to identifying different micro- or macro-evolutionary processes operating within geographically separated populations within a species.

Six out of the 13 species of anemonefish with cross-region distributions exhibit less genetic distance among populations than between their most closely related species: *A. bicinctus*, *A. chagosensis*, *A. nigripes*, *A. perideraion*, *A.*

akallopisos, and *A. melanopus*. In other words, the genetic data from Cytb sequences support the species boundaries currently recognized for these anemonefishes. Two of these species, *A. akallopisos* and *A. bicinctus*, have been the subject of studies of gene flow within their distribution.

Huyghe and Kochzius (2017) examined the case of *A. akallopisos* in further detail. This species is distributed across regions A (Western Indian region) and C (Central Indo-Pacific region) (Figure 3.1). However, none of the haplotypes of the mitochondrial marker CRA/CRE was shared between samples of *A. akallopisos* collected from these two regions (Huyghe and Kochzius 2017). A single mutation segregates samples from these two region samples that is not sufficient to show reciprocally monophyletic patterns. Genetic diversity (calculated based on the CRA/CRE marker) was much higher in samples from region B than samples in region A, suggesting that *A. akallopisos* geographically originated from region B. Given the large distance between the disjunct populations and the short pelagic larval duration, long-distance dispersal is rather unlikely. A stepping stone model involving islands in the central Indian Ocean is a more likely scenario for the historical colonization emanating from the region B populations. Prevailing ocean currents create (or created in the past) irregular but unidirectional gene flow from region B. Alternatively, there could be recent barriers to dispersal among populations that formed too recently to establish sufficient genetic distance to be considered as separate species. A "young" barrier would result in some genetic distance but only at the level expected for population differences within a species (Huyghe and Kochzius 2017).

Nanninga et al. (2014) investigated the population-level genetic patterns of *A. bicinctus* along the Saudi Arabian coast of the Red Sea using microsatellite markers. Gene flow among the northern and central Red Sea populations appeared to follow a stepping-stone model, which was disrupted by a distinct genetic break at a latitude of approximately 19°N. This break corresponds with the pronounced environmental changes in the Red Sea between the southernmost end of the Farasan Banks reef complex and the northernmost end of the Farasan Islands, roughly separating the northern and central Red Sea from the southern basin (Nanninga et al. 2014). Interestingly, a similar genetic structure was found for a species of sponge (Giles et al. 2015), suggesting that a similar phenomenon may occur across a wide range of taxa. These patterns were further confirmed by a more comprehensive ddRAD-based analysis (Saenz-Agudelo et al. 2015). The ddRAD data also confirmed the previous findings of interspecific hybrids, including hybrids of *A. bicinctus* and *A. omanensis*, at an intermediate suture zone with the Indian Ocean island of Socotra (DiBattista et al. 2015). Samples of anemonefishes from the Gulf of Aden indicated complex patterns of genomic admixture with evidence of introgression between species. Nonetheless, the SNPs analysis suggests that the species boundary between *A. bicinctus* and *A. omanensis* is strong (despite the lack of differences in mtDNA previously reported; DiBattista et al. 2015; Litsios and Salamin 2014).

Among the 13 anemonefish species analyzed previously, two species (*A. polymnus* and *A. sebae*) show no genetic difference within the species or between the sister species. This indicates that the Cytb gene pool is shared among these sister species and this marker shows no segregation abilities for these species. Litsios et al. (2014) were able to resolve the two species using several nuclear markers. Although there is no *p*-distance difference detected by the Cytb mitochondrial marker, the genetic structure has been found between *A. polymnus* populations based on microsatellite markers (Saenz-Agudelo et al. 2009). A low rate of gene flow was indicated between two locations separated by 1,500 km: a metapopulation in Bootless Bay and a population at Schumann Island, both in Papua New Guinea (region C) (Saenz-Agudelo et al. 2009).

Three of the 13 examined anemonefish species, *A. chrysopterus*, *A. clarkii*, and *A. percula*, have higher genetic distances among individuals within the species than compared to the genetic distance with their closely related sister species. For these three, the species boundaries are not supported by Cytb sequence data. Interspecific hybridization among these species might have contributed to the high genetic variation among individuals. Hybridization may transfer unique alleles from a different species into the gene pool and the resultant genetic distance estimation may be larger than the species boundaries for the purebreds. Natural hybridization events between *A. chrysopterus* and *A. sandaracinos* in Kimbe Bay, Papua New Guinea, are well documented (Gainsford et al. 2015; He et al. 2018). Potential nuclear DNA introgression was also evident through distinct intermediate hybrid genotypes identifiable in both parent species (Gainsford et al. 2015). As was the case for *A. polymnus* and *A. sebae*, Litsios et al. (2014) were able to resolve clear species relationships for these species using several nuclear markers. Further population genetics investigations on these three species, involving substantial sampling sizes and multiple genetic markers from different regions across their entire distribution ranges, could evaluate the potential existence of cryptic speciation or at least the existence of distinct subspecies.

Finally, although not a species included in our *p*-distance assessment, Simpson et al. (2014) assessed long-distance dispersal in the Oman anemonefish (*A. omanensis*). This species occurs in region B. Genetic assignment tests (based on a multi-marker database) demonstrated the bidirectional exchange of first-generation migrants of *A. omanensis*, with subsequent social and reproductive integration, between two populations separated by over 400 km (Simpson et al. 2014). The species integrity of *A. omanensis* appears to have well-maintained local connectivity of populations via the dispersal of planktonic larvae within region B. The biogeography and availability of suitable habitats on the southern end of the Arabian peninsula indicate that some larvae must disperse 100–1,000 kilometres (Simpson et al. 2014).

3.3 MICRO- TO MACRO-EVOLUTIONARY PROCESSES THAT BOOST OR POTENTIALLY CONSTRAIN SPECIATION

As reviewed in Chapter 1, anemonefishes form a monophyletic lineage within damselfishes (Litsios et al. 2012; Tang et al. 2021) and are known to have settled with success across a large range of ecological habitats and marine ecoregions. The macroevolutionary divergence of anemonefishes is relatively recent (McCord et al. 2021). Damselfishes seem to have evolved rapidly in response to their environment, which is an opportunity to question the similarity of macroevolutionary and microevolutionary drivers of anemonefish diversity. From a macroevolutionary biogeographic perspective, parallel evolution of damselfishes in different marine ecoregions resulted in similar assemblages of ecomorphs formed by different species within habitats (Gajdzik et al. 2018). This process was rendered possible by means of "reticulate adaptive radiation" (Cooper and Westneat 2009) or "iterative evolution" (Frédérich et al. 2013). Network-like contact and hybridization followed by local divergence combined with gradual adaptive evolutionary change, therefore shaping the diversity of anemonefish species.

In terms of microevolutionary mechanisms, it is widely acknowledged that both connectivity (i.e., successful dispersal events from one island source population to a destination population on another island) and self-recruitment (i.e., successful settlement of an offspring into the population on its natal island) shape the genetic diversity of anemonefish populations (Berumen et al. 2012; Jones et al. 2005). For example, in the orange clownfish, *A. percula*, connectivity and self-recruitment have been found to reach a near fifty-fifty balance in some populations (Salles et al. 2016). Both at the scale of anemonefish populations and species, locally specific evolutionary dynamics are made possible by the presence of self-recruitment. Furthermore, species hybridization and population contact enhance diversity. Macro- and microevolutionary mechanisms, therefore, seem to be tightly linked and likely produce coherent changes in anemonefish diversity at the biogeographic and local scales.

Whether macro- and microevolutionary mechanisms (drift, migration, mutation, and/or drift) are shaping the diversity of species and populations equally remains unknown to date in most species. Little is known about the anemonefish selection-mutation balance at the scale of species and populations. From the global evolutionary biogeographic knowledge that can be gathered on anemonefish species, it is likely that the environmental demand of the local habitat gradually selected for a similar suite of functional traits in different marine ecoregions. We discuss in the following section various possibilities for further examination of anemonefish ecology in terms of better understanding links and influences on their evolutionary ecology.

In wild clownfish populations, recent quantitative genetic results indicate the presence of a microevolutionary equilibrium in terms of adaptation rate, as illustrated by little to no genetic variation for fitness (Salles et al. 2020). In other words, selection and mutation appeared balanced in the populations. It remains, however, difficult to link this finding with macroevolutionary mechanisms.

3.4 CONCLUSIONS AND FUTURE DIRECTIONS

For studies of larval dispersal and connectivity, anemonefishes have been the focus of some pioneering work (see Chapter 20). In particular, parentage analysis techniques have been applied in anemonefish as a proof of concept and, subsequently, anemonefish have been the source of valuable empirical measurements of larval connectivity patterns. However, parentage approaches can be extremely labour-intensive and expensive. Further, parentage studies typically yield limited temporal information (e.g., a field campaign may capture only a single year's or a single season's recruitment). Pinsky et al. (2017) use a large parentage dataset generated for *A. percula* in Papua New Guinea to test indirect methods (such as isolation-by-distance metrics commonly used in population genetics studies). Encouragingly, the results indicate that certain aspects of connectivity may be derived from traditional population genetics approaches (requiring much less sampling effort and sequencing costs). Biogeography patterns are, in part, the result of larval dispersal patterns over evolutionary timescales, so insights into the drivers and maintenance of connectivity patterns are also useful in understanding broader principles underpinning speciation processes.

The evolutionary biogeography of anemonefish in marine ecoregions and our knowledge of their population ecology and population genetics collectively point towards a coherent impact of macro- and microevolutionary mechanisms on anemonefish diversity. Successive divergence and contact between populations and/or species are widely acknowledged to have enhanced adaptive evolution in many plant and animal species, terrestrial and marine alike. This scenario only partly covers how macro- and microevolutionary mechanisms shape anemonefish diversity. Anemonefish populations and species provide a remarkable example of a strong link between macro- and microevolutionary mechanisms where connectivity, self-recruitment, local evolutionary changes, and convergent evolution act across ecoregions, ecological timescales, and evolutionary timescales.

As in so many other aspects of marine biology and marine ecology, anemonefish have great potential as model systems for understanding biogeography and speciation. Many aspects of their life history make them amenable to in-situ observations and aquaria experiments (including breeding and larval rearing studies). Carefully designed studies, particularly if considering the various biogeographic regions inhabited by anemonefish, could shed light on principles of evolutionary ecology with broad implications for other marine organisms. For example, future work could target questions about the maintenance of species boundaries, especially in biodiversity "hotspots" where many species occur in sympatry. Do behaviors or diets play a role? Are there subtle timing differences in spawning behaviors that serve as reproductive barriers? How does specialization in the use of or selection of host anemone species influence a species' ability to colonize new areas? These are among the many tractable questions that may be more feasible to address in anemonefish studies than in many other types of marine organisms.

ACKNOWLEDGEMENTS

S. H. and M. L. B. acknowledge financial support from the King Abdullah University of Science and Technology.

REFERENCES

Berumen, M. L., G. R. Almany, S. Planes, G. P. Jones, P. Saenz-Agudelo, and S. R. Thorrold. 2012. Persistence of self-recruitment and patterns of larval connectivity in a marine protected area network. *Ecology and Evolution* 2(2): 444–452.

Chen, J., Z. Ye, Z. Yu, J. Wang, P. Li, X. Chen, Y. Liu, et al. 2017. The complete mitochondrial genome of the hybrid grouper (*Cromileptes altivelis*♀ × *Epinephelus lanceolatus*♂) with phylogenetic consideration. *Mitochondrial DNA Part B* 2(1): 171–172.

Cooper, W. J., and N. W. Westneat. 2009. Form and function of damselfish skulls: Rapid and repeated evolution into a limited number of trophic niches. *BMC Evolutionary Biology* 9(1): 1–17.

DiBattista, J. D., L. A. Rocha, J.-P. A. Hobbs, S. He, M. A. Priest, T. H. Sinclair-Taylor, B. W. Bowen, et al. 2015. When biogeographical provinces collide: Hybridization of reef fishes at the crossroads of marine biogeographical provinces in the Arabian Sea. *Journal of Biogeography* 42(9): 1601–1614.

Frédérich, B., L. Sorenson, F. Santini, G. J. Slater, and M. E. Alfaro. 2013. Iterative ecological radiation and convergence during the evolutionary history of damselfishes (Pomacentridae). *The Ameican Naturalist* 181(1): 94–113.

Gainsford, A., L. Herwerden, and G. Jones. 2015. Hierarchical behaviour, habitat use and species size differences shape evolutionary outcomes of hybridization in a coral reef fish. *Journal of Evolutionary Biology* 28(1): 205–222.

Gajdzik, L., E. Parmentier, L. N. Michel, N. Sturaro, K. Soong, G. Lepoint, B. Frédérich. 2018. Similar levels of trophic and functional diversity within damselfish assemblages across Indo-Pacific coral reefs. *Functional Ecology* 32(5): 1358–1369.

Giles, E. C., P. Saenz-Agudelo, N. E. Hussey, T. Ravasi, and M. L. Berumen. 2015. Exploring seascape genetics and kinship in the reef sponge *Stylissa carteri* in the Red Sea. *Ecology and Evolution* 5(13): 2487–2502.

He, S., J. L. Johansen, A. S. Hoey, M. K. Pappas, and M. L. Berumen. 2017. Molecular confirmation of hybridization between *Dascyllus reticulatus* × *Dascyllus aruanus* from the Great Barrier Reef. *Marine Biodiversity* 49(1): 395–404.

He, S., S. Planes, T. H. Sinclair-Taylor, and M. L. Berumen. 2018. Diagnostic nuclear markers for hybrid Nemos in Kimbe Bay, PNG-*Amphiprion chrysopterus* x *Amphiprion sandaracinos* hybrids. *Marine Biodiversity* 49(3): 1261–1269.

Huyghe, F., and M. Kochzius. 2017. Highly restricted gene flow between disjunct populations of the skunk clownfish (*Amphiprion akallopisos*) in the Indian Ocean. *Marine Ecology* 38(1): e12357.

Jones, G. P., S. Planes, and S. R. Thorrold. 2005. Coral reef fish larvae settle close to home. *Current Biology* 15(14): 1314–1318.

Litsios, G., P. B. Pearman, D. Lanterbecq, N. Tolou, and N. Salamin. 2014. The radiation of the clownfishes has two geographical replicates. *Journal of Biogeography* 41(11): 2140–2149.

Litsios, G., L. Pellissier, F. Forest, C. Lexer, P. B. Pearman, N. E. Zimmermann and N. Salamin. 2012. Trophic specialization influences the rate of environmental niche evolution in damselfishes (Pomacentridae). *Proceedings of the Royal Society B* 279(1743): 3662–3669.

Litsios, G., and N. Salamin. 2014. Hybridisation and diversification in the adaptive radiation of clownfishes. *BMC Evolutionary Biology* 14(1): 245.

McCord, C. L., C. M. Nash, W. J. Cooper, and M. W. Westneat. 2021. Phylogeny of the damselfishes (Pomacentridae) and patterns of asymmetrical diversification in body size and feeding ecology. *PLOS ONE* 16(10): e0258889.

Nanninga, G. B., P. Saenz-Agudelo, A. Manica, and M. L. Berumen. 2014. Environmental gradients predict the genetic population structure of a coral reef fish in the Red Sea. *Molecular Ecology* 23(3): 591–602.

Pinsky, M. L., P. Saenz-Agudelo, O. C. Salles, G. R. Almany, M. Bode, M. L. Berumen, S. Andréfouët et al. 2017. Marine dispersal scales are congruent over evolutionary and ecological time. *Current Biology* 27(1): 149–154.

Saenz-Agudelo, P., J. D. Dibattista, M. J. Piatek, M. R. Gaither, H. B. Harrison, G. B. Nanninga, and M. L. Berumen. 2015. Seascape genetics along environmental gradients in the Arabian Peninsula: Insights from ddRAD sequencing of anemonefishes. *Molular Ecology* 24(24): 6241–6255.

Saenz-Agudelo, P., G. P. Jones, S. R. Thorrold, and S. Planes. 2009. Estimating connectivity in marine populations: An empirical evaluation of assignment tests and parentage analysis under different gene flow scenarios. *Molecular Ecology* 18(8): 1765–1776.

Salles, O. C., G. R. Almany, M. L. Berumen, G. P. Jones, P. Saenz-Agudelo, M. Srinivasan, S. R. Thorrold, et al. 2020. Strong habitat and weak genetic effects shape the lifetime reproductive success in a wild clownfish population. *Ecology Letters* 23(2): 265–273.

Salles, O. C., B. Pujol, J. A. Maynard, G. R. Almany, M. L. Berumen, G. P. Jones, P. Saenz-Agudelo, et al. 2016. First genealogy for a wild marine fish population reveals multigenerational philopatry. *The Proceedings of the National Academy of Sciences* 113(46): 13245–13250.

Simpson, S. D., H. B. Harrison, M. R. Claereboudt, and S. Planes. 2014. Long-distance dispersal via ocean currents connects Omani clownfish populations throughout entire species range. *PLOS ONE* 9(9): e107610.

Tamura, K., G. Stecher, D. Peterson, A. Filipski, and S. Kumar. 2013. MEGA6: Molecular evolutionary genetics analysis version 6.0. *Molecular Biology and Evolution* 30(12): 2725–2729.

Tang, K. L., M. L. Stiassny, R. L. Mayden, and R. DeSalle. 2021. Systematics of damselfishes. *Ichthyology and Herpetology* 109: 258–318.

APPENDIX AND SUPPLEMENTAL MATERIALS

Appendix 3.1 Cytb Sequence Alignments, and additional supporting material, can be accessed at:

https://drive.google.com/drive/folders/1TK47PcfqtzqDJYwQFAbhEmI4zBTdoW2q?usp=sharing

4 Genomic Evidence of Hybridization during the Evolution of Anemonefishes

Anna Marcionetti, Sarah Schmid, and Nicolas Salamin

CONTENTS

4.1 INTRODUCTION

Adaptive radiation is defined as the rapid diversification of an ancestral population into several ecologically different species, associated with adaptive morphological or physiological divergence (Schluter 2000). This process is considered to play a central role in creating the spectacular diversity of life on Earth (Simpson 1953; Schluter 2000). For decades, researchers have been investigating the causes and consequences of adaptive radiations (e.g., Givnish and Sytsma 1997; Schluter 2000; Seehausen 2004; Glor 2010; Yoder et al. 2010; Givnish 2015; Soulebeau et al. 2015; Stroud and Losos 2016; Martin and Richards 2019), with the ultimate goal to broaden our understanding of the mechanisms governing species diversification and, ultimately, the buildup of biodiversity.

The intrinsic genomic factors that may promote adaptive radiation have started to be studied using the recent genomic data for non-model systems made available. Recent work included cichlids (Brawand et al. 2014; Faber-Hammond et al. 2019; McGee et al. 2020; Xiong et al. 2021), threespine sticklebacks (Jones et al. 2012; Verta and Jones 2019), *Anolis* lizards (Feiner 2016), Darwin's finches (Lamichhaney et al. 2015), *Heliconius* butterflies (Dasmahapatra et al. 2012; Supple et al. 2013; Edelman et al. 2019), and *Dysdera* spiders (Vizueta et al. 2019). What becomes evident is that a wide array of genome-wide changes, including chromosomal duplications, expansions of gene families, bursts of transposable elements (TEs), and accelerated evolution of coding and non-coding sequences, could predispose particular lineages to radiate adaptively (Jones et al. 2012; Brawand et al. 2014; Fan and Meyer 2014; Feiner 2016; Berner and Salzburger 2015; Faber-Hammond et al. 2019; Verta and Jones 2019; Xiong et al. 2021). Besides these overall genomic features potentially being linked with the rapid diversification, there is a clear and essential role played by ancient polymorphism and hybridization events in shaping adaptive radiations (e.g., Dasmahapatra et al. 2012; Berner and Salzburger 2015; Lamichhaney et al. 2015; Meier et al. 2017; Edelman et al. 2019; Svardal et al. 2020; Kozak et al. 2021).

The exchange of genetic material between evolutionary-distinct entities can lead to different processes according to the level of divergence between the two parents. Shortly after population differentiation, when divergence is weak, gene flow between two sister populations will often hinder further diversification due to its homogenization effect on allele frequencies (Cutter and Gray 2016; Weir and Price 2011). However, at an intermediate level of divergence, the genomes are often thought to be only partially permeable to introgression. When exchanges do occur, they might involve genomic regions associated with adaptive traits or neutral regions in linkage with those (Payseur and Rieseberg 2016). This process can promote the divergence between populations thanks to the recruitment of new alleles (Poelstra et al. 2018). Therefore, adaptive introgression has the potential to occur at an optimal degree of divergence between the parental species, bound between a minimum divergence required for the evolution of new and advantageous genotype combinations and a maximum divergence beyond which the level of genetic incompatibilities is too high, repressing the potential hybridization benefits. In such circumstances, adaptive introgression could lead to ecological adaptation (Jones et al. 2018), expansion of geographic range, or even adaptive radiation (Meier et al. 2017; Marques et al. 2019). Two hypotheses have been put forward to explain the role of hybridization in adaptive radiations: the *hybrid swarm* and the *syngameon* hypotheses (Seehausen 2004). The *hybrid swarm* hypothesis predicts hybridization events between distantly related species that have evolved independently for a substantial amount of time (Seehausen 2004). In this case, the onset of the entire adaptive radiation is promoted

DOI: 10.1201/9781003125365-5

by the resulting new combination of old alleles that were not previously found in either of the parental species (Seehausen 2004; Marques et al. 2019). In contrast, in the *syngameon* hypothesis, hybridization occurs among species during an adaptive radiation, potentially facilitating further speciation events within the radiation, as demonstrated in the *Heliconius* butterflies (Dasmahapatra et al. 2012; Pardo-Diaz et al. 2012) or Darwin's finches (Lamichhaney et al. 2015).

An interesting group to extend our understanding of these processes are anemonefishes (genera *Amphiprion* and *Premnas*). These iconic coral reef fishes consist of 28 recognized species and two natural hybrids (Fautin and Allen, 1997; Ollerton et al. 2007; Gainsford et al. 2015), and their distribution spans the whole tropical belt of the Indo-West Pacific Ocean (Figure 4.1). One distinctive characteristic of this group is the mutualistic interaction they maintain with sea anemones. Indeed, anemonefishes live unharmed within the toxic tentacles of sea anemones, and they benefit from this relationship with shelter from predators (Buston 2003) and reproductive benefits gained through the protection of eggs (Saenz-Agudelo 2011). While all species are associated with sea anemones, there is a large variability in host usage within the group. Indeed, some species are strictly specialists and can interact with a single species of sea anemones, while others are generalists and can live within up to ten hosts (Figure 4.1; Fautin and Allen 1997; Ollerton et al. 2007; Gainsford et al. 2015). Within the sea anemones, anemonefishes live in a size-based social hierarchy (Fricke 1979; Ochi 1989; Buston 2003) and are sequential hermaphrodites (Fricke and Fricke 1977; Moyer and Nakazono 1978; Fricke 1979).

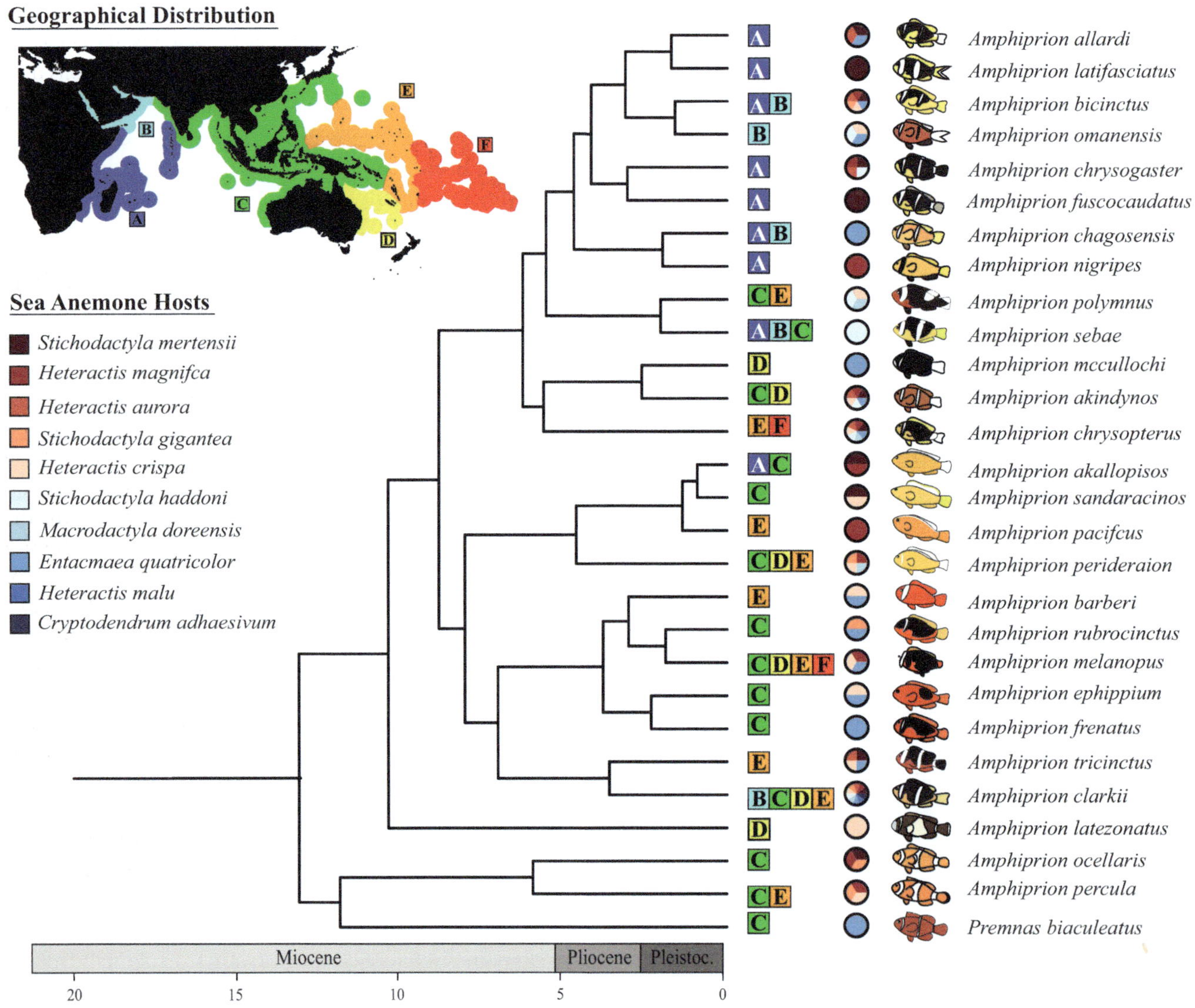

FIGURE 4.1 Dated phylogeny of anemonefishes based on the 20 most informative genes using BEAST v2.6.2. (Schmid et al. 2022). Geographical distributions, sea anemone hosts and representative phenotypes are schematized for each species. Letters from A to F represent the geographic regions. A. Western Indian Ocean; B. North-Western Indian Ocean; C. Central Indo-Pacific Ocean; D. South-Western Pacific Ocean; E. Central Pacific Ocean; F. Polynesian Ocean (modified from Litsios et al. 2014). Colored circles represent the sea anemones species with which each anemonefish can interact (Fautin and Allen 1997).

The mutualism with sea anemones is particularly interesting as it acted as the trigger of anemonefish adaptive radiation (Litsios et al. 2012). Indeed, it worked as a key innovation, opening new habitats with additional exploitable resources by anemonefishes and thus providing new ecological opportunities (Yoder et al. 2010; Litsios et al. 2012). Anemonefishes then diversified into several ecological niches associated with host and habitat usage, resulting in the variety of interactions observed today (generalists-specialists gradient; Figure 4.1). During their diversification, anemonefishes developed phenotypes correlated with their ecological niches, resulting in the phenotypic convergence of species inhabiting similar hosts and habitats (Figure 4.1; Litsios et al. 2012). Although a detailed description of anemonefish ecotypes is still lacking, it has been proposed that, for instance, more generalist species are likely to eat more planktonic food (and thus have more gill rakers) and be better swimmers than specialists, which never leave the close vicinity of their sea anemone host (Litsios et al. 2012). The primary radiation of anemonefish occurred in the Indo-Australian Archipelago, where the group originated (Litsios et al. 2014). Following a colonization event along the eastern coast of Africa, a second geographically independent radiation occurred in the Western Indian Ocean (Litsios et al. 2014). Indeed, the seven species that diversified there span the whole range of possible mutualistic interaction with sea anemones, from specialist to generalist species, and display the phenotypes associated with their ecological niche (Figure 4.1; Litsios et al. 2012; Litsios et al. 2014). This ecological sorting along the generalist to specialist axis allows different species to coexist in sympatry (Elliott and Mariscal 2001). Nevertheless, the genomics underlying this adaptive radiation are yet to be investigated.

Previous work suggested that the rapid diversification of anemonefishes was linked with hybridization events among members of the radiation. Indeed, the cytonuclear inconsistency observed in clownfishes was associated with a substantial increase in diversification rate (Litsios and Salamin 2014). Additionally, hybridization events are still happening in the group, as shown by the presence of two natural hybrids, *A. leucokranos* and *A. thiellei* (Ollerton et al. 2007; Gainsford et al. 2015). Hybridization may be facilitated by the occasional cohabitation of different species within the same sea anemone hosts (Songploy et al. 2021), which is mainly observed when host availability is limited (Camp et al. 2016). These observations open the question of the role that hybridization played and is still playing in anemonefish diversification. While previous analyses were based on very limited genetic data and did not include all the species (Litsios and Salamin 2014), the recent increase in genomic data available for the group provides a unique opportunity to clarify the role played by hybridization in the evolution of anemonefishes and to evaluate the impacts of hybridization on the genealogical relationships across the genome of clownfishes. Here, we build on two recent studies extending the genomic analyses in anemonefishes (Marcionetti et al. 2022; Schmid et al. 2022) to illustrate potential genomic signatures of past introgression. Such analyses might be valuable to potentially highlight parts of the genome exhibiting patterns of adaptive introgression and thus playing a key role in the adaptive radiation process.

4.2 WHOLE-GENOMIC DATA FOR ALL ANEMONEFISH SPECIES AND THE DETECTION OF HYBRIDIZATION EVENTS

The current genomic resources for anemonefishes (Lehmann et al. 2019; Marcionetti et al. 2018, 2019) cover whole-genome assemblies for 11 species of the group (*Premnas biaculeatus*, *Amphiprion ocellaris*, *A. percula*, *A. perideraion*, *A. akallopisos*, *A. polymnus*, *A. sebae*, *A. melanopus*, *A. bicinctus*, *A. nigripes*, and *A. frenatus*) and one of their sister species, the lemon damselfish (*Pomacentrus moluccensis*). The approaches used to build the genome assemblies were very different, with high coverage obtained mostly by short Illumina reads for all species except *A. percula* (from Marcionetti et al. 2018, 2019), which led to a high number of scaffolds despite the assembled genomes containing all the essential genes. The genome of *A. percula* (Lehmann et al. 2019) was achieved with a very thorough data collection combining short and long reads with a high coverage combined with HiC scaffolding that enabled the reconstruction of chromosome level assembly. However, the main summary statistics obtained by the two studies were congruent (Marcionetti et al. 2019, Chapter 2), which gave the first hint that the overall genomic architecture within the genus is conserved.

This data was recently augmented with whole-genome sequencing data for all 28 species of clownfishes (Schmid et al. 2022), following the same procedure as the one described in Marcionetti et al. (2019). The genomic reads of each clownfish species were mapped to the chromosome-level assembly obtained for *A. percula* to obtain comparable alignments across all the species sampled (see Schmid et al. 2022 for the details).

This genomic data available for clownfishes allows assessing the presence of potential hybridization events during the diversification of the group. In the case of hybridization, cytoplasmic (chloroplast or mitochondrial) and nuclear DNA have distinct evolutionary histories. Thus, phylogenetic reconstructions based on cytoplasmic or nuclear DNA should result in inconsistent trees showing different relationships between species (i.e., cytonuclear discordance). Similarly, topological inconsistencies should also be observed when reconstructing phylogenetic trees along the nuclear genome. Nevertheless, the presence of incongruent phylogenetic trees is not, alone, evidence for hybridization. Indeed, topological inconsistency can also result from the retention of ancestral polymorphism because of incomplete lineage sorting (ILS). Because these mechanisms leave similar genetic signatures, distinguishing between introgressive hybridization and ILS has been notoriously difficult (e.g., Holder et al. 2001; Qu et al. 2012;

Sousa and Hey 2013). While approaches to testing for gene flow by fitting models using maximum-likelihood or Bayesian methods have been developed (Pinho and Hey 2010), these methods are computationally intensive. An alternative approach that has been widely used consists of testing for an excess of shared derived polymorphism using a four-taxon (or ABBA-BABA) test and its related statistics (e.g., Patterson's D, f_4 -ratio, f_d; Kulathinal et al. 2009; Green et al. 2010; Durand et al. 2011; Martin et al. 2015). This simpler and more computationally efficient method can be more easily applied on a genomic scale to distinguish which evolutionary process – between hybridization and ILS – is responsible for the disparities in the phylogenetic relationships between species.

4.3 INCONGRUENCE IN PHYLOGENOMIC INFERENCE

The presence of potential hybridization across the evolution of anemonefishes was first investigated by comparing the phylogenetic relationships obtained from the whole mitochondrial and nuclear genomes. The mitochondrial phylogenetic tree was reconstructed based on the whole-mitochondrial sequence (16,647 bp aligned across all species) using RAxML v8.2.4 (GTR+Gamma model of substitution using ascertainment bias option; Stamatakis 2014). The nuclear phylogenetic tree was based on whole-genome data for all chromosomes combined, which produced 65,378,772 SNPs that were combined in genomic windows of 10 kb to estimate gene trees using RAxML v8.2.4 (same parameters as for the mitochondrial tree). Over 13,000 gene trees were then combined to obtain the nuclear phylogenetic tree using ASTRAL v5.7.3 (Zhang et al. 2018). The full details of the analyses are found in Schmid et al. (2022). Ancestral nodes for both mitochondrial and nuclear phylogenetic trees are well supported, while more recent splits display lower node support, particularly in the tree inferred with the mitochondrial genome (Figure 4.2). The two genomic datasets display differences in the topologies obtained, which is similar to what was previously shown with limited genetic data (Litsios and Salamin 2014).

The first difference between the two phylogenetic trees consists in the position of *P. biaculeatus*. Indeed, the nuclear phylogenetic tree places this species as the basal lineage of all other anemonefishes, while the mitochondrial genome supports a basal group containing the three species *P. biaculeatus*, *A. ocellaris*, and *A. percula*. This disparity mirrors the conflicting phylogenies reported in the literature (e.g., Frédérich et al. 2013; Mirande 2017; na Ayudhaya et al. 2017; Lobato et al. 2014; DiBattista et al. 2016). While *Premnas* has been recently recovered within *Amphiprion* (Tang et al. 2021), these inconsistencies suggest a complex evolutionary history of *P. biaculeatus* – such as potential hybridization events with species outside the *Amphiprion* genus or with the ancestor of *A. percula*/*A. ocellaris*.

Additional topological differences between the mitochondrial and nuclear phylogenetic trees are observed when looking at the relationship between the clades. First, in the nuclear phylogenetic tree, the *clarkii* clade is sister to the *ephippium* complex, and both are clustered with the *akallopisos* clade (Figure 4.2). In the mitochondrial phylogenetic tree, however, the *clarkii* and the *akallopisos* clades are both monophyletic, and the *ephippium* clade is sister to the Australian clade (containing *A. akindynos* and *A. mccullochi*; Figure 4.2). Second, the *polymnus* clade is interspersed inside the Indian clade in the mitochondrial dataset, while the nuclear genome resolves this clade as the sister clade to the whole Indian clade (Figure 4.2). These topological disparities are consistent with those previously described (Litsios and Salamin 2014), and nodes in the nuclear and the mitochondrial phylogenetic trees obtained with whole-genomic data are strongly supported, thus validating cytonuclear discordance in anemonefishes.

These results suggest that hybridization events occurred during the diversification of the group. Topology inconsistencies are mainly observed in basal branches, indicating that hybridization events may have occurred between ancestral species and potentially fuelled the radiation of anemonefishes (Litsios and Nicolas 2014; Schmid et al. 2022). Indeed, hybridization has the potential to increase the amount of standing genetic variation, which can, in turn, generate adaptive novelty, corresponding to two key mechanisms in adaptive radiation (Grant and Grant 1997; Seehausen 2004). Nevertheless, besides hybridization and introgression, cytonuclear discordance may also result from incomplete lineage sorting (Ballard and Whitlock 2004). Further analyses are therefore necessary to assess the contribution of these different processes in shaping phylogenetic relationships of anemonefishes and, more generally, their role in the diversification of the group.

4.4 GENE TREE INCONGRUENCE ACROSS THE GENOME

In addition to the cytonuclear discordances highlighting nodes potentially affected by hybridization, the genomic-wide data available for anemonefishes allows investigating inconsistencies between gene trees across the nuclear genome. This can give additional hints of potential hybridization events or incomplete lineage sorting across the nuclear genome of anemonefishes and allows to identify genomic regions affected by these processes that may be involved in anemonefish diversification.

In order to explore gene tree incongruence across the genomes, the assemblies of the ten anemonefish species shown in Marcionetti et al. (2019) were aligned, and the phylogenetic relationship between species was reconstructed with PhyML (GTR+Gamma model, 100 bootstraps: v3.3; Guindon et al. 2010) for non-overlapping sliding-windows of 100 kb. A total of 5,936 topologies estimated across the genomes were summarized by calculating the Robinson-Foulds distances between them and applying Metric Multi-Dimensional Scaling to group similar trees (Figure 4.3A–B), obtaining five main topologies (Figure 4.3C) distributed

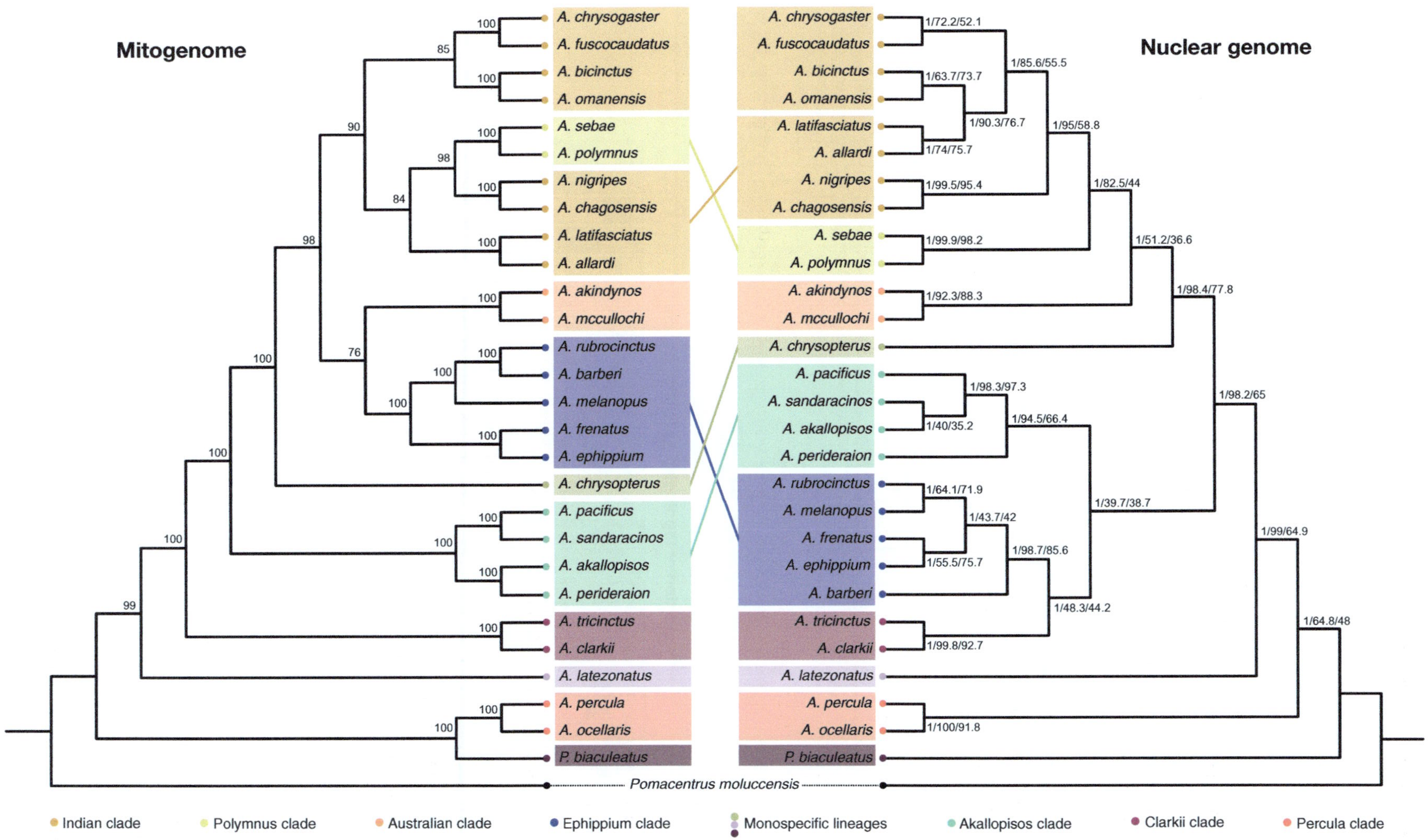

FIGURE 4.2 Comparison of mitochondrial (left) and nuclear (right) phylogenetic trees for the 28 anemonefish species. Phylogenetic trees are represented as cladograms. Colors correspond to the anemonefish clades, as defined in Litsios and Salamin (2014). Lines between the two trees indicate different placement of particular clades. Node labels of the mitochondrial phylogeny correspond to the bootstrap support based on 100 bootstraps resampling. For the nuclear phylogeny, node labels correspond to the posterior probability. *A.* stands for *Amphiprion* and *P.* for *Premnas* (adapted from Schmid et al. 2022). Geographical distribution of all species is reported in Figure 4.1.

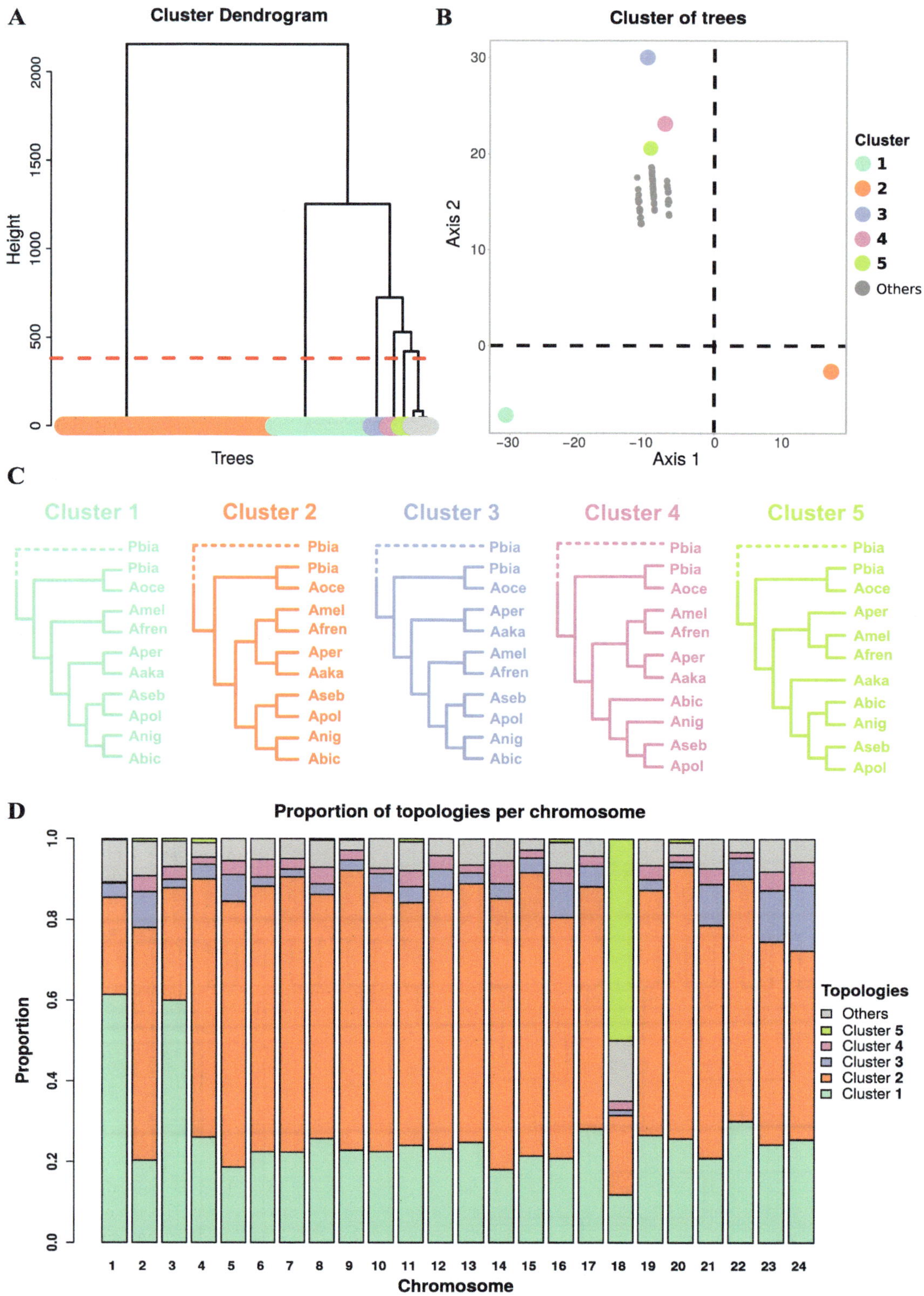

FIGURE 4.3 Topological inconsistencies across anemonefish nuclear genome. A. Hierarchical clustering based on the Robinson-Foulds distance between the trees reconstructed with PhyML along the genome (non-overlapping windows of 100 kb). The red line corresponds to the cutoff set to define the five major clusters of trees, represented by the different colors. B. Multidimensional scaling plot of the five clusters of trees. C. The topologies of the five main clusters. Pbia, Aoce, Amel, Afre, Aper, Aaka, Abic, Anig, Aseb, and Apol correspond to *P. biaculeatus*, *A. ocellaris*, *A. melanopus*, *A. frenatus*, *A. perideraion*, *A. akallopisos*, *A. bicinctus*, *A. nigripes*, *A. sebae*, and *A. polymnus*, respectively. The position of *P. biaculeatus* in the trees was variable (represented with dotted lines), and the numbers on the top of the branches represent the proportion of trees showing the corresponding topology. The trees were rooted with the outgroup *P. moluccensis* (not shown). D. Distribution of the five topologies across anemonefish chromosomes. The additional topologies, reported in grey, were not considered as they are variable, with small rearrangements of the leaves. Figure adapted from Marcionetti et al. (2022).

along the 24 chromosomes (Figure 4.3D). The full details of the analyses can be found in Marcionetti et al. (2022).

The most frequent topology observed along the genome (57% of the windows) is associated with cluster 2 (in orange in Figure 4.3A–D), which is consistent with the nuclear phylogenetic tree (Figure 4.2). The species relationship obtained for the mitochondrial phylogenetic tree (Figure 4.2) is also observed along the nuclear genome, but it is less frequent (5% of the windows; topology of cluster 3 in Figure 4.3A–D). By contrast, the second most common topology (27% of the windows) is associated with cluster 1 (Figure 4.3A–D), characterized by the species-pair *A. akallopisos–A. perideraion* being basal to the *A. bicinctus–A. nigripes–A. polymnus–A. sebae* complex. This topology is prevailing in chromosomes 1 and 3 (Figure 4.3D), which therefore showed a different pattern than the other chromosomes. The topology of cluster 4 is similar to cluster 2, but with *A. bicinctus* being basal to *A. nigripes*, and it is observed at low frequencies in all chromosomes (Figure 4.3A–D). Finally, the topology of cluster 5 is characterized by *A. perideraion* and *A. akallopisos* not branching as sister species but being basal to the *A. frenatus–A. melanopus* pair and to the *A. bicinctus–A. nigripes–A. polymnus–A. sebae* complex, respectively. This topology is almost exclusively observed on chromosome 18, where it is the most prevalent topology (Figure 4.3D).

The topological disparities observed across the anemonefish nuclear genome mainly concern the deep nodes of the trees, while the five pairs of closely related species mostly branch together. These observations further indicate that potential hybridization events and/or incomplete lineage sorting have occurred during the diversification of the group, potentially facilitating the radiation of anemonefishes. Nevertheless, incongruences in the terminal branches of the tree are also observed, suggesting further hybridization events outside the main radiation of anemonefish. First, in all five clusters of trees, *P. biaculeatus* mostly branches as sister species of *A. ocellaris*, but it is also frequently placed as basal to the whole *Amphiprion* clade (represented by dotted lines in the topologies in Figure 4.3C). This disparity mirrors the inconsistencies observed in the mitochondrial and nuclear phylogenetic trees (Figure 4.2) and is in line with the revised systematic position of the genus *Premnas* (Tang et al. 2021), which we discussed in the previous section. Second, *A. perideraion* is not branching with its sister species *A. akallopisos* in cluster 5 but is basal to the *A. melanopus–A. frenatus* pair. This observation could suggest past gene flow between *A. perideraion* and the *A. melanopus–A. frenatus* ancestor. This topology is almost exclusively observed on this chromosome and clusters in two large regions, potentially indicating that the introgression signal was removed from the rest of the genome by extensive backcrossing but persisted on chromosome 18, likely through the disruption of recombination (Marcionetti et al. 2022).

Such disruption of recombination can be achieved, for instance, through genomic inversion (Stevison et al. 2011). Large genomic inversions that break recombination, creating clusters of loci controlling ecologically important traits, consequently fixed by natural selection, are observed in the case of supergenes (e.g., Joron et al. 2011; Zinzow-Kramer et al. 2015; Branco et al. 2018). In anemonefish, a total of 331 functionally annotated genes are located in the two regions of chromosome 18 whose topology is associated with cluster 5 (Marcionetti et al. 2022). Among them, genes with functions associated with epithelium morphogenesis, fertilization, axis elongation, and retinal vasculature development were overrepresented in these regions compared to the rest of the genome. However, a link between these functions and important ecological traits for anemonefishes cannot be easily drawn. Thus, a potential role of selection in fixing and maintaining the two regions of alternative topology on chromosome 18, and the importance of those two regions in the anemonefish diversification, cannot be clearly established without further studies. It is worth mentioning that the evolution of sex chromosomes may also result in particular patterns as those observed on chromosome 18 (e.g., Natri et al. 2019). However, this might not be relevant in anemonefishes because of their sequential hermaphrodites with no sex chromosomes (Fricke and Fricke 1977; Fricke 1979; Arai 2011), and genes involved in the sex change are scattered throughout the genome (Casas et al. 2018).

4.5 ESTIMATING PAST INTROGRESSION IN ANEMONEFISHES

In order to confirm past introgression and exclude the possibility that the topological inconsistencies observed in anemonefishes were uniquely due to ILS, we formally tested for the presence of ancestral hybridization events using a four-taxon test. Introgression between all combinations of the anemonefishes clades shown in Figure 4.4 was estimated using f_4-ratio statistics as implemented in DSUITE (Malinsky et al. 2020), and each of the 11 lineages was alternatively considered as a potential donor (P_3) or recipient (P_1 and P_2). The f_4-ratio statistics provide an estimation of introgression between non-sister species, and a positive f_4-ratio indicates introgression between P_3 and P_2, whereas a negative value is the result of introgression between P_3 and P_1. Statistical significance and standard errors of the resulting f_4-ratio statistics were calculated using a standard block jackknife procedure (Green et al. 2010; Durand et al. 2011). The full details of the analyses can be found in Schmid et al. (2022).

Hybridization is pervasive among all anemonefishes clades and chromosomes, with significant f_4-ratios varying between 0.0008 and 0.024 (Figure 4.4). The f_4-ratio values in two portions of chromosome 18 are up to four times higher than in the rest of the genome (Figure 4.4). This result is consistent with the pattern of topological inconsistency observed in Marcionetti et al. (2022) and described in the previous section.

These results further suggest that events of ancestral hybridization might have taken place at the base of the

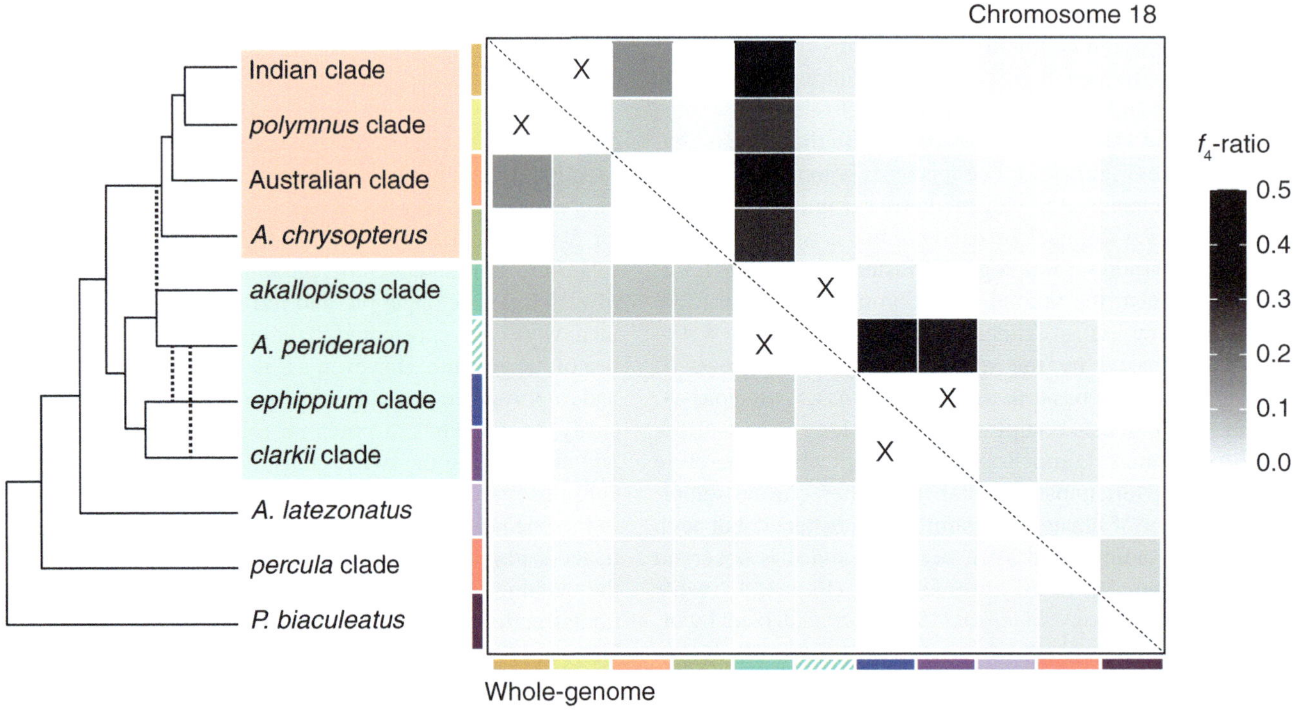

FIGURE 4.4 Introgression between all the different anemonefish clades/species. The lower triangle represents significant f_4-ratio values calculated over the whole genome, while the upper triangle represents significant f_4-ratio values calculated only over chromosome 18. Each grey square stands for a specific combination of P_2 (left of the square) and P_3 (bottom of the square). Introgression between sister taxa cannot be assessed and is marked by an *x*. Dashed lines on the cladogram represent the strong introgression events highlighted in chromosome 18. Although introgression is pervasive among all anemonefish clades, remarkable results consist of significant statistics between the three most ancestral lineages of anemonefishes (i.e., *Premnas biaculeatus*, *A. percula* clade, and *A. latezonatus*) and each one of the most recent clades. Furthermore, chromosome 18 displays f_4-ratio up to four times higher for combinations involving *A. latezonatus* and the members of the *akallopisos-A. perideraion-ephippium-clarkii* clade. Chromosome 18 also exhibits a strong signal of introgression between the members of the *Indian-polymnus-Australian-A. chrysopterus* clade and the *akallopisos clade*, as well as between the *ephippium-clarkii clade* and *A. perideraion*, with f_4-ratio reaching values of 0.5 (adapted from Schmid et al. 2022).

anemonefish radiation. Hybridization at the onset of adaptive radiation was already described in various species (e.g., Barrier et al.1999; Meier et al. 2017). Furthermore, it was recently suggested that hybridization has the potential to reshuffle old genetic variations into new combinations and eventually lead to speciation or adaptive radiation (Marques et al. 2019). Contrary to new mutations, old variations have already been through a selection filter and are optimized for their genomic and ecological context (Abbott et al. 2013). Several examples of species arising from new combinations of old alleles exist (see Marques et al. 2019 for a review), and many of them are well-known cases of adaptive radiation. Anemonefish are well-suited candidates for this "combinatorial view of speciation", but further evidence is still needed, such as a comparison of the age of the ancient alleles and the species splitting time as well as the linkage disequilibrium pattern of such alleles, which will be possible to investigate when a better knowledge of the genes involved in the anemonefish radiation will be acquired.

Additionally, we cannot rule out the possibility of one or several introgression events with an ancestral species that was not sampled or extinct, and which generated the observed pattern of introgression. Such events known as ghost introgression have the potential to enhance adaptation and speciation (Ottenburghs 2020). Highlighting archaic introgressed tracts opens new horizons to disentangle the impact of those ancestral variations on the recipient lineage, thus improving our understanding of the role of introgression in the evolutionary trajectory of species (Jacobs and Therkildsen 2019). But up to now, only a few studies investigated in detail archaic tracts in a non-hominid genome and were done, among others, in the sea bass (Duranton et al. 2019), the killer whale (Foote et al. 2019), and the bonobo (Kuhlwilm et al. 2019).

4.6 DISCUSSION AND PERSPECTIVES

The patterns observed across the genomic windows analyzed, together with the topological inconsistencies and the estimates of f_4-ratio, suggest ancestral hybridization events in the diversification of anemonefishes. Gene flow spreading ancient genetic variation among species has been proposed to facilitate adaptive radiation (Berner and Salzburger 2015; Marques et al. 2019). For instance, ancestral hybridization between distinct lineages has fueled the adaptive radiation of cichlids (i.e., the *hybrid swarm* hypothesis, Seehausen 2004; Meier et al. 2017; Svardal et al. 2020), while introgressive hybridization among members of the radiating lineages (i.e., the *syngameon* hypothesis, Seehausen 2004) has facilitated ecological speciation in *Heliconius* butterflies (Dasmahapatra et al. 2012; Pardo-Diaz et al. 2012) and Darwin's finches (Lamichhaney et al. 2015). While hybridization events may have also participated in anemonefish diversification, we cannot yet exclude that the mosaic genomes observed in anemonefishes are, at least partially, the result of other processes.

The extent of introgression between anemonefish species across the genome does not inform us whether the process is mainly neutral or promoted by a selective advantage. It is known that many introgression events have involved adaptation and resulted in some well-known cases of adaptive radiation (Dasmahapatra et al. 2012; Grant and Grant 2014). Moreover, a number of known cases of adaptive introgression consist of a few important genes that confer a specific advantage to the species and display a clear phenotype–genotype association (e.g., Pardo-Diaz et al. 2012; Huerta-Sânchez et al. 2014; Fontaine et al. 2015). Adaptive introgression can not only transfer advantageous loci between species but can also lead to new adaptive combinations (Seehausen 2004). It was previously reported that anemonefish rapid radiation coincides with their mutualism with sea anemones (Litsios et al. 2012). Yet, knowledge about the genetic basis of the mutualism between the anemonefish and the sea anemone is still at an early stage. Seventeen positively selected genes at the basis of the anemonefish radiation were previously highlighted, some coding for function associated with *N*-acetylated sugars, molecules that are known to play a role in sea anemone discharge of toxins (Marcionetti et al. 2019). But the causal link between those genes and the ability to interact with the sea anemone remains to be validated with further experimental approaches. Thus, the mutualism with sea anemones appears to have complex genetic bases only partially understood, which impede the identification of adaptive introgressed segments in the anemonefish genome.

The genomic resources currently available for the anemonefishes have extended our understanding of the mechanisms driving the adaptive radiation of this group. The results presented here show clear signs of ancient hybridization in the group. Taken together with the presence of current hybrid species such as *A. leucokranos* and *A. thiellei*, they illustrate the important role of these processes in the evolution of anemonefishes. Further work on the genomic architecture of these current hybrid species as well as detailed population genomic studies of key species complexes potentially hybridizing – such as the *akallopisos* clade – would provide important cues to understand the genomic regions involved in the hybridization and their exact role in the diversification and adaptation of anemonefishes. This seems to be within the reach of current research and will open new questions to fully comprehend the evolution of anemonefishes.

ACKNOWLEDGEMENTS

We would like to thank Vincent Laudet and Timothy Ravassi for the invitation to participate in this book. We would like to thank as well Baptiste Micheli, Benjamin Titus, and Aline Sondra Adler for advice and discussion during the analyses of the genomic data. The bioinformatic analyses were done on the Computing Infrastructure of DCSR of the University of Lausanne while the genomic data were obtained in the Lausanne Genomic Facility of the University of Lausanne. The work was funded by a Swiss National Science Foundation grant to N. S. (grant 310030_185223) as well as by funds from the University of Lausanne.

REFERENCES

Abbott, R., D. Albach, S. Ansell, J. W. Arntzen, S. J. E. Baird, N. Bierne, J. Boughman, et al. 2013. Hybridization and speciation. *Journal of Evolutionary Biology* 26: 229–246.

Arai, R. 2011. *Fish karyotypes: A check list.* Springer.

Ballard, J. W. O., and M. C. Whitlock. 2004. The incomplete natural history of mitochondria. *Molecular Ecology* 13(4): 729–744.

Barrier, M., B. G. Baldwin, R. H. Robichaux, and M. D. Purugganan. 1999. Interspecific hybrid ancestry of a plant adaptive radiation: Allopolyploidy of the Hawaiian silversword alliance (Asteraceae) inferred from floral homeotic gene duplications. *Molecular Biology and Evolution* 16(8): 1105–1113.

Berner, D., and W. Salzburger. 2015. The genomics of organismal diversification illuminated by adaptive radiations. *Trends in Genetics* 31(9): 491–499.

Branco, S., F. Carpentier, R. C. R. de la Vega, H. Badouin, A. Snirc, S. Le Prieur, M. A. Coelho, et al. 2018. Multiple convergent supergene evolution events in mating-type chromosomes. *Nature Communications* 9(1): 1–13.

Brawand, D., C. E. Wagner, Y. I. Li, M. Malinsky, I. Keller, S. Fan, O. Simakov, et al. 2014. The genomic substrate for adaptive radiation in African cichlid fish. *Nature* 513: 375–381.

Buston, P. 2003. Size and growth modification in clownfish. *Nature* 424: 145–146.

Camp, E. F., J.-P. A. Hobbs, M. De Brauwer, A. J. Dumbrell, and D. J. Smith. 2016. Cohabitation promotes high diversity of clownfishes in the Coral Triangle. *Proceedings of the Royal Society B: Biological Sciences* 283: 20160277.

Casas, L., P. Saenz-Agudelo, and X. Irigoien. 2018. High-throughput sequencing and linkage mapping of a clownfish genome provide insights on the distribution of molecular players involved in sex change. *Scientific Reports* 8(1): 1–12.

Cutter, A. D. and J. C. Gray. 2016. Ephemeral ecological speciation and the latitudinal biodiversity gradient. *Evolution* 70: 2171–2185.

Dasmahapatra, K. K., J. R. Walters, A. D. Briscoe, J. W. Davey, A. Whibley, N. J. Nadeau, A. V. Zimin, et al (Heliconius Genome Consortium). 2012. Butterfly genome reveals promiscuous exchange of mimicry adaptations among species. *Nature* 487: 94–98.

DiBattista, J. D., J. Howard Choat, M. R. Gaither, J. P. A. Hobbs, D. F. Lozano-Cortés, R. F. Myers, and M. L. Berumen. 2016. On the origin of endemic species in the Red Sea. *Journal of Biogeography* 43(1): 13–30.

Durand, E. Y., N. Patterson, D. Reich, and M. Slatkin. 2011. Testing for ancient admixture between closely related populations. *Molecular Biology and Evolution* 28(8): 2239–2252.

Duranton, M., F. Allal, S. Valière, O. Bouchez, F. Bonhomme, and P.-A. Gagnaire. 2019. The contribution of ancient admixture to reproductive isolation between European sea bass lineages, *Evolutionary Biology* 4(3): 226–242.

Edelman, N. B., P. B. Frandsen, M. Miyagi, B. Clavijo, J. Davey, R. B. Dikow, G. Garcia-Accinelli, S. M. Van Belleghem, et al. 2019. Genomic architecture and introgression shape a butterfly radiation. *Science* 366: 594–599.

Elliot, J. K. and R. N. Mariscal. 2001. Coexistence of nine anemonefish species: Differential host and habitat utilization, size and recruitment. *Marine Biology* 138: 23–36.

Faber-Hammond, J. J., E. Bezault, D. H. Lunt, D. A. Joyce, and S. C. Renn. 2019. The genomic substrate for adaptive radiation: Copy number variation across 12 tribes of African cichlid species. *Genome Biology and Evolution* 11(10): 2856–2874.

Fan, S., and A. Meyer. 2014. Evolution of genomic structural variation and genomic architecture in the adaptive radiations of African cichlid fishes. *Frontiers in Genetics* 5: 163.

Fautin, D. G., and G. R. Allen. 1997. *Anemone fishes and their host sea anemones*. Western Australian Museum.

Feiner, N. 2016. Accumulation of transposable elements in Hox gene clusters during adaptive radiation of Anolis lizards. *Proceedings of the Royal Society B: Biological Sciences* 283: 20161555.

Fontaine, M. C., J. B. Pease, A. Steele, R. M. Waterhouse, D. E. Neafsey, I. V. Sharakhov, X. Jiang, et al. 2015. Extensive introgression in a malaria vector species complex revealed by phylogenomics. *Science* 347: 1258524.

Foote, A. D., M. D. Martin, M. Louis, G. Pacheco, K. M. Robertson, M-H. S. Sinding, A. R. Amaral, et al. 2019. Killer whale genomes reveal a complex history of recurrent admixture and vicariance. *Molecular Ecology* 28(14): 3427–3444.

Frédérich, B., L. Sorenson, F. Santini, G. J. Slater, and M. E. Alfaro. 2013. Iterative ecological radiation and convergence during the evolutionary history of damselfishes (Pomacentridae). *The American Naturalist* 181(1): 94–113.

Fricke, H. W. 1979. Mating system, resource defence and sex change in the anemonefish Amphiprion akallopisos. *Zeitschrift für Tierpsychologie* 50(3): 313–326.

Fricke, H., and S. Fricke. 1977. Monogamy and sex change by aggressive dominance in coral reef fish. *Nature* 266: 830–832.

Gainsford, A., L. Van Herwerden, and G. P. Jones. 2015. Hierarchical behaviour, habitat use and species size differences shape evolutionary outcomes of hybridization in a coral reef fish. *Journal of Evolutionary Biology* 28(1): 205–222.

Givnish, T., and K. Sytsma. 1997. *Molecular evolution and adaptive radiation*. Cambridge University Press.

Givnish, T. J. 2015. Adaptive radiation versus 'radiation' and 'explosive diversification': Why conceptual distinctions are fundamental to understanding evolution. *New Phytologist* 207(2): 297–303.

Glor, R. E. 2010. Phylogenetic insights on adaptive radiation. *Annual Review of Ecology, Evolution, and Systematics* 41: 251–270.

Grant, P. R. and B. R. Grant. 1997. Hybridization, sexual imprinting, and mate choice. *The American Naturalist* 149: 1–28.

Grant, P. R. and B. R. Grant. 2014. Synergism of natural selection and introgression in the origin of a new species. *The American Naturalist* 183: 671–681.

Green, R. E., J. Krause, A. W. Briggs, T. Maricic, U. Stenzel, M. Kircher, N. Patterson, et al. 2010. A draft sequence of the Neandertal genome. *Science* 328: 710–722.

Guindon, S., J. F. Dufayard, V. Lefort, M. Anisimova, W. Hordijk, and O. Gascuel. 2010. New algorithms and methods to estimate maximum-likelihood phylogenies: Assessing the performance of PhyML 3.0. *Systematic Biology* 59(3): 307–321.

Holder, M. T., J. A. Anderson, and A. K. Holloway. 2001. Difficulties in detecting hybridization. *Systematic Biology* 50(6): 978–982.

Huerta-Sánchez, E., X. Jin, Asan, Z. Bianba, B. M. Peter, N. Vinckenbosch, Y. Liang, et al. 2014. Altitude adaptation in Tibetans caused by introgression of Denisovan-like DNA. *Nature* 512: 194–197.

Jacobs, A., and N. O. Therkildsen. 2019. Excavating ghost footprints and tangled trees from modern genomes. *Molecular Ecology* 28: 3287–3290.

Jones, F. C., M. G. Grabherr, Y. F. Chan, P. Russell, E. Mauceli, J. Johnson, and D. M. Kingsley. 2012. The genomic basis of adaptive evolution in threespine sticklebacks. *Nature* 484: 55–61.

Jones, M. R., L. S. Mills, P. C. Alves, C. M. Callahan, J. M. Alves, D. J. R. Lafferty, F. M. Jiggins, et al. 2018. Adaptive introgression underlies polymorphic seasonal camouflage in snowshoe hares. *Science* 360: 1355–1358.

Joron, M., L. Frezal, R. T. Jones, N. L. Chamberlain, S. F. Lee, C. R. Haag, A. Whibley, et al. 2011. Chromosomal rearrangements maintain a polymorphic supergene controlling butterfly mimicry. *Nature* 477: 203–206.

Kozak, K. M., M. Joron, W. O. McMillan, and C. D. Jiggins. 2021. Rampant genome-wide admixture across the Heliconius radiation. *Genome Biology and Evolution* 13(7): evab099.

Kuhlwilm, M., S. Han, V. C. Sousa, L. Excoffier, and T. Marques-Bonet. 2019. Ancient admixture from an extinct ape lineage into bonobos. *Nature Ecology and Evolution* 3: 957–965.

Kulathinal, R. J., L. S. Stevison, and M. A. Noor. 2009. The genomics of speciation in Drosophila: Diversity, divergence, and introgression estimated using low-coverage genome sequencing. *PLoS Genetics* 5(7): e1000550.

Lamichhaney, S., J. Berglund, M. S. Almén, K. Maqbool, M. Grabherr, A. Martinez-Barrio, M. Promerová et al. 2015. Evolution of Darwin's finches and their beaks revealed by genome sequencing. *Nature* 518: 371–375.

Lehmann, R., D. J. Lightfoot, C. Schunter, C. T. Michell, H. Ohyanagi, K. Mineta, K.S. Foret, et al. 2019. Finding Nemo's Genes: A chromosome-scale reference assembly of the genome of the orange clownfish Amphiprion percula. *Molecular Ecology Resources* 19(3): 570–585.

Litsios, G., P. B. Pearman, D. Lanterbecq, N. Tolou, and N. Salamin. 2014. The radiation of the clownfishes has two geographical replicates. *Journal of Biogeography* 41(11): 2140–2149.

Litsios, G., and N. Salamin. 2014. Hybridisation and diversification in the adaptive radiation of clownfishes. *BMC Evolutionary Biology* 14(1): 245.

Litsios, G., C. A. Sims, R. O. Wüest, P. B. Pearman, N. E. Zimmermann, and N. Salamin. 2012. Mutualism with sea anemones triggered the adaptive radiation of clownfishes. *BMC Evolutionary Biology* 12(1): 212.

Lobato, F. L., D. R. Barneche, A. C. Siqueira, A. M. Liedke, A. Lindner, M. R. Pie, D. R. Bellwood, et al. 2014. Diet and diversification in the evolution of coral reef fishes. *PLoS One* 9(7): e102094.

Malinsky, M., M. Matschiner, and H. Svardal. 2020. Dsuite – Fast *D*-statistics and related admixture evidence from VCF files. *Molecular Ecology Resources* 21(2): 584–595.

Marcionetti, A., V. Rossier, J. A. Bertrand, G. Litsios, and N. Salamin. 2018. First draft genome of an iconic clownfish species (Amphiprion frenatus). *Molecular Ecology Resources* 18(5): 1092–1101.
Marcionetti, A., V. Rossier, N. Roux, P. Salis, V. Laudet, N. Salamin. 2019. Insights into the genomics of clownfish adaptive radiation: Genetic basis of the mutualism with sea anemones. *Genome Biology and Evolution* 11(3): 869–882.
Marcionetti, A., and N. Salamin. 2022. Insights into the genomics of clownfish adaptive radiation: The genomic substrate of the diversification. *bioRxiv* 2022.05.12.491701, doi: https://doi.org/10.1101/2022.05.12.491701
Marques, D. A., J. I. Meier, and O. Seehausen. 2019. A combinatorial view on speciation and adaptive radiation. *Trends in Ecology and Evolution* 34(6): 531–544.
Martin, S. H., J. W. Davey, C. D. Jiggins. 2015. Evaluating the use of ABBA–BABA statistics to locate introgressed loci. *Molecular Biology and Evolution* 32(1): 244–257.
Martin, C. H., and E. J. Richards. 2019. The paradox behind the pattern of rapid adaptive radiation: How can the speciation process sustain itself through an early burst? *Annual Review of Ecology, Evolution, and Systematics* 50: 569–593.
McGee, M. D., S. R. Borstein, J. I. Meier, D. A. Marques, S. Mwaiko, A. Taabu, M. A. Kishe, et al. 2020. The ecological and genomic basis of explosive adaptive radiation. *Nature* 586: 75–79.
Meier, J. I., D. A. Marques, S. Mwaiko, C. E. Wagner, L. Excoffier, and O. Seehausen. 2017. Ancient hybridization fuels rapid cichlid fish adaptive radiations. *Nature Communications* 8(1): 1–11.
Mirande, J. M. 2017. Combined phylogeny of ray-finned fishes (Actinopterygii) and the use of morphological characters in large-scale analyses. *Cladistics* 33(4): 333–350.
Moyer, J. T., and A. Nakazono. 1978. Protandrous hermaphroditism in six species of the anemonefish genus Amphiprion in Japan. *Japanese Journal of Ichthyology* 25(2): 101–106.
na Ayudhaya, P. T., N. Muangmai, N. Banjongsat, W. Singchat, S. Janekitkarn, S. Peyachoknagul, and K. Srikulnath. 2017. Unveiling cryptic diversity of the anemonefish genera Amphiprion and Premnas (Perciformes: Pomacentridae) in Thailand with mitochondrial DNA barcodes. *Agriculture and Natural Resources* 51(3): 198–205.
Natri, H. M., J. Merilä, and T. Shikano. 2019. The evolution of sex determination associated with a chromosomal inversion. *Nature Communications* 10(1): 1–13.
Ochi, H. 1989. Mating behavior and sex change of the anemonefish, Amphiprion clarkii, in the temperate waters of southern Japan. *Environmental Biology of Fishes* 26(4): 257–275.
Ollerton, J., D. McCollin, D. G. Fautin, and G. R. Allen. 2007. Finding NEMO: Nestedness engendered by mutualistic organization in anemonefish and their hosts. *Proceedings of the Royal Society B: Biological Sciences* 274: 591–598.
Ottenburghs, J. 2020. Ghost introgression: Spooky gene flow in the distant past. *BioEssays*, 42(6): 2000012.
Pardo-Diaz, C., C. Salazar, S. W. Baxter, C. Merot, W. Figueiredo-Ready, M. Joron, W. O. McMillan, and C. D. Jiggins. 2012. Adaptive introgression across species boundaries in heliconius butterflies. *PloS Genetics* 8: e1002752.
Payseur, B. A., and L. H. Rieseberg. 2016. A genomic perspective on hybridization and speciation. *Molecular Ecology* 25: 2337–2360.
Pinho, C., and J. Hey. 2010. Divergence with gene flow: Models and data. *Annual Review of Ecology, Evolution, and Systematics* 41: 215–230.
Poelstra, J. W., E. J. Richards, and C. H. Martin. 2018. Speciation in sympatry with ongoing secondary gene flow and a potential olfactory trigger in a radiation of Cameroon cichlids. *Molecular Ecology* 27: 4270–4288.
Qu, Y., R. Zhang, Q. Quan, G. Song, S. H. Li, and F. Lei. 2012. Incomplete lineage sorting or secondary admixture: Disentangling historical divergence from recent gene flow in the Vinous-throated parrotbill (*Paradoxornis webbianus*). *Molecular Ecology* 21(24): 6117–6133.
Saenz-Agudelo, P., G. P. Jones, S. R. Thorrold, and S. Planes. 2011. Detrimental effects of host anemone bleaching on anemonefish populations. *Coral Reefs* 30(2): 497–506.
Schluter, D. 2000. *The ecology of adaptive radiation*. Oxford University Press.
Schmid, S., B. Micheli, F. Cortesi, G. Donati, and N. Salamin. 2022. Extensive hybridisation throughout clownfishes evolutionary history. *bioRxiv* 2022.07.08.499304. doi: https://doi.org/10.1101/2022.07.08.499304.
Seehausen, O. 2004. Hybridization and adaptive radiation. *Trends in Ecology and Evolution* 19(4): 198–207.
Simpson, G. G. 1953. *The major features of evolution*. Columbia University Press.
Songploy, S., S. Chavanich, Y. Mariasingarayan, and V. Viyakarn. 2021. The sharing of the same host of two species of anemonefish in the Gulf of Thailand, one of which is possibly introduced. *Diversity* 13(7): 304.
Soulebeau, A., X. Aubriot, M. Gaudeul, G. Rouhan, S. Hennequin, T. Haevermans, J-Y. Dubuisson, et al. 2015. The hypothesis of adaptive radiation in evolutionary biology: Hard facts about a hazy concept. *Organisms Diversity and Evolution* 15(4): 747–761.
Sousa, V., and J. Hey. 2013. Understanding the origin of species with genome-scale data: Modelling gene flow. *Nature Reviews Genetics* 14(6): 404–414.
Stamatakis, A. 2014. RaxML version 8: A tool for phylogenetic analysis and post-analysis of large phylogenies. *Bioinformatics* 30(9): 1312–1313.
Stevison, L. S., K. B. Hoehn, M. A. Noor. 2011. Effects of inversions on within-and between-species recombination and divergence. *Genome Biology and Evolution* 3: 830–841.
Stroud, J. T., and J. B. Losos. 2016. Ecological opportunity and adaptive radiation. *Annual Review of Ecology, Evolution, and Systematics* 47: 507–532.
Supple, M. A., H. M. Hines, K. K. Dasmahapatra, J. J. Lewis, D. M. Nielsen, C. Lavoie, D. A. Ray, et al. 2013. Genomic architecture of adaptive color pattern divergence and convergence in Heliconius butterflies. *Genome Research* 23(8): 1248–1257.
Svardal, H., F. X. Quah, M. Malinsky, B. P. Ngatungmaiera, E. A. Miska, W. Salzburger, M. J. Genner, et al. 2020. Ancestral hybridization facilitated species diversification in the Lake Malawi cichlid fish adaptive radiation. *Molecular Biology and Evolution* 37(4): 1100–1113.
Tang, K. L., M. L. Stiassny, R. L. Mayden, and R. DeSalle. 2021. Systematics of Damselfishes. *Ichthyology and Herpetology* 109(1): 258–318.
Verta, J. P., and F. C. Jones. 2019. Predominance of cis-regulatory changes in parallel expression divergence of sticklebacks. *Elife* 8: e43785.
Vizueta, J., N. Macías-Hernández, M. A. Arnedo, J. Rozas, and A. Sánchez-Gracia. 2019. Chance and predictability in evolution: The genomic basis of convergent dietary specializations in an adaptive radiation. *Molecular Ecology* 28(17): 4028–4045.

Weir, J. T. and M. Price. 2011. Andean uplift promotes lowland speciation through vicariance and dispersal in Dendrocincla woodcreepers. *Molecular Ecology* 20: 4550–4563.

Xiong, P., C. D. Hulsey, C. Fruciano, W. Y. Wong, A. Nater, A. F. Kautt, O. Simakov, et al. 2021. The comparative genomic landscape of adaptive radiation in crater lake cichlid fishes. *Molecular Ecology* 30(4): 955–972.

Yoder, J. B., E. Clancey, S. Des Roches, J. M. Eastman, L. Gentry, W. Godsoe, T. J. Hagey, et al. 2010. Ecological opportunity and the origin of adaptive radiations. *Journal of Evolutionary Biology* 23(8): 1581–1596.

Zhang, C., M. Rabiee, E. Sayyari, and S. Mirarab. 2018. ASTRAL-III: Polynomial time species tree reconstruction from partially resolved gene trees. *BMC Bioinformatics* 19: 153.

Zinzow-Kramer, W. M., B. M. Horton, C. D. McKee, J. M. Michaud, G. K. Tharp, J. W. Thomas, and D. J. Maney. 2015. Genes located in a chromosomal inversion are correlated with territorial song in white-throated sparrows. *Genes, Brain and Behavior* 14(8): 641–654.

5 The Use of Modern Genetic Tools in Anemonefishes

Laurie John Lee Mitchell, Sakuto Yamanaka, Masato Kinoshita, and Fabio Cortesi

CONTENTS

5.1 INTRODUCTION

Anemonefishes present scientists with a wealth of fascinating biological traits such as their natural symbiosis with sea anemones, sex change, dominance hierarchy, socially controlled growth rate, and conserved skin pattern development. Combined with their ease of upkeep and culturing in captivity, this has led to their emergence as a model group for studying ecology, evolution, and developmental biology in a reef fish. To identify the genes underlying a given trait and examine its molecular evolution, it is often necessary to conduct an initial general screening using forward genetic approaches (i.e., seeking the genetic basis of a phenotype or trait) or exploratory analyses of transcriptomes and/or genomes. This can then inform reverse genetic studies (i.e., seeking what phenotypes are controlled by specific genetic sequences) by performing more targeted in-vivo manipulations of genetic sequences. In this chapter, an overview is first given of the discoveries made by comparative genomic and transcriptomic studies on anemonefishes. Next, the potential usefulness of forward genetic approaches already applied in more established teleost models is discussed. Finally, an in-depth explanation is given on the application of the CRISPR/Cas9 genome-editing system in the false clown anemonefish (*Amphiprion ocellaris*), including current injection protocols for gene knockouts, notes on possible modifications for alternative edits (e.g., gene knock-ins), and the challenges to be addressed before achieving a reliable production of mutant anemonefish.

5.2 INSIGHTS FROM COMPARATIVE TRANSCRIPTOMICS AND GENOMICS

Transcriptome analysis and/or results from quantitative real-time polymerase chain reaction (qRT-PCR) experiments have yielded insights into changes in gene expression levels under different environmental contexts that pertain to anemonefish development (e.g., embryogenesis, larval metamorphosis, sex change) (Casas et al. 2016; Salis et al. 2019; Schalm et al. 2021) and adaptive responses towards external stressors (e.g., pollutants, social stress, oxidative stress) (Ryu et al. 2019; Khamkaew et al. 2020; Zhang et al. 2020). A wealth of transcriptomic sequences is available for multiple tissue and organ samples taken from multiple anemonefish species and reproductive/developmental states, including for different colored skin, the brain, gonads, gill, and retina (see Table 5.1 for a detailed summary of assembled transcriptomes).

Comparative gene expression between different phenotypes using whole-tissue bulk transcriptomic analyses have highlighted individual or clusters of genes expressed in anemonefish tissues that correlate with a variety of functions, including skin and colouration (Maytin et al. 2018;

DOI: 10.1201/9781003125365-6

TABLE 5.1
Summary of Assembled Transcriptomes for Anemonefish Species

Species	Tissue/organ	Sex/developmental state	NCBI bioproject no.	Associated study
A. ocellaris	All organs and muscle	F, M	PRJNA374650	NA
A. ocellaris	Orange and white skins	F, M, J	PRJNA482393	Salis et al. (2019)
A. ocellaris	Retina	F, M	PRJNA547682	Mitchell et al. (2021)
A. percula	Skins (orange, white, black)	J	PRJNA471968	Maytin et al. (2018)
A. percula	Brain	J	PRJEB27750	Sahm et al. (2019)
A. akindynos	Retina	F, M, J	PRJNA547682	Stieb et al. (2019)
A. bicinctus	Brain and gonads	F, M, J	PRJNA261388	Casas et al. (2016)
A. clarkii	Brain	J	PRJEB27750	Sahm et al. (2019)
A. melanopus	Gill	–	PRJNA398732	Sun et al. (2016)
A. sebae	Brain	F, J	PRJNA285007	NA

– = non-disclosed information; NA = not applicable; F = female, M = male, J = juvenile.

Salis et al. 2019), retina and color vision (Stieb et al. 2019; Mitchell et al. 2021), and gonads and brain during sex change (Casas et al. 2016). Experiments using qRT-PCR analyses have further revealed the differential expression of genes which strongly correlate positively or negatively across different developmental and reproductive states. For example, transitional and non-transitional sex states indicate the importance of the aromatase gene (*cyp19a1*) in modulating estrogen production and its regulation in the brain by *sox6* and *foxp4* expression, and by *foxl2* and *dmrt1* expression in the gonads (Casas et al. 2016). Exogenous steroid (i.e., estradiol and cortisol) treatments combined with qRT-PCR analysis have further demonstrated the role that *cyp19a1* serves in feminization and promoting dominant behavior (Iwata and Suzuki 2020). Other applications of qRT-PCR analysis in anemonefish studies have shown the role of at least three genes (*fhl2a*, *fhl2b*, and *apoD1a*) in skin iridophore development during metamorphosis (Salis et al. 2019), and the rhythmic expression of internal clock genes (*bmal*, *clocka*, *cry1b*, *per1b*, *per2*, and *per3*) (Schalm et al. 2021). Moreover, the latter study also established, for the first time, an anemonefish embryonic cell line to study clock gene expression which could feasibly be applied to in-vitro studies comparing gene expression in various biological functions. Localisation of gene expression at the cellular level has also been achieved in anemonefishes by the fluorescent labelling of mRNA transcripts using fluorescent in-situ hybridization. This approach has highlighted specific gene activity in anemonefish embryogenesis (Ghosh et al. 2009), olfactory epithelium (Veilleux et al. 2013), gonadal development (Kobayashi et al. 2017), and retinal photoreceptors (Stieb et al. 2019; Mitchell et al. 2021).

Continual improvements in the cost and accessibility of genome sequencing technologies have extended its application to non-model organisms, including anemonefishes. The year 2018 saw the first draft genome assembly for an anemonefish (*Amphiprion frenatus*) (Marcionetti et al. 2018), which was followed soon after by the two sister species, *A. ocellaris* (Tan et al. 2018) and *A. percula* (Lehmann et al. 2019), the latter of which is one of the most contiguous and complete teleost fish genome assemblies currently available. The draft genome assemblies of a further eight anemonefish species have since been made publicly available (Marcionetti et al. 2019), giving a total of 11 species for which we have genomic data (see Table 5.2 for a summary of anemonefish genomes). These resources are extremely valuable for studying the molecular evolution and adaptation of common anemonefish traits (Lehmann et al. 2019; Marcionetti et al. 2019; Mitchell et al. 2021). Moreover, this raises the exciting prospect of applying modern genome-editing tools in anemonefishes to perform targeted gene knockouts/knock-ins with presumably low off-target activity to directly study individual gene function.

The availability of assembled genomes for multiple anemonefish species has enabled the testing of evolutionary theories pertaining to their adaptive radiation by conducting comparative genomics and molecular evolutionary analyses. The identification of genes under positive selection has revealed 17 genes at the origin of anemonefish radiation, including *HOG1437* and *HOG16500* which remove or mask *N*-acetylated sugars in the mucus coating of anemonefish skin. *N*-acetylated sugars have an important role in stimulating chemoreceptors surrounding sea anemone cnidocytes which triggers their discharge and release of toxins (i.e., stings), and therefore, their absence or reduction in anemonefish skin likely had a crucial role in the evolution of their symbiosis (Marcionetti et al. 2019). Studying anemonefish genomes has also enabled the in-silico identification of highly similar paralogous gene sequences, and analysis of their synteny in the contiguous genome of *A. percula* (Lehmann et al. 2019). This approach has identified visual genes including those involved in the visual transduction pathway, and multiple subclasses of visual opsin genes that can support color vision (Figure 5.1a). Interestingly, across anemonefishes, there is evidence of two tandem duplication events involving the ultraviolet-sensitive (*SWS1*) opsin gene subclass, of which two functionally-coding *SWS1* opsin genes (*SWS1α* and *SWS1β*) are retained in the genomes of

TABLE 5.2
Summary of Whole Genome Assemblies for Anemonefish Species

Species	Sequencing technology and depth	Insert size (bp)	NCBI bioproject no.	Associated study
A. ocellaris	Paired-end Illumina HiSeq2000, 54x coverage; Oxford Nanopore MinION, 11x coverage	300 bp short read 8–30 kbp long read	PRJNA407816	Tan et al. (2018)
A. ocellaris	Paired-end Illumina HiSeq2000, ~130x coverage; mate-pair Illumina HiSeq2500, ~50x coverage	350 bp short read 3 kbp long read	PRJNA515163	Marcionetti et al. (2019)
A. percula	SMRTbell PacBio RS II, 121x coverage	10 and 15 kbp long read	PRJNA436093	Lehmann et al. (2019)
A. frenatus	Paired-end Illumina HiSeq2000, 126x coverage; mate-pair Illumina HiSeq2500, 36x coverage	350 bp short read 3 kbp long read	SRP132439	Marcionetti et al. (2018)
A. bicinctus	Paired-end Illumina HiSeq2000, ~49x coverage	350 bp short read	PRJNA515163	Marcionetti et al. (2019)
A. melanopus	Paired-end Illumina HiSeq2000, 44.5x coverage	350 bp short read	PRJNA515163	Marcionetti et al. (2019)
A. polymnus	Paired-end Illumina HiSeq2000, ~56x coverage	350 bp short read	PRJNA515163	Marcionetti et al. (2019)
A. sebae	Paired-end Illumina HiSeq2000, ~37x coverage	350 bp short read	PRJNA515163	Marcionetti et al. (2019)
A. akallopisos	Paired-end Illumina HiSeq2000, ~55x coverage	350 bp short read	PRJNA515163	Marcionetti et al. (2019)
A. nigripes	Paired-end Illumina HiSeq2000, ~55x coverage	350 bp short read	PRJNA515163	Marcionetti et al. (2019)
A. perideraion	Paired-end Illumina HiSeq2000, 38x coverage	350 bp short read	PRJNA515163	Marcionetti et al. (2019)
A. biaculeatus	Paired-end Illumina HiSeq2000, ~47x coverage	350 bp short read	PRJNA515163	Marcionetti et al. (2019)

bp = base pairs.

all anemonefishes, an exceptionally rare finding in teleost fishes (Figure 5.1b; Mitchell et al. 2021). It remains unclear whether conserving the duplicate pairs conveys any functional benefit to vision in anemonefishes (more information pertaining to anemonefish vision can be found in Chapter 9).

5.3 THE POTENTIAL FOR FORWARD GENETIC STUDIES OF ANEMONEFISH BIOLOGY

Forward genetic investigations aim to identify the genes or genetic elements such as transcription factors, which underly a given phenotype or biological process. Commonly applied techniques in model animals include quantitative trait locus (QTL) analysis or genome-wide association (GWAS) studies, along with forward mutagenesis and subsequent genetic screening. This section describes these approaches in the context of their application in more common teleost models and discusses their potential usefulness in anemonefishes.

As of yet, no QTL or GWAS studies have been performed on anemonefishes. The intent of these methods is to test phenotypic and/or genetic variants from across the genomes of numerous individuals to detect genotype–phenotype associations. These approaches have been of immense usefulness in providing new directions of study for various traits in commonly used genetic models in biomedical research such as in zebrafish (*Danio rerio*) by identifying key genetic loci that are worthwhile to functionally validate (reviewed by Cano-Gamez and Trynka 2020). One suggested application of GWAS in anemonefishes is to identify and map the genetic loci underlying natural color pattern differentiation and skin pigment mutations observed in captive strains, of which some are also observed in wild anemonefish populations (Roux et al. 2021). Combining GWAS with non-targeted mutagenesis and mass mutant screening protocols could also prove to be highly useful for the large-scale mapping of numerous genetic loci with their associated traits in anemonefishes.

Because no forward mutagenesis approach has yet been applied in anemonefishes, a good start would be adapting and trialling techniques already commonly used in well-established teleost models such as zebrafish and medaka (*Oryzias latipes*). The standard choice for chemical mutagenesis in the aforementioned teleost models is alkylating agents such as N-ethyl-N-nitrosourea (ENU), which most often induces AT to TA transversions and AT to GC transitions in male germline cells but can also cause rare small base pair deletions or other chromosomal changes (de Bruijn et al. 2009; Kegel et al. 1936). Gamma radiation treatment can also efficiently induce mutations directly in embryos, but the often-excessive size of deletions and other sequence changes raises difficulty in the accurate identification of specific genes that underly mutant phenotypes (reviewed by Lawson and Wolfe 2011). The point mutations predominantly induced by ENU make it a more favourable technique (Loosli et al. 2003; Furutani-Seiki et al. 2004); however, the propensity of ENU to induce mutations at multiple genetic loci can also complicate distinguishing the gene(s) or genetic elements responsible for an observed phenotype (Lawson and Wolfe 2011). This issue can be circumvented by the alternative use of replication-deficient retroviruses for mutagen delivery in 1,000-cell stage embryos (Nagayoshi et al. 2008; Sivasubbu et al. 2006) at a cost of substantially lower mutagenic efficiency, reportedly one-ninth the efficiency of ENU in zebrafish (Amsterdam et al. 1999).

One important consideration when applying these forward mutagenesis approaches is the mutation screening which usually requires multi-generational breeding

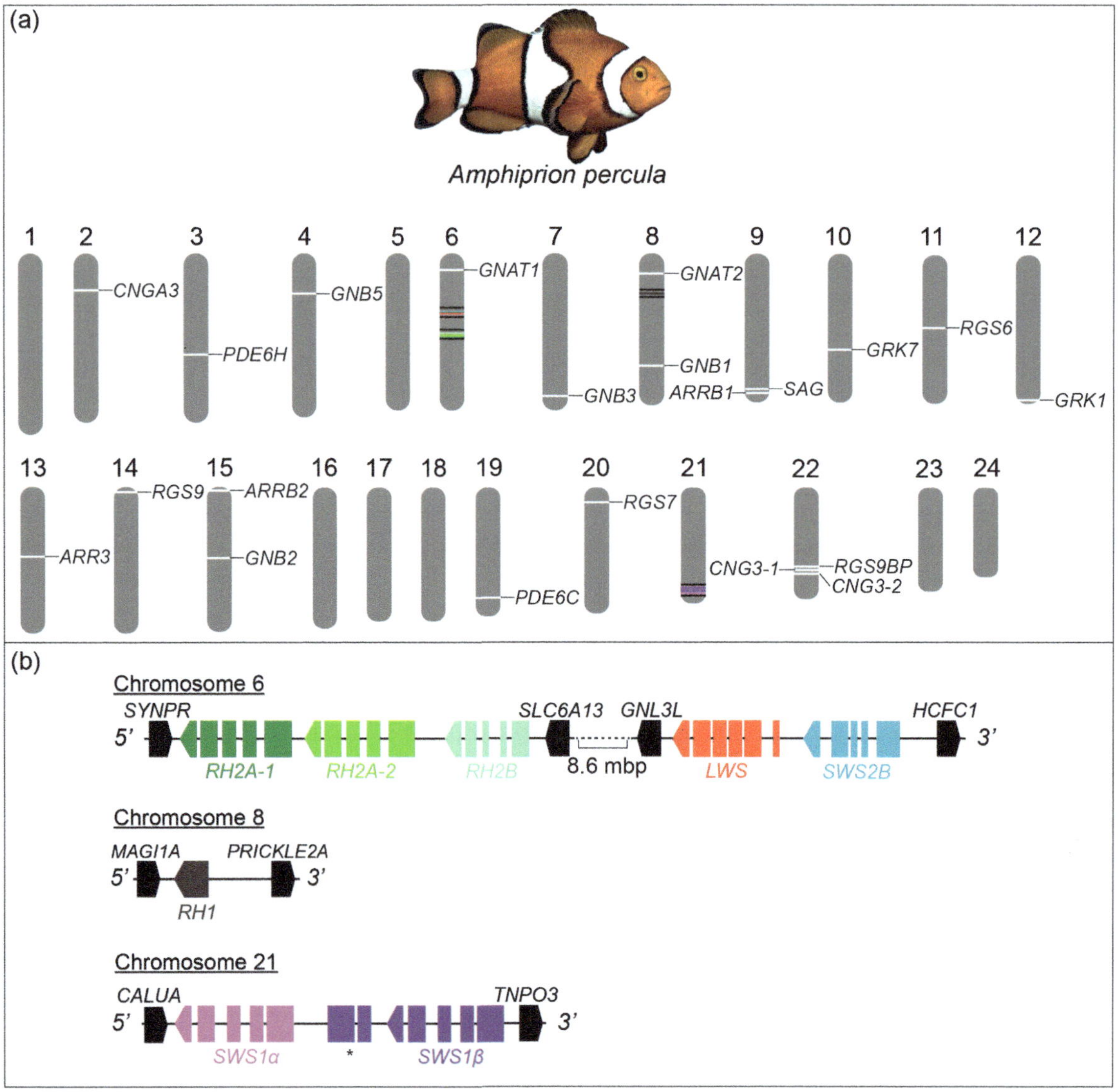

FIGURE 5.1 Chromosomal assembly of the *Amphiprion percula* genome and synteny of visual genes. (a) Organisation of the *A. percula* genome in 24 chromosomal pairs (one shown per pair) according to the assembly by Lehmann et al. (2019). Highlighted are visual genes mapped against the *A. percula* genome, including genes involved in the phototransduction pathway (white lines) and visual opsin genes (colored lines) with immediate flanking genes (black lines). (b) Closeup views of syntenic regions for different subclasses of opsin genes that form teleost visual pigments including green-sensitive (*RH2s*), yellow-red sensitive (*LWS*), violet-sensitive (*SWS2B*), ultraviolet-sensitive (*SWS1s*) opsin, and rod opsin (*RH1*). Flanking genes found immediately upstream (5') and downstream (3') of each syntenic region are in black. "*" denotes an *SWS1* pseudogene. Figure is modified from Mitchell et al. (2021). *A. percula* image taken with permission of Valerio Tettamanti.

schemes to generate results. This process involves crossing mutated males with wildtype females to produce first-generation (F1) offspring that are raised and outcrossed to produce second-generation (F2) progeny, which can be screened to identify homozygous mutations that are recoverable in the backcrossed, third-generation (F3) offspring. The relatively long-generation time of anemonefishes (~9–18 months) would require the planning of long-term experiments over a timescale which may be prohibitive in some circumstances. To achieve a single F3 screening scheme with anemonefishes would take a minimum of 27 months, a substantially longer period of time compared to zebrafish and medaka which take a minimum of six months (both species take approximately two to four months to reach maturation; Lawrence et al. 2012). Regardless, where possible such forward genetic experiments would yield valuable insight into the genetics underlying complex biological processes in anemonefishes (e.g., skin pattern formation and sex change).

Another approach for linking a phenotype of interest with its genetic basis is by the experimental disruption (i.e., knockdown or enhancement) of normal gene expression. In anemonefishes, this can be induced using various pharmacological treatments that often serve to increase or block hormonal pathway activity (e.g., Nakamura et al. 2015; Salis et al. 2018; Iwata and Suzuki 2020). Using drugs can help deduce the genetic basis of a phenotype or trait but require caution and careful consideration of controls due to their often wide-ranging/whole-organism effects.

5.4 REVERSE GENETIC STUDIES OF ANEMONEFISH BIOLOGY

Reverse genetics is a powerful method to understand the functions of the gene of interest by analyzing its phenotypic effects caused by introducing mutations at a specific DNA sequence, while forward genetics is a method to search and identify the gene responsible for the phenotype. In reverse genetic studies, gene targeting methods using ECS (embryonic stem cells) and chemical mutagenesis approaches (for example, Targeting Induced Local Lesions in Genomes: TILLING) have been used for some model organisms (Amsterdam and Hopkins 2006; Dahm and Geisler 2006; Moens et al. 2008). However, these approaches have rarely been applied to non-model organisms including anemonefishes because ECS has not been established (Jasin and Rothstein 2013) and the chemical mutagenesis approach requires great labour of breeding and screening to obtain desired mutants. RNA interference, another reverse genetic approach that uses small interfering RNAs to silence the function of the target gene (Wilson and Doudna 2013), is relatively easy to use for non-model organisms. However, this method also has not been applied to anemonefishes because of a lack of established transfection technologies. In the last decade, genome editing technologies such as Transcription Activator-Like Effector Nuclease (TALEN) and Clustered Regularly Interspaced Short Palindromic Repeats (CRISPR)/CRISPR-associated protein (CRISPR/Cas) have been established. These systems that recognize specific sequences of DNA and induce double-strand break (DSB) greatly contribute to genetic engineering not only in model organisms but also in non-model organisms (Hsu et al. 2014; Joung and Sander 2013; Knott and Doudna 2018, Maraffini 2016). In particular, CRISPR/Cas9 is widely used in many organisms because of its simplicity and commercial availability. The production of mutant strains using genome editing technologies requires the precise DNA sequence information of the target gene and well-established breeding and transfection (microinjection) techniques. Among these requirements, breeding technologies and genome/transcriptome information have been studied quite extensively in anemonefishes (Buston and Elith 2011; Maytin et al. 2018; Roux et al. 2021; Salis et al. 2018, 2019, 2021; Tan et al. 2018). However, only recently several research groups have started to generate genetic mutants of anemonefish using newly developed microinjection methods in combination with the CRISPR/Cas9 technology (Mitchell et al. 2021; Yamanaka et al. 2021). This section first describes general genome editing protocols used in teleost fishes and then provides examples of applied research on genome editing in anemonefishes (mainly *A. ocellaris*).

5.4.1 General Protocol for Gene Knockout in Fishes

In this section, we introduce a general genome editing protocol for teleost fishes based on CRISPR/Cas9 derived from *Streptococcus pyogenes*, which is the most universal genome editing system. For detailed information on TALEN and other CRISPR systems, we refer the readers to the topic-specific literature including studies from Christian et al. (2010), Makarova et al. (2011), Carroll (2014), and Pickar-Oliver and Gersbach (2019).

The CRISPR/Cas9 system consists of a nuclease protein, Cas9, and a guide RNA (gRNA) that recognizes the target DNA sequence. In this system, Cas9 protein forms Cas9/gRNA ribonucleoprotein complexes (CRISPR-RNPs). The RNPs recognize the 5'-NGG (protospacer adjacent motif: PAM) directly downstream of the gRNA sequence (Shah et al. 2013; Zhang et al. 2014) and induce DSB at 3–4 bp upstream of the PAM sequence (Gasiunas et al. 2012; Jinek et al. 2012). The intrinsic DNA repair processes after DSB is prone to mistakes (Chiruvella et al. 2013; Roth et al. 1992), allowing for various types of experimental modifications such as targeted mutagenesis with small insertions or deletions (indels) and gene knock-in.

5.4.1.1 Step 1: Design and Preparation of CRISPR-gRNAs

First, the DNA sequence of the target gene must be confirmed before designing the gRNAs. Since there are sometimes population-specific mutations in non-model organisms, the sequence of the target region may differ from the database due to the presence of mutations such as SNVs (single nucleotide variant). The DNA sequence followed by PAM is critical for the recognition of targets by CRISPR-RNP. Then, using web tools such as CRISPR-Direct (Naito et al. 2015) and CRISPR gRNA Design Software in Geneious (Hsu et al. 2014), several candidates for the target sequence are selected with the following criteria. In the case of CRISPR-based gene knockout, gRNAs are generally designed on the downstream of the translation start codon or on the upstream of important domains. Small insertions or deletions (in/del) induced by the CRISPR-RNP are expected to cause frameshift mutations resulting in nonsense amino acid sequences in transcript and/or newly generated stop codons.

Usually, the in/del patterns are random. However, frameshift mutations can be efficiently generated by utilizing the microhomology-mediated end-joining (MMEJ) repair mechanism. The MMEJ is an intrinsic DNA repair mechanism for DSBs, which relies on two short homologous sequences (microhomologies) flanking the DSB site and results in the deletion of the intervening sequence (Qi et al. 2013; Bae et al. 2014; Grajcarek et al. 2019; McVey and Lee 2008). After the target sequence is determined, gRNA is synthesized as described previously (for example, Ansai and Kinoshita 2014) or ordered from vendors.

5.4.1.2 Step 2: Preparation of Cas9

Cas9 can be commercially acquired as protein or can be synthesized in vitro as mRNA. Detailed information about synthesis of Cas9-mRNA is described in Ansai and Kinoshita (2014). It is reported that Cas9 is more efficient when used as a protein to form RNP complexes with gRNA compared to when used as an expression plasmid or mRNA (Kagita et al. 2021; Kotani et al. 2015).

5.4.1.3 Step 3: Microinjection

The microinjection method in teleost fishes was first developed for medaka and zebrafish (Rosen et al. 2009; Kinoshita et al. 2009) and has since been applied to many other fish species such as red sea bream (Kishimoto et al. 2018, 2019), tiger puffer (Kato-Unoki et al. 2018; Kishimoto et al. 2019), and Atlantic salmon (Edvardsen et al. 2014). The equipment required for microinjection consists of a stereo microscope, a needle made from a glass capillary (GD-1, NARISHIGE, Tokyo, Japan), a puller (PC-10, NARISHIGE, Tokyo, Japan) to draw the needle, a micromanipulator (M-152, NARISHIGE, Tokyo, Japan), a stand (GJ-1, NARISHIGE, Tokyo, Japan) for fixing the manipulator, an injection holder (HI-7, NARISHIGE, Tokyo, Japan), an injector (IM-12, NARISHIGE, Tokyo, Japan) for controlling the pressure in the needle, and acrylic plates or agar plates with trenches for holding the eggs (Figure 5.2).

In fishes, the egg envelope (chorion) gradually hardens after fertilization (Yamagami et al. 1992; Sano et al. 2017), making it difficult for the needle to penetrate. The injection period, needle size, and injection target site depend on the time until the first cleavage, the nature of the yolk cell, and the hardness of the chorion (Goto et al. 2019). Therefore, the parameters for microinjection need to be optimized for each fish species. A brief and general procedure of microinjection in fishes is described below (for detailed information see Kinoshita et al. 2009; Goto et al. 2019) and videos of injection techniques and tips are introduced in detail in Murakami and Kinoshita (2018).

5.4.1.3.1 Step 3.1

Prepare the injection solution containing genome editing reagents (for example, 100 ng/μl of Cas9-mRNA and 50 ng/μl of gRNA) and incubate on ice until just before use. The concentration of the injection solution (or rather the amount of solution to be injected) depends on the species and the activity of the gRNA.

5.4.1.3.2 Step 3.2

Pour the solution into the glass needle with a micro-loading tip by back-filing (loading from the opposite side to the tip). For pneumatic injectors, centrifuge the needle filled with the solution to remove air bubbles in the needle, while for oil pressure injectors, after centrifugation to remove air bubbles, fill up the needle with mineral oil.

5.4.1.3.3 Step 3.3

Set up the instruments including injector, manipulator, and glass needles. Then, slowly touch the surface of the acrylic plate with the tip of the needle, apply pressure and ensure that the solution gradually comes out from the tip of the needle. Alternatively, the tip of the needle can be broken with fine tweezers.

5.4.1.3.4 Step 3.4

Collect fertilized eggs after artificial insemination or natural spawning and align them in the trench on an acrylic (for red sea bream and tiger puffer) or agar plate (for zebrafish and medaka).

5.4.1.3.5 Step 3.5

Inject the solution into eggs at the one-cell stage under a stereomicroscope. The amount of solution to be injected into

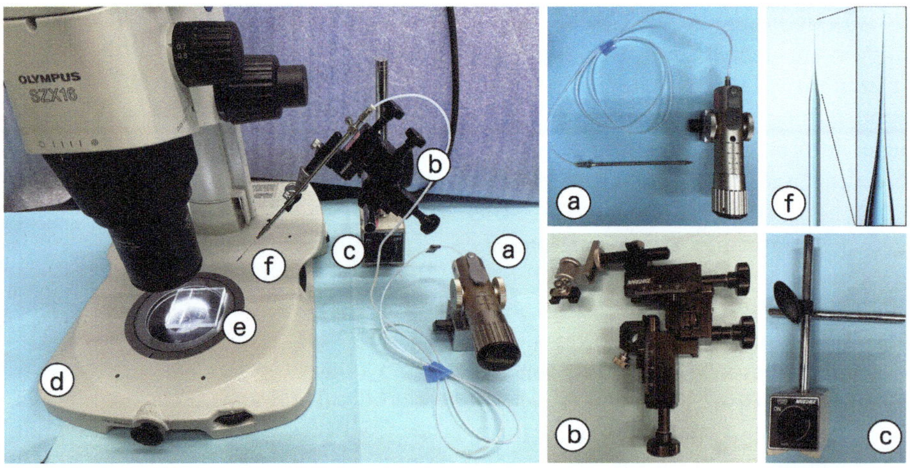

FIGURE 5.2 Set up for microinjection instruments. The injection system consists of a pneumatic injector (a), a manipulator (b), a magnetic stand (c), a stereomicroscope (d), a trench acryl plate (e), and glass needles (f). The fine end of the glass needle is less than 10 um in diameter (the needle tip was observed using a microscopic micrometer and blue dye is filled at the end of the glass needle for illustration purposes).

an egg is roughly 1–10 pl, depending on the size of the eggs. The pressure into the needle tip depends on the injection system and the diameter of the needle tip. The number of eggs to be injected depends on the genome-editing activity of the injected solution and the fish-specific embryo survival rate.

5.4.1.4 Step 4: Detection of Induced Mutation by Heteroduplex Mobility Assay

To assess the mutation-inducing activity of the CRISPR-RNP, Heteroduplex Mobility Assay (HMA) is simpler, quicker, and less expensive than analyzing DNA sequences or utilizing enzymes (for example, Cel 1) that recognize mismatches in DNA duplexes. HMA is performed with PCR amplification of genomic DNA containing the target site and subsequent electrophoresis (Ota et al. 2013; Ansai et al. 2014; Chenouard et al. 2016). HMA is based on the phenomenon that homoduplexes and heteroduplexes (wild-type strand and mutated strand) exhibit different mobilities in electrophoresis. The CRISPR-RNP with high mutation-induce activity produces a multiband pattern in HMA while a single band is observed without mutation.

5.4.2 Application and Tips for Genome Editing in Anemonefishes

The application of microinjection to anemonefish eggs, as the first step of gene editing, faces several difficulties. One of the biggest is how to prepare the fertilized eggs for microinjection. Anemonefishes are benthic spawners, which lay sticky eggs on rough surfaces (Roux et al. 2020). The adhesive eggs make rapid handling and injection difficult. Recently, two research groups overcame this issue using *A. ocellaris* (Mitchell et al. 2021; Yamanaka et al. 2021). In this section, their protocols and tips for microinjection and the application of CRISPR for anemonefish genome editing are introduced.

5.4.2.1 Step 1: Preparation of Fertilized Eggs

A. ocellaris parents usually start cleaning the surface of the spawning substrate (a stone plate or a terracotta pot) with their mouths one or two days prior to spawning. On the morning of the spawning day, the female's ovipositor visibly protrudes.

Spawning occurs in the late afternoon when the female lays sticky eggs on the substrate, and then the male fertilizes the eggs. This process is repeated, and about 300–1,000 eggs are spawned within 30–60 minutes.

When performing microinjection, how the fertilized eggs are handled and collected depends on the type of substrate on which they are spawned. Mitchell et al. (2021) used a terracotta pot as a substrate. Immediately after spawning, they broke the terracotta pot containing egg clutches apart into multiple shards (2.0 × 4.0 cm) using a hammer and chisel for the microinjection procedure.

On the other hand, Yamanaka et al. (2021) used a stone plate as a substrate and carefully detached the eggs from the stone plate using tweezers under a microscope. The use of either substrate may depend on individual preferences by anemonefish breeding pairs, and while switching substrates is possible this often requires an adjustment period (and disrupted spawning cycle).

5.4.2.2 Step 2: Microinjection of Anemonefish Eggs

Using the injector and manipulator described above, insert a glass needle into the cytoplasm on the animal pole side of the egg (i.e., the side that is attached to the substrate), and inject the solution containing the genome editing tools. The volume of the injection solution is 3–5 pl which is equivalent to a droplet of about 0.1 mm in diameter.

When the blast disk (cytoplasm) is not developed enough to be recognized just after fertilization, microinjection into the yolk is also effective and easier (Figure 5.3) (Yamanaka et al. 2021).

Microinjection usually ceases 40–90 minutes after fertilization, about when the embryos reach the two-cell stage and chorion becomes too hard to penetrate by glass needle (Mitchell et al. 2021; Yamanaka et al. 2021).

In order to inject more eggs, it is useful to exchange the substrate while the pair is spawning. After about 100–200 eggs are spawned, remove the substrate with eggs for

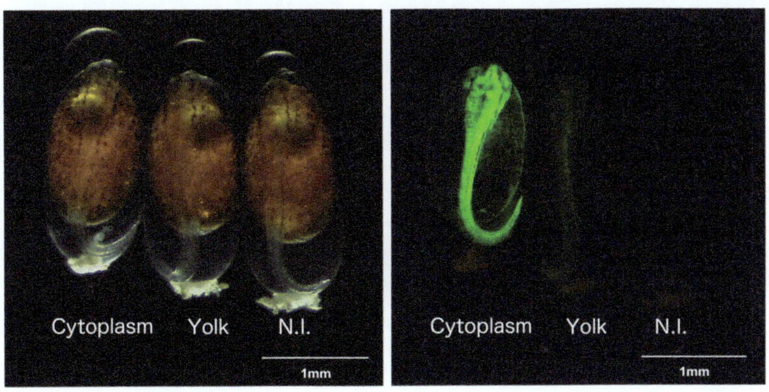

FIGURE 5.3 Microinjection into cytoplasm or yolk. Embryos on day 2 after injection in brightfield (left) and fluorescent light (right). "Cytoplasm" and "Yolk" are the embryos injected with GFP mRNA into cytoplasm and yolk, respectively. "N.I." is the embryo without injection.

injection and replace it with a new one. When the substrate is removed, the parents stop spawning temporarily; after a while they start to spawn again on the new substrate. In this way, a new clutch can be obtained for the next round of microinjection.

5.4.2.3 Step 3: Tips for Anemonefish Microinjection

Co-injection of GFP-mRNA is useful for selecting embryos that have been successfully injected into the cytoplasm. That is, successfully injected embryos exhibit green fluorescence under a fluorescence microscope several hours after injection. From our experience, almost all embryos exhibiting green fluorescence harbour in/del mutations in the target site. As for the preparation of GFP-mRNA, please refer to Yamanaka et al. (2021).

5.4.3 Examples of Gene KO in Anemonefishes

So far there are two reports showing the efficiency of CRISPR/Cas9 for genome editing in anemonefishes. Both reports targeted the genes which concern black pigmentation because the efficiency of gene knockout is easy to recognize by lacking black pigment on the skin and retina in early embryos.

Mitchell et al. (2021) targeted the *tyrosinase* gene (*tyr*) involved in the initial step of melanin production (Cal et al. 2017) and *rhodopsin-like 2B* opsin gene (*RH2B*) encoding a mid-wavelength-sensitive visual pigment (Bowmaker 2008). After confirming the mutation inducing-activity of the gRNAs, the CRISPR/Cas9 mixtures were injected into the fertilized eggs on pieces of terracotta pot as described earlier. Up to half of the injected embryos (range of positive mutants out of four injection rounds, 12.2–53.8%) showed complete or partial deletion of black pigment (Figure 5.4a) suggesting that gene knockout in both alleles has occurred at high efficiency. Mutations in the target sequence were confirmed by cloning the PCR products of mutants and subsequent Sanger DNA sequencing, revealing that most mutations were situated 4–14 bp upstream of the PAM sequence.

Yamanaka et al. (2021) targeted the *solute carrier family 45 member 2* gene (*slc45a2*), which is a transporter protein concerning melanin synthesis (Lamason et al. 2005). They injected the CRISPR/Cas9 mixture with GFP-mRNA, which was the indicator of successful microinjection. After injection, embryos with green fluorescence were selected and then the genome editing activity was investigated by PCR and subsequent HMA as described previously (Ansai et al. 2014). All the embryos exhibited multi-bands which indicated the targeted gene had been mutated (Figure 5.4c). The remaining embryos with green fluorescence were further cultured and visually confirmed to lack black pigmentation in the skin and retina (Figure 5.4b).

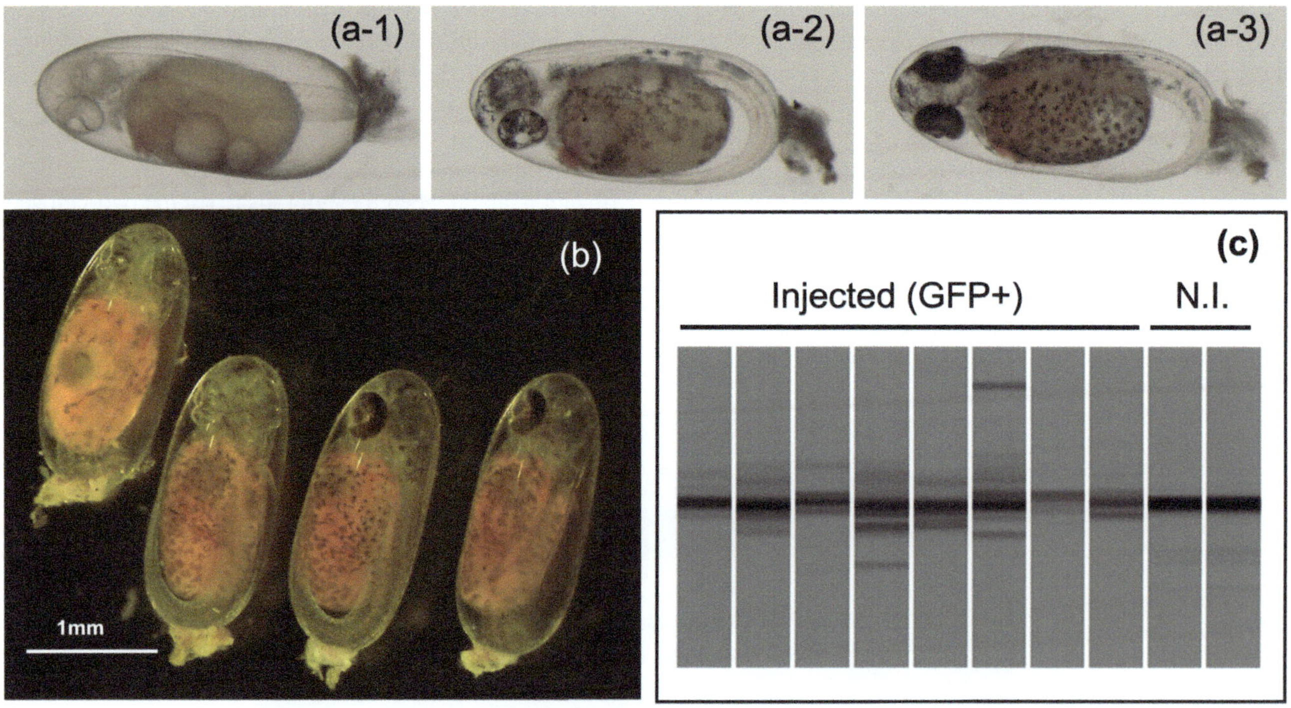

FIGURE 5.4 Effect of gene knockout by CRISPR/Cas9. Embryos of *tyr* mutant *Amphiprion ocellaris* lacking black pigment (a-1) completely and (a-2) partially on the retina and surface and (a-3) a wildtype embryo for comparison (Mitchell et al. 2021). (b) Embryos of *slc45a2* mutant *A. ocellaris* partially lacking black pigments on retina and surface (left two images) and wildtype embryos (right two images) (Yamanaka et al. 2021). (c) Heteroduplex mobility assay image of 4-dpf embryos exhibiting green fluorescence. Each column represents one embryo. Injected (GFP+): the embryos with green fluorescence four days after injection of CRISPR/Cas9 targeted for *slc45a2* and GFP-RNA. N.I.: embryos without injection.

5.5 CONCLUSION

In this chapter, we provide an overview of how forward and reverse genetics are being used and further developed to study anemonefish ecology, evolution, and development. Albeit challenges such as relatively long generation times and a complex larval cycle remain, the use of -omics approaches in anemonefishes promises exciting insights into various aspects of animal biology including the evolution of mutualism, social hierarchy, sex change, vision, and colouration. Future work in the forward genetic realm is likely to focus on long-read technology, resolution at the single-cell level, integration with high-end microscopy and forays into the epigenetic control of gene function. The newly developed microinjection protocols in connection with the CRISPR/Cas9 technology make it possible to manipulate any genetic element of interest and showed that mutations can be introduced with high efficiency in the first generation, thus at least partially overcoming the need for long-term multi-generational experiments. Other reverse-genetic approaches such as *Tol2*-mediated knock-ins are now also well within our grasp. This opens exciting new avenues of research such as the potential to manipulate color patterns involved in species differentiation or the use of optogenetics to control neuronal networks involved in dominance behaviors. We hope that our contribution will serve as a resource and an inspiration for the anemonefish and wider research communities and that it will drive discussion and future planning to coordinate efforts in this space.

REFERENCES

Amsterdam, A., S. Burgess, G. Golling, W. Chen, Z. Sun, K. Townsend, and N. Hopkins. 1999. A large-scale insertional mutagenesis screen in zebrafish. *Genes and Development* 13: 2713–2724.

Amsterdam, A., and N. Hopkins. 2006. Mutagenesis strategies in zebrafish for identifying genes involved in development and disease. *Trends in Genetics* 22: 473–478.

Ansai, S., K. Inohaya, Y. Yoshiura, M. Schartl, N. Uemura, R. Takahashi, and M. Kinoshita. 2014. Design, evaluation, and screening methods for efficient targeted mutagenesis with TALENs in medaka. *Development, Growth & Differentiation* 56: 98–107.

Ansai, S., and M. Kinoshita. 2014. Targeted mutagenesis using CRISPR/Cas system in medaka. *Biology Open* 3: 362–371.

Bae, S., J. Kweon, H. S. Kim, and J. S. Kim. 2014. Microhomology-based choice of Cas9 nuclease target sites. *Nature Methods* 11: 705–706.

Bowmaker, J. K. 2008. Evolution of vertebrate visual pigments. *Vision Research* 48: 2022–2041.

Buston, P. M., and J. Elith. 2011. Determinants of reproductive success in dominant pairs of clownfish: A boosted regression tree analysis. *Journal of Animal Ecology* 80: 528–538.

Cal, L., P. Suarez-Bregua, J. M. Cerdá-Reverter, I. Braasch, and J. Rotllant. 2017. Fish pigmentation and the melanocortin system. *Comparative Biochemistry and Physiology Part A: Molecular and Integrative Physiology* 211: 26–33.

Cano-Gamez, E., and G. Trynka. 2020. From GWAS to function: Using functional genomics to identify the mechanisms underlying complex diseases. *Frontiers in Genetics* 11: 424.

Carroll, D. 2014. Genome engineering with targetable nucleases. *Annual Review of Biochemistry* 83: 409–439.

Casas, L., F. Saborido-Rey, T. Ryu, C. Michell, T. Ravasi, and C. Irigoien. 2016. Sex change in clownfish: Molecular insights from transcriptome analysis. *Scientific Reports* 6: 35461.

Chenouard, V., L. Brusselle, J. M. Heslan, S. Remy, S. Ménoret, C. Usal, L. H. Ouisse, et al. 2016. A rapid and cost-effective method for genotyping genome-edited animals: A heteroduplex mobility assay using microfluidic capillary electrophoresis. *Journal of Genetics and Genomics* 43: 341–348.

Chiruvella, K. K., Z. Liang, and T. E. Wilson. 2013. Repair of double-strand breaks by end joining. *Cold Spring Harbor Perspectives in Biology* 5: a012757.

Christian, M., T. Cermak, E. L. Doyle, C. Schmidt, F. Zhang, A. Hummel, A. J. Bogdanove, et al. 2010. Targeting DNA double-strand breaks with TAL effector nucleases. *Genetics* 186: 756–761.

Dahm, R., and R. Geisler. 2006. Learning from small fry: The zebrafish as a genetic model organism for aquaculture fish species. *Marine Biotechnology* 8: 329–345.

de Bruijn, E., E. Cuppen, and H. Feitsma. 2009. Highly efficient ENU mutagenesis in zebrafish. *Methods in Molecular Biology* 546: 3–12.

Edvardsen, R. B., S. Leininger, L. Kleppe, K. O. Skaftnesmo, and A. Wargelius. 2014. Targeted mutagenesis in Atlantic salmon (*Salmo salar L.*) using the CRISPR/Cas9 system induces complete knockout individuals in the F0 generation. *PLoS ONE* 9(9): e108622.

Furutani-Seiki, M., T. Sasado, C. Morinaga, H. Suwa, K. Niwa, H. Yoda, T. Deguchi, et al. 2004. A systematic genome-wide screen for mutations affecting organogenesis in Medaka, *Oryzias latipes*. *Mechanisms of Development* 121(7-8): 647–658.

Gasiunas, G., R. Barrangou, P. Horvath, and V. Siksnys. 2012. Cas9-crRNA ribonucleoprotein complex mediates specific DNA cleavage for adaptive immunity in bacteria. *Proceedings of the National Academy of Sciences of the United States of America* 109: E2579–E2586.

Ghosh, J., R. W. Wilson, and T. Kudoh. 2009. Normal development of the tomato clownfish *Amphiprion frenatus*: Live imaging and in situ hybridization analyses of mesodermal and neuroectodermal development. *Journal of Fish Biology* 75: 2287–2298.

Goto, R., T. Saito, T. Matsubara, and E. Yamaha. 2019. Microinjection of marine fish eggs. In *Microinjection. Methods in Molecular Biology*, eds C. Liu and Y. Du, 475–487. New York: Humana Press.

Grajcarek, J., J. Monlong, Y. Nishinaka-Arai, M. Nakamura, M. Nagai, S. Matsuo, D. Lougheed, et al. 2019. Genome-wide microhomologies enable precise template-free editing of biologically relevant deletion mutations. *Nature Communications* 10: 4856.

Hsu, P. D., E. S. Lander, and F. Zhang. 2014. Development and applications of CRISPR-Cas9 for genome engineering. *Cell* 157: 1262–1278.

Iwata, E., and N. Suzuki. 2020. Steroidal regulation of the aromatase gene and dominant behavior in the false clown anemonefish *Amphiprion ocellaris*. *Fisheries Science* 86: 457–463.

Jasin, M., and R. Rothstein. 2013. Repair of strand breaks by homologous recombination. *Cold Spring Harbor Perspectives in Biology* 5(11): a012740.

Jinek, M., K. Chylinski, I. Fonfara, M. Hauer, J. A. Doudna, and E. Charpentier. 2012. A programmable dual-RNA-guided DNA endonuclease in adaptive bacterial immunity. *Science* 337: 816–821.

Joung, J. K., and J. D. Sander. 2013. TALENs: A widely applicable technology for targeted genome editing. *Nature Reviews Molecular Cell Biology* 14: 49–55.

Kagita, A., M. S. Y. Lung, H. Xu, Y. Kita, N. Sasakawa, T. Iguchi, M. Ono, et al. 2021. Efficient ssODN-mediated targeting by avoiding cellular inhibitory RNAs through precomplexed CRISPR-Cas9/sgRNA ribonucleoprotein. *Stem Cell Reports* 16: 985–996.

Kato-Unoki, Y., Y. Takai, M. Kinoshita, T. Mochizuki, R. Tatsuno, Y. Shimasaki, and Y. Oshima. 2018. Genome editing of pufferfish saxitoxin- and tetrodotoxin-binding protein type 2 in *Takifugu rubripes*. *Toxicon* 153: 58–61.

Kegel, L., M. Rubio, R. G. Almeida, S. Benito, A. Klingseisen, and D. A. Lyons. 1936. Forward genetic screen using zebrafish to identify new genes involved in myelination. In *Oligodendrocytes Methods and Protocols*, eds D. A. Lyons and L. Kegel, 185–209. Edinburgh, UK: Humana Press.

Khamkaew, A., J. Thamnawasolos, C. Boonphakdee, T. Boonphakdee, and K. Hrimpeng. 2020. Effects of bisphenol A on the expression of *CYP1A* transcripts in juvenile false clown anemonefish (*Amphiprion ocellaris*). *Genomics and Genetics* 13: 69–78.

Kinoshita, M., K. Murata, K. Naruse, M. Tanaka. 2009. Chapter 7 Transgenesis. In *Medaka: Biology, Management, and Experimental Protocols*, eds M. Kinoshita, K. Murata, K. Naruse and M. Tanaka, 277–296. New Jersey: Wiley-Blackwell.

Kishimoto, K., Y. Washio, Y. Murakami, T. Katayama, M. Kuroyanagi, K. Kato, Y. Yoshiura, and M. Kinoshita. 2019. An effective microinjection method for genome editing of marine aquaculture fish: Tiger pufferfish *Takifugu rubripes* and red sea bream *Pagrus major*. *Fisheries Science* 85: 217–226.

Kishimoto, K., Y. Washio, Y. Yoshiura, A. Toyoda, T. Ueno, H. Fukuyama, K. Kato, et al. 2018. Production of a breed of red sea bream *Pagrus major* with an increase of skeletal muscle muss and reduced body length by genome editing with CRISPR/Cas9. *Aquaculture* 495: 415–427.

Knott, G. J., and J. A. Doudna. 2018. CRISPR-Cas guides the future of genetic engineering. *Science* 361: 866–869.

Kobayashi, Y., R. Nozu, M. Nakamura. 2017. Expression and localization of two gonadotropin receptors in gonads of the yellowtail clownfish. *Amphiprion Clarkii. Journal of Aquaculture & Marine Biology* 5(3): 00120.

Kotani, H., K. Taimatsu, R. Ohga, S. Ota, and A. Kawahara. 2015. Efficient multiple genome modifications induced by the crRNAs, tracrRNA and Cas9 protein complex in zebrafish. *PLoS ONE* 10(5): e0128319.

Lamason, R. L., M.-A. P. Mohideen, J. R.Mest, A. C. Wong, H. L. Norton, M. C. Aros, M. J. Jurynec, et al. 2005. SLC24A5, a putative cation exchanger, affects pigmentation in zebrafish and humans. *Science* 310: 1782–1786.

Lawrence, C., I. Adatto, J. Best, A. James, and K. Maloney. 2012. Generation time of zebrafish (*Danio rerio*) and medakas (*Oryzias latipes*) housed in the same aquaculture facility. *Lab Animal* 41: 158–165.

Lawson, N. D., and S. A. Wolfe. 2011. Forward and reverse genetic approaches for the analysis of vertebrate development in the zebrafish. *Developmental Cell* 21: 48–64.

Lehmann, R., D. J. Lightfoot, C. Schunter, C. T. Michell, H. Ohyanagi, K. Mineta, and T. Ravasi. 2019. Finding Nemo's genes: A chromosome-scale reference assembly of the genome of the orange clownfish *Amphiprion percula*. *Molecular Ecology Resources* 19: 570–585.

Loosli, F., W. Staub, K. C. Finger-Baier, E. A. Ober, H. Verkade, J. Wittbrodt, and H. Baier. 2003. Loss of eyes in zebrafish caused by mutation of *chokh/rx3*. *EMBO Reports* 4: 894.

Makarova, K. S., D. H. Haft, R. Barrangou, S. J. J. Brouns, E. Charpentier, P. Horvath, S. Moineau, et al. 2011. Evolution and classification of the CRISPR-Cas systems. *Nature Reviews Microbiology* 9: 467–477.

Marcionetti, A., V. Rossier, J. A. M. Bertrand, G. Litsios, and N. Salamin. 2018. First draft genome of an iconic clownfish species (*Amphiprion frenatus*). *Molecular Ecology Resources* 18: 1092–1101.

Marcionetti, A., V. Rossier, N. Roux, P. Salis, V. Laudet, and N. Salamin. 2019. Insights into the genomics of clownfish adaptive radiation: Genetic basis of the mutualism with sea anemones. *Genome Biology and Evolution* 11: 869–882.

Marraffini, L. A. 2016. The CRISPR-Cas system of *Streptococcus pyogenes*: Function and applications. In *Streptococcus Pyogenes: Basic Biology to Clinical Manifestations*, eds J. J. Ferretti, D. L. Stevens and V. A. Fischetti, 1–17. Oklahoma City: University of Oklahoma Health Sciences Center.

Maytin, A. K., S. W. Davies, G. E. Smith, S. P. Mullen, and P. M. Buston. 2018. De novo transcriptome assembly of the clown anemonefish (*Amphiprion percula*): A new resource to study the evolution of fish color. *Frontiers in Marine Science* 5: 1–11.

McVey, M., and S. E. Lee. 2008. MMEJ repair of double-strand breaks (director's cut): Deleted sequences and alternative endings. *Trends in Genetics* 24: 529–538.

Mitchell, L. J., K. L. Cheney, M. Luehrmann, N. J. Marshall, K. Michie, and F. Cortesi. 2021. Molecular evolution of ultraviolet visual opsins and spectral tuning of photoreceptors in anemonefishes (*Amphiprioninae*). *Genome Biology and Evolution* 13: evab184.

Mitchell, L. J., V. Tettamanti, J. S. Rhodes, J. N. Marshall, K. L. Cheney, and F. Cortesi. 2021. CRISPR/Cas9-mediated generation of biallelic F0 anemonefish (*Amphiprion ocellaris*) mutants. *PLoS ONE* 16: e0261331.

Moens, C. B., T. M. Donn, E. R. Wolf-Saxon, and T. P. Ma. 2008. Reverse genetics in zebrafish by TILLING. *Briefings in Functional Genomics and Proteomics* 7: 454–459.

Murakami, Y., and M. Kinoshita. 2018. Medaka-microinjection with an upright microscope. *Bio-Protocol* 8: 1–8.

Nagayoshi, S., E. Hayashi, G. Abe, N. Osato, K. Asakawa, A. Urasaki, K. Horikawa, et al. 2008. Insertional mutagenesis by the Tol2 transposon-mediated enhancer trap approach generated mutations in two developmental genes: *tcf7* and *synembryn*-like. *Development* 135: 159–169.

Naito, Y., K. Hino, H. Bono, and K. Ui-Tei. 2015. CRISPRdirect: Software for designing CRISPR/Cas guide RNA with reduced off-target sites. *Bioinformatics* 31: 1120–1123.

Nakamura, M., S. Miura, R. Nozu, and Y. Kobayashi. 2015. Opposite-directional sex change in functional female protandrous anemonefish, *Amphiprion clarkii*: Effect of aromatase inhibitor on the ovarian tissue. *Zoological Letters* 1: 30.

Ota, S., Y. Hisano, M. Muraki, K. Hoshijima, T. J. Dahlem, D. J. Grunwald, Y. Okada, et al. 2013. Efficient identification of TALEN-mediated genome modifications using heteroduplex mobility assays. *Genes to Cells* 18: 450–458.

Pickar-Oliver, A., and C. A. Gersbach. 2019. The next generation of CRISPR–Cas technologies and applications. *Nature Reviews Molecular Cell Biology* 20: 490–507.

Qi, Y., Y. Zhang, F. Zhang, J. A. Baller, S. C. Cleland, Y. Ryu, C. G. Starker, et al. 2013. Increasing frequencies of site-specific mutagenesis and gene targeting in *Arabidopsis* by manipulating DNA repair pathways. *Genome Research* 23: 547–554.

Rosen, J. N., M. F. Sweeney, and J. D. Mably. 2009. Microinjection of zebrafish embryos to analyze gene function. *Journal of Visualized Experiments* 25: e1115.

Roth, D. B., P. B. Nakajima, J. P. Menetski, M. J. Bosma, and M. Gallert. 1992. Recombination in mouse thymocytes: Double-strand breaks near T cell receptor gamma rearrangement signals. *Cell* 69: 41–53.

Roux, N., V. Logeux, N. Trouillard, R. Pillot, K. Magré, P. Salis, D. Lecchini, et al. 2021. A star is born again: Methods for larval rearing of an emerging model organism, the false clownfish *Amphiprion ocellaris. Journal of Experimental Zoology Part B: Molecular and Developmental Evolution* 336: 376–385.

Roux, N., P. Salis, S-H. Lee, L. Besseau, and V. Laudet. 2020. Anemonefish, a model for Eco-Evo-Devo. *EvoDevo* 11: 20.

Ryu, H. S., C. Y. Choi, J. A. Song, and G. S. Kil. 2019. Effects of UV radiation on oxidative Stress in yellowtail clownfish *Amphiprion clarkii. Ocean Science Journal* 54: 205–212.

Sahm, A., P. Almaida-Pagán, M. Bens, M. Mutalipassi, A. Lucas-Sánchez, J. de Costa Ruiz, M. Görlach, et al. 2019. Analysis of the coding sequences of clownfish reveals molecular convergence in the evolution of lifespan. *BMC Evolutionary Biology* 19: 89.

Salis, P., S. Lee, N. Roux, D. Lecchini, and V. Laudet. 2021. The real Nemo movie: Description of embryonic development in *Amphiprion ocellaris* from first division to hatching. *Developmental Dynamics* 250: 1651–1667.

Salis, P., T. Lorin, V. Lewis, C. Rey, A. Marcionetti, M. L. Escande, N. Roux, et al. 2019. Developmental and comparative transcriptomic identification of iridophore contribution to white barring in clownfish. *Pigment Cell and Melanoma Research* 32: 391–402.

Salis, P., N. Roux, O. Soulat, D. Lecchini, V. Laudet, and B. Frédérich. 2018. Ontogenetic and phylogenetic simplification during white stripe evolution in clownfishes. *BMC Biology* 16: 1–13.

Sano, K., M. Kawaguchi, K. Katano, K. Tomita, M. Inokuchi, T. Nagasawa, J. Hiroi, et al. 2017. Comparison of egg envelope thickness in teleosts and its relationship to the sites of ZP protein synthesis. *Journal of Experimental Zoology Part B: Molecular and Developmental Evolution* 328: 240–258.

Schalm, G., K. Bruns, N. Drachenberg, N. Geyer, N. S. Foulkes, C. Bertolucci, and G. Gerlach. 2021. Finding Nemo's clock reveals switch from nocturnal to diurnal activity. *Scientific Reports* 11: 1–11.

Shah, S. A., S. Erdmann, F. J. M. Mojica, and R. A. Garrett. 2013. Protospacer recognition motifs: Mixed identities and functional diversity. *RNA Biology* 10: 891–899.

Sivasubbu, S., D. Balciunas, A. E. Davidson, M. A. Pickart, S. B. Hermanson, K. J. Wangensteen, S. C. Ekker. 2006. Gene-breaking transposon mutagenesis reveals an essential role for histone H2afza in zebrafish larval development. *Mechanisms of Development* 123: 513–529.

Stieb, S., F. de Busserolles, K. L. Carleton, F. Cortesi, W. Chung, B. E. Dalton, and N. J. Marshall. 2019. A detailed investigation of the visual system and visual ecology of the Barrier Reef anemonefish, *Amphiprion akindynos. Scientific Reports* 9: 16459.

Sun, Y., Y. Huang, X. Li, C. C. Baldwin, Z. Zhou, Z. Yan, K. A. Crandall, et al. 2016. Fish-T1K (transcriptomes of 1,000 fishes) project: Large-scale transcriptome data for fish evolution studies. *GigaScience* 5: s13742–016–0124–7.

Tan, M. H., C. M. Austin, M. P. Hammer, Y. P. Lee, L. J. Croft, and H. M. Gan. 2018. Finding Nemo: Hybrid assembly with Oxford Nanopore and Illumina reads greatly improves the clownfish (*Amphiprion ocellaris*) genome assembly. *GigaScience* 7: 1–6.

Veilleux, H. D., L. Van Herwerden, N. J. Cole, E. K. Don, C. De Santis, D. L. Dixson, P. L. Munday. 2013. *Otx2* expression and implications for olfactory imprinting in the anemonefish, *Amphiprion percula. Biology Open* 2: 907–915.

Wilson, R. C., and J. A. Doudna. 2013. Molecular mechanisms of RNA interference. *Annual Review of Biophysics* 42: 217–239.

Yamagami, K., T. S. Hamazaki, S. Yasumasu, K. Masuda, I. Iuchi. 1992. Molecular and cellular basis of formation, hardening, and breakdown of the egg envelope in fish. *International Review of Cytology* 136: 51–92.

Yamanaka, S., Y. Okada, T. Furuta, M. Kinoshita. 2021. Establishment of culture and microinjection methods for false clownfish embryos without parental care. *Development, Growth & Differentiation* 63: 459–466.

Zhang, F., Y. Wen, and X. Guo. 2014. CRISPR/Cas9 for genome editing: Progress, implications and challenges. *Human Molecular Genetics* 23: 40–46.

Zhang Y, H, Zhang, J. Wang, X. Zhang, S. Bu, X. Liu, Q. Wand, et al. 2020. Molecular characterization and expression patterns of glucocorticoid receptor (GR) genes in protandrous hermaphroditic yellowtail clownfish, Amphiprion clarkii. *Gene* 745: 144651

Part II

Life History and Development

6 The Post-Embryonic Period of Anemonefishes

Natacha Roux, David Lecchini, and Vincent Laudet

CONTENTS

6.1 INTRODUCTION

The role of thyroid hormone (TH) as the main trigger and coordinator of distinct biological processes during metamorphosis (transitions from larval to juvenile states) has been extensively studied in vertebrates with spectacular transformations such as anuran amphibians and flatfishes (Power et al. 2008; Laudet 2011; Denver 2013). However, as interesting and fascinating as they are, these organisms are exceptions to the rule, as morphological changes are often more subtle than spectacular in other vertebrate species. Indeed, TH-regulated metamorphosis has been shown to be ancestral to chordates with most vertebrates passing through a period of morphological, physiological, and ecological changes, often during post-embryonic development (Laudet 2011). Even in amniotes a remodelling period similar to metamorphosis has been observed and was shown to be very important for the future life of the emanated organism (Holzer and Laudet 2013; Buchholz 2015). A particular challenge for the organism is to align and coordinate this transformation with environmental conditions as well as with its metabolic and physiological status. This critical step is triggered by TH, which ensures the coordination of these changes in various organs and cell types with the environment and the physiological status of the organism (Denver 2013). Ultimately, the ability to successfully undergo metamorphosis will impact the quality and the survival of the juvenile emanating from this complex process, with direct consequences on its ability to perform its ecological function and to grow to reach adulthood (Holzer et al. 2017; Besson et al. 2020). However, the precise underlying mechanisms that facilitate the global action of TH during metamorphosis remain poorly understood.

In teleost fishes, TH action is particularly diverse and is critical for the future juvenile fish (McMenamin and Parichy 2013; Lazcano and Orozco 2018; Deal and Volkoff 2020). TH trigger massive morphological changes enabling, for example, the transition from oceanic bilateral symmetric larvae to benthic asymmetric juveniles in flatfishes (Power et al. 2008). It also triggers metamorphosis when larvae are still in the ocean; however, it is currently not clear how larvae determine the right time to initiate this transition, as early or late metamorphosis may jeopardize recruitment processes. It has been shown in a coral reef fish (the convict surgeonfish) that TH disruption alters metamorphosis and its signalling pathway, resulting in sensory ability impairments, decreased grazing activity, and higher predation. Thus, the juvenile quality was altered (Holzer et al. 2017; Besson et al. 2020). It is therefore important to identify the environmental factors involved in TH regulation. We know that in the teleost brain the hypothalamo-pituitary axis is responsible for the integration of both internal and external stimuli to ensure TH synthesis, but it is currently unknown which stimuli in the ocean trigger this action at the correct time (Deal and Volkoff, 2020; see Chapter 11 by Dussenne et al.).

Understanding how TH control metamorphosis is not only important to better understand the pleiotropic action of the hormone, but also to determine how the correct integration of the transformative process controls the ecological function and quality of the future juvenile. However, understanding all of these steps requires a genetic model organism, and currently, most teleost fish models used in evo/devo labs (e.g., zebrafish, medaka) are freshwater fish. In fact, most marine teleost species used for research purposes are aquaculture models such as sea bass, sea bream, salmon, flatfish, tuna, and grouper. Unfortunately, the size of adult individuals, difficulties identifying their sex, low spawning frequency, the minute size of the larvae, and the challenges associated with their husbandry make these species difficult to maintain and reproduce at the required scale. Recently, owing to several remarkable biological traits, anemonefish have received special attention, allowing studies to tackle a series of scientific questions in several disciplines: ecology, evolutionary sciences, and developmental biology (eco-evo-devo), including the investigation of metamorphosis (Roux et al. 2020).

DOI: 10.1201/9781003125365-8

Anemonefishes have a biphasic life cycle with an oceanic dispersal during the larval phase followed by a reef phase during which newly metamorphosed juveniles recruit in a sea anemone species (Figure 6.1). Due to their ecological features and their ability to be reared in aquaria, anemonefish recently became an emerging model in the field of eco-evo-devo (Roux et al. 2020; Roux et al. 2021). As they can be used for both ecological and functional experiments in the lab, anemonefish have been used to provide insight into the central role of TH in orchestrating metamorphosis. In this chapter, we will first describe the morphological changes that occur during post-embryonic development and metamorphosis. We will then outline the role of TH in the regulation of this process, before finally discussing the possible role of environmental cues in metamorphosis regulation and their integration with the TH system.

6.2 THE POST-EMBRYONIC DEVELOPMENT OF ANEMONEFISH

Several descriptions of the post-embryonic development of anemonefish species (*Amphiprion ocellaris*, *A. perideraion*, *A. ephippium*, *A. clarkii*, *A. sebae*, *A. frenatus*) are available (Ghosh et al. 2009; Putra et al. 2012; Adams et al. 2014; Gunasekaran et al. 2017; Krishna 2018); however, only one description has categorized larval development into distinct stages based on morphological criteria (Roux et al. 2019). Developmental stages are essential for model organisms as they allow the standardization of development. Such standardization is critical as various factors such as temperature, photoperiod, and food nutritional value may cause heterogeneity in developmental timing, meaning that time alone cannot be used to predict developmental changes in anemonefishes (Arvedlund et al. 2000; Avella et al. 2007; Olivotto et al. 2011; Ye et al. 2011).

The first developmental table of anemonefishes has been made on the false clownfish *A. ocellaris* reared under laboratory conditions at 25°C with a 12 hours light:12 hours dark photoperiod (Roux et al. 2019). A total of seven developmental stages have been identified for this species using eight morphological criteria visible under a binocular or a stereomicroscope: notochord, dorsal fin soft rays, anal fin soft rays, dorsal fin spines, anal fin spines, pelvic fins, head white bar, and caudal white bar (Table 6.1). At stages 1, 2 and 3, larvae have an overall elongated shape, caudal, anal, and dorsal fin soft rays and spines are slowly developing and reach their final state at stage 3, meaning that soft rays and spines are fully visible. The notochord develops with three different states: first a pre-flexion state (stage 1, Figure 6.2A), followed by a flexion state during which it bends dorsally (stage 2, Figure 6.2A, B1–B7), and finally the post-flexion state characterized by the vertical position of the bones supporting caudal rays (stage 3, Figure 6.2A). At these stages, the pigmentation pattern is composed of aggregated black cells (melanophores) forming two lines on the body and stellate black and yellow cells (xanthophores) scattered on the head and the body.

Stage 4 marks a turn in the developmental dynamic of *A. ocellaris* as larvae slowly begin their transformation into adult-like individuals (Figure 6.2A). The overall shape is

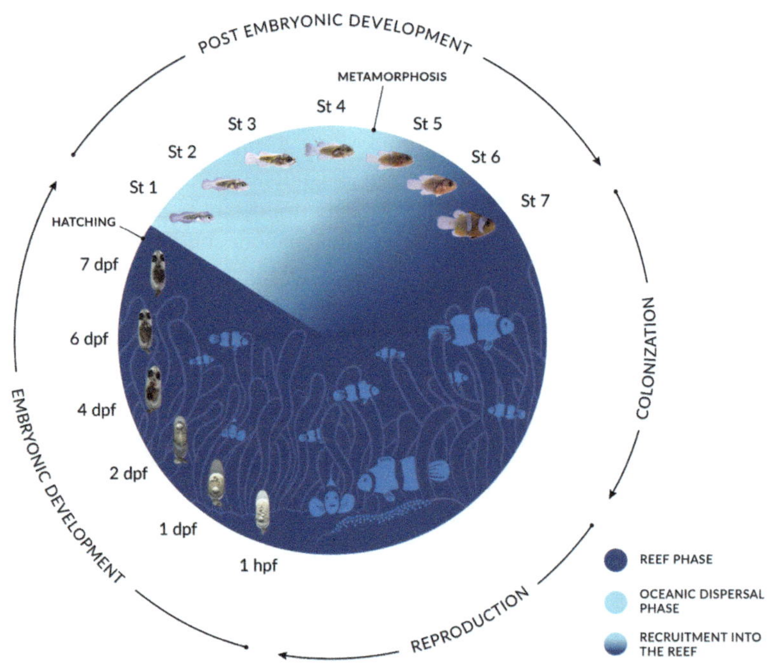

FIGURE 6.1 Illustration of the life cycle of anemonefishes from Roux et al. (2020). Anemonefish lay their eggs close to their sea anemone, where they will develop for six to ten days, depending on temperature and species. After hatching, larvae are directly dispersed into the open ocean, where they will grow for 10 to 15 days before returning to the reef. Larval development is characterized by seven distinct stages. The transition between the ocean and the reef is associated with the metamorphosis of larvae into juveniles. Juveniles will then settle in a sea anemone. Picture from Roux et al. (2020).

TABLE 6.1
Qualitative Traits Characterizing Each Developmental Stage

Developmental Stages	Notochord	Anal soft rays	Anal spines	Dorsal soft rays	Dorsal spines	Pelvic fin	Head and body white bars	Caudal fin white bar
1	**Pre-flexion**	Absent	Absent	Absent	Absent	Absent/Bud	Absent	Absent
2	**Flexion**	Present	Absent	Present	Absent	Bud	Absent	Absent
3	**Post-flexion**	Present	Present	Present	Absent	Bud/**Fin**	Absent	Absent
4	Post-flexion	Present	Present	Present	Present	**Spines**	Absent	Absent
5	Post-flexion	Present	Present	Present	Present	Spines	**Transparent**	Absent
6	Post-flexion	Present	Present	Present	Present	Spines	**White**	Absent
7	Post-flexion	Present	Present	Present	Present	Spines	White	**Present**

Source: from Roux et al. (2019).
Note: bold terms correspond to criteria used to differentiate each stage from the other.

becoming more ovoid, the pelvic fin spines are present, and the pigmentation becomes progressively orange. At stages 5 and 6 the characteristic white bars begin to appear on the head and the trunk, being transparent at the beginning of their formation (stage 5, Figure 6.2A) and then intensively whitish at stage 6 (Figure 6.2A, B12 andB17). This white hue is due to specific reflective cells that have been morphological and genetically identified as iridophores (Figure 1C of Salis et al. 2019). Stage 7 marks the end of the larval development as the third and last white bar is appearing on the caudal peduncle (before the caudal fin), the overall pigmentation pattern is bright orange, and individuals now resemble mini adult-like individuals, namely juveniles.

The other studies describing the larval development of *A. clarkii* (Ghosh et al. 2009), *A. frenatus* (Putra et al. 2012), *A. nigripes* (Anil et al. 2012), *A. sebae* (Gunasekaran et al. 2017), *A. ephippium* (Krishna 2018), and *A. perideraion* (Salis et al. 2018) reveal some similarities in the development of anemonefish larvae. As for *A. ocellaris*, larvae from four of these species have an elongated body shape after hatching, are surrounded by a transparent fin fold that differentiates into the caudal, dorsal, and anal, and have a pigmentation pattern composed of stellar melanophores and xanthophores on the head and body. The pelvic fins are always the last to differentiate with massive pigmentation and body shape changes occurring once the notochord has reached the post-flexion stage. From these studies, it seems that the developmental timing of morphological and pigmentation changes occurring during the larval development of anemonefish is relatively similar (see Table 6.2 and Figure 6.3); however, fully confirming this requires additional and more precise comparative studies similar to what has been done in *A. ocellaris* in Roux et al. (2019). Similar methodology using qualitative criteria can be applied to other anemonefish species to identify distinct stages during the post-embryonic development. This way it will allow assessing the developmental differences existing between all anemonefish species. It is also worth mentioning that the sequence of appearance of developmental modifications, as well as stage-to-stage transitions, may vary from one rearing condition to another depending on the abiotic factors mentioned earlier, but also larval density in rearing tanks and food quality and availability (Arvedlund et al. 2000; Avella et al. 2007; Olivotto et al. 2008; Olivotto et al. 2011; Ye et al. 2011; Dhaneesh et al. 2012; Chambel et al. 2015).

One post-development difference between anemonefishes and other pomacentrids, and even other coral reef fishes, is that instead of opting for a long larval development and thus a long oceanic phase, anemonefish have evolved a shorter oceanic phase (10 to 15 days) (Kavanagh and Frederich 2016; Salis et al. 2021). This developmental strategy is linked to the fact that anemonefish embryonic development is longer than the other species, seven to ten days instead of two to three days, which consequently results in bigger larvae at hatching, and more importantly, more developed larvae that already have a functional digestive tract and are able to capture live prey within a few hours of hatching (Önal et al. 2008; Putra et al. 2012; Kavanagh and Frederich 2016). They are also already equipped with pectoral fins that improve swimming ability, allowing them to swim against currents and move towards live prey (Bellwood and Fisher 2001; Leis 2006). The reason for such difference is still poorly understood and it also poses interesting questions about the mechanisms involved in regulating the timing of larval development and the transition between the larval and juvenile period (metamorphosis), compared to other species (Kavanagh and Frederich 2016; Salis et al. 2021). The question of why metamorphosis occurs sooner in anemonefishes is therefore still open and will require study integrating ecological as well as developmental approaches.

6.3 INTERPLAY BETWEEN ENVIRONMENTAL CUES AND LIFE HISTORY TRANSITIONS

Marine ecologists studying teleosts have always been fascinated by how marine fish larvae succeed in finding their reef habitat following development in the open ocean (Leis 2006). Anemonefishes are no exception to the rule as they

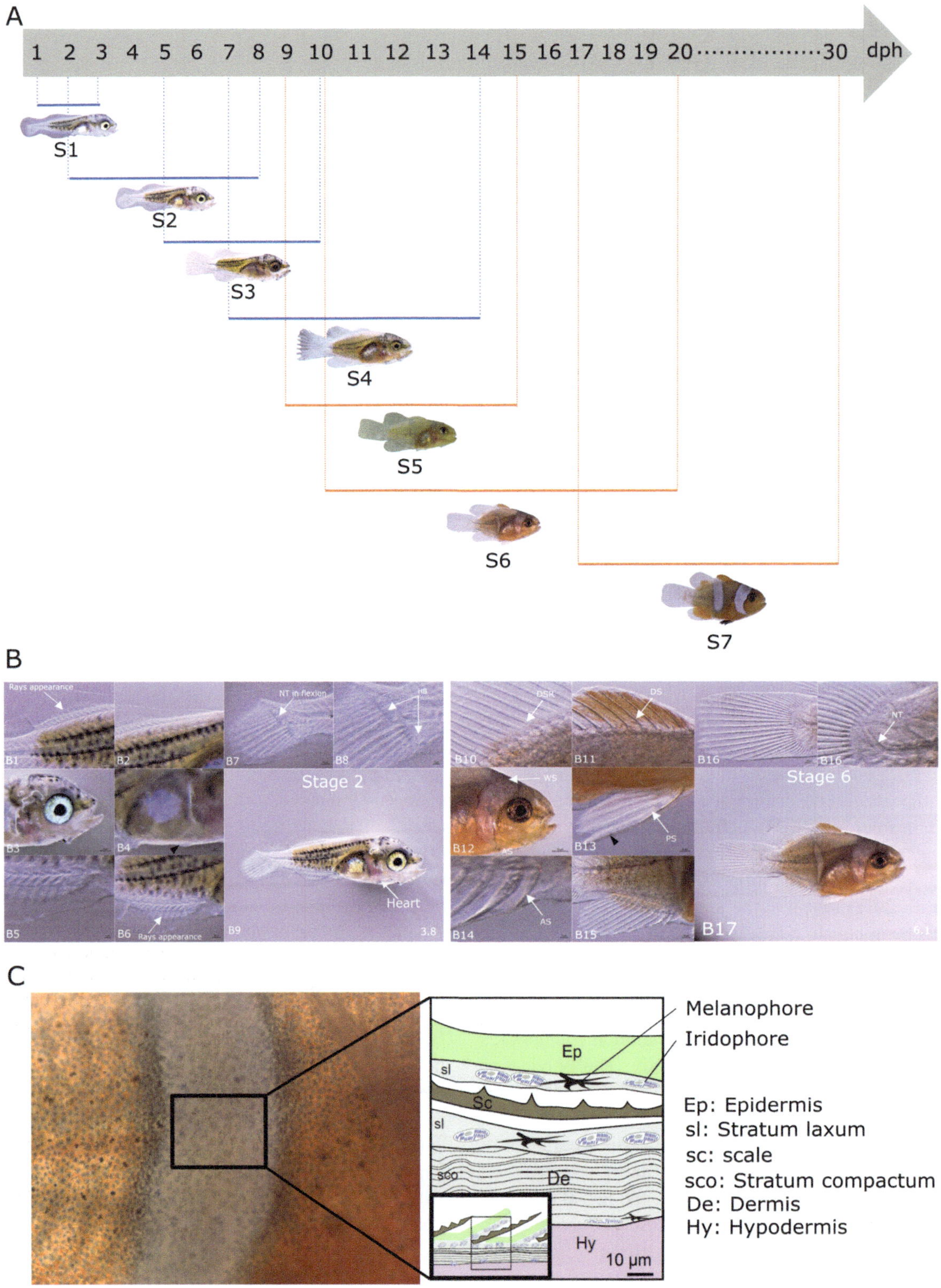

FIGURE 6.2 Developmental stages and white bars cell organization of the false clownfish *Amphiprion ocellaris*. A. Sequence of appearance of the developmental stage, dph: days post-hatching. B. Developmental stage 2 and 6 illustrations from Roux et al. (2019). Black arrowheads indicate pelvic bud in B4 (DSR: dorsal soft rays; DS: dorsal spines; WS: white stripe; PS: pelvic spines; AS: anal spines; HB: hypural bones; NT: notochord). C. Close up on the body white bar and cartoon from Salis et al. (2019) showing a transverse section of the white skin. Melanophores (black pigment cells) and iridophores (white pigment cells) are mostly encountered in the stratum laxum of the dermis.

TABLE 6.2
Anemonefish Timing of Pigmentation Changes

Species	Rearing temperature	Hatchling range size (SL)	White bars appearance	Completion of pigmentation changes	Band lost	Reference
Amphiprion ephippium	27°C	3.9–4	10–15 dph	Around 45 dph	Yes, starts at 45 dph	Krishna et al. 2018
Amphiprion clarkii	–	3.5–3.8	15–17 dph	Around 22–25 dph	No	Ghosh et al. 2011
Amphiprion sebae	–	2.5–3.5	–	–	–	Gunasekaran et al. 2017
Amphiprion percula	26°C (+/– 2°C)	3.79	–	–	No	Onal et al. 2008
Amphiprion perideraion	27°C	4.4	15 dph	–	No	Salis et al. 2019
Amphiprion ocellaris	25°C	3.8	14 dph	30 dph		Roux et al. 2019
Amphiprion frenatus	27.8–28.9°C	4.6	10 dph	–	Yes	Putra et al. 2012

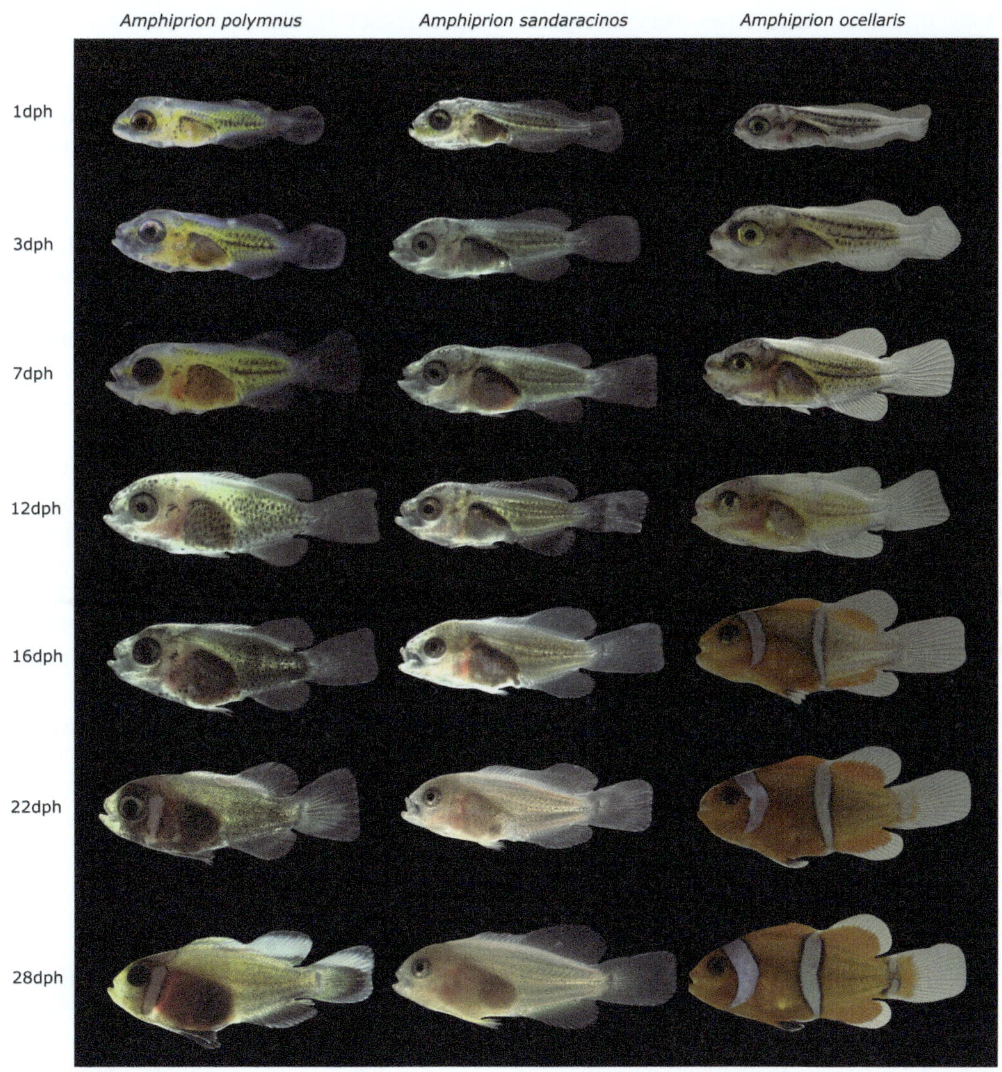

FIGURE 6.3 Pictures showing the larval development of *Amphiprion polymnus*, *A. sandaracinos*, and *A. ocellaris* illustrating similarities in terms of morphological changes but differences in the timing of the formation of the pigmentation pattern. Picture credits: Ken Maeda, Pollina Pilieva, and Natacha Roux.

also have an oceanic larval dispersal followed by a reef phase. Their larval recruitment into reef environments is not a random outcome, rather it is due to their amazing swimming and sensory abilities that allow them to orientate themselves in the big blue, find a reef, and locate their sea anemone host (Bellwood and Fisher 2001; Leis et al. 2011). Considering the relatively short duration of their larval development, it was initially thought that anemonefish larvae were not able to travel long distances in the ocean before settling in a sea anemone on a reef, but it has been proven otherwise. By marking *A. polymnus* embryo with tetracycline in Schuman island (Kimbe Bay, Papua New Guinea), Jones et al. (2005) showed that 32% of the new *A. polymnus* recruits returned to their birth island. They also observed that many of them (68%) may have travelled a considerable distance (more than 10 km) before ending up on Schuman Island. This study showed that despite their short larval duration, anemonefish larvae are fully capable of travelling long distances. This poses very intriguing questions such as: how are they able to locate their habitat? Are they integrating environmental cues to regulate the timing of metamorphosis? How do they know when it is the right time to start this process? To answer these questions, we need to first understand the sensitivity of anemonefishes to various olfactory cues.

Murata et al. (1986) demonstrated that both *A. perideraion* and *A. ocellaris* were sensitive and responded to chemical compounds isolated from symbiotic sea anemone host (*Heteractis crispa* and *Stichodactyla gigantea*). This was the starting point of numerous studies investigating the olfactory sensitivity of anemonefish. Elliot et al. (1995) tested the ability of nine anemonefish species (*Premnas biaculeatus*, *A. percula*, *A. perideraion*, *A. polymnus*, *A. sandaracinos*, *A. leucokranos*, *A. melanopus*, *A. clarkii*, *A. akindynos*, *A. chrysopterus*) to locate sea anemones in the field. All species were reared in the laboratory (breeding pair held with a sea anemone) and released into the environment when they reached the stage of recruitment competency. According to the authors, this stage was reached when the larvae started to display an adult-like body pattern with white bars and adopted benthic swimming behavior. All larvae aged between 7 to 16 dph that were released in the field were chemically attracted to their sea anemone host. They were able to detect the odour of their host up to 8 m away when they were facing the current, but this ability was reduced when there was no current. These results have been confirmed by the work of Dixson et al. who revealed that newly settled *A. percula* juveniles collected from the field were sensitive to olfactory cues of their sea anemone host, reef island water, and even rainforest leaves encountered on reef islands (Dixson et al. 2008, 2011, 2014; Scott and Dixson 2016). Additionally, the authors demonstrated that the olfactory sensitivity of *A. percula* and *A. melanopus* reared in the laboratory started at 7 dph and that they were both attracted to sea anemone and coral cues. *A. percula* larvae also demonstrated olfactory sensitivity towards tropical plant cues. More surprisingly, Scott and Dixson (2016) observed that the olfactory abilities of anemonefish larvae not only help them to locate a suitable sea anemone host but also help them to distinguish between bleached sea anemones and healthy sea anemones.

Altogether, these studies allow us to suggest a hypothesis: anemonefish integrate environmental cues that help them to adjust the timing of their metamorphosis. In vertebrates, environmental cues (temperature, photoperiod, chemicals, etc.) as well as internal information (hormones, metabolic signals) are integrated by the central nervous system, resulting in the release of neuroendocrine signals from a brain region called the hypothalamus. This triggers the production of hormones from another brain region, the pituitary gland, which induces peripheral endocrine secretions such as thyroid hormones or cortisol. Thyroid hormones then regulate morphological, physiological, and behavioral changes, such as metamorphosis, and cortisol regulates stress responses (Power et al. 2001; Blanton and Specker 2007). These cascades are called the hypothalamo-pituitary-thyroid axis (HPT) and the hypothalamo-pituitary-internal axis (HPI). The HPT and HPI are the keys to perceiving, processing, and transducing environmental cues into neural and hormonal signals (see a complete review in Chapter 12). There is evidence from frogs that the environment is triggering metamorphosis by interacting with these two axes, but this is less studied in teleost fishes (Blanton and Specker 2007; Denver 2013; Deal and Volkoff 2020).

To demonstrate the role of environmental cues in the metamorphosis of marine teleosts, we must have access to the entire larval developmental period and be able to expose these young stages to various environmental cues. Interestingly, anemonefish are the perfect candidates to test the effects of environmental cues on TH-regulated metamorphosis as ecological data collected to date demonstrate that anemonefish larvae are sensitive to a variety of chemical cues. It would be extremely interesting to test the effects of such cues on the timing of metamorphosis and on the HPT and HPI axis using the tools that are now available for anemonefishes (see Chapters 5 and 23). We may assume that environmental factors perceived by anemonefish larvae are responsible for increasing the production of TH, thus triggering metamorphosis. In the final part of this chapter, we will discuss the role of TH in anemonefish metamorphosis.

6.4 TH AND ANEMONEFISH METAMORPHOSIS

Compared to flatfishes whose metamorphosis is morphologically spectacular as the larvae become asymmetric and one of the eyes migrates to the other side of the body, anemonefishes undergo subtle morphological changes. Anemonefish metamorphosis is mainly noticeable visually by a change in the overall shape observed at stage 4 in *A. ocellaris*, and the acquisition of an adult-like body pattern with the appearance of white bars and a change of color that starts at stage 5 in *A. ocellaris* (Roux et al. 2019). This change of pigmentation is characterized molecularly by a change in the expression levels of several genes involved in pigment cell specification, xanthophore development, pteridine pigment synthesis of xanthophores, melanophore development, melanogenesis regulation, melanosomes biogenesis, and more interestingly genes specifically expressed in iridophores (Salis et al. 2021). If we compare such information with data obtained in zebrafish, they support the hypothesis that TH signalling has an evolutionarily conserved role in regulating the timing of pigmentation pattern development in teleosts as both species have markedly different adult pigment patterns. Indeed, several studies in zebrafish demonstrated the central role of TH in promoting the maturation of melanophores and xanthophores (McMenamin et al. 2014; Saunders et al. 2019). In *A. ocellaris*, TH is also implicated in adult pigment pattern formation and is responsible for the maturation of iridophores, in particular (Salis et al. 2021).

Whilst their pigmentation pattern changes, anemonefishes also undergo a change in swimming behavior during metamorphosis. They adopt a more benthic lifestyle, swimming closer to the wall and bottom of their tank when they are reared in captivity, as observed in *P. biaculeatus*, *A. percula*, *A. perideraion*, *A. polymnus*, *A. sandaracinos*, *A. leucokranos*, *A. melanopus*, *A. clarkii*, *A. akindynos*, *A. chrysopterus*, and *A. ocellaris* (Elliott et al. 1995; Roux et al. 2019). While there is still no direct evidence on the role of TH in this change of behavior in anemonefishes, we can expect that these hormones will also be involved in such modification as they seem to be involved in the regulation of different types of behavior in zebrafish such as feeding, anxiety response, and social interaction (McMenamin et al. 2017; Park et al. 2019).

In the natural environment, metamorphosis is associated with an environmental change; anemonefish larvae actively swimming in the open ocean adopt a sedentary lifestyle when they settle in their sea anemone host. This change of behavior seems associated with metabolic changes as observed by Paul and Kunzmann (2019) who noticed changes in the enzymatic activities of aerobic and anaerobic metabolism during *A. ocellaris* metamorphosis which could be explained by this change in swimming activities. Such metabolic changes often occur during teleost metamorphosis (Darias et al. 2008; Ferraresso et al. 2013).

In addition to the metabolic changes that allow anemonefishes to be more suited to a benthic lifestyle, an important but cryptic modification undergone by metamorphosing larvae concerns their visual system. Indeed, when evolving in an oceanic environment during the beginning of their larval development, anemonefishes are exposed to a relatively blue/green environment that only requires the use of specific protein in their retina sensitive to these wavelengths, namely the short-wavelength opsins (Figure 6.4A, B). Contrastingly, during metamorphosis and whilst transitioning to a reef habitat, larvae need to be sensitive to a more colourful habitat that requires the use of proteins sensitive to long wavelengths (such as orange/red): the long-wavelength opsin (Figure 6.4A, B). As mentioned in Chapter 8, anemonefish possess both short- and long-wavelength opsins in their retina at the adult stage. But what is interesting is that the ontogeny of the expression levels of the genes encoding for these proteins shows a pattern with a clear separation between the early larval stage and the metamorphosing stage in *A. ocellaris* (Figure 6.4C). By conducting transcriptomic analysis of all the developmental stages identified by Roux et al. (2019), it has been observed that short-wavelength opsin genes are highly expressed during the early larval stage (stages 1, 2 and 3) and then strongly decrease, whereas the long-wavelength opsin gene starts to be highly expressed from stage 4, and is thus associated with all the changes we mentioned before: shape modification, pigmentation changes, swimming behavior, and metabolic changes. All these changes appeared to be under the regulation of TH (Roux et al. 2022; Salis et al., 2021).

According to Salis et al. (2021), TH is involved in the formation of white bars in *A. ocellaris*. Indeed, by exposing un-metamorphosed larvae (stage 3) to various concentrations of TH, Salis et al. observed accelerated formation of white bars as well as the up-regulation of genes specifically expressed in iridophores. Furthermore, when larvae were exposed to a pharmacological treatment blocking TH synthesis (mix of methimazole perchlorate, potassium perchlorate, and Iopanoic acid), white bars formation was delayed (Figure 6.4D). Considering these results, it would be interesting to conduct similar studies that investigate the effects of TH on the opsin shift observed in *A. ocellaris* and on the metabolic and swimming changes observed during metamorphosis. This will allow us to gain insights into how TH orchestrate the numerous diverse processes that ensure the success of metamorphosis in marine teleosts. For example, few data are available on the role of TH in fin and skeletal development in marine fishes whereas these two processes are essential to ensure larvae swim actively towards their recruitment habitat but also to successfully catch prey (jaw formation). Recent studies in zebrafish showed that TH affects the development of several bone structures, highlighting the need to gather similar information in marine fishes such as anemonefish to assess the conservation of TH regulation regarding these processes (Galindo et al. 2019; Keer et al. 2019) Additionally, anemonefish are relevant candidates to investigate the role of TH on sensory organs development also regulated by TH in zebrafish (Hu et al. 2019), but also the role of other hormones produced by the HPI axis such as corticoids that are believed to interfere with the thyroid system, but for which we currently have limited information.

Overall, these data suggest that, as in other teleost fish, thyroid hormones trigger metamorphosis and control its progression, thus ensuring the coordination of changes that take place in different organs. As in *Xenopus* and zebrafish which remain the most advanced models for studying the molecular mechanisms of metamorphosis, the way in which this coordination is ensured is still poorly understood

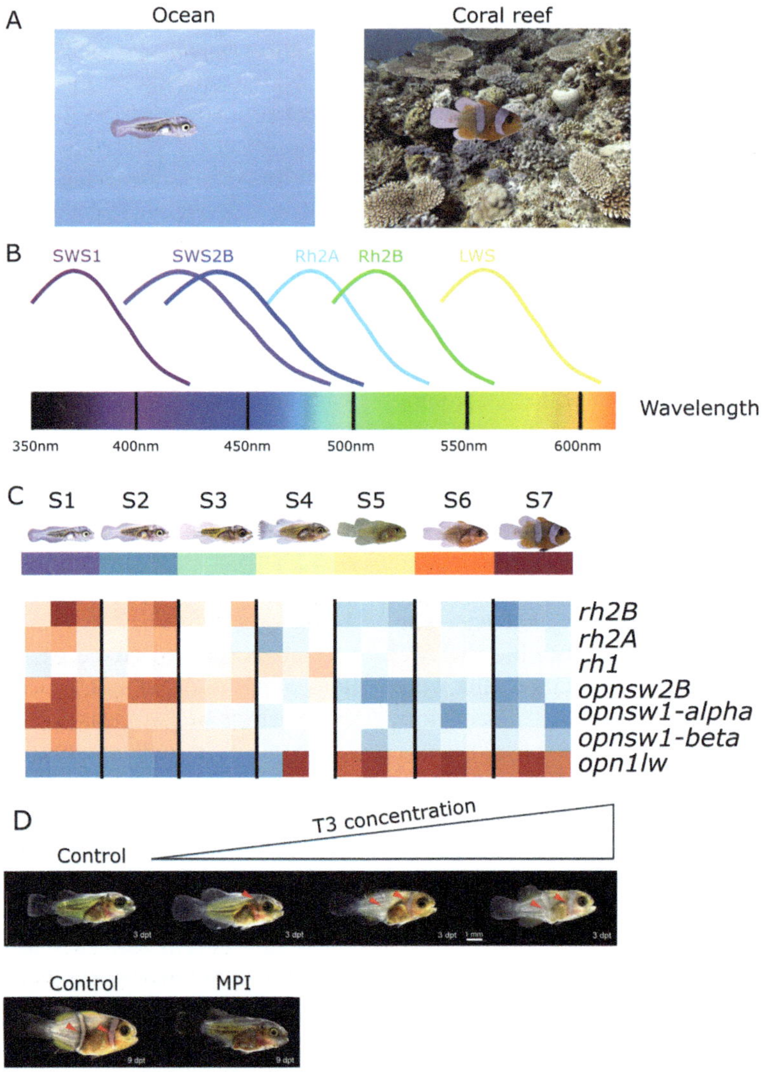

FIGURE 6.4 Opsin gene expression during metamorphosis and effects of thyroid hormones on white bar formation in the false clownfish *Amphiprion ocellaris*. A. Pictures illustrating the color differences between the environment in which anemonefish larvae live before metamorphosis (in the open ocean) and after (in reef environment). B. Wavelength of the six visual opsins encountered in anemonefishes. C. Heatmap showing the expression levels of the genes encoding for the opsins during *Amphiprion ocellaris* post-embryonic development (blue for low expression levels, red for high expression levels). D. Pharmacological treatment with the thyroid hormone T3 which accelerates white bar formation and with MPI which slow down white bar appearance (Salis et al. 2021).

in other marine teleost species and should be the subject of future studies.

It is striking that virtually all organs are affected in one way or another by this transformation. In that sense, even if the metamorphosis of anemonefish is less spectacular than, for example, the transformation of a symmetrical flatfish larvae to an asymmetrical adult, the two processes of metamorphosis are similar. However, for anemonefishes, this transformation corresponds to a less spectacular transformation characterized by changes in shape and pigmentation and a clear ecological transition between an oceanic habitat and a reef. The genomes of anemonefishes, like those of other teleost fishes, must therefore integrate these two different and equally important constraints as the ultimate reproductive success can only be achieved if both the larval and juveniles succeed.

6.5 CONCLUSION

The metamorphosis is a crucial yet understudied step in the life cycle of all marine fish. The renewal of fish populations on a reef is highly dependent on the regular influx of larvae that transform and settle in their reef habitat. If this influx is interrupted or decreased, the fish population will rapidly collapse. In this context, the global and local anthropogenic stressors impacting this critical step must be better understood (Lowe et al. 2021). Noise pollution which can affect the capacity of larvae to locate reefs from a distance, and the urbanization of coasts which can modify the odorous bouquet of the reef, further disrupting its localization by the larvae, can both affect this crucial step (Dixson et al. 2008). Likewise, chemical pollution by endocrine disruptors targeting the thyroid hormone signalling pathway can have disastrous effects on the completion of metamorphosis (Holzer et al. 2017). Studying the effects of these stressors on the metamorphosis and the settlement of young juveniles in their final habitat is required. Anemonefish, which can be studied both in the laboratory but also at sea (only young recruits at the moment as larvae are nearly impossible to capture before recruitment in a sea anemone), offer a promising model for this type of study and it is certainly an important research direction in the future.

Another important research direction is to try to reconcile the ecological study of metamorphosis (the ecological factors that trigger it, the processes that guide the larvae, and the population dynamics in situ) and the more mechanistic approaches in the laboratory. For the moment, the metamorphosis itself, as well as the role of thyroid hormones, but also of other hormones such as corticosteroids, has mainly been studied in laboratory (*Xenopus*, zebrafish) or aquaculture (flatfish) models for which it is difficult to make ecologically relevant conclusions. These models therefore do not allow integrating the ecological dimension into our understanding of metamorphosis. We are convinced that anemonefish can make a relevant contribution to the eco-evo-devo analysis of post-embryonic development of coral reef fish.

REFERENCES

Adams J. A., T. S. Galloway, D. Mondal, S. C. Esteves, and F. Mathews. 2014. Effect of mobile telephones on sperm quality: A systematic review and meta-analysis. *Environment International* 70: 106–112.

Anil M., S. B. Santhosh, B. Prasad, and R. George. 2012. Broodstock development and breeding of black-finned anemone fish *Amphiprion nigripes* Regan, 1908 under captive conditions. *Indian Journal of Fisheries* 59: 77–82.

Arvedlund M., M. I. McCormick, and T. Ainsworth. 2000. Effects of photoperiod on growth of larvae and juveniles of the anemonefish *Amphiprion melanopus. Naga, The ICLARM Quarterly* 23(2): 18–23.

Avella M. A., I. Olivotto, G. Gioacchini, F. Maradonna, and O. Carnevali. 2007. The role of fatty acids enrichments in the larviculture of false percula clownfish *Amphiprion ocellaris. Aquaculture* 273(1): 87–95.

Bellwood D., and R. Fisher. 2001. Relative swimming speeds in reef fish larvae. *Marine Ecology Progress Series* 211: 299–303.

Besson M., W. E. Feeney, I. Moniz, L. François, R. M. Brooker, G. Holzer, M. Metian, N. Roux, V. Laudet, and D. Lecchini. 2020. Anthropogenic stressors impact fish sensory development and survival via thyroid disruption. *Nature Communications* 11(1): 3614.

Blanton M. L., and J. L. Specker. 2007. The hypothalamic-pituitary-thyroid (HPT) axis in fish and its role in fish development and reproduction. *Critical Reviews in Toxicology* 37(1–2): 97–115.

Buchholz D. R. 2015. More similar than you think: Frog metamorphosis as a model of human perinatal endocrinology. *Developmental Biology* 408(2): 188–195.

Chambel J., V. Severiano, T. Baptista, S. Mendes, and R. Pedrosa. 2015. Effect of stocking density and different diets on growth of Percula Clownfish, *Amphiprion percula* (Lacepede, 1802). *SpringerPlus* 4(1): 183.

Darias M. J., J. L. Zambonino-Infante, K. Hugot, C. L. Cahu, and D. Mazurais. 2008. Gene expression patterns during the larval development of European sea bass (Dicentrarchus labrax) by microarray analysis. *Marine Biotechnology* 10(4): 416–428.

Deal C. K., and H. Volkoff. 2020. The role of the thyroid axis in fish. *Frontiers in Endocrinology* 11: 861.

Denver R. J. 2013. Neuroendocrinology of amphibian metamorphosis. *Current Topics in Developmental Biology* 103: 195–227.

Dhaneesh K. V., K. N. Devi, T. Ajith Kumar, T. Balasubramanian, and K. Tissera. 2012. Breeding, embryonic development and salinity tolerance of Skunk clownfish *Amphiprion akallopisos. Journal of King Saud University - Science* 24(3): 201–209.

Dixson D. L., G. P. Jones, P. L. Munday, S. Planes, M. S. Pratchett, M. Srinivasan, C. Syms, and S. R. Thorrold. 2008. Coral reef fish smell leaves to find island homes. *Proceedings of the Royal Society B: Biological Sciences* 275(1653): 2831–2839.

Dixson, D.L., Jones, G.P., Munday, P.L., Planes, S., Pratchett, M.S., and Thorrold, S.R. 2014. Experimental evaluation of imprinting and the role innate preference plays in habitat selection in a coral reef fish. *Oecologia* 174: 99–107. https://doi.org/10.1007/s00442-013-2755-z.

Dixson D. L., P. L. Munday, M. Pratchett, and G. P. Jones. 2011. Ontogenetic changes in responses to settlement cues by Anemonefish. *Coral Reefs* 30(4): 903–910.

Elliott J. K., J. M. Elliot, and R. N. Mariscal. 1995. Host selection, location, and association behaviors of anemonefishes in field settlement experiments. *Marine Biology* 122(3): 377–389.

Ferraresso S., A. Bonaldo, L. Parma, S. Cinotti, P. Massi, L. Bargelloni, and P. P. Gatta. 2013. Exploring the larval transcriptome of the common sole (*Solea solea* L.). *BMC Genomics* 14: 315.

Galindo D., E. Sweet, Z. DeLeon, M. Wagner, A. DeLeon, C. Carter, S. K. McMenamin, and W. J. Cooper. 2019. Thyroid hormone modulation during zebrafish development recapitulates evolved diversity in danionin jaw protrusion mechanics. *Evolution & Development* 21(5): 231–246.

Ghosh J., R. W. Wilson, and T. Kudoh. 2009. Normal development of the tomato clownfish *Amphiprion frenatus*: Live imaging and *in situ* hybridization analyses of mesodermal and neuroectodermal development. *Journal of Fish Biology* 75(9): 2287–2298.

Ghosh, S., T. T. A. Kumar, K. Nanthinidevi, and T. Balasubranananian. 2011. Hatchery Production of Clark's Clownfish, Amphiprion clarkii (Bennett, 1830) Using Brackishwater. *IPCBEE* 22: 51–56.

Gunasekaran K., A. Sarvanakumar, D. Selvam, and R. Mahesh. 2017. Embryonic and larval developmental stages of sebae clownfish *Amphiprion sebae* (Bleeker 1853) in captive condition. *Indian Journal of Marine Science* 46(05): 9.

Holzer G., M. Besson, A. Lambert, L. François, P. Barth, B. Gillet, S. Hughes, et al. 2017. Fish larval recruitment to reefs is a thyroid hormone-mediated metamorphosis sensitive to the pesticide chlorpyrifos. *eLife* 6: e27595.

Holzer G., and V. Laudet. 2013. Thyroid hormones and post-embryonic development in amniotes. *Current Topics in Developmental Biology* 103: 397–425.

Hu Y., A. Mauri, J. Donahue, R. Singh, B. Acosta, and S. McMenamin. 2019. Thyroid hormone coordinates developmental trajectories but does not underlie developmental truncation in danionins. *Developmental Dynamics* 248(11): 1144–1154.

Jones G. P., S. Planes, and S. R. Thorrold. 2005. Coral reef fish larvae settle close to home. *Current Biology* 15(14): 1314–1318.

Kavanagh K. D., and B. Frederich. 2016. Ontogeny and early life stages of damselfishes. In *Biology of Damselfishes*, eds B. Frederich and E. Parmentier, 167–182. Boca Raton: CRC-Press/Taylor&Francis.

Keer S., K. Cohen, C. May, Y. Hu, S. McMenamin, and L. P. Hernandez. 2019. Anatomical assessment of the adult skeleton of zebrafish reared under different thyroid hormone profiles. *The Anatomical Record* 302(10): 1754–1769.

Krishna R. 2018. Larval development and growth of Red Saddleback Anemonefish, Amphiprion ephippium,(Bloch, 1790) under captive conditions. *Indian Journal of Marine Sciences* 47(12): 9.

Laudet V. 2011. The origins and evolution of vertebrate metamorphosis. *Current Biology* 21(18): R726–R737.

Lazcano I., and A. Orozco. 2018. Revisiting available knowledge on teleostean thyroid hormone receptors. *General and Comparative Endocrinology* 265: 128–132.

Leis J. M. 2006. Are larvae of demersal fishes plankton or nekton? *Advances in Marine Biology* 51: 57–141.

Leis J. M., U. Siebeck, and D. L. Dixson. 2011. How Nemo finds home: The neuroecology of dispersal and of population connectivity in larvae of marine fishes. *Integrative and Comparative Biology* 51(5): 826–843.

Lowe W. H., T. E. Martin, D. K. Skelly, and H. A. Woods. 2021. Metamorphosis in an era of increasing climate variability. *Trends in Ecology & Evolution* 36(4): 360–375.

McMenamin S., C. Carter, and W. J. Cooper. 2017. Thyroid hormone stimulates the onset of adult feeding kinematics in zebrafish. *Zebrafish* 14(6): 517–525.

McMenamin S. K., E. J. Bain, A. E. McCann, L. B. Patterson, D. S. Eom, Z. P. Waller, J. C. Hamill, et al. 2014. Thyroid hormone-dependent adult pigment cell lineage and pattern in zebrafish. *Science* 345(6202): 1358–1361.

McMenamin S. K., and D. M. Parichy. 2013. Metamorphosis in teleosts. *Current Topics in Developmental Biology* 103: 127–165.

Murata M., K. Miyagawa-Koshima, and K. Nakanishi. 1986. Characterization of compounds that induce symbiosis between sea anemone and anemone fish. *Science* 234(4776): 585–587.

Olivotto I., F. Capriotti, I. Buttino, A. M. Avella, V. Vitiello, F. Maradonna, and O. Carnevali. 2008. The use of harpacticoid copepods as live prey for Amphiprion clarkii larviculture: Effects on larval survival and growth. *Aquaculture* 274(2): 347–352.

Olivotto I., M. Di Stefano, S. Rosetti, L. Cossignani, A. Pugnaloni, F. Giantomassi, and O. Carnevali. 2011. Live prey enrichment, with particular emphasis on HUFAs, as limiting factor in false percula clownfish (*Amphiprion ocellaris*, Pomacentridae) larval development and metamorphosis: Molecular and biochemical implications. *Comparative Biochemistry and Physiology Part A: Molecular & Integrative Physiology* 159(3): 207–218.

Önal U., C. Langdon, and I. Çelik. 2008. Ontogeny of the digestive tract of larval percula clownfish, *Amphiprion percula* (Lacépède 1802): A histological perspective: Digestive tract ontogeny of larval clownfish, Amphiprion percula. *Aquaculture Research* 39(10): 1077–1086.

Park J.-S., T.-I. Choi, O.-H. Kim, T. I. Hong, W.-K. Kim, W.-J. Lee, and C.-H. Kim. 2019. Targeted knockout of duox causes defects in zebrafish growth, thyroid development, and social interaction. *Journal of Genetics and Genomics* 46(2): 101–104.

Paul N., and A. Kunzmann. 2019. Glycolytic capacities depend on developmental stage in the clownfish *Amphiprion ocellaris*. *Aquatic Biology* 28: 187–195.

Power D. M., I. E. Einarsdóttir, K. Pittman, G. E. Sweeney, J. Hildahl, M. A. Campinho, N. Silva, et al. 2008. The molecular and endocrine basis of flatfish metamorphosis. *Reviews in Fisheries Science* 16(supl): 95–111.

Power D. M., L. Llewellyn, M. Faustino, M. A. Nowell, B. T. Björnsson, I. E. Einarsdottir, A. V. Canario, and G. E. Sweeney. 2001. Thyroid hormones in growth and development of fish. *Comparative Biochemistry and Physiology Part C: Toxicology & Pharmacology* 130(4): 447–459.

Putra D. F., A. B. Abol-Munafi, Z. A. Muchlisin, and J.-C. Chen. 2012. Preliminary studies on morphology and digestive tract development of tomato clownfish, *Amphiprion frenatus* under captive condition. *Aquaculture, Aquarium, Conservation & Legislation* 5(1): 29–35.

Roux N., V. Logeux, N. Trouillard, R. Pillot, K. Magré, P. Salis, D. Lecchini, L. Besseau, V. Laudet, and P. Romans. 2021. A star is born again: Methods for larval rearing of an emerging model organism, the False clownfish *Amphiprion ocellaris*. *Journal of Experimental Zoology B: Molecular and Developmental Evolution* 336(4): 376–385.

Roux N., S. Miura, M. Dussenne, Y. Tara, F. Lee, S. de Bernard, M. Reynaud, et al. 2022. The multi-level regulation of clownfish metamorphosis by thyroid hormones. *BioRxiv* 2022.03.04.482938.

Roux N., P. Salis, A. Lambert, V. Logeux, O. Soulat, P. Romans, B. Frédérich, et al. 2019. Staging and normal table of postembryonic development of the clownfish (*Amphiprion ocellaris*). *Developmental Dynamics* 248(7): 545–568.

Roux N., P. Salis, S.-H. Lee, L. Besseau, and V. Laudet. 2020. Anemonefish, a model for Eco-Evo-Devo. *EvoDevo* 11(1): 20.

Salis P., T. Lorin, V. Lewis, C. Rey, A. Marcionetti, M.-L. Escande, N. Roux, et al. 2019. Developmental and comparative transcriptomic identification of iridophore contribution to white barring in clownfish. *Pigment Cell and Melanoma Research* 32: 391–402.

Salis P., N. Roux, D. Huang, A. Marcionetti, P. Mouginot, M. Reynaud, O. Salles, et al. 2021. Thyroid hormones regulate the formation and environmental plasticity of white bars in clownfishes. *Proceedings of National Academy of Sciences* 118(23): e2101634118.

Salis P., N. Roux, D. Lecchini, and V. Laudet. 2018. The postembryonic development of Amphiprion perideraion reveals a decoupling between morphological and pigmentation changes. *Cybium* 42(4): 309–312.

Saunders L. M., A. K. Mishra, A. J. Aman, V. M. Lewis, M. B. Toomey, J. S. Packer, X. Qiu, et al. 2019. Thyroid hormone regulates distinct paths to maturation in pigment cell lineages. *eLife* 8: e45181.

Scott A., and D. L. Dixson. 2016. Reef fishes can recognize bleached habitat during settlement: Sea anemone bleaching alters anemonefish host selection. *Proceedings of the Royal Society B: Biological Sciences* 283(1831): 20152694.

Ye L., S.-Y. Yang, X.-M. Zhu, M. Liu, J.-Y. Lin, and K.-C. Wu. 2011. Effects of temperature on survival, development, growth and feeding of larvae of Yellowtail clownfish *Amphiprion clarkii* (Pisces: Perciformes). *Acta Ecologica Sinica* 31(5): 241–245.

7 Color Patterns in Anemonefish
Development, Role, and Diversity

Pauline Salis, Marleen Klann, and Vincent Laudet

CONTENTS

7.1 INTRODUCTION

Body colouration, or pigmentation, plays an essential part in every animal's survival. It is not only important for predator avoidance (for example through camouflage or mimicry), protection against UV radiation or antiparasitic defence, but also for reproductive success, and more generally social interactions (Kelley et al. 2013; Dale et al. 2015; Marshall 2000; Cuthill et al. 2017; Marshall et al. 2019). Pigmentation patterning systems are of great interest not only in ecology but also for development and evolution, all aiming to understand the wide variety of patterns seen in living animals. These patterns are often conspicuous, but their origins remain elusive for many species. Coral reef fishes with their brilliant colors and their astonishing diversity in terms of ecology are of great interest at that level (reviewed in Salis et al. 2019a). Anemonefish that can be studied both in the wild and in the lab are becoming valuable models to decipher the ecological, evolutionary, and developmental forces acting on pigmentation (Roux et al. 2020).

In fish, skin pigmentation depends on the distribution of pigment cells, also called chromatophores, which are derivatives of the neural crest and are typically classified based on the colored pigment they are bearing and their ultrastructure (Fujii 1993; Schartl et al. 2016). The four major chromatophore subtypes are (1) melanin-containing melanophores, which appear brown or blackish, (2) yellow/orange/red xanthophores/erythrophores that are distinguished by color and contain carotenoids and/or pteridines, (3) iridophores which contain guanine crystals, which usually gives them a silver or iridescent appearance, and (4) uric acid bearing white leucophores.

Our understanding of color pattern formation and development in fishes has progressed very much thanks to studies performed on zebrafish and to a lesser extent medaka (Patterson and Parichy 2019; Lamoreux et al. 2005). Zebrafish possess three chromatophore cell types (melanophores, xanthophores, and iridophores), which interact and communicate to establish the adult color pattern of dark stripes and bright interstripes following a Turing-like model according to which the number of stripes increases with the growth of the fish (Nakamasu et al. 2009; Kondo et al. 2009). These Turing-like patterns have been observed in the wild in many other fish and among coral reef fish, in particular angelfish (Kondo and Asai 1995) and pufferfish (Miyazawa 2020). However, not all pigmentation patterns follow this model, and these alternative systems provide interesting opportunities to explore color pattern development.

Anemonefish offer such an alternative system in which different mechanisms controlling vertical bar patterns can be analyzed. This color pattern comprises zero to three vertical white bars outlined with a black edge, that are visible on a darker body background (red, orange, or black) (Figure 7.1a–e), with five exceptions displaying one horizontal dorsal white stripe associated or not with a thin vertical white bar on the head (*Amphiprion akallopisos*, *A. leucokranos*, *A.*

DOI: 10.1201/9781003125365-9

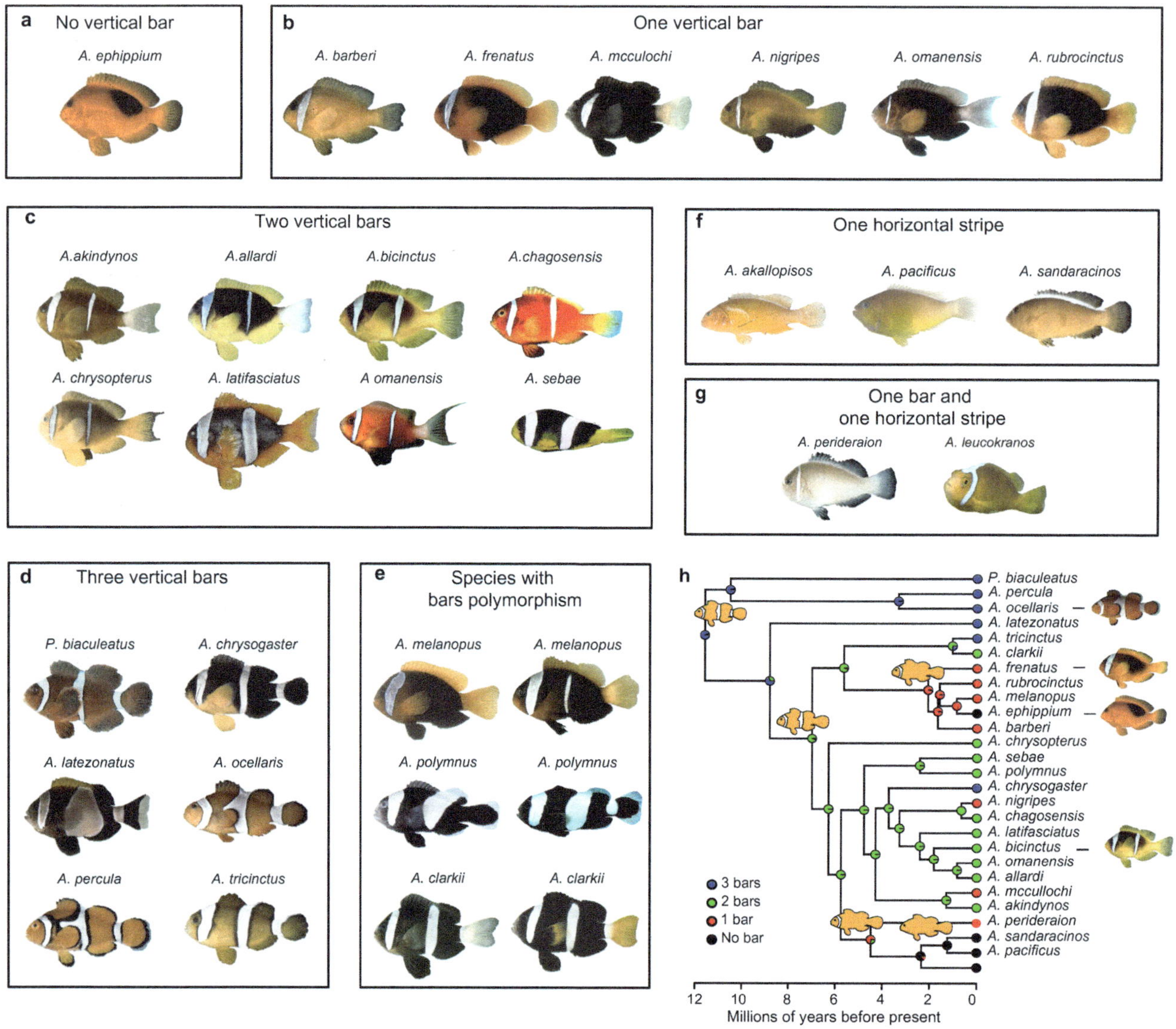

FIGURE 7.1 Adult color patterns of anemonefishes species and successive caudo-rostral loss of vertical white bars during evolution. Pictures of adult anemonefishes classified depending on their color patterns. a) No vertical white bar, b) one vertical white bar on the head, c) two vertical white bars (one on the head, the other on the body), d) three vertical white bars (one on the head, one on the body trunk, and the last one on the peduncle), e) fishes having white bar number polymorphism, f) fishes having one horizontal white stripe, and g) anemonefishes having one vertical white bar on the head and a horizontal white stripe. h) Phylogenetic tree of anemonefishes from Litsios et al. (2014) with a summary map of white vertical bar number histories generated through stochastic character mapping. This trait mapping shows that the diversification of white bar pattern is a history of loss from an ancestral anemonefish having three white vertical bars and that these losses occurred in a progressive and sequential fashion from caudal to rostral. Circles at the tips of the tree indicate each species' white vertical bar pattern and circles at all internal nodes give probabilities of ancestral striped pattern.

pacificus, *A. perideraion*, and *A. sandaracinos*) (Figure 7.1f and g) (Salis et al. 2018a). The pattern observed is clearly stereotyped as only three patterns of bars are observed over all anemonefishes: one bar on the head; one bar on the head and the other on the trunk; or on the head, the trunk, and the tail. There is not a single species with a posterior bar without a bar on the head, for example. This suggests a strong antero-posterior coupling of bar formation.

The evolutionary mapping of the various patterns (zero, one, two, or three white vertical bars) has revealed that the ancestral anemonefish that arose ca. 12 million years ago almost certainly had three white bars and that there was a successive loss of those bars from posterior to anterior during evolutionary history (Figure 7.1h) (Salis et al. 2018b). This phylogenetic analysis also revealed that these losses occurred several times independently during anemonefish diversification, suggesting this may have played an important role in the evolutionary radiation. This observation reinforces the notion that there is a conserved mechanism responsible for color patterning shared by all anemonefish and that this patterning mechanism is somehow connected to the antero-posterior axis of the fish.

The pattern system is still elusive but as this chapter will show, available information on anemonefish color patterns, including contributing pigment cells, ontogenesis, functions, and variation both in natural and in domesticated stocks are all providing clues to how this pattern arises.

7.2 PIGMENT CELLS

Anemonefishes exhibit three types of chromatophores (Salis et al. 2019b; Yasir and Qin 2007; Maytin et al. 2018): white iridophores (forming the white bars), orange/reddish xanthophores (that provide the main body pigmentation), and black melanophores (forming a black border between the bars and the main body pigmentation but also contributing to the main body colouration) (Figure 7.2a).

Microscopic analysis (transmission electron microscopy, TEM), as well as gene expression analysis of the white skin of *A. ocellaris* and *A. frenatus*, confirmed that the white bar is composed of iridophores (Salis et al. 2019b). Moreover, these TEM experiments showed that anemonefish iridophores have a stereotyped morphology (Figure 7.2b) (round-shaped cells with stacks of flattened guanine platelets) and are very similar in some respects to a subtype of iridophores in zebrafish (so-called S-type iridophores), which are characterized by densely stacked platelets (Hirata et al. 2003; Gur et al. 2020). We have preliminary evidence, however, that there may be several additional types of iridophores in anemonefish as in zebrafish. Moreover, a pharmacological approach that targets iridophore genes (such as the compound TAE684 that inhibits the activity of *Ltk* and *Alk* expressed in iridophores) in *A. ocellaris* larvae reduced the number of presumptive iridophores at an ultrastructural level (TEM analysis) and concomitantly reduced the white hue of the bars at the organismal level, confirming that white colouration of bars is due to iridophores and not leucophores (Figure 7.2e–f) (Salis et al. 2019b).

Similar sets of experiments have been done on orange skin, and the ultrastructural organization of orange cells and gene expression analysis confirmed that xanthophores are responsible for orange colouration (Salis et al. 2019b) (Figure 7.2c). Moreover, black cells correspond to melanophores as observed by the presence of melanosomes in TEM experiments (Figure 7.2d).

7.3 COLOR PATTERN ONTOGENESIS

Similar to ontogenetic changes in the color pattern of zebrafish, anemonefishes have two main color patterns, occurring at different stages of the life cycle: a larval color pattern that consists of a yellowish body with two horizontal black stripes (see Figure 7.3g), and an adult color pattern that differs among anemonefish species, but generally comprises zero to three vertical white bars (see Figure 7.3i). Whereas the larval pattern develops during embryogenesis (Figure 7.3a–f), the adult color pattern develops during a critical step named metamorphosis, and during subsequent juvenile

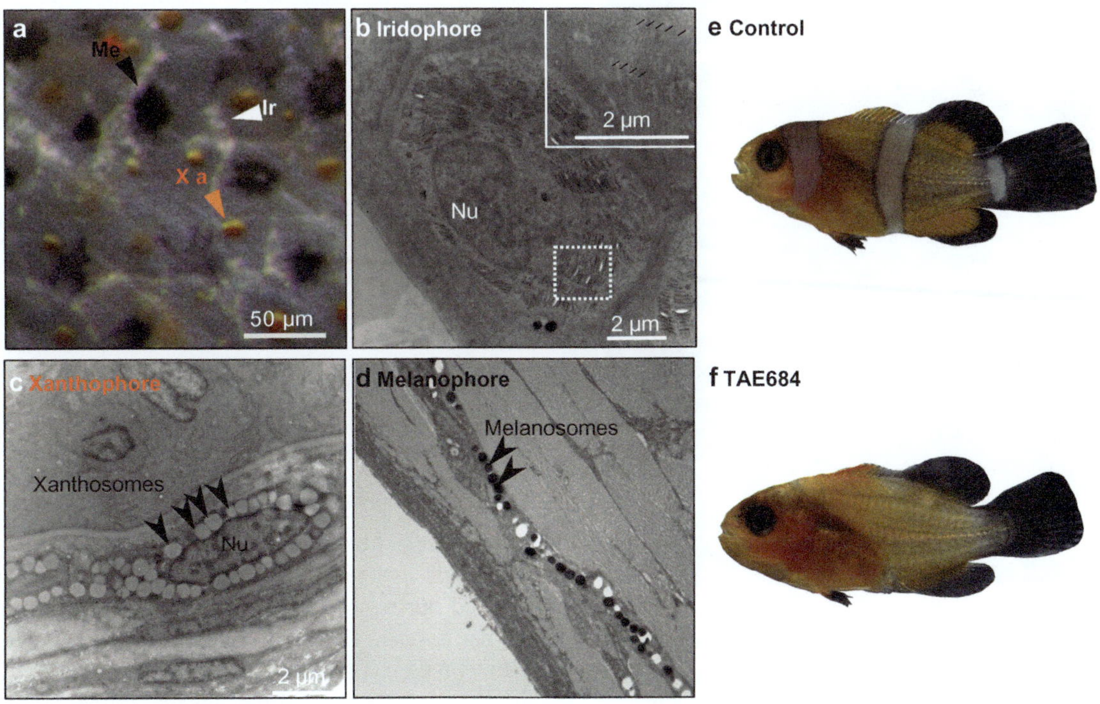

FIGURE 7.2 Chromatophores identity for white, orange, and black skin. a) Stereomicroscope pictures showing the three types of chromatophores within the trunk of juvenile *A. ocellaris*. b) Transmission electron microscopy (TEM) image of an iridophore of the middle white bar of *A. ocellaris*: please note the parallel platelets (inset). c) TEM image of a xanthophore of the body of *A. ocellaris*: please note the xanthosomes. d) TEM of black edge of *A. ocellaris* showing the presence of melanophores. e–f) Modifications of color pattern after 13 days of TAE684 drug treatment of *A. ocellaris* at 18 dph at 0.3 µM (f) compared to DMSO (control, e). Ir, iridophore; Xa, xanthophore; Me, melanophore.

FIGURE 7.3 Ontogeny of white bar formation reveals a rostro-caudal stereotyped pattern. Stereomicroscope pictures of *A. ocellaris* color pattern ontogenesis before hatching (at 28 hpf [a], 29 hpf [b], 32 hpf [c], 40 hpf [d], 4 dpf [e], 7 dpf [f]) and after hatching during metamorphosis (7 dph [g], 11 dph [h], 14 dph [i]). Circles indicate migration of the same melanophores over the yolk at 28 hpf and 29 hpf (a and b). Black and orange arrows point respectively melanophores and xanthophores.

development (Figure 7.3g–i) (Salis et al. 2021b; Roux et al. 2021).

7.3.1 Pigmentation during Embryonic Development

The emergence of pigmentation during embryonic development has been well described in *A. ocellaris* (Salis et al. 2021a) and occurs in three main steps. First, it starts with the appearance of melanophores over the yolk at the 12-somites stages (around 28 hpf – hours post-fertilization – in *A. ocellaris*) (Figure 7.3a). These cells then migrate towards and invade the entire yolk between 28–30 hpf (Figure 7.3a and b). Second, melanophores appear in the vicinity of the eyes and later the tail and in the posterior trunk at the border with the yolk (black arrows in Figure 7.3c and d). Around 4 dpf (days post-fertilization), pale orange xanthophores are visible on the ventral side of the trunk and posterior to the eye (Figure 7.3e). Lastly, at 7 dpf, some melanophores localize along the myosepta to form a stripe and ventrally concentrate at the border between the yolk and the ventral side of the embryo (Figure 7.3f). At that stage xanthophores form a stripe that runs through the eye. This pattern is very similar among most anemonefishes, including *Premnas biaculeatus* and *A. perideraion*, in which lateral black marks of melanophores over a yellow pale body are clearly visible at hatching (Madhuet al. 2012; Salis et al. 2018a). Comparison of embryonic pigmentation processes with other fish, for example, zebrafish, is impeded by different timing of embryonic development. However, in zebrafish first melanophores will appear much later (after all somites have been formed already) and they appear first posterior to the otic vesicle and will later on migrate towards the yolk (Kimmel et al. 1995). Both anemonefish and zebrafish embryos hatch with two longitudinal stripes, one dorsal and one ventral. In contrast to zebrafish, iridophores are not discernible during anemonefish embryonic development.

7.3.2 Development of Pigmentation during Metamorphosis

During metamorphosis, larvae acquire their white bars, white stripe, or both (juvenile color pattern). The bars develop in a very stereotyped manner from the anterior to the posterior region (Figure 7.3g–i) (Dhaneesh et al. 2012; Kumar et al. 2012; Roux et al. 2019; Salis et al. 2018a; Salis et al. 2018b; Madhu et al. 2006). In *A. ocellaris*, the timing of the formation of these white bars is controlled by thyroid hormones (TH). Treating *A. ocellaris* larvae with higher doses of thyroid hormones leads to a faster development of their white bars (Salis et al. 2021b). Conversely, when treating larvae with drugs that impair TH production, bar formation is delayed (Salis et al. 2021b) indicating that in *A. ocellaris* TH are important for the timing of white bar formation.

Interestingly, some differences can be observed between species of anemonefishes in the development timing of white bar formation. Whereas larvae of *A. ocellaris* and *A. frenatus* for example acquire their white bars concomitantly on the head and body at around 11 dph (Figure 7.3h), we observed that *A. percula* acquire first their head bar during metamorphosis and then their second bar as juveniles after about two to three months (Salis et al. 2021b; Salis et al. 2018b). This suggests a different mechanism controlling the timing of white bar formation that may be very interesting to understand. It is particularly critical to explore the differences between *A. ocellaris* and *A. percula* since both are sister species and it is rather unexpected for them to employ differing mechanisms for white bar formation.

Although the exact mechanism of white bar formation has not been described yet, our observations suggest that a coordinated interplay of melanophores and iridophores is likely to be required for normal patterning. In zebrafish, color patterns composed of periodic horizontal stripes form in a spontaneous way, with dynamics that resemble those of a Turing model (Nakamasu et al. 2009; Yamaguchi et al. 2007). In fact, it has been shown that cell-cell communication between the various types of pigment cells is instrumental in controlling pattern formation with the consequence that stripe numbers depend on the size of the fish (Patterson and Parichy 2019). In anemonefishes, the developmental mechanism at the origin of the formation of white bars must be different since the number of bars does not depend on the size of the fish: new bars do not form when the distance between two previous ones increases but following an ordered anterior-to-posterior sequence and likely roles for positional information in the tissue environment. For example, whatever the size of the fish (small *A. ocellaris* or large *P. biaculeatus*) the trunk white bar will always be at the level of the spine/soft ray boundary of the dorsal fin. Clearly, a Turing-like model alone cannot explain the evolutionary history and stage-specificity of white bar formation during anemonefish ontogeny. This suggests that when and where the bars are formed is controlled by specific patterning mechanisms that remain to be elucidated. This is emphasized by the fact that the number of white bars is significantly correlated to the geometry and size of the dorsal fin, strongly suggesting this landmark may act as a spatial reference (Salis et al. 2018b).

7.3.3 Maturation of Pigmentation Patterns in Juveniles

After metamorphosis, juveniles of at least eight species have supplementary bars that disappear later caudo-rostrally. These eight species are placed in three different areas of the evolutionary tree suggesting that this ability to lose white bars during late ontogeny occurred several times independently during anemonefish evolution (Salis et al. 2018b). However, the developmental processes at the origin of the disappearance of such bars remain still unknown (Salis et al. 2018b). The reduction of bar numbers during the ontogeny matches the sequence of bar loss across evolution (Figure 7.1h), demonstrating that diversification in color patterns among anemonefish lineages may result from changes in developmental processes.

7.4 FUNCTIONAL ASPECT OF ANEMONEFISH SKIN COLOR AND PATTERN

Whatever the genetic and developmental systems are that control the formation of the conspicuous color pattern of anemonefish, the function of these patterns is a central question. It is clear that there will be no full understanding of these patterns if we do not understand the proximal and the ultimate causes that underlie their occurrence (Laland et al. 2011).

Several authors have suggested contrasting hypotheses that are interesting to discuss briefly. These can be grouped into two main classes: the brilliant colors of clownfish could be linked to predator avoidance or could have a social function. It is important to note that these two classes of explanation are not mutually exclusive.

7.4.1 Predator Avoidance

In a recent study, Merilaita and Kelley (2018) tested the association between anemonefish color pattern and several ecological features such as the nature of the sea anemone host, habitat, and depth. They observed that fish with zero to one bars are associated with a more limited number of giant sea anemone species (that is, specialists *senso*; Litsios et al. 2014) in comparison with anemonefish having two to three bars. In addition, they suggest that there was a negative relationship between sea anemone host toxicity (as defined by Nedosyko et al. 2014), tentacle length, and bar evolution. They therefore propose that ancestral anemonefish had multiple bars (as also observed by Salis et al. 2018b) that may have served for hiding and camouflage among the long tentacles of *Heteractis* or *Entacmaea* types of sea anemone (Titus et al. 2019; Nguyen et al. 2020). Later, when anemonefish entered carpet sea anemone (*Stichodactyla*-type) they used the colouration in a different way, as an aposematic signal that could inform potential predators of their host's toxicity. However, there are still many uncertainties around this intriguing model. First, the taxonomy and evolution of giant sea anemones are far from being entirely understood and there may be several cryptic species (Nguyen et al. 2020; Titus et al. 2019). In addition, some of the sea anemone species previously believed to belong to the weakly toxic group are more closely related to the highly toxic carpet sea anemone and share their toxic gene repertoire (Kashimoto et al. 2022). So, it is probably too soon to be able to fully understand the link between sea anemone hosts and pigment patterns of anemonefishes. Second, this camouflage/aposematism model comes up against our almost complete ignorance of anemonefish predators. According to early studies, it is clear that outside the protection conferred by the sea anemone, anemonefish are at risk of being attacked (Mariscal 1970). But to our knowledge, no direct attack by a predator on an anemonefish living in their anemone has ever been observed and it is therefore not easy to understand the role of an aposematism pigmentation signal in this context. If the predators are nocturnal, as some authors have suggested, the role of pigmentation would have to be minimal (Mariscal 1970; Allen 1975).

7.4.2 Social Function

The brilliant colouration of anemonefish makes them highly visible, resulting in an easy observation by divers in the field. This may therefore suggest that the pigmentation is here to be seen and as anemonefish live in a complex hierarchical social system, the other main class of explanation for their pigmentation pattern is to play a key role in their social life.

This has been proposed by several authors which suggest that these patterns may be used to advertise social rank (Fautin and Allen 1997; Militz et al. 2016), signal individual identity (Fricke 1973; Nelson et al. 1994), or be employed for species recognition (Salis et al. 2018b). We have, for example, tested the hypothesis that anemonefish species living in sympatry may have a distinct number of bars in order to facilitate species recognition. We effectively observed that similar bar patterns were found less often than expected in communities living in sympatry than in a random set of species. This suggests that the bar system could be used to discourage the association of non-conspecific and/or encourage the association of conspecific. This role in species recognition may also explain why in many species very young recruits are colored distinctly different than older juveniles. We suspect this is a message conveyed to the larger juveniles by the new recruit to avoid potentially harmful antagonistic aggressive behavior.

This role of pigmentation as a social signal implies of course that the colors are seen by the conspecific fishes and therefore suggests a coupling between pigmentation and color vision as suggested in other coral reef fishes (Hench et al. 2019). Recent research on *A. akindynos* has shown that anemonefish can perceive UV light (Stieb et al. 2019). The same authors also showed that white as well as orange skin reflects light in the UV range, which might indeed be used for communication (Stieb et al. 2019). Therefore, future experiments that address the function of anemonefish skin color and pattern should account for UV signalling, particularly regarding the white bars.

7.5 COLOR PATTERN POLYMORPHISM

As discussed earlier the mechanisms that preside over the formation of anemonefish color patterns are still far from clear and the functions of these patterns also remain elusive. To gain insight into these fundamental questions we believe that it is interesting to exploit the variability of these patterns, that is, their polymorphisms observed in natural conditions but also the many types of variants which are known. Whether they are rare natural variants or the colored mutants found in pet shops, they offer the opportunity to better understand pigmentation patterning in anemonefish. Below we will discuss polymorphisms and in the next section the variants and mutants.

Color polymorphism, such as differences in the number of white bars or differences in background color or melanism, is known to occur frequently in anemonefishes (Figure 7.1e and Figure 7.4). However, they are mainly restricted to some anemonefish species including *A. clarkii*, *A. polymnus*, *A. melanogaster*, *A. chrysopterus*, *A. percula*, and *A. nigripes*. It is important to note that a change in bar appearance can be also linked with senescence. Older individuals of *A. omanensis*, *A chrysopterus*, or *P. biaculeatus*, for example, exhibit narrowing bars, that slowly regress from ventral to dorsal and eventually might disappear completely.

FIGURE 7.4 Polymorphism in anemonefishes. Polymorphic melanistic variations in *A. clarkii* (a–c) and *A. melanopus* (d–f). All images are taken by John E. Randall.

An obvious intraspecific polymorphism is the geographical variation that is very common among widely distributed species. This is well exemplified by *A. clarkii* in which continuous variation of traits (respective amount of orange/black, width of body bar) have been observed along a south-to-north cline in the Ryukyu archipelago in Japan (Bell et al. 1982; Fautin and Allen 1997; Moyer 1976). In other cases, like *A. melanopus*, geographically isolated populations with specific patterns occur: for example, in the Fiji islands *A. melanopus* harbour a strong orange-reddish morph which might be devoid of melanophores (Drew et al. 2008) (Figure 7.4).

But polymorphism occurs also in fishes sharing the same habitat, that is in sympatry. Two causes are well identified behind these cases: age and sex differences. In anemonefish, there is no generalized sexual dimorphism apart from the size. However, some specific cases can be observed, such as *A. clarkii*, that show differences in caudal fin colouration with males having an orange caudal fin whereas females have a yellow one (Fautin and Allen 1997; Moyer 1976). As discussed previously, early young recruits are frequently different from older juveniles. This has been observed in *A. percula* where young recruits only have one white bar and are yellow whereas older juveniles have three bars and are orange, similar to the adults (Salis et al. 2021b; Maytin et al. 2018). In *A. frenatus* a reduction of bars is observed as juvenile fish mature (Salis et al. 2018b; Fautin and Allen 1997).

The developmental mechanisms at the origin of such variations have only been explored in the context of variation of white bar developmental timing in juvenile *A. percula*. It has been reported that juvenile fish acquire their white bars at a different speed depending on the sea anemone species in which they recruit (Figure 7.5): white bars form earlier when *A. percula* develop in *Stichodactyla gigantea* in comparison with *A. percula* developing in *Heteractis magnifica*. This earlier formation of white bars in young *A. percula* living in *S. gigantea* is associated with an increase of thyroid hormone (TH) levels, which is interesting given that administration of exogenous TH effectively speeds up white bar formation in *A. ocellaris* (Salis et al. 2021b). Interestingly in the *A. percula* juveniles living in *Stichodactyla* a higher expression of *duox* (Salis et al. 2021b), a dual oxidase known to be important for formation of TH (Chopraet al. 2019; Park et al. 2019), has been observed. Further experiments in zebrafish have revealed that *duox* (and TH abundance downstream of *duox*) also regulates the rate of iridophore development in this species (Salis et al. 2021b). Taken together these data suggest a model in which the juvenile fish present in *Stichodactyla* and *Heteractis* detect that they are in different microenvironments and adapt their physiology and development by regulating their level of TH to fit to the demand of each microenvironment (here, sea anemone species). It remains to be established which specific features in these various microenvironments trigger the differences in TH levels. Another important question is to know if the increased level of TH observed in *Stichodactyla* young recruits only affects pigmentation, or if this is only the visible part of a wider series of differences.

Another important pigmentation polymorphism frequently observed in anemonefish is melanism. This has

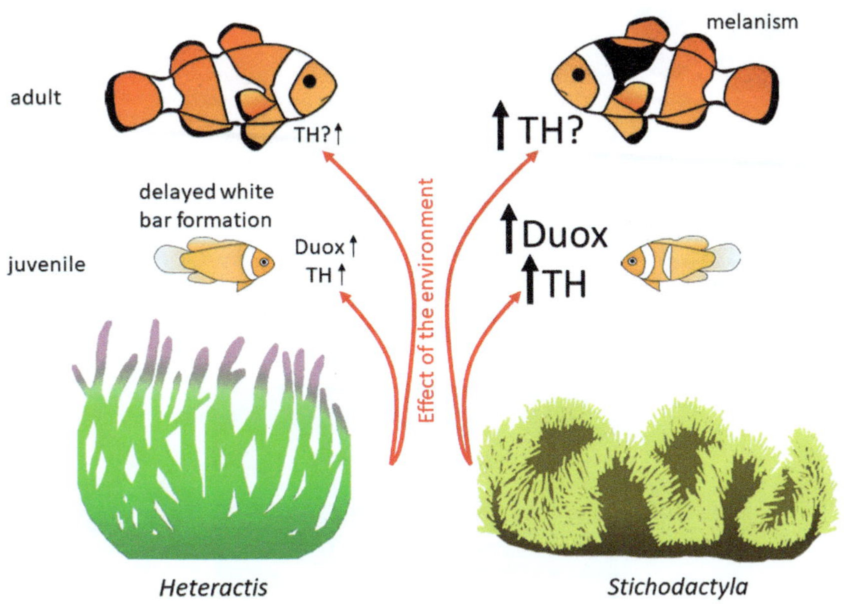

FIGURE 7.5 Developmental mechanisms at the origin of color pattern polymorphism in *A. percula*. Two different polymorphic traits have been noted with *A. percula* either living in *Heteractis* or *Stichodactyla* sea anemones: first, juvenile *A. percula* living in *Heteractis* exhibit a delayed white bar formation that has been shown to be associated with a decrease in thyroid hormones and a decreased expression of *duox* (a gene known to control TH synthesis). Second, adult *A. percula* living in *Stichodactyla* have a higher tendency to show melanism, which might also be connected to differences in thyroid hormone levels.

been described in several species, like *A. chrysopterus*, *A. clarkii* (Figure 7.4), *A. percula* (Figure 7.5), and *A. polymnus* (Fautin and Allen 1997; Militz et al. 2016; Moyer 1976). A study by Militz et al. (2016) identified a suite of interacting and conditional ecological factors encompassing host anemone species, social rank, and location effects as the primary factors predicting the distribution of melanistic morphs. This suggests a more complex mechanism involved in melanistic anemonefish polymorphism.

Furthermore, it is known that sometimes melanistic polymorphic traits can be genetically fixed in restricted geographical populations, such as the black *A. ocellaris* from Darwin, North Australia (see the following). These fish have been in the aquarium trade for many years, and adults display a permanent black color pattern, that has become independent of ecological and environmental factors (Klann et al. 2021).

7.6 COLOR PATTERN VARIANTS AND PET SHOP MUTANTS

Color pattern variants – alterations from the normal barred pattern – in anemonefishes are known from the wild as well as the aquarium trade. Anemonefish with abnormal colouration coming from the aquarium trade are mutants and can therefore be used to identify genes underlying various color traits. Sometimes, rare color variants can be found in the wild and while their genetics is unknown, the collective analysis of those fish can provide valuable insights into biases and constraints associated with developmental pathways controlling pigmentation, as well as their ecological importance. In general, pigmentation abnormalities can be categorized into (1) imbalance of chromatophore subtypes and (2) irregular patterning mechanisms (Klann et al. 2021).

7.6.1 Imbalance of Chromatophores

Geographically restricted wild melanistic populations of *A. clarkii* (Japan and Philippines), *A. frenatus* (Japan and Philippines), *A. ocellaris* (Darwin, North Australia), and *A. percula* (Solomon Islands) can be found (Figure 7.6a). Even though adults are melanistic, larvae and juveniles display an orange/dark orange body colouration that is slowly replaced by black color. This is most likely due to an increasing number of melanophores, which ultimately dominate the color of nearly the entire fish, apart from the white bars. Many individuals from various species (e.g., *A. bicinctus*, *A. chrysopterus*, *A. clarkii*, *A. melanopus*, *A. ocellaris*, *A. percula*, *A. polymnus*, and *P. biaculeatus*) with reduced numbers of iridophores and decreased black edges are known, the so-called "Misbar" phenotype (Figure 7.6b). This phenotype is most likely caused by a genetic defect in a gene responsible for iridophore specification or development, such as *Foxd3*, *ltk*, *Edn3b*, or *Ednr3b* for example (Lister et al. 2006; Fadeev et al. 2016; Krauss et al. 2014).

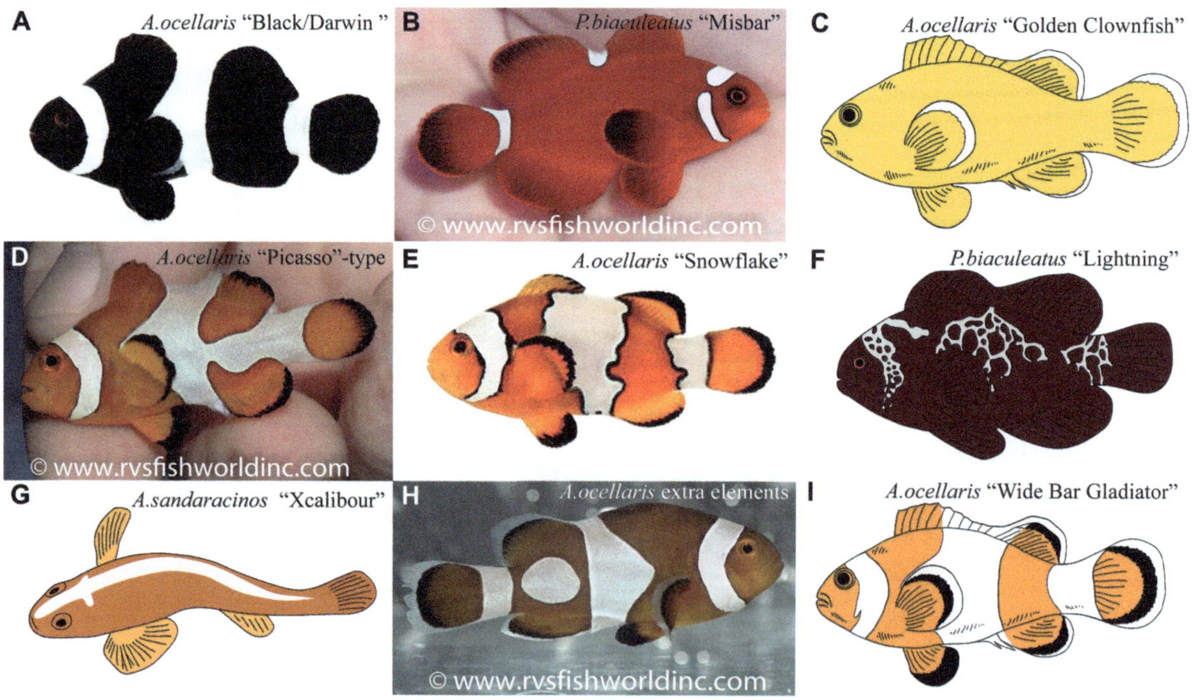

FIGURE 7.6 Color pattern variation. a) Black *A. ocellaris* "Darwin" retail line, b) wild "Misbar" *P. biaculeatus* from the Philippines, c) wild "Golden Clownfish", d) wild "Picasso"-type *A. ocellaris* from the Philippines, e) "Snowflake" *A. ocellaris* retail line, f) "Lighting" *P. biaculeatus*, g) "Xcalibour" *A. sandaracinos*, h) wild *A. ocellaris* from the Philippines with extra elements, i) "Wide Bar Gladiator" *A. ocellaris* retail line. Images were kindly provided by RVSFishworld (b, d, h) or ORA (a, e). Schematic drawings were made by M. Klann (c, f, g, i).

There is only one known wild-caught individual that lacks both melanophores and iridophores – known as the "Golden Clownfish" (Figure 7.6c). In contrast, hypomelanic *A. ocellaris* individuals are only known from the aquarium trade. This phenotype represents albinism in which melanophores are present in their normal arrangement and distribution, but they are hypopigmented (highly reduced melanin levels). Potential candidate genes responsible for albinism are genes involved in melanin synthesis, like *MCR1* or *oca2* (Tezuka et al. 2011; Richardson et al. 2008; Klaassen et al. 2018).

7.6.2 Irregular Patterning Mechanisms

Variations in bar shape are highly variable between individuals and can vary from a small bulge to extensive shape deformations. Both in the wild and aquarium trade anemonefish display bars that are interconnected with irregular but smooth outlines; they are called "Picasso"-typed in aquaculture (Figure 7.6d). This phenotype has been observed in at least eight different anemonefish species and striking similarities between heterozygous and homozygous appearances of three (*A. ocellaris*, *A. percula*, and *P. biaculeatus*) of these eight species suggest that the genetic mechanisms underlying this phenotype are very similar. Other patterning defects include (1) irregular white bars with exaggerated and jagged edges, called "Snowflake" in aquaculture (Figure 7.6e), (2) network-like connections of all three white bars, so-called "Lightning" and seen only in *P. biaculeatus* (Figure 7.6f), (3) a cross-like mark, formed by the horizontal branching of the dorsal stripe, posterior of the eyes, which is only known for *A. sandaracinos* so far, called "Xcalibour" (Figure 7.6g), (4) display of extra elements, like dots (Figure 7.6h), and (5) increased width of the white bars, called "Wide Bar (Gladiator)" (Figure 7.6i).

Pet-shop mutants are available for at least eight different species of anemonefish: *A. ocellaris* (most variants available), *P. biaculeatus* (second most variants available), *A. percula* (third most variants available), *A. clarkii*, *A. bicinctus*, *A. melanopus*, *A. polymnus*, and *A. sandaracinos*. Continuing breeding efforts of existing and new anemonefish species will most likely result in increased numbers of commercially available color mutants in the future. In the wild more species have been observed showing color variations, in addition to the eight species listed above: *A. akindynos*, *A. chrysopterus*, *A. frenatus*, *A. nigripes*, *A. omanensis*, and *A. perideraion*. Accurate estimates of abnormal colouration in anemonefish are obscured by highly limited scientific field observations and by the bias to restrict observation (scientific or private) to a few species as well as to visit a few geographical regions only.

7.7 CONCLUSION

Pigmentation patterning systems are of great interest not only in ecology but also for development and evolution, all aiming to understand the wide variety of patterns seen in living animals. Our understanding of color pattern formation and development in fishes is mainly restricted to a single species: zebrafish. While it has been shown that stripe formation resembles a Turing pattern in which cell–cell interactions among pigment cells result in the addition of stripes as fish size is increasing, other models have been postulated that explain pigmentation processes very well (Volkening and Sandstede 2018; Owen et al. 2020). However, not all pigmentation patterns follow similar models and thus provide interesting models to explore color pattern development. One such example is anemonefishes. Once matured, the color pattern (one to three vertical white bars with a black outline on an orange to reddish body) will not change or adjust when the fish is growing, hinting that different mechanisms are at play. Anemonefishes have and always will remain prominent models for ecological studies, but now those can be linked with lab-based evo-devo approaches. Since anemonefish are open to developmental and genetic manipulation, researchers can aim to identify genes underlying various color traits, manipulate them in laboratory environments, test the effect in complex social systems of captive anemonefish colonies, and study the polymorphism and evolutionary patterns of these genes in wild populations. Moreover, detailed analysis of closely related anemonefishes with differing color patterns or color development will certainly shed light on how and why brilliant pigmentation patterns emerged in coral reef fishes.

REFERENCES

Allen, G. R. 1975. *The Anemonefishes: Their Classification and Biology*. Hong Kong: T.F.H. Publications.

Bell, L. J., J. T. Moyer, and K. Numachi. 1982. Morphological and genetic variation in Japanese populations of the anemonefish *Amphiprion clarkii*. *Marine Biology* 72 (2): 99–108.

Chopra, K., S. Ishibashi, and E. A.-O. Amaya. 2019. Zebrafish duox mutations provide a model for human congenital hypothyroidism. *Biology Open* 8 (2): bio037655.

Cuthill, C. I., L. W. Allen, K. Arbuckle, B. Caspers, G. Chaplin, E. M. Hauber, E. G. Hill, et al. 2017. The biology of color. *Science* 357 (6350): eaan0221.

Dale, J., C. J. Dey, K. Delhey, B. Kempenaers, and M. Valcu. 2015. The effects of life history and sexual selection on male and female plumage colouration. *Nature* 527 (7578): 367–370.

Dhaneesh, K. V., T. T. A. Kumar, G. Swagat, and T. Balasubramanian. 2012. Breeding and mass scale rearing of clownfish *Amphiprion percula*: Feeding and rearing in brackishwater. *Chinese Journal of Oceanology and Limnology* 30 (4): 528–534.

Drew, J., G. R. Allen, L. E. S. Kaufman, and P. H. Barber. 2008. Endemism and regional color and genetic differences in five putatively cosmopolitan reef fishes. *Conservation Biology* 22 (4): 965–975.

Fadeev, A., J. Krauss, A. P. Singh, and C. Nüsslein-Volhard. 2016. Zebrafish leucocyte tyrosine kinase controls iridophore establishment, proliferation and survival. *Pigment Cell & Melanoma Research* 29 (3): 284–296.

Fautin, D. G., and G. R. Allen. 1997. *Anemone Fishes and Their Host Sea Anemones: A Guide for Aquarists and Divers*. Revised Edition. Perth: Western Australian Museum.

Fricke, H. W. 1973. Individual partner recognition in fish: Field studies on *Amphiprion bicinctus*. *Naturwissenschaften* 60 (4): 204–205.

Fujii, R. 1993. Cytophysiology of fish chromatophores. *International Review of Cytology* 143: 191–255.

Gur, D., E. J. Bain, K. R. Johnson, A. J. Aman, H. A. Pasoili, J. D. Flynn, M. C. Allen, et al. 2020. In situ differentiation of iridophore crystallotypes underlies zebrafish stripe patterning. *Nature Communications* 11 (1): 6391.

Hench, K., M. Vargas, M. P. Höppner, W. O. McMillan, and O. Puebla. 2019. Inter-chromosomal coupling between vision and pigmentation genes during genomic divergence. *Nature Ecology & Evolution* 3 (4): 657–667.

Hirata, M., K. Nakamura, T. Kanemaru, Y. Shibata, and S. Kondo. 2003. Pigment cell organization in the hypodermis of zebrafish. *Developmental Dynamics* 227 (4): 497–503.

Kashimoto, R., M. Tanimoto, S. Miura, N. Satoh, L. Vincent, and K. Khalturin. 2022. Transcriptomes of giant sea anemones from Okinawa as a tool for understanding their phylogeny and symbiotic relationships with anemonefish. *Zoological Science* 39 (4): 374–387.

Kelley, J. L., J. L. Fitzpatrick, and S. Merilaita. 2013. Spots and stripes: Ecology and colour pattern evolution in butterflyfishes. *Proceedings of the Royal Society B: Biological Sciences* 280 (1757): 20122730.

Kimmel, C. B., W. W. Ballard, S. R. Kimmel, B. Ullmann and T. F. Schilling. 1995. Stages of embryonic development of the zebrafish. *Developmental Dynmics* 203: 253–310.

Klaassen, H., Y. Wang, K. Adamski, N. Rohner, and J. E. Kowalko. 2018. CRISPR mutagenesis confirms the role of *oca2* in melanin pigmentation in *Astyanax mexicanus*. *Developmental Biology* 441 (2): 313–318.

Klann, M., M. Mercader, L. Carlu, K. Hayashi, J. D. Reimer, and V. Laudet. 2021. Variation on a theme: Pigmentation variants and mutants of anemonefish. *EvoDevo* 12 (1): 8.

Kondo, S., and R. Asai. 1995. A reaction–diffusion wave on the skin of the marine angelfish Pomacanthus. *Nature* 376 (6543): 765–768.

Kondo, S., M. F. Iwashita, and M. Yamaguchi. 2009. How animals get their skin patterns: Fish pigment pattern as a live Turing wave. *International Journal of Developmental Biology* 53: 851–856

Krauss, J., H. G. Frohnhöfer, B. Walderich, H. M. Maischein, C. Weiler, U. Irion, and C. Nüsslein-Volhard. 2014. Endothelin signalling in iridophore development and stripe pattern formation of zebrafish. *Biology Open* 3 (6): 503–509.

Kumar, T. T., M. Gopi, K. V. Dhaneesh, R. Vinoth, S. Ghosh, T. Balasubramanian, and T. Shunmugaraj. 2012. Hatchery production of the clownfish *Amphiprion nigripes* at Agatti Island, Lakshadweep, India. *Journal of Environmental Biology* 33 (3): 623–628.

Laland, K. N., K. Sterelny, J. Odling-Smee, W. Hoppitt, and T. Uller. 2011. Cause and effect in biology revisited: Is Mayr's proximate-ultimate dichotomy still useful? *Science* 334 (6062): 1512–1516.

Lamoreux, M. L., R. N. Kelsh, Y. Wakamatsu, and K. Ozato. 2005. Pigment pattern formation in the medaka embryo. *Pigment Cell Research* 18 (2): 64–73.

Lister, J. A., C. Cooper, K. Nguyen, M. Modrell, K. Grant, and D. W. Raible. 2006. Zebrafish *Foxd3* is required for development of a subset of neural crest derivatives. *Developmental Biology* 290 (1): 92–104.

Litsios, G., A. Kostikova, and N. Salamin. 2014. Host specialist clownfishes are environmental niche generalists. *Proceedings of the Royal Society B: Biological sciences* 281 (1795): 20133220.

Litsios, G., P. B. Pearman, D. Lanterbecq, N. Tolou, and N. Salamin. 2014. The radiation of the clownfishes has two geographical replicates. *Journal of Biogeography* 41 (11): 2140–2149.

Madhu, K., R. Madhu, G. Gopakumar, and C. S. Sasidharan. 2006. Breeding, larval rearing and seed production of maroon clown *Premnas biaculeatus* under captive conditions. *Marine Fishereies Information Service, Technical and Extension Series* 190: 1–5.

Madhu, K., R. Madhu, and T. Retheesh. 2012. Broodstock development, breeding, embryonic development and larviculture of spine-cheek anemonefish, *Premnas biaculeatus* (Bloch, 1790). *Indian Journal of Fisheries* 59 (1): 65–75.

Mariscal, R. N. 1970. The nature of the symbiosis between Indo-Pacific anemone fishes and sea anemones. *Marine Biology* 6 (1): 58–65.

Marshall, N. J. 2000. Communication and camouflage with the same 'bright' colours in reef fishes. *Philosophical Transactions of the Royal Society of London B: Biological Science* 355 (1401): 1243–1248.

Marshall, N. J., F. Cortesi, F. de Busserolles, U. E. Siebeck, and K. L. Cheney. 2019. Colours and colour vision in reef fishes: Past, present and future research directions. *Journal of Fish Biology* 95 (1): 5–38.

Maytin, A. K., S. W. Davies, G. E. Smith, S. P. Mullen, and P. M. Buston. 2018. De novo transcriptome assembly of the clown anemonefish (*Amphiprion percula*): A new resource to study the evolution of fish color. *Frontiers in Marine Science* 5: 284.

Merilaita, S., and J. L. Kelley. 2018. Scary clowns: Adaptive function of anemonefish coloration. *Journal of Evolutionary Biology* 31 (10): 1558–1571.

Militz, T. A., M. I. McCormick, D. S. Schoeman, J. Kinch, and P. C. Southgate. 2016. Frequency and distribution of melanistic morphs in coexisting population of nine clownfish species in Papua New Guinea. *Marine Biology* 163 (10): 200.

Miyazawa, S. 2020. Pattern blending enriches the diversity of animal colorations. *Science Advances* 6 (49): eabb9107.

Moyer, J. T. 1976. Geographical variation and social dominance in Japanese populations of the anemonefish *Amphiprion clarkii*. *Japanese Journal of Ichthyology* 23 (1): 12–22.

Nakamasu, A., G. Takahashi, A. Kanbe, and S. Kondo. 2009. Interactions between zebrafish pigment cells responsible for the generation of Turing patterns. *PNAS* 106 (21): 8429–8434.

Nedosyko, A. M., J. E. Young, J. W. Edwards, and K. Burke da Silva. 2014. Searching for a toxic key to unlock the mystery of anemonefish and anemone symbiosis. *PLOS ONE* 9 (5): e98449.

Nelson, J. S., L. M. Chou, and V. P. E. Phang. 1994. Pigmentation variation in the anemonefish Amphiprion ocellaris (teleostei, pomacentridae) – type, stability and its usefulness for individual identification. *Raffles Bulletin of Zoology* 42: 927–930.

Nguyen, H-T. T., B. T. Dang, H. Glenner, and A. J. Geffen. 2020. Cophylogenetic analysis of the relationship between anemonefish Amphiprion (Perciformes: Pomacentridae) and their symbiotic host anemones (Anthozoa: Actiniaria). *Marine Biology Research* 16 (2): 117–133.

Owen, J. P., R. N. Kelsh, and C. A. Yates. 2020. A quantitative modelling approach to zebrafish pigment pattern formation. *eLife* 9: e52998.

Park, J. S., T. I. Choi, O. H. Kim, T. I. Hong, W. K. Kim, W. J. Lee, and C. H. Kim. 2019. Targeted knockout of duox causes defects in zebrafish growth, thyroid development, and social interaction. *Journal of Genetics and Genomics* 46 (2): 101–104.

Patterson, L. B., and D. M. Parichy. 2019. Zebrafish pigment pattern formation: Insights into the development and evolution of adult form. *Annual Review of Genetics* 53: 505–530.

Richardson, J., P. R. Lundegaard, N. L. Reynolds, J. R. Dorin, D. J. Porteous, I. J. Jackson, and E. E. Patton. 2008. *mc1r* pathway regulation of zebrafish melanosome dispersion. *Zebrafish* 5 (4): 289–295.

Roux, N., V. Logeux, N. Trouillard, R. Pillot, K. Magré, P. Salis, D. Lecchini, L. Besseau, V. Laudet, and P. Romans. 2021. A star is born again: Methods for larval rearing of an emerging model organism, the False Clownfish *Amphiprion ocellaris. Journal of Experimental Zoology Part B: Molecular and Developmental Evolution* 336 (4): 376–385.

Roux, N., P. Salis, A. Lambert, V. Logeux, O. Soulat, P. Romans, B. Frédérich, et al. 2019. Staging and normal table of post-embryonic development of the clownfish (*Amphiprion ocellaris*). *Developmental Dynamics* 248 (7): 545–568.

Roux, N., P. Salis, S.-H. Lee, L. Besseau, and V. Laudet. 2020. Anemonefish, a model for Eco-Evo-Devo. *EvoDevo* 11 (1): 20.

Salis, P., S. H. Lee, N. Roux, D. Lecchini, and V. Laudet. 2021a. The real Nemo movie: Description of embryonic development in *Amphiprion ocellaris* from first division to hatching. *Developmental Dynamics* 250 (11): 1651–1667.

Salis, P., T. Lorin, V. Laudet, and B. Frédérich. 2019a. Magic traits in magic fish: Understanding color pattern evolution using reef fish. *Trends in Genetics* 35 (4): 265–278.

Salis, P., T. Lorin, V. Lewis, C. Rey, A. Marcionetti, M. L. Escande, N. Roux et al. 2019b. Developmental and comparative transcriptomic identification of iridophore contribution to white barring in clownfish. *Pigment Cell & Melanoma Research* 32 (3): 391–402.

Salis, P., N. Roux, D. Huang, A. Marcionetti, P. Mouginot, M. Reynaud, O. Salles, et al. 2021b. Thyroid hormones regulate the formation and environmental plasticity of white bars in clownfishes. *Proceedings of the National Academy of Sciences* 118 (23): e2101634118.

Salis, P., N. Roux, D. Lecchini, and V. Laudet. 2018a. The post-embryonic development of *Amphiprion perideraion* reveals a decoupling between morphological and pigmentation changes. *Cybium* 42 (4): 309–312.

Salis, P., N. Roux, O. Soulat, D. Lecchini, V. Laudet, and B. Frédérich. 2018b. Ontogenetic and phylogenetic simplification during white stripe evolution in clownfishes. *BMC Biology* 16 (1): 90.

Schartl, M., L. Larue, M. Goda, M. W. Bosenberg, H. Hashimoto, and R. N. Kelsh. 2016. What is a vertebrate pigment cell? *Pigment Cell & Melanoma Research* 29 (1): 8–14.

Stieb, S. M., F. de Busserolles, K. L. Carleton, F. Cortesi, W-S. Chung, B. E. Dalton, L. A. Hammond, et al. 2019. A detailed investigation of the visual system and visual ecology of the Barrier Reef anemonefish, *Amphiprion akindynos. Scientific Reports* 9 (1): 16459.

Tezuka, A., H. Yamamoto, J. Yokoyama, C. van Oosterhout, and M. Kawata. 2011. The MC1R gene in the guppy (*Poecilia reticulata*): Genotypic and phenotypic polymorphisms. *BMC Research Notes* 4: 31.

Titus, B. M., C. Benedict, R. Laroche, L. C. Gusmão, V. Van Deusen, T. Chiodo, C. P. Meyer, et al. 2019. Phylogenetic relationships among the clownfish-hosting sea anemones. *Molecular Phylogenetics and Evolution* 139: 106526.

Volkening, A., and B. Sandstede. 2018. Iridophores as a source of robustness in zebrafish stripes and variability in Danio patterns. *Nature Communications* 9 (1): 3231.

Yamaguchi, M., E. Yoshimoto, and S. Kondo. 2007. Pattern regulation in the stripe of zebrafish suggests an underlying dynamic and autonomous mechanism. *Proceedings of the National Academy of Sciences* 104 (12): 4790–4793.

Yasir, I., and J. G. Qin. 2007. Embryology and early ontogeny of an anemonefish *Amphiprion ocellaris. Journal of the Marine Biological Association of the United Kingdom* 87 (4): 1025–1033.

8 Age and Longevity

Mirko Mutalipassi, Eva Terzibasi Tozzini, and Alessandro Cellerino

CONTENTS

8.1 INTRODUCTION

8.1.1 Senescence in Teleosts

Teleost fishes are a large and diversified class of vertebrates comprising more than 20,000 existing species (Patnaik et al. 1994). Within this biodiversity, some species (zebrafish, killifish, and clownfish; Schartl 2014) show some interesting traits that make them ideal model species for ecological, physiological, genetic, and biotechnological investigations (Holtze et al. 2021; Mutalipassi 2019; Bahls et al. 2003; Levy and Currie 2015; Bodnar 2016; Goldstein and King 2017). Among these aquatic models, some species occupy the extremes in the spectrum of ageing processes, with lifespan differences of two orders of magnitude. Some examples include short lifespan species, such as the turquoise killifish *Nothobranchius furzeri* which presents a lifespan of few months (Cellerino et al. 2015; Terzibasi et al. 2007), and species with a life expectancy of more than a century, as in the case of some rockfishes of the genus *Sebates* (Mangel et al. 2007) and the Greenland shark *Somniosus microcephalus* (Nielsen et al. 2016). In teleost fishes, we can identify three types of senescence (Finch 1998). The first is the rapid senescence, generally linked with the sudden death at first spawning, with examples in species such as lamprey and salmon. Rapid senescence is usually triggered by hormonal inductions (Finch 1998). The second one is the gradual senescence characterized by an age-related decline in reproduction, loss of compact bone, endothelial proliferation, collagen oxidation, and accumulation of brain amyloid and other forms of protein aggregation, etc. (Kishi 2004). This second typology of senescence has been observed in many other vertebrates including humans, and it is present in teleosts such as *Poecilia* spp. (Reznick et al. 2006) or zebrafish (Kishi et al. 2003). The third typology of senescence is observed in fishes characterized by an indeterminate growth, where senescence is supposed to be slow or negligible, with evidence of undiminished functions during ageing and with reproductive activities maintained in old age (Finch 1998).

8.1.2 Theories Concerning the Evolution of Senescence

Evolutionary theories of ageing correlate low extrinsic mortality conditions to the evolution of slow senescence and increased lifespan. Extrinsic mortality is one of the most important factors which contribute to the accumulation of deleterious mutations by limiting the exposure of the late-acting mutation to selection. It has been theorized that extrinsic mortality is the principal determinant of the senescence rate in age-structured populations (Kirkwood 2000). Three main theories were proposed to explain the evolution of ageing. The mutation accumulation theory postulates that ageing is caused by the accumulation of mutations with late-life phenotypes that behave like quasi-neutral mutations as they have marginal effects on whole-life fitness (Medawar 1952); the antagonistic pleiotropy theory postulates that alleles that are positively selected due to their effects on growth and fertility early in life reduce fitness later in life (Williams 1957); and the disposable soma theory postulates that ageing is caused by the trade-off in the energetic resources devoted to growth and maintenance (Kirkwood 2002). All three theories have as a corollary that reduced extrinsic mortality should drive the

DOI: 10.1201/9781003125365-10

evolution of longevity as observed by the exceptional longevity of several vertebrate species living in predator-free or protected environments, such as caves (Voituron et al. 2011) or arboreal habitats (Shattuck and Williams 2010), or in species that evolved the ability to produce or bioaccumulate antipredator chemical compounds, as in the case of several lineages of amphibians and snakes (Hossie et al. 2013).

The aforementioned killifish *Nothobranchius furzeri* is a case study that provides an example of relationships between extrinsic mortality and the evolution of senescence. *Nothobranchius* species, characterized by accelerated senescence and short lifespan, are subjected to high extrinsic mortality (Terzibasi et al. 2013; Blažek et al. 2017; Cellerino et al. 2015) and this selection revealed a correlation between the evolution of mitochondrial biogenesis genes and lifespan (Sahm et al. 2017). In *N. furzeri*, investigations on positively selected genes, within three evolutionary lineages, demonstrated that genes under positive selection were significantly enriched for functions involved in all steps of mitochondrial biogenesis, as in the case of mitochondrial proteins and respiratory chain complex I (Sahm and Cellerino 2017). Under the influence of such a peculiar ecology, protein expression has evolved to sustain fast growth and early maturation. Yet, through the process of antagonistic pleiotropy, they drive at the same time an accelerated ageing process (Sahm and Cellerino 2017). For these reasons, *N. furzeri* is one of the most interesting and emerging model species for ageing studies, due to its unique short life expectancy which is extremely reduced in those strains living in the most arid regions of its distribution area (Terzibasi et al. 2008; Holtze et al. 2021).

8.1.3 Searching for a New Long-Lived Model Organism

In contrast to *Nothobranchius* which is widely used as an experimental model species in ageing for its lifespan of less than one year, the research community still lacks an established model organism with exceptional longevity that can be easily cultured in captivity. Current field models are inadequate for common search purposes because of various difficulties related to their culturing. The scientific community is asking for a model that can be easy to rear and manage and that can, at the same time, answer scientific questions, for example in the ageing research field. For instance, the Greenland shark (*Somniosus microcephalus*), a model species for field studies, is quite complex to culture and breed in captivity due to its size, the unusual and remote habitat, and the extremely slow generation time. In addition, further studies are needed to explore and take advantage of the full potential of these remote and/or poorly studied species. Especially in comparative studies dealing with the ageing field, it is needed to improve our knowledge not only of transcriptome and genome sequences of these target species, but also of their distinctive physiology (Holtze et al. 2021). Several other species that demonstrated extreme lifespan, such as the olm (*Proteus anguinus*), an aquatic cave-dwelling salamander, are rare in nature and exhibited undesirable features, such as an age at sexual maturity of 15.6 years and the fact that they lay, on average, 35 eggs every 12.5 years (Voituron et al. 2011), making them not suitable for laboratory investigations. Similar problems are faced when trying to use long-lived mammals that have been demonstrated to be difficult or impossible to be kept in captivity and manipulated experimentally (Sahm et al. 2019; Holtze et al. 2021). These constraints create the necessity for a long-lived vertebrate that has all the characteristics required in a model organism (Ankeny and Leonelli 2021).

8.2 ANEMONEFISHES: IDEAL MODELS OF LONGEVITY?

8.2.1 Anemonefish: The Answer to the Search for a New Marine Model

Fishes belonging to the subfamily of Amphiprioninae (Pomacentridae family) could provide such a model. Amphiprioninae comprises two genera, *Amphiprion* and the monospecific *Premnas*, and all the species in these two genera are commonly known as anemonefish or clownfish. Clownfishes evolved a peculiar adaptation, probably inherited by a common ancestor, that enables a symbiosis with sea anemones and this symbiosis can be considered iconic of coral reefs. These species, originally used only in ecological investigations, are gaining interest as a more flexible model with potential application in various research fields. Consequently, the interest in clownfishes as model organisms increased in the last decade, for example in ageing (Sahm et al. 2019) and eco-evo-devo (Roux et al. 2020) studies, with several publications that described the unique and distinctive characteristics as well as the advantages of this model compared to the standard ones. They possess several characteristics that are precious in model organisms that make. They are phylogenetically related to damselfishes, such as *Chromis* and *Dascyllus* (subfamily Chrominae), *Chrysiptera* (subfamily Pomacentrinae), and *Lepidozygus* (subfamily Lepidozyginae) (Quenouille, Bermingham, and Planes 2004). The complex phylogeny of anemonefishes has been resolved using mtDNA and nuclear markers, as well as whole-genome sequencing (Litsios et al. 2012; Marcionetti et al. 2019).

8.2.2 Anemonefish Peculiarities

Clownfish species are socially controlled sequential protandrous hermaphrodites (Olivotto and Geffroy 2017) and their assemblages are characterized by a strong social hierarchy based on size that behaves as queues for reproduction (Casas et al. 2016). In fact, the two largest individuals are the dominant female and dominant male, respectively; this breeding couple is surrounded by a variable number of immature males of smaller size (Fricke and Fricke 1977). Apart from the aforementioned physiological and genetic peculiarities, clownfishes are relatively small, with some species that achieve a maximum size of less than ten centimetres. In addition, several *Amphiprion* species are cheap, common as well as robust aquarium fishes, easy to culture, feed, and breed in large numbers (Roux et al. 2021). All these characteristics allow the use of these species and in particular the false anemonefish *Amphiprion ocellaris* or the orange clownfish *Amphiprion percula*, to investigate a wide range of scientific questions in field or mesocosms, ranging from sex changes to social behaviors, sound production, as well as the symbiotic relationship with sea anemones (Marcionetti et al. 2019; Litsios et al. 2012; Dixson et al. 2014; Mebs 2009; Buston 2003).

8.2.3 Anemonefish and Predation

The search for long-lived vertebrates focused on those species that live in predator-free environments, or in stable habitats being two of the aforementioned environmental constraints correlated with the evolution of extreme lifespan (Wilkinson and South 2002; Rose 1991). Anemonefish belongs to this group of species that occupy a predator-free environment thanks to the symbiosis with sea anemones that provides for protection from predation. *Amphiprion* fishes are not hit by the lethal nematocysts present in the epithelium of the sea anemones tentacles thanks to a protective mucous coat that prevents the discharge of these cnidarian organelles (Mebs 2009). When facing danger these small reef fishes instantly search for protection in anemones' tentacles and it has been proven that if deprived of their symbiont anemone, the predation rate on clownfishes drastically increases (Elliott, Elliott, and Mariscal 1995; Mariscal 1970). Field observations have demonstrated that intimate relationships with anthozoans are not unique to clownfishes but can be found in other species of Pomacentridae, such as *Chromis viridis*, that showed interesting relationships with scleractinian corals (Ben-Tzvi et al. 2008; Lecchini et al. 2006). These species use branching corals as shelters (Garcia-Herrera et al. 2017; Holbrook et al. 2008), and the lack of available refuges exposes them to high predation (Hixon et al. 1997). Nevertheless, with the presence of favourable microhabitats and shelters, Chrominae are intensively predated by a wide range of carnivorous organisms ranging from resident-benthic to generalist-pelagic ones (Hixon et al. 1997).

8.3 LIFESPAN DATA

Lifespan data in the wild or even in captivity are not available for many Pomacentridae species, since these species build huge schools and have few interspecific differences making it impossible to identify them individually. Despite this, indirect evidence such as high adult mortality and very rapid growth (80% of maximum size reached within the first year) clearly indicates that these animals are short-lived in the wild and they are considered, by definition, a model for short-lived reef inhabitants (Wantiez and Thollot 2000). On the contrary, clownfishes live in small assemblages associated with a distinctive sea anemone, making their identification easy. The presence of the interspecific relationship between clownfish and anemones proved to have a significant impact on the population mortality rate that is lower than the one observed in other coral reef fishes or in the aforementioned species, of the same size, belonging to Pomacentridae (Munro and Williams 1985; Eckert 1987; Aldenhoven 1986; Buston and García 2007). Field investigations in several study sites on wild populations of clownfish demonstrated a low annual mortality rate ranging from 12.9% (Salles et al. 2015) to 13.7% (Buston 2003). In populations of *A. percula* in Madang Lagoon, Papua New Guinea, it was demonstrated that the low mortality, and consequently the predatory pressure on local populations, was not equally distributed according to the different stages of adulthood, with a mortality up to five times higher in non-breeding males (low-rank individuals) if compared to breeding couples (high-rank individuals) (Buston 2003). In the population of the same species living at Kimbe Island in Papua New Guinea, the mortality rate among the various social ranks did not produce statistical evidence although it was possible to determine that the annual mortality remained quite low compared to the one described in other reef fishes. In the case of Kimbe Island, the biannual mortality rate of local populations ranged from 18% to 49% for juveniles or immature males, from 9% to 44% in mature males, and from 19% to 55% in dominant females (Salles et al. 2015). In addition to mortality studies, an in-situ investigation performed using the recapture probability techniques demonstrated a lower bound of about 30 years in the estimated maximum lifespan of *A. percula* (Buston and García 2007). The long lifespan of clownfishes in the wild has been confirmed by a survey questionnaire (Table 8.1) distributed to researchers working with clownfishes and to public aquariums across Europe (Sahm et al. 2019).

The results of this survey demonstrated that a) for several clownfish species, the lifespan in captivity is more than a decade, with species living more than 20 years, as in the case of *A. melanopus* and *A. ocellaris*, and b) all the considered individuals in the survey were not approaching the limit of their lifespan, being actively spawning and showing no reproductive senescence (Sahm et al. 2019) (Figure 8.1).

8.4 TRANSCRIPTOMIC ANALYSIS OF ANEMONEFISH FOR LONGEVITY STUDIES

8.4.1 Amphiprioninae vs Chrominae: Positively Selected Genes

The low intrinsic mortality described earlier is correlated to anemonefishes' low predatory pressure, leading to an extraordinarily long life for all the species belonging to the subfamily Amphiprioninae. These observations are quite interesting, especially comparing the ageing of clownfishes with the short life span that characterizes the other

TABLE 8.1
Maximum Lifespan Registered by Clownfish Survey

Amphiprion species	Maximum size (cm)	Maximum lifespan registered	Status at census
clarkii in wild	15	12	Alive
clarkii (private aquarium)	15	16	Alive
clarkii	15	9	Alive
frenatus	14	18	Dead
melanopus	12	21	Alive
ocellaris (private aquarium)	11	22	Alive
ocellaris	11	17	Alive
perideraion	10	18	Alive
akydinos	9	13	Dead

Source: as described by Sahm et al. (2019).

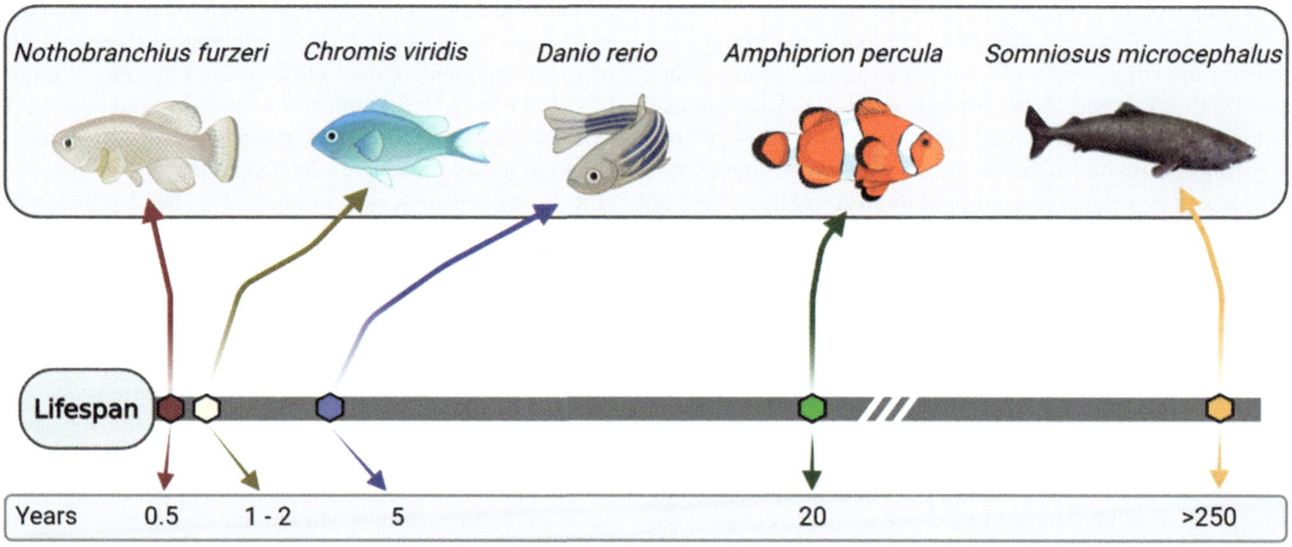

FIGURE 8.1 Lifespan in four teleost species and one elasmobranch. In established model species, such as *Nothobranchius furzeri* and *Danio rerio*, the lifespan ranges from six months to five years respectively. In the common damselfish such as blue damsel *Chromis viridis*, the lifespan in wild population could reach two years. On the contrary, clownfishes (*Amphiprion percula*) can easily reach a lifespan of 20 years and *Somniosus microcephalus* can live for more than 250 years. Created with BioRender.com.

species of the Pomacentridae family. In Amphiprioninae vs Chrominae, a total of 157 positively selected genes were identified belonging to 19 biological processes, several of them interesting for ageing research. In particular, nine of these are associated with the metabolism of xenobiotics, detoxification, and glutathione metabolism. These processes are up-regulated in experimental conditions promoting long life, such as dietary restriction, manipulation of mitochondrial translation (Houtkooper et al. 2013), or somatotropic axis inhibition, using common model organisms such as the nematode *Caenorhabditis elegans* and the fruit fly *Drosophila melanogaster* (McElwee et al. 2007), various mice laboratory strains (McElwee et al. 2007; Plank et al. 2012; Amador-Noguez et al. 2007; Steinbaugh et al. 2012), rats, pigs, and rhesus monkeys (Plank et al. 2012). Clownfishes, and mole rats too, show a positive selection of two lysosomal membrane proteins LAMP2 and CD63 (LAMP3) (Sahm et al. 2019) playing an important role in chaperone-mediated autophagy, lysosomal protein degradation in response to starvation (Berditchevski and Odintsova 2007; Eskelinen 2006), and a still unknown role in adaptive immune response and apoptosis (Tanaka et al. 2020). Lysosomal dysfunction is one of the key hallmarks of ageing (Carmona-Gutierrez et al. 2016). When the function of the lysosomal pumps is impaired, it leads to an increase in lysosomal pH (Colacurcio and Nixon 2016) reducing the activity of lyases and leading to the widespread age-dependent accumulation of lysosomal aggregates (Sacramento et al. 2020) such as lipofuscin (Brunk and Terman 2002) and ubiquitin-positive inclusions (Gray et al. 2003). A marker of lysosomal dysfunction is also the conserved up-regulation across tissues and species of genes coding for proteins of lysosomal pathways that probably is an effort for a compensatory response (Aramillo Irizar et al. 2018; de Magalhães, Curado, and Church 2009; Kurz et al. 2008). Earlier findings associated selection on lysosomal genes with evolution of mammalian longevity (Li and De Magalhães 2013). Considering these results, it is reasonable to think that positive selection related to lysosomal function is one of the processes that trigger the evolution of extraordinarily long life in clownfishes. Analysis of age-dependent protein aggregation would be important to further investigate the lysosomal function in the clownfish. Other biological processes that have been observed to be under positive selection in clownfish are translation, inflammation, and autophagy. Inflammation and autophagy impairments are considered evolutionary-conserved key hallmarks of ageing (López-Otín et al. 2013) and reduction of translation rates is associated with lifespan extension in nematodes and mice (Hofmann et al. 2015; Steffen and Dillin 2016).

8.4.2 Ageing and Anti-Parallel Evolution

Transcriptomic analysis performed on various Amphiprioninae vs Chrominae (Sahm et al. 2019) and killifishes (Baumgart et al. 2016) showed signs of anti-parallel evolution, *id est* a process by which the same genetic pathways show signatures of positive selection in two lineages that evolved lifespan in opposite directions, as in the case of GSTK1, a protein involved in glutathione metabolism and protection from oxidative stress. This gene was demonstrated to be positively selected both in clownfishes and in very short-lived annual killifishes. Signs of convergence were observed in genes linked to the biogenesis of mitochondrially encoded proteins, as in the case of FASTKD2 and FASTKD5, involved in the biogenesis of mitochondrial ribosomes (Sahm et al. 2019). It is remarkable to observe that those signs of positive selection detected in both short-lived species (*Nothobranchius furzeri*) and long-lived species (clownfish and mole-rat) (Sahm et al. 2019) have been corroborated by analysing the expression of MTERF, a gene that acts as a negative regulator of mitochondrial transcription (Roberti et al. 2009). These positively selected genes involved in functions like "Mitochondrial large/small ribosomal subunit" (GO:0005762/GO:0005763) and "Mitochondrial respiratory chain complex I" (GO:0005747) have been detected in numerous species with unique lifespan, like ants (Roux et al. 2014) and African mole-rats (Sahm and Cellerino 2017; Sahm et al. 2018), as well as in Amphiprionidae (Figure 8.2). Since detailed structures of this protein are available (Ladner et al. 2004; Wang et al. 2011), homology modelling was possible, and it strongly indicates that positive selection targeted positions that are implicated in the enzymatic activity and function of the encoded protein.

Paradoxically, the short lifespan of annual killifishes of the Nothobranchiidae family, of the genus *Callopanchax* from West Africa (Cui et al. 2019) and the species *Austrofundulus limnaeus* from South America (Wagner et al. 2018) was associated with an enrichment of positively selected genes for a gene-set that stands explicitly for mitochondrial biogenesis (Sahm et al. 2017). Mitochondrial biogenesis and mitonuclear balance were related to the increase in longevity in experimental studies in several

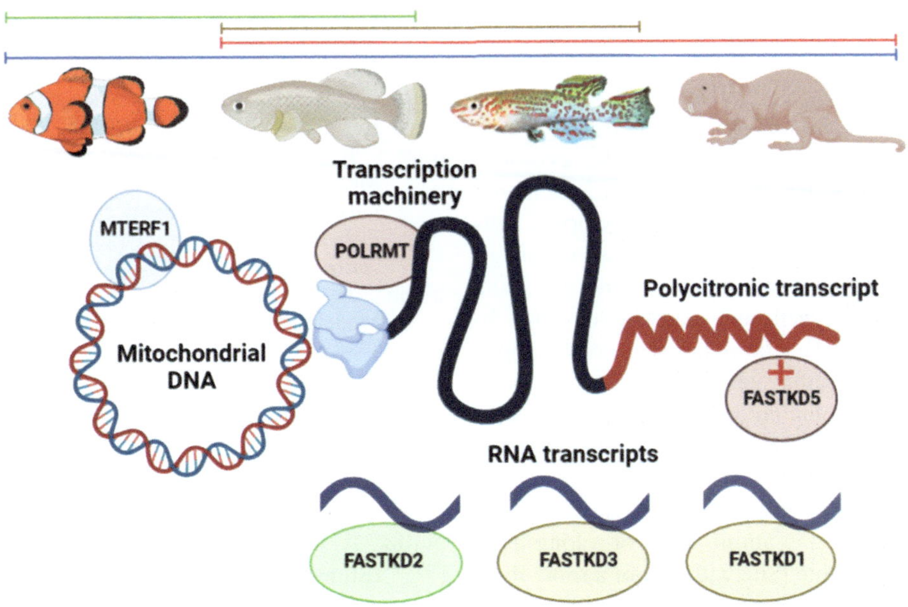

FIGURE 8.2 Convergent evolution of positively selected genes involved in mitonuclear balance in species characterized by a very short (*Nothobranchius furzeri* and *Callopanchax occidentalis*) and long lifespan (clownfishes and mole rats). In particular, the process depicted corresponds to the transcription and transcript processing of mitochondrially encoded genes. The color code of the genes corresponds to the upper bars indicates the species where the positive selection is observed: blue (MTERF1) indicates all four species, red (FASTKD5) the two killifishes and the mole rat, green (FASTKD2) *N. furzeri* and clownfishes, and brown the two killifishes. Created with BioRender.com.

model organisms (Karpac and Jasper 2013) and demonstrated to be key pathways in the regulation of ageing and lifespan (Houtkooper et al. 2013). Further investigations demonstrated that the same pathway is under positive selection in both short- and long-lived species. For example, the termination factor mTERF1 is under positive selection in clownfishes, killifishes, and mole rats (Sahm et al. 2019), and genes coding for mitochondrial ribosomal proteins and for members of complex I of the respiratory chain are under positive selection both in killifishes (Sahm et al. 2018; Cui et al. 2019; Sahm et al. 2017) and in long-lived rodents (Sahm et al. 2018) indicating that the same genetic design triggers both evolution of longevity and reduced life expectancy (Holtze et al. 2021). Therefore, some biological processes as in the case of mitochondrial biogenesis could be considered as a core genetic substrate in the evolution of lifespan. It has been probably recruited multiple times independently, in various species for various ecological adaptations, causing a modulation of lifespan in the opposite direction (short-lived vs long-lived species) depending on the life-history strategy that was selected for each evolutive clade (Holtze et al. 2021; Sahm et al. 2019).

8.5 CONCLUSION

In the 20th century, biological investigations faced a transition from descriptive to a mechanistic understanding of the biological processes leading to the conscious decision to employ model organisms as effective tools to study life. Although experimental organisms do not necessarily have to be representative of species other than themselves, in many cases model organisms should assure a wide representation of biological diversity and should allow researchers to observe phenomena that are arguably not directly observable using other target organisms, for various reasons. Since the proposal of anemonefish as model organisms suitable for ecological and, afterwards, evo-devo studies, several investigations have made use of anemonefish experimental and unique advantages. Anemonefish is an organism so convenient to study a wide range of biological phenomena that researchers are developing tools and resources specifically designed, such as collections of techniques and methods and genetic databases. For example, processes, genes, and specific sites of genome under positive selection represent potential and promising targets for follow-up studies in various scientific fields. One example of the potential application of these studies on the positively selected sites is given by the use of CRISPR/Cas9 technology in order to substitute the amino acids of a long-lived species at a positively selected site with that of a short-lived species. Considering that genomes of 12 different clownfish species are available (Pryor et al. 2020), that they can be easily cultured in captivity (Roux et al. 2021) and that several species are currently used as models to experimentally induce sex reversal (Casas et al. 2016) and pigmentation phenotypes (Salis et al. 2019), we can affirm that anemonefishes are efficient model organisms that assure a representation of biological phenomena. As a model, they represent a larger group of organisms beyond themselves and serve as the basis for articulating processes thought to be shared across several other types of organisms. For this reason, this model is powerful and gives an effective repertoire of answers to scientific questions in modern investigation not only related to ageing theories but also considering other research fields. Anemonefish are no doubt the first long-living experimental fish model for ageing studies, but they represent a fundamental model for many scientific fields due to their biological, physiological, and genetic peculiarities.

REFERENCES

Aldenhoven, J. M. 1986. Local Variation in Mortality Rates and Life-Expectancy Estimates of the Coral-Reef Fish *Centropyge bicolor* (Pisces: Pomacanthidae). *Marine Biology: International Journal on Life in Oceans and Coastal Waters* 92 (2): 237–44.

Amador-Noguez, D., A. Dean, W. Huang, K. Setchell, D. Moore, and G. Darlington. 2007. Alterations in Xenobiotic Metabolism in the Long-Lived Little Mice. *Aging Cell* 6 (4): 453–70.

Ankeny, R. A., and S. Leonelli. 2021. *Model Organisms (Elements in the Philosophy of Biology)*. Cambridge: Cambridge University Press.

Bahls, C., J. Weitzman, and R. Gallagher. 2003. Biology's Models. *The Scientist* 17 (11): S5.

Baumgart, M., S. Priebe, M. Groth, N. Hartmann, U. Menzel, L. Pandolfini, P. Koch, et al. 2016. Longitudinal RNA-Seq Analysis of Vertebrate Aging Identifies Mitochondrial Complex I as a Small-Molecule-Sensitive Modifier of Lifespan. *Cell Systems* 2 (2): 122–32.

Ben-Tzvi, O., A. Abelson, O. Polak, and M. Kiflawi. 2008. Habitat Selection and the Colonization of New Territories by *Chromis viridis*. *Journal of Fish Biology* 73 (4): 1005–18.

Berditchevski, F., and E. Odintsova. 2007. Tetraspanins as Regulators of Protein Trafficking. *Traffic* 8 (2): 89–96.

Blažek, R., M. Polačik, P. Kačer, A. Cellerino, R. Řežucha, C. Methling, O. Tomášek, et al. 2017. Repeated Intraspecific Divergence in Life Span and Aging of African Annual Fishes along an Aridity Gradient. *Evolution; International Journal of Organic Evolution* 71 (2): 386–402.

Bodnar, A. 2016. Lessons from the Sea: Marine Animals Provide Models for Biomedical Research. *Environment* 58 (2): 16–25.

Brunk, U. T., and A. Terman. 2002. Lipofuscin: Mechanisms of Age-Related Accumulation and Influence on Cell Function. *Free Radical Biology and Medicine* 33 (5): 611–19.

Buston, P. 2003. Mortality Is Associated with Social Rank in the Clown Anemonefish (*Amphiprion Percula*). *Marine Biology* 143 (4): 811–15.

Buston, P., and M. B. García. 2007. An Extraordinary Life Span Estimate for the Clown Anemonefish *Amphiprion percula*. *Journal of Fish Biology* 70 (6): 1710–19.

Carmona-Gutierrez, D., A. L. Hughes, F. Madeo, and C. Ruckenstuhl. 2016. The Crucial Impact of Lysosomes in Aging and Longevity. *Ageing Research Reviews* 32: 2–12.

Casas, L., F. Saborido-Rey, T. Ryu, C. Michell, T. Ravasi, and X. Irigoien. 2016. Sex Change in Clownfish: Molecular Insights from Transcriptome Analysis. *Scientific Reports* 6: 35461.

Cellerino, A., D. Valenzano, and M. Reichard. 2015. From the Bush to the Bench: The Annual *Nothobranchius* Fishes as a New Model System in Biology. *Biological Reviews of the Cambridge Philosophical Society* 91 (2): 511–33.

Colacurcio, D.J., and R.A. Nixon. 2016. Disorders of Lysosomal Acidification-The Emerging Role of v-ATPase in Aging and Neurodegenerative Disease. *Ageing Research Reviews* 32: 75–88.

Cui, R., T. Medeiros, D. Willemsen, L. N. M. Iasi, G. E. Collier, M. Graef, M. Reichard, et al. 2019. Relaxed Selection Limits Lifespan by Increasing Mutation Load. *Cell* 178 (2): 385–99.e20.

de Magalhães, J. P., J. Curado, and G. M. Church. 2009. Meta-Analysis of Age-Related Gene Expression Profiles Identifies Common Signatures of Aging. *Bioinformatics* 25 (7): 875–81.

Dixson, D. L., G. P. Jones, P. L. Munday, S. Planes, M. S. Pratchett, and S. R. Thorrold. 2014. Experimental Evaluation of Imprinting and the Role Innate Preference Plays in Habitat Selection in a Coral Reef Fish. *Oecologia* 174 (1): 99–107.

Eckert, G. J. 1987. Estimates of Adult and Juvenile Mortality for Labrid Fishes at One Tree Reef, Great Barrier Reef. *Marine Biology* 95 (2): 167–71.

Elliott, J. K., J. M. Elliott, and R. N. Mariscal. 1995. Host Selection, Location, and Association Behaviors of Anemonefishes in Field Settlement Experiments. *Marine Biology* 122 (3): 377–89.

Eskelinen, E-L. 2006. Roles of LAMP-1 and LAMP-2 in Lysosome Biogenesis and Autophagy. *Molecular Aspects of Medicine* 27: 495–502.

Finch, C. E. 1998. Variations in Senescence and Longevity Include the Possibility of Negligible Senescence. *Journals of Gerontology - Series A Biological Sciences and Medical Sciences* 53 (4): 235–39.

Fricke, H., and S. Fricke. 1977. Monogamy and Sex Change by Aggressive Dominance in Coral Reef Fish. *Nature* 266 (5605): 830–32.

Garcia-Herrera, N., S. C. A. Ferse, A. Kunzmann, and A. Genin. 2017. Mutualistic Damselfish Induce Higher Photosynthetic Rates in Their Host Coral. *The Journal of Experimental Biology* 220 (10): 1803–11.

Goldstein, B., and N. King. 2017. The Future of Cell Biology: Emerging Model Organisms Reasons to Turn to Nontraditional Models *Trends in Cell Biology* 26 (11): 818–24.

Gray, D. A., M. Tsirigotis, and J. Woulfe. 2003. Ubiquitin, Proteasomes, and the Aging Brain. *Science of Aging Knowledge Environment: SAGE KE* 4 (34): 65.

Hixon, M. A., M. H. Carr, V. B. O. Riordan, and A. M. Burnell. 1997. Synergistic Predation, Density Dependence, and Population Regulation in Marine Fish. *Science* 277 (5328): 946–49.

Hofmann, J. W., X. Zhao, M. De Cecco, A. L. Peterson, L. Pagliaroli, J. Manivannan, G. B. Hubbard, et al. 2015. Reduced Expression of MYC Increases Longevity and Enhances Healthspan. *Cell* 160 (3): 477–88.

Holbrook, S. J., A. J. Brooks, R. J. Schmitt, and H. L. Stewart. 2008. Effects of Sheltering Fish on Growth of Their Host Corals. *Marine Biology* 155 (5): 521–30.

Holtze, S., E. Gorshkova, S. Braude, A. Cellerino, P. Dammann, T. B. Hildebrandt, A. Hoeflich, et al. 2021. Alternative Animal Models of Aging Research. *Frontiers in Molecular Biosciences* 8: 660959.

Hossie, T. J., C. Hassall, W. Knee, and T. N. Sherratt. 2013. Species with a Chemical Defence, but Not Chemical Offence, Live Longer. *Journal of Evolutionary Biology* 26 (7): 1598–602.

Houtkooper, R. H., L. Mouchiroud, D. Ryu, N. Moullan, E. Katsyuba, G. Knott, R. W. Williams, et al. 2013. Mitonuclear Protein Imbalance as a Conserved Longevity Mechanism. *Nature* 497 (7450): 451–57.

Irizar, P. A., S. Schäuble, D. Esser, M. Groth, C. Frahm, S. Priebe, M. Baumgart, et al. 2018. Transcriptomic Alterations during Ageing Reflect the Shift from Cancer to Degenerative Diseases in the Elderly. *Nature Communications* 9 (1): 327.

Karpac, J., and H. Jasper. 2013. Aging: Seeking Mitonuclear Balance. *Cell* 154 (2): 271–73.

Kirkwood, T. B. L. 2000. Why Do We Age? *Nature* 408 (6809): 233–38.

Kirkwood, T. B. L. 2002. Evolution of Ageing. *Mechanisms of Ageing and Development* 123 (7): 737–45.

Kishi, S. 2004. Functional Aging and Gradual Senescence in Zebrafish. *Annals of the New York Academy of Sciences* 1019: 521–26.

Kishi, S., J. Uchiyama, A. M. Baughman, T. Goto, M. C. Lin, and S. B. Tsai. 2003. The Zebrafish as a Vertebrate Model of Functional Aging and Very Gradual Senescence. *Experimental Gerontology* 38 (7): 777–86.

Kurz, T., A. Terman, B. Gustafsson, and U. T. Brunk. 2008. Lysosomes and Oxidative Stress in Aging and Apoptosis. *Biochimica et Biophysica Acta – General Subjects* 1780 (11): 1291–303.

Ladner, J. E., J. F. Parsons, C. L. Rife, G. L. Gilliland, and R. N. Armstrong. 2004. Parallel Evolutionary Pathways for Glutathione Transferases: Structure and Mechanism of the Mitochondrial Class Kappa Enzyme rGSTK1-1. *Biochemistry* 43 (2): 352–61.

Lecchini, D., Y. Nakamura, J. Grignon, and M. Tsuchiya. 2006. Evidence of Density-Independent Mortality in a Settling Coral Reef Damselfish, *Chromis viridis*. *Ichthyological Research* 53 (3): 298–300.

Levy, A., and A. Currie. 2015. Model Organisms Are Not (Theoretical) Models. *The British Journal for the Philosophy of Science* 66 (2): 327–48.

Litsios, G., C. A. Sims, R. O. Wüest, P. B. Pearman, N. E. Zimmermann, and N. Salamin. 2012. Mutualism with Sea Anemones Triggered the Adaptive Radiation of Clownfishes. *BMC Evolutionary Biology* 12: 212.

Li, Y., and J. P. De Magalhães. 2013. Accelerated Protein Evolution Analysis Reveals Genes and Pathways Associated with the Evolution of Mammalian Longevity. *Age* 35 (2): 301–14.

López-Otín, C., M. A. Blasco, L. Partridge, M. Serrano, and G. Kroemer. 2013. The Hallmarks of Aging. *Cell* 153 (6): 1194–217.

Mangel, M., H. K. Kindsvater, and M. B. Bonsall. 2007. Evolutionary Analysis of Life Span, Competition, and Adaptive Radiation, Motivated by the Pacific Rockfishes (Sebastes). *Evolution; International Journal of Organic Evolution* 61 (5): 1208–24.

Marcionetti, A., V. Rossier, N. Roux, P. Salis, V. Laudet, and N. Salamin. 2019. Insights into the Genomics of Clownfish Adaptive Radiation: Genetic Basis of the Mutualism with Sea Anemones. *Genome Biology and Evolution* 11 (3): 869–82.

Mariscal, R. N. 1970. The Nature of the Symbiosis between Indo-Pacific Anemone Fishes and Sea Anemones. *Marine Biology* 6 (1): 58–65.

McElwee, J. J., E. Schuster, E. Blanc, et al. 2007. Evolutionary Conservation of Regulated Longevity Assurance Mechanisms. *Genome Biology* 8: R132.

Mebs, D. 2009. Chemical Biology of the Mutualistic Relationships of Sea Anemones with Fish and Crustaceans. *Toxicon: Official Journal of the International Society on Toxinology* 54 (8): 1071–74.

Medawar, P. B. 1952. An Unsolved Problem of Biology. In *The Uniqueness of the Individual*, ed. H. K. Lewis, (pp. 44-70). New York: Basic Books Inc.

Munro, J. L., and D. Mcb Williams. 1985. Assessment and Management of Coral Reef Fisheries: Biological, Environmental and Socio-Economic Aspects. In *Proceedings of the Fifth International Coral Reef Symposium*, eds. C. Gabrie and B. Salvat 4: 543–81. Tahiti, French Polynesia: Antenne Museum.

Mutalipassi, M. 2019. *Re-Defining the Concept of Model Species: An Experimental Approach on a Range of Marine Animals*. PhD thesis, The Open University.

Nielsen, J., R. B. Hedeholm, J. Heinemeier, et al. 2016. Eye Lens Radiocarbon Reveals Centuries of Longevity in the Greenland Shark (*Somniosus Microcephalus*). *Science* 353 (6300): 702–4.

Olivotto, I., and B. Geffroy. 2017. Clownfish. In *Marine Ornamental Species Aquaculture*, eds. R. Calado, I. Olivotto, M. P. Oliver, and G. J. Holt, 177–99. Chichester: Wiley-Blackwell.

Patnaik, B. K., N. Mahapatro, and B. S. Jena. 1994. Ageing in Fishes. *Gerontology* 40 (2–4): 113–32.

Plank, M., D. Wuttke, S. Van Dam, S. A. Clarke, and J. P. De Magalhães. 2012. A Meta-Analysis of Caloric Restriction Gene Expression Profiles to Infer Common Signatures and Regulatory Mechanisms. *Molecular bioSystems* 8 (4): 1339–49.

Pryor, S. H., R. Hill, D. L. Dixson, N. J. Fraser, B. P. Kelaher, and A. Scott. 2020. Anemonefish Facilitate Bleaching Recovery in a Host Sea Anemone. *Scientific Reports* 10 (1): 18586.

Quenouille, B., E. Bermingham, and S. Planes. 2004. Molecular Systematics of the Damselfishes (Teleostei: Pomacentridae): Bayesian Phylogenetic Analyses of Mitochondrial and Nuclear DNA Sequences. *Molecular Phylogenetics and Evolution* 31 (1): 66–88.

Reznick, D., M. Bryant, and D. Holmes. 2006. The Evolution of Senescence and Post-Reproductive Lifespan in Guppies (*Poecilia reticulata*). *PLoS Biology* 4 (1): 136–43.

Roberti, M., P. L. Polosa, F. Bruni, C. Manzari, S. Deceglie, M. N. Gadaleta, and P. Cantatore. 2009. The MTERF Family Proteins: Mitochondrial Transcription Regulators and Beyond. *Biochimica et Biophysica Acta - Bioenergetics* 1787 (5): 303–11.

Rose, M. R. 1991. *The Evolutionary Biology of Aging*. Vol. 68. New York: Oxford University Press.

Roux, J., E. Privman, S. Moretti, J. T. Daub, M. Robinson-Rechavi, and L. Keller. 2014. Patterns of Positive Selection in Seven Ant Genomes. *Molecular Biology and Evolution* 31 (7): 1661–85.

Roux, N., V. Logeux, N. Trouillard, R. Pillot, K. Magré, P. Salis, D. Lecchini, et al. 2021. A Star Is Born Again: Methods for Larval Rearing of an Emerging Model Organism, the False Clownfish *Amphiprion ocellaris*. *Journal of Experimental Zoology. Part B, Molecular and Developmental Evolution* 336 (4): 376–85.

Roux, N., P. Salis, S. H. Lee, L. Besseau, and V. Laudet. 2020. Anemonefish, a Model for Eco-Evo-Devo. *EvoDevo* 11 (1): 1–11.

Sacramento, E. K., J. M. Kirkpatrick, M. Mazzetto, M. Baumgart, A. Bartolome, S. Di Sanzo, C. Caterino, et al. 2020. Reduced Proteasome Activity in the Aging Brain Results in Ribosome Stoichiometry Loss and Aggregation. *Molecular Systems Biology* 16 (6): e9596.

Sahm, A., P. Almaida-Pagán, M. Bens, M. Mutalipassi, A. Lucas-Sánchez, J. De Costa Ruiz, M. Görlach, et al. 2019. Analysis of the Coding Sequences of Clownfish Reveals Molecular Convergence in the Evolution of Lifespan. *BMC Evolutionary Biology* 19 (1): 89.

Sahm, A., M. Bens, Y. Henning, C. Vole, M. Groth, M. Schwab, S. Hoffmann, et al. 2018. Higher Gene Expression Stability during Aging in Long-Lived Giant Mole-Rats than in Short-Lived Rats. *Aging* 10 (12): 3938–56.

Sahm, A., M. Bens, M. Platzer, and A. Cellerino. 2017. Parallel Evolution of Genes Controlling Mitonuclear Balance in Short-Lived Annual Fishes. *Aging Cell* 16 (3): 488–96.

Sahm, A., M. Bens, K. Szafranski, S. Holtze, M. Groth, M. Görlach, C. Calkhoven, et al. 2018. Long-Lived Rodents Reveal Signatures of Positive Selection in Genes Associated with Lifespan and Eusociality. *Plos Genetics* 14 (3): e1007272.

Sahm, A., and A. Cellerino. 2017. (Anti-)parallel Evolution of Lifespan. *Aging* 9 (10): 2018–19.

Salis, P., T. Lorin, V. Lewis, et al. 2019. Developmental and Comparative Transcriptomic Identification of Iridophore Contribution to White barring in Clownfish. *Pigment Cell & Melanoma Research* 32 (3): 391–402.

Salles, O. C., J. A. Maynard, M. Joannides, C. M. Barbu, P. Saenz-Agudelo, G. R. Almany, M. L. Berumen, et al. 2015. Coral Reef Fish Populations Can Persist without Immigration. *Proceedings of the Royal Society B: Biological Sciences* 282 (1819): 20151311.

Schartl, M. 2014. Beyond the Zebrafish: Diverse Fish Species for Modeling Human Disease. *Disease Models & Mechanisms* 7 (2): 181–92.

Shattuck, M. R., and S. A. Williams. 2010. Arboreality Has Allowed for the Evolution of Increased Longevity in Mammals. *Proceedings of the National Academy of Sciences of the United States of America* 107 (10): 4635–39.

Steffen, K. K., and A. Dillin. 2016. A Ribosomal Perspective on Proteostasis and Aging. *Cell Metabolism* 23 (6): 1004–12.

Steinbaugh, M. J., L. Y. Sun, A. Bartke, and R. A. Miller. 2012. Activation of Genes Involved in Xenobiotic Metabolism Is a Shared Signature of Mouse Models with Extended Lifespan. *American Journal of Physiology - Endocrinology and Metabolism* 303 (4): E488–E495.

Tanaka, T., B. M. Warner, T. Odani, Y. Ji, Y-Q. Mo, H. Nakamura, S-I. Jang, et al. 2020. LAMP3 Induces Apoptosis and Autoantigen Release in Sjögren's Syndrome Patients. *Scientific Reports* 10 (1): 1–17.

Terzibasi, E., A. Dorn, E. Ng'Oma, M. Polačik, R. Blažek, K. Reichwald, A. Petzold, et al. 2013. Parallel Evolution of Senescence in Annual Fishes in Response to Extrinsic Mortality. *BMC Evolutionary Biology* 13: 77.

Terzibasi, E., D. R. Valenzano, and A. Cellerino. 2007. The Short-Lived Fish *Nothobranchius furzeri* as a New Model System for Aging Studies. *Experimental Gerontology* 42 (1–2): 81–89.

Terzibasi, E., D. R. Valenzano, P. Roncaglia, A. Cattaneo, L. Domenici, and A. Cellerino. 2008. Large Differences in Aging Phenotype between Strains of the Short-Lived Annual Fish *Nothobranchius furzeri*. *PloS One* 3 (12): e3866.

Voituron, Y., M. De Fraipont, J. Issartel, O. Guillaume, and J. Clobert. 2011. Extreme Lifespan of the Human Fish (*Proteus anguinus*): A Challenge for Ageing Mechanisms. *Biology Letters* 7 (1): 105–7.

Wagner, J. T., P. P. Singh, A. L. Romney, C. L. Riggs, P. Minx, S. C. Woll, J. Roush, et al. 2018. "The Genome of *Austrofundulus limnaeus* Offers Insights into Extreme Vertebrate Stress Tolerance and Embryonic Development." *BMC Genomics* 19 (1): 1–21. doi: 10.1186/s12864-018-4539-7.

Wang, J., J. Zhu, S. Liu, B. Liu, Y. Gao, and Z. Wu. 2011. Generation of Reactive Oxygen Species in Cyanobacteria and Green Algae Induced by Allelochemicals of Submerged Macrophytes. *Chemosphere* 85 (6): 977–82.

Wantiez, L., and P. Thollot. 2000. Settlement, Post-Settlement Mortality and Growth of the Damselfish *Chromis fumea* (Pisces: Pomacentridae) on Two Artificial Reefs in New Caledonia (South-West Pacific Ocean). *Journal of the Marine Biological Association of the United Kingdom* 80 (6): 1111–18.

Wilkinson, G. S., and J. M. South. 2002. Life History, Ecology and Longevity in Bats. *Aging Cell* 1 (2): 124–31.

Williams, G. C. 1957. Pleiotropy, Natural Selection, and the Evolution of Senescence. *Evolution; International Journal of Organic Evolution* 11 (4): 398–411.

9 The Visual Ecology of Anemonefishes

Fabio Cortesi, Valerio Tettamanti, and Fanny de Busserolles

CONTENTS

9.1 INTRODUCTION

> If you want beauty and wildlife, you want a coral reef. Put on a mask and stick your head under the water. The sight is mind-blowing. And that, actually, is still a mystery: why are coral reefs so beautiful and colorful?
>
> **(Sir David Attenborough, 28 October 2012, *The Guardian*, UK)**

Sir Attenborough's question is still plaguing (vision) scientists who are trying to decipher the language of color and vision in reef critters. Anemonefishes are a fitting example; they are easily distinguishable from other reef fishes due to their characteristic orange, black, and white stripes, and their association with anemones (Fautin and Allen 1997) (Figure 9.1A, B). However, between and within the 28 anemonefish species, the patterns vary considerably, including species with none, one, two, and three vertical bars and others that have horizontal bars on their head and dorsum (Klann et al. 2021; Roux et al. 2020) (see Chapter 7 for details). Whether anemonefishes use these patterns to communicate, what messages they might be sending, and who the intended receivers are, remain mostly unknown. Hence, if we want to understand anemonefish visual communication and how this influences their behavior, we need to see the world through their eyes. This involves characterizing their visual systems' molecular, anatomical, and physiological properties and setting them in the context of the visual environment anemonefishes experience during their daily tasks.

9.2 THE LIGHT ENVIRONMENT OF ANEMONEFISHES

As opposed to the broad light spectrum found in most terrestrial habitats, light underwater is absorbed and scattered through particles leading to a decrease in intensity and a narrowing of the spectrum with increasing depth and distance (Jerlov 1976). Put simply, the deep ocean appears dark blue to us, because only blue wavelengths ~480–490 nm remain at depth. On the contrary, bodies of water that are rich in dissolved organic matter, phytoplankton or that contain silt in the water column such as many lakes, rivers, and inshore reefs are prone to excessive scatter, shifting the available light to longer red-dominated wavelengths > 550 nm (Jerlov 1976, Munz, and McFarland 1977). Consequently, differences in aquatic light environments exert strong selection on the visual systems of their inhabitants, with marine fishes generally having more blue-green centred visual systems and freshwater fishes having red-shifted vision (reviewed in Carleton et al. 2020; Musilova et al. 2021).

Most anemonefishes are found on tropical coral reefs, which are typically nutrient-poor, clear water habitats that, at least for the first ten meters, are flooded with a broad spectrum of light ranging from the ultraviolet (UV, < 400 nm) to the red (> 600 nm) (McFarland 1991; Jerlov 1976) (Figure 9.1C). Although marked differences in photic environments between reef microhabitats exist (Marshall et al. 2003) and these have been shown to drive visual system evolution in some reef fish families (e.g., cardinalfishes, Luehrmann et al. 2020; soldierfishes, Busserolles et al. 2021), anemones or more specifically their photosynthetic algal symbionts thrive in broad-spectrum light (Foo et al. 2020). As we will explore in detail in this chapter, anemonefish take advantage of this broad spectrum of light, which together with various ecological and behavioral traits shape the specifics of their visual systems (Stieb et al. 2019; Mitchell et al. 2021a).

9.3 THE MORPHOLOGICAL BASIS FOR VISION IN ANEMONEFISHES

Anemonefishes, like all vertebrates, possess camera-type eyes, where light enters the eye through a pupillary aperture and is focused onto the retina by a lens. Like most bony fishes, they have laterally placed eyes providing a large monocular field of view on either side of the head (Figure 9.2A). Although the extent of their visual field is currently unknown, anemonefishes likely possess some degree

DOI: 10.1201/9781003125365-11

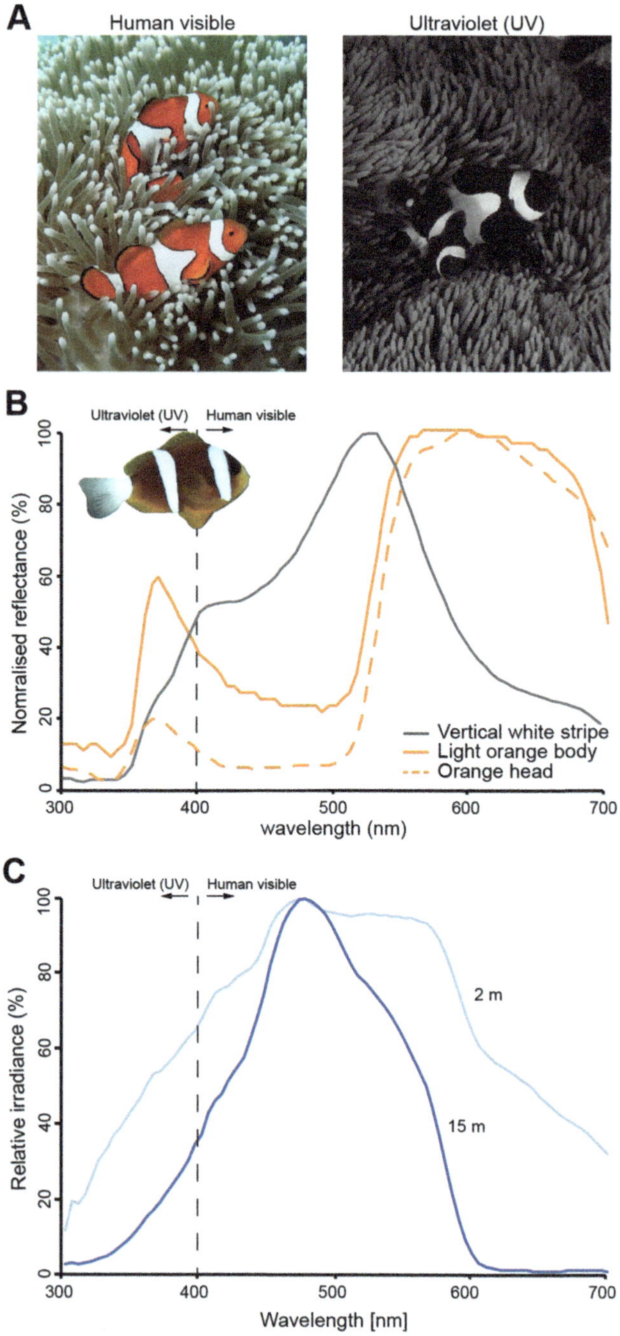

FIGURE 9.1 Anemonefish color and underwater light environment. A) Photographs of *Amphiprion percula* in the human visible (left) and ultraviolet (right). B) Normalized spectral reflectance of various body parts in *Amphiprion akindynos*. C) Relative sidewelling spectral irradiance at 2 m (light blue) and 15 m (dark blue) from reefs around Lizard Island, Australia, where the fishes in A and B were observed. Note how with increasing depth, wavelengths at either end of the light spectrum are attenuated. Images in A, courtesy of Justin Marshall. Panels B and C, adapted with permission from (Stieb et al. 2016, 2019).

of binocular vision. Such a frontward-looking, overlapping visual field is likely as we can see both eyes when looking at the fish head-on and because of the presence of small gaps between their iris and lens, called aphakic gaps, which allow extra light to enter the eye frontally (Figure 9.2B).

For vision to occur, light needs to reach the retina, located at the back of the eye. As such, light must first travel through a series of ocular media composed of the cornea, the lens, and the vitreous humour (Figure 9.2C). These ocular media can have filtering properties, removing specific wavelengths of light before reaching the retina. In most fishes, it is the UV part of the spectrum, which can be damaging to the visual system (Ivanov et al. 2018; Zigman 1993) but also causes excessive scattering (Muntz 1973), which is filtered out (Carleton et al. 2020). However, for anemonefishes and damselfishes more generally, ocular media are UV-transparent at all stages of their lives, therefore enabling the entire light spectrum to reach the retina and, as we will see later, allowing for UV vision to occur (Siebeck and Marshall 2001, 2007; Stieb et al. 2019).

Once light reaches the retina, it must pass through several neuronal layers before reaching the light-sensitive visual pigments located in the photoreceptor cells (Figure 9.2C). Like most vertebrates, anemonefishes possess two morphological types of photoreceptors, rods, and cones. Each type mediates vision under specific light conditions, dim-light vision for the rods, and bright light and color vision for the cones. In bony fishes, cone photoreceptors can further be divided into several morphological subtypes such as single, double, triple, and quadruple cones, the latter three composed of several cones fused together (Walls 1942). Anemonefishes possess rods and single and double cones (Mitchell et al. 2021a; Stieb et al. 2019). In addition, their cones are organized in a regular fashion, called a square mosaic, a conformation commonly found in teleosts where every single cone is surrounded by four double cones, resulting in a double to single cone ratio of 2:1 (Figure 9.2E) (Collin and Shand 2003). These photoreceptors, which constitute the first stage of visual processing, will convert the light information into an electrical signal via a process called phototransduction. This electrical signal will then be further processed and transmitted to the retinal ganglion cells via a series of interneurons (Figure 9.2C). Ultimately, it is the ganglion cells, which constitute the last stage of visual processing in the retina, that transmit the visual information to the central nervous system, and therefore set the upper limit of anatomical acuity and sensitivity (Warrant and Locket 2004) (Figure 9.2D).

Anatomical visual acuity was estimated in the Barrier Reef anemonefish, *Amphiprion akindynos*, based on the highest density of ganglion cells in the retina. Acuity is commonly measured in cycles per degree of visual angle, which refers to the number of black and white line pairs (one black and one white line = 1 cycle) that an animal can distinguish in one degree of their visual field. Anemonefish acuity varied from 3.6 to 5.8 cycles per degree in individuals measuring around 4 cm and 8 cm in standard length, respectively (Stieb et al. 2019). Accordingly, an 8 cm *A. akindynos* may be able to perceive another individual of similar size from several meters away (Figure 9.3). While anemonefish acuity appears relatively low compared to other damselfish species (four to eight cycles per degree; Collin and Pettigrew 1989; Parker et al. 2017) and coral reef fishes in general (four to

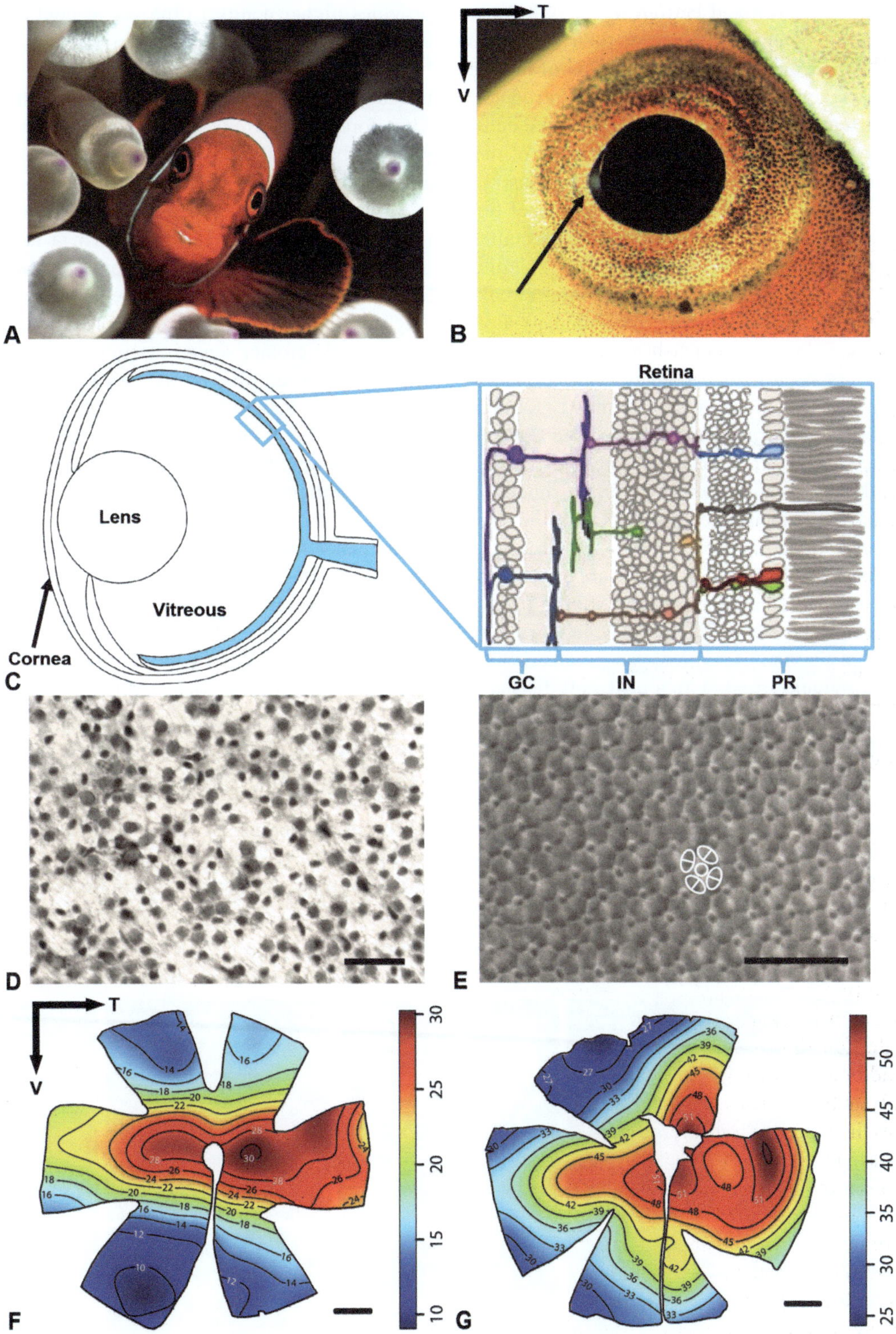

FIGURE 9.2 Anatomy of the visual system of anemonefishes. A) Frontal picture of *Premnas biaculeatus* showing the position of the eyes in the head with some degree of binocular overlap. B) Photograph of the nasal aphakic gap (black arrow) in *Amphiprion ocellaris*. The black arrow indicates the orientation of the head. T = temporal, V = ventral. C) Schematic of an anemonefish eye and its retina. The schematic of the retina was adapted with permission from (Baden 2020). GC = ganglion cells, IN = interneurons, PR = photoreceptors. D) Wholemount view of the ganglion cell layer stained with Nissl stain in *Amphiprion akindynos*. E) Wholemount view of the photoreceptor layer in *A. akindynos* showing the square mosaic pattern. Scale bars in D, E = 25 μm. F) and G) Topographic distribution of ganglion cells and total cone photoreceptors, respectively, in *A. akindynos* (reproduced from Stieb at al. 2019). The black lines represent iso-density contours and values are expressed in densities × 10^3 cells/mm^2. The black arrow indicates the orientation of the retinas. T = temporal, V = ventral. Scale bars: 1 mm.

27 cycles per degree; Collin and Pettigrew 1989), it seems well suited to the relatively small size of their territories, feeding at close range, and the close-range type of interactions that they partake in with conspecifics and intruders.

As in most vertebrates, *A. akindynos* ganglion cell densities vary greatly across the retina, forming regions of high cell density, often referred to as retinal specializations, that provide greater acuity in a specific part of the visual field of the animal (Figure 9.2F). Two main types of specializations are usually identified in vertebrates: areas, defined by a concentric increase of cells, and horizontal streaks, characterized by an elongated increase in cell density along the retinal meridian. In coral reef fishes, the type of specialization found in a particular species often correlates with the structure of its habitat. Areas are mostly found in species living in highly structured and enclosed environments where the sand-water interface is interrupted while horizontal streaks are usually found in species inhabiting open environments with an uninterrupted view of the sand-water interface (Collin and Pettigrew 1988a, 1988b). In the case of anemonefishes, area specializations are expected as their habitat is relatively small and enclosed, living in symbiosis with their host anemone. Surprisingly, in *A. akindynos*, irrespective of sex and life stages, ganglion cells form a well-defined horizontal streak with a peak cell density in the temporo-central part of the retina (Figure 9.2F). Since anemonefish are highly territorial, a horizontal streak may allow them to scan a broad range of their environment without distinctive head movements and within the safety of their anemone to facilitate the detection of intruders from a relatively long distance (Figure 9.3). Furthermore, the peak density of ganglion cells found in the central retina, which provides higher acuity to the visual field on either side of the fish, may be advantageous in assessing a conspecific's body size, and therefore social status, while swimming side by side at the beginning of an encounter (Stieb et al. 2019). Therefore, this streak specialization likely reflects the peculiar behavior and highly structured social hierarchy of anemonefishes.

Cone photoreceptor density in *A. akindynos* also varies greatly across the retina to form a horizontal streak. However, compared to the ganglion cells, the streak is wider and extends vertically into the temporal part of the retina (Figure 9.2G). This vertical component in cone distribution may provide greater sensitivity in the field of view situated directly above and below the fish to potentially help with the detection of predators/intruders situated above or below them. It may also facilitate the visualization of the vertical body stripes used for species recognition (Stieb et al. 2019).

It is worth noting that most of the work mentioned in this section derives from a single anemonefish species, *A. akindynos*. Since most anemonefishes have similar ecologies in terms of habitat, diet, and behavior we expect little interspecific variation in the general morphology of their visual systems. However, as discussed next, since species differ in color and patterns it is likely that their vision differs at the molecular and physiological levels.

9.4 THE MOLECULAR BASIS FOR VISION IN ANEMONEFISHES

As for all vertebrates, at the core of the anemonefish visual system are a class of membrane-bound G-protein-coupled opsin receptors that are situated in the outer segments of the

FIGURE 9.3 Coral reef and anemone habitat (bottom row) and an individual *Amphiprion akindynos* (top row) as seen by an 8 cm *A. akindynos* with a spatial resolving power of 5.8 cycles per degree from several viewing distances. The R package AcuityView (Caves and Johnsen 2018) was used to generate the images using a Fourier transformation approach to adjust the spatial resolution of the image according to the visual acuity of the receiver from a given distance. In AcuityView, these images were assumed to be 86.24 cm (bottom row) and 11.47 cm across (top row). Images courtesy of Justin Marshall.

photoreceptors (Hunt et al. 2014). Opsin proteins together with a vitamin A–derived chromophore retinal (the light-sensitive unit) form the visual pigment, which absorbs photons of light from the environment and initiates the phototransduction cascade ultimately propagating the signal to the central nervous system. The type of chromophore and the angle at which it is bound to the opsin protein determines the peak spectral sensitivity of the visual pigment (λ_{max}) (Cronin et al. 2014). Five different visual opsin types, which are likely to have evolved in the vertebrate ancestor (Lamb 2020; Larhammar et al. 2009; Fain 2020), exist in anemonefishes (Mitchell et al. 2021a; Stieb et al. 2019). These five types can be separated based on their genealogy, their λ_{max} values, and their photoreceptor specificity into one rod-specific opsin (rhodopsin, RH1, λ_{max} for anemonefishes 491–499 nm) and four cone-specific opsins: two short-wavelength-sensitive opsins, SWS1 (ultraviolet, 356–370 nm) and SWS2 (violet-blue, 406–407 nm), middle-wavelength-sensitive rhodopsin-like opsin RH2 (blue-green, 497–528 nm), and long-wavelength-sensitive opsin LWS (red, 553–561 nm) (Mitchell et al. 2021a). Opsin genes in anemonefishes, like in most teleosts (Musilova et al. 2021), have continued to evolve through gene duplication, gene loss, mutation, and gene conversion. Indeed, a detailed search in eleven anemonefish species revealed a single rod opsin and between six to seven cone opsins within their genomes (Mitchell et al. 2021a) (Figure 9.4A).

Interestingly, in all anemonefish species, UV-sensitive *SWS1* duplications (*SWS1α*, *SWS1β*) can be found (Mitchell et al. 2021a). *SWS1* duplications are rare in bony fishes and have previously only been found in a few species (Lin et al. 2017; Musilova et al. 2019; Rennison et al. 2012). The significance of the anemonefish *SWS1* duplicates remains to be investigated. However, as was suggested for the Ayu, *Plecoglossus altivelis*, the paralogs might be differentially expressed between seasons or during development (Minamoto and Shimizu 2005). As discussed in more detail in the following, they may also assist in UV communication or during feeding (Mitchell et al. 2021a; Stieb et al. 2019).

A duplication of *RH2A* (*RH2A-1*, *RH2A-2*) was also found in four anemonefish species: *A. percula*, *A. ocellaris*, *A. perideraion*, and *A. akallopisos* (Mitchell et al. 2021a) (Figure 9.4A). *RH2* genes show the highest evolutionary dynamics in fishes, which is likely due to their sensitivity to the centre blue-green part of the light spectrum; the most common light underwater (Musilova and Cortesi 2021; Musilova et al. 2021). However, the reason for having an extra duplicate in these four species and what function the paralogs may have remains unknown.

Opsin gene expression data is available for *A. akindynos* (wild-caught, i.e., under natural lighting) and *A. ocellaris* (captive bred, i.e., under artificial aquarium lighting), representing the two major anemonefish clades (Figure 9.4B). Interestingly, for both species, no difference in expression was found between the dominant female and male, and between the breeding fish and immature individuals in the case of *A. akindynos*. Both species express *RH1* and five to six cone opsins: *SWS1*, *SWS2B*, *RH2A* (predominantly *RH2A-1* for *A. ocellaris*), *RH2B*, and *LWS* (Mitchell et al. 2021a; Stieb et al. 2019). *RH2B* and *RH2A* are highly expressed in opposite members of the double cones (*RH2A* is sometimes also coexpressed with *LWS*), and maximally sensitive to blue (RH2B, ~498 nm λ_{max}) and green (RH2A, ~516–528 nm λ_{max}) light (Mitchell et al. 2021a; Stieb et al. 2019) (Figure 10.4C, D). This is a common characteristic amongst coral reef fish double cones that match the blue-green-light-dominated environment found on coral reefs (Marshall et al. 2003; Losey et al. 2003; Marshall et al. 2019; Cortesi et al. 2020) (Figure 9.1C).

The single cone gene expression varied more substantially between the captive-reared *A. ocellaris* and the wild-caught *A. akindynos*. Under artificial aquarium light, *A. ocellaris* coexpresses *SWS1β* with *SWS2B* throughout its retina, while under natural lighting *A. akindynos* expresses *SWS1* (the two paralogs were not differentiated) throughout the retina with only a small area located dorso-temporally that coexpresses *SWS2B* in the same cones (Stieb et al. 2019; Mitchell et al. 2021a) (Figure 9.4E). UV light is common on shallow coral reefs and almost all damselfishes have UV-reflecting skin patterns, which is likely to serve as a secret communication channel hidden away from UV-blind predators (Stieb et al. 2017; Siebeck 2004; Siebeck et al. 2010). The white bars of anemonefishes are highly UV-reflective (Figure 9.1A, B), and thus are potentially very conspicuous to a fish with UV vision when seen from up close. Moreover, when modelling the world through the eyes of an *A. akindynos*, the coexpression of *SWS1* with *SWS2B* is predicted to further increase the color contrast between orange and white stripes when seen against the water column or the host anemone (Stieb et al. 2019). As anemonefishes readily feed on UV absorbing or reflecting zooplankton, having a UV receptor is also likely to assist in spotting their favourite meal (Stieb et al. 2017). UV-sensitivity has previously been shown to increase feeding success in several freshwater fishes (e.g., Novales Flamarique 2013; Yoshimatsu et al. 2020; Loew et al. 1993; Rick et al. 2012), and indeed larval *Premnas biaculeatus* are able to feed on zooplankton when only UV light is available (Job and Bellwood 2007).

9.5 ANEMONEFISH VISUAL BEHAVIOR AND WHERE TO GO FROM HERE

Although we are slowly learning about the molecular and anatomical features of coral reef fish visual systems, much is still unknown about how vision and color guide specific behaviors on the reef (Cortesi et al. 2020; Marshall et al. 2019). Anemonefishes are ideal to investigate the visual ecology of reef fishes as they have well-developed visual senses that are representative of many smaller reef fishes including their ability to see in the UV. They vary greatly in body patterns, diet, and social network structure between species. They also perform several interesting behaviors

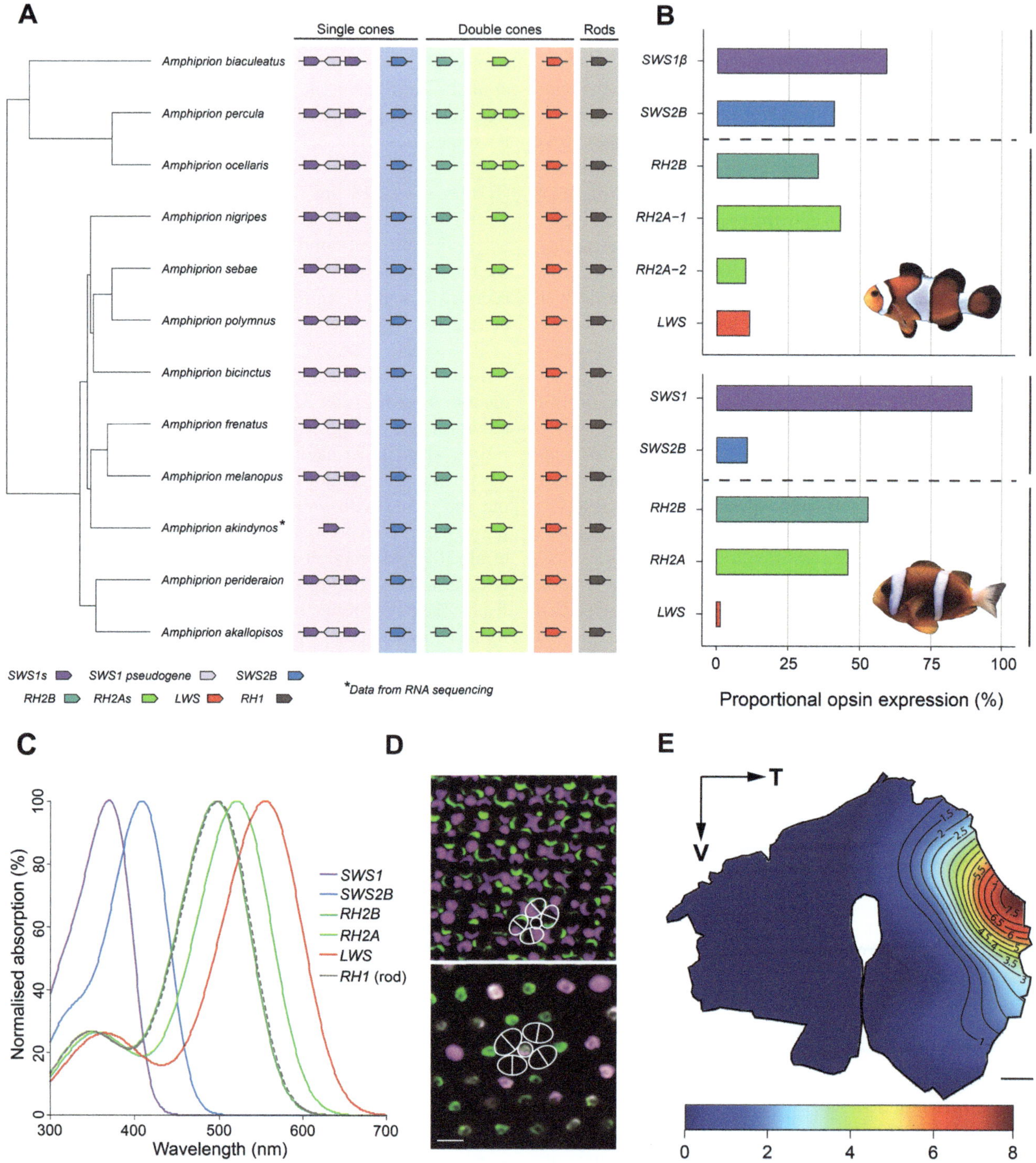

FIGURE 9.4 Opsin gene evolution, gene expression and spectral sensitivity in anemonefishes. A) Anemonefish genomes contain members of all four vertebrate cone opsin types and a single-rod opsin (*RH1*). B) Cone opsin gene expression in *Amphiprion ocellaris* (top) and *A. akindynos* (bottom) shows that members of all cone opsins are expressed to different degrees in the two species. Note that, the proportional single and double cone opsin expression is displayed separately. C) Chromophore A_1-based spectral absorbance curves in *A. akindynos* for cone- and rod opsin-based visual pigments. D) Double label fluorescent in-situ hybridization (FISH) in the *A. akindynos* retina. Top, *RH2A* (green) and *RH2B* (magenta) are always expressed in opposite members of double cones. Bottom, *SWS1* (green) is expressed in all single cones and sometimes coexpressed with *SWS2B* (magenta). E) Retinal wholemount of the FISH labelled *SWS2B* expressing single cones in *A. akindynos* shows a small dorso-temporal area of coexpression (i.e., the area of the eye that looks forward and downward). The black lines represent iso-density contours and values are expressed in densities $\times 10^3$ cells/mm^2. The black arrow indicates the orientation of the retinas. T = temporal, V = ventral. Scale bar: 1 mm. Opsin gene/protein acronyms stand for: *SWS1 or 2*, short-wavelength-sensitive 1 or 2; *RH2*, rhodopsin-like 2; *LWS*, long-wavelength-sensitive; *RH1*, rhodopsin (rod opsin). Panels A, B (top), adapted with permission from Mitchell et al. (2021a); B (bottom)–F, adapted with permission from Stieb et al. (2019).

including the characteristic up and down "bobbing" movement that might serve to advertise stinging anemones. Anemonefishes are also ideal to study the visual development of reef fishes as they are relatively easy to keep and breed in captivity. While more observational and experimental studies in the field are clearly needed, the recent development of a UV-Violet-RGB screen has shown that *A. ocellaris* is trainable in the laboratory to peck at colorful displays including in the UV (Powell et al. 2021). This enables future work to test specific aspects of anemonefish vision including their chromatic and achromatic thresholds to determine their color vision space, i.e., their visual capabilities in the UV and human visible range. Recently, reverse-genetic methods have also been developed for *A. ocellaris* (Mitchell et al. 2021b; Yamanaka et al. 2021), enabling the manipulation of specific photoreceptors (e.g., the ablation of *SWS1* paralogs, or genetic elements of the visual pathway to test their function in anemonefish vision) (see Chapter 5 for details). Together with an increased understanding of their evolutionary history and biology in general, this opens exciting new avenues to study all aspects of vision in anemonefishes and reef fishes in general.

ACKNOWLEDGEMENTS

We would like to acknowledge the Traditional Owners of the lands on which much of the reported work has been conducted. This work was supported by two Australian Research Council DECRA Fellowships to F.C. (DE200100620) and F.d.B (DE180100949).

REFERENCES

Baden, T. 2020. Vertebrate Vision: Lessons from Non-Model Species. *Seminars in Cell & Developmental Biology* 106:1–4.

Carleton, K. L., D. Escobar-Camacho, S. M. Stieb, F. Cortesi, and N. J. Marshall. 2020. Seeing the Rainbow: Mechanisms Underlying Spectral Sensitivity in Teleost Fishes. *Journal of Experimental Biology* 223(8):jeb193334.

Caves, E. M., and S. Johnsen. 2018. AcuityView: An R Package for Portraying the Effects of Visual Acuity on Scenes Observed by an Animal. *Methods in Ecology and Evolution* 9(3):793–97.

Collin, S. P., and J. D. Pettigrew. 1988a. Retinal Topography in Reef Teleosts I. Some Species with Well-Developed Areae but Poorly-Developed Streaks. *Brain, Behavior and Evolution* 31(5):269–82.

Collin, S. P., and J. D. Pettigrew. 1988b. Retinal Topography in Reef Teleosts II. Some Species with Prominent Horizontal Streaks and High-Density Areae. *Brain, Behavior and Evolution* 31(5):283–95.

Collin, S. P., and J. D. Pettigrew. 1989. Quantitative Comparison of the Limits on Visual Spatial Resolution Set by the Ganglion Cell Layer in Twelve Species of Reef Teleosts. *Brain, Behavior and Evolution* 34(3):184–92.

Collin, S. P., and J. Shand. 2003. Retinal Sampling and the Visual Field in Fishes. In *Sensory Processing in Aquatic Environments*, eds. S. P. Collin and N. J. Marshall, 139–69. New York: Springer.

Cortesi, F., L. J. Mitchell, V. Tettamanti, L. G. Fogg, F. de Busserolles, K. L. Cheney, and N. J. Marshall. 2020. Visual System Diversity in Coral Reef Fishes. *Seminars in Cell & Developmental Biology* 106:31–42.

Cronin, T. W., S. Johnsen, N. J. Marshall, and E. J. Warrant. 2014. *Visual Ecology*. Princeton: Princeton University Press.

de Busserolles, F., F. Cortesi, L. Fogg, S. M. Stieb, M. Luehrmann, and N. J. Marshall. 2021. The Visual Ecology of Holocentridae, a Nocturnal Coral Reef Fish Family with a Deep-Sea-Like Multibank Retina. *Journal of Experimental Biology* 224(1):jeb233098.

Fain, G. L. 2020. Lamprey Vision: Photoreceptors and Organization of the Retina. *Seminars in Cell & Developmental Biology* 106:5–11.

Fautin, D. Gail, and G. R. Allen. 1997. *Anemonefishes and Their Host Sea Anemones: A Guide for Aquarists and Divers*. Perth: Western Australian Museum.

Foo, S. A., L. Liddell, A. Grossman, and K. Caldeira. 2020. Photo-Movement in the Sea Anemone Aiptasia Influenced by Light Quality and Symbiotic Association. *Coral Reefs* 39(1):47–54.

Hunt, D. M., M. W. Hankins, S. P. Collin, and N. J. Marshall. 2014. *Evolution of Visual and Non-Visual Pigments*. Vol. 4. New York: Springer.

Ivanov, I. V., T.Mappes, P. Schaupp, C. Lappe, and S. Wahl. 2018. Ultraviolet Radiation Oxidative Stress Affects Eye Health. *Journal of Biophotonics* 11(7):e201700377.

Jerlov, N. G. 1976. *Marine Optics*. Amsterdam: Elsevier.

Job, S., and D. R. Bellwood. 2007. Ultraviolet Photosensitivity and Feeding in Larval and Juvenile Coral Reef Fishes. *Marine Biology* 151(2):495–503.

Klann, M., M. Mercader, L. Carlu, K. Hayashi, J. D. Reimer, and V. Laudet. 2021. Variation on a Theme: Pigmentation Variants and Mutants of Anemonefish. *EvoDevo* 12(1):1–16.

Lamb, T. D. 2020. Evolution of the Genes Mediating Phototransduction in Rod and Cone Photoreceptors. *Progress in Retinal and Eye Research* 76:100823.

Larhammar, D., K. Nordström, and T. A. Larsson. 2009. Evolution of Vertebrate Rod and Cone Phototransduction Genes. *Philosophical Transactions of the Royal Society B: Biological Sciences* 364(1531):2867–80.

Lin, J.-J., F.-Y. Wang, W.-H. Li, and T.-Y. Wang. 2017. The Rises and Falls of Opsin Genes in 59 Ray-Finned Fish Genomes and Their Implications for Environmental Adaptation. *Scientific Reports* 7(1):15568.

Loew, E. R., W. N. McFarland, E. L. Mills, and D. Hunter. 1993. A Chromatic Action Spectrum for Planktonic Predation by Juvenile Yellow Perch, *Perca flavescens*. *Canadian Journal of Zoology* 71(2):384–86.

Losey, G. S., W. N. McFarland, E. R. Loew, J. P. Zamzow, P. A. Nelson, and N. J. Marshall. 2003. Visual Biology of Hawaiian Coral Reef Fishes. I. Ocular Transmission and Visual Pigments. *Copeia* 2003(3):433–54.

Luehrmann, M., F. Cortesi, K. L. Cheney, F. de Busserolles, and N. J. Marshall. 2020. Microhabitat Partitioning Correlates with Opsin Gene Expression in Coral Reef Cardinalfishes (Apogonidae). *Functional Ecology* 34(5):1041–52.

Marshall, N. J., F. Cortesi, F. de Busserolles, U. E. Siebeck, and K. L. Cheney. 2019. Colours and Colour Vision in Reef Fishes: Past, Present and Future Research Directions. *Journal of Fish Biology* 95(1):5–38.

Marshall, N. J., K. Jennings, W. N. McFarland, E. R. Loew, and G. S. Losey. 2003. Visual Biology of Hawaiian Coral Reef Fishes. III. Environmental Light and an Integrated Approach to the Ecology of Reef Fish Vision. *Copeia* 2003 (3):467–80.

McFarland, W. N. 1991. The Visual World of Coral Reef Fishes. In *The Ecology of Fishes on Coral Reefs*, ed. P. F. Sale. San Diego, CA: Academic Press, Inc.

Minamoto, T., and I. Shimizu. 2005. Molecular Cloning of Cone Opsin Genes and Their Expression in the Retina of a Smelt, Ayu (*Plecoglossus altivelis*, Teleostei). *Comparative Biochemistry and Physiology Part B: Biochemistry and Molecular Biology* 140(2):197–205.

Mitchell, L. J., K. L. Cheney, M. Luehrmann, N. J. Marshall, K. Michie, and F. Cortesi. 2021a. Molecular Evolution of Ultraviolet Visual Opsins and Spectral Tuning of Photoreceptors in Anemonefishes (Amphiprioninae). *Genome Biology and Evolution* 13(10):evab184.

Mitchell, L. J., V. Tettamanti, J. S. Rhodes, N. J. Marshall, K. L. Cheney, and F. Cortesi. 2021b. CRISPR/Cas9-Mediated Generation of Biallelic F0 Anemonefish (*Amphiprion ocellaris*) Mutants. *PLoS ONE* 16(12):e0261331.

Muntz, W. R. A. 1973. Yellow Filters and the Absorption of Light by the Visual Pigments of Some Amazonian Fishes. *Vision Research* 13(12):2235–54.

Munz, F. W., and W. N. McFarland. 1977. Evolutionary Adaptations of Fishes to the Photic Environment. In *The Visual System in Vertebrates*, eds. F. Crescitelli, C. A. Dvorak, D. J. Eder, A. M. Granda, D. Hamasaki, K. Holmberg, A. Hughes, et al. 193–274. *Handbook of Sensory Physiology*. Berlin, Heidelberg: Springer.

Musilova, Z., and F. Cortesi. 2021. Multiple Ancestral and a Plethora of Recent Gene Duplications during the Evolution of the Green Sensitive Opsin Genes (*RH2*) in Teleost Fishes. *BioRxiv*, May. Cold Spring Harbor Laboratory, 2021.05.11.443711.

Musilova, Z., F. Cortesi, M. Matschiner, W. I. L. Davies, J. Suresh Patel, S. M. Stieb, F. de Busserolles, et al. 2019. Vision Using Multiple Distinct Rod Opsins in Deep-Sea Fishes. *Science* 364(6440):588–92.

Musilova, Z., W. Salzburger, and F. Cortesi. 2021. The Visual Opsin Gene Repertoires of Teleost Fishes: Evolution, Ecology, and Function. *Annual Review of Cell and Developmental Biology* 37(1):441–68.

Novales, F. I. 2013. Opsin Switch Reveals Function of the Ultraviolet Cone in Fish Foraging. *Proceedings of the Royal Society B: Biological Sciences* 280(1752):20122490.

Parker, A. N., K. A. Fritsches, C. Newport, G. Wallis, and U. E. Siebeck. 2017. Comparison of Functional and Anatomical Estimations of Visual Acuity in Two Species of Coral Reef Fish. *Journal of Experimental Biology* 220(13):2387–96.

Powell, S., L. J. Mitchell, A. M. Phelan, F. Cortesi, J. N. Marshall, and K. L. Cheney. 2021. A 5-Channel LED Display to Investigate UV Perception. *Methods in Ecology and Evolution* 12(4):602–607.

Rennison, D. J., G. L. Owens, and J. S. Taylor. 2012. Opsin Gene Duplication and Divergence in Ray-Finned Fish. *Molecular Phylogenetics and Evolution* 62(3):986–1008.

Rick, I. P., D Bloemker, and T. C. M. Bakker. 2012. Spectral Composition and Visual Foraging in the Three-Spined Stickleback (Gasterosteidae: *Gasterosteus aculeatus* L.): Elucidating the Role of Ultraviolet Wavelengths. *Biological Journal of the Linnean Society* 105(2):359–68.

Roux, N., P. Salis, S-H. Lee, L. Besseau, and V. Laudet. 2020. Anemonefish, a Model for Eco-Evo-Devo. *EvoDevo* 11(1):1–11.

Siebeck, U. E. 2004. Communication in Coral Reef Fish: The Role of Ultraviolet Colour Patterns in Damselfish Territorial Behaviour. *Animal Behaviour* 68(2):273–82.

Siebeck, U. E. 2007. Potential Ultraviolet Vision in Pre-Settlement Larvae and Settled Reef Fish - A Comparison across 23 Families. *Vision Research* 47(17):2337–52.

Siebeck, U. E., and N. J. Marshall. 2001. Ocular Media Transmission of Coral Reef Fish - Can Coral Reef Fish See Ultraviolet Light? *Vision Research* 41(2):133–49.

Siebeck, U. E., A. N. Parker, D. Sprenger, L. M. Mäthger, and G. Wallis. 2010. A Species of Reef Fish that Uses Ultraviolet Patterns for Covert Face Recognition. *Current Biology: CB* 20(5):407–10.

Stieb, S. M., F. de Busserolles, K. L. Carleton, F. Cortesi, W.-S. Chung, B. E. Dalton, L. A. Hammond, and N. J. Marshall. 2019. A Detailed Investigation of the Visual System and Visual Ecology of the Barrier Reef Anemonefish, *Amphiprion akindynos*. *Scientific Reports* 9(1):1–14.

Stieb, S. M., K. L. Carleton, F. Cortesi, N. J. Marshall, and W. Salzburger. 2016. Depth Dependent Plasticity in Opsin Gene Expression Varies between Damselfish (Pomacentridae) Species. *Molecular Ecology* 25(15):3645–61.

Stieb, S. M., F. Cortesi, L. Sueess, K. L. Carleton, W.Salzburger, and N. J. Marshall. 2017. Why UV Vision and Red Vision Are Important for Damselfish (Pomacentridae): Structural and Expression Variation in Opsin Genes. *Molecular Ecology* 26(5):1323–42.

Walls, G. 1942. *The Vertebrate Eye and Its Adaptive Radiation*. Bloomfield, MI: Cranbrook Science.

Warrant, E. J., and N. A. Locket. 2004. Vision in the Deep Sea. *Biological Reviews* 79(3):671–712.

Yamanaka, S., Y. Okada, T. Furuta, and M. Kinoshita. 2021. Establishment of Culture and Microinjection Methods for False Clownfish Embryos without Parental Care. *Development, Growth & Differentiation* 63(9):495–66.

Yoshimatsu, T., C. Schröder, N. E. Nevala, P. Berens, and T. Baden. 2020. Fovea-Like Photoreceptor Specializations Underlie Single UV Cone Driven Prey-Capture Behavior in Zebrafish. *Neuron* 107(2):320–37.

Zigman, S. 1993. Ocular Light Damage. *Photochemistry and Photobiology* 57(6):1060–68.

10 Sound Communication

Eric Parmentier and David Lecchini

CONTENTS

10.1 INTRODUCTION

Today, fish acoustic communication is considered an important aspect of teleost social behavior across a wider taxonomic spectrum since fish sounds have been reported in many different unrelated taxa (Parmentier et al. 2021; Lobel et al. 2010). Acoustic signals mediate fish social interactions in a wide range of activities such as distress or alarm situations, conspecific identification, courtship and agonistic interactions, mate choice, mate quality assessment, and coordination of gamete release (Amorim et al. 2015).

Damselfishes are a well-known vocal species from coral reefs. Some species are not only able to make sounds; they can also emit different kinds of sounds that are produced in various behavioral contexts (Mann and Lobel 1998; Parmentier et al. 2010; Parmentier et al. 2016). To date, sounds have been recorded and analysed in various species from the genera *Abudefduf*, *Amphiprion*, *Chromis*, *Dascyllus*, *Plectroglyphidodon*, *Pomacentrus*, and *Stegastes* (Parmentier et al. 2016). Sounds were also reported but not analyzed in *Hypsypops* (Limbaugh 1964; Fish and Mowbray 1970), *Microspathodon* (Emery 1973), and *Chrysiptera* (Graham 1992). Most pomacentrid sounds are a series of short-duration pulses. In that kind of series, one can measure the duration and number of pulses in the series, pulse period (time between the start of one pulse and the next), the related pulse repetition rate (number of pulses per unit time), interpulse interval (the silent period between pulses), pulse duration, and the frequency or power spectrum (Myrberg et al. 1978; Spanier 1979; Lobel and Mann 1995). All these characters do not carry the same kind of information and the physical properties of the acoustic environment can affect the cues in different ways during sound propagation. Features such as pulse periods are the most important in order to discriminate the pomacentrid sounds (Mann and Lobel 1997). Other parameters may not be relevant for interspecific comparisons. For example, the dominant frequency and the pulse duration are only related to fish size (Colleye et al. 2009; Myrberg et al. 1993), not to the species. Moreover, the number of pulses in a sound could be simply owing to its motivational state (Parmentier et al. 2010).

10.2 PRODUCTION OF DIFFERENT TYPES OF SOUNDS

Collingwood (1868) was the first to report in a scientific textbook the unusual phenomenon of a fish living in association with tropical sea anemones. As this intimate relationship between anemonefishes and their invertebrate hosts is the more glamorous aspect of their general biology, the considerable emphasis placed on this topic has tended to obscure other equally interesting specificities about the behaviors of these fish. In recent years, attention has turned to other aspects of the life history of anemonefishes. Any diver that has attempted to approach an anemonefish has experienced how it can rush toward intruders, making rapid nodding movements with the head, opening and closing jaws with convulsive jerks. These movements are related to sound production, audible to human ears at a distance of a meter or more (Moyer and Sawyers 1973) showing that these fish can produce volitional sounds. The present chapter aims at synthesizing knowledge about acoustic communication in anemonefishes. Previous reviews can be found in the book entitled *Biology of Damselfishes* (Frédérich and Parmentier 2016).

Literature on sound production in anemonefishes can be traced back to as early as 1930 when Verwey stated that *A. akallopisos* and *A. polymnus* were able to produce sounds (Verwey 1930). The sounds, which were clearly audible to the human ear, were mainly associated with agonistic activity. They were emitted by the fish in conjunction with both threat and submissive postures. Then, Schneider studied sound production in *A. clarkii*, *A. polymnus*, *A. frenatus*, and *A. percula* (Schneider 1964). He documented three types of sounds: threatening, fighting, and shaking sounds. Threatening sounds were used to intimidate other fish from a large distance, while fighting sounds were produced when attacking other specimens. Both types of sounds are the same but fighting sounds usually possess a single unit. According to the author, shaking sounds are "by contrast very different and were produced by the attacked fish" (Schneider 1964).

Unfortunately, this first study by Schneider revealed few detailed data about the acoustic features of vocalizations. Later, Allen recorded sounds for *A. chrysopterus* and *A.*

DOI: 10.1201/9781003125365-12

perideraion, both in the field and in the laboratory (Allen 1972). He differentiated two distinct sounds he called "clicks" and "grunts". Allen postulated clicks probably correspond to threatening sounds and can be emitted alone or in series of three to 15 pulses. However, he also noted that grunts are emitted in conjunction with threat postures by resident fish after new *Amphiprion* specimens were released in the tank. He did not record grunts in the field and did not detect the shaking sound observed by Schneider. His description and sonagraph do not provide enough data to distinguish grunts and clicks. We suspect the clicks and grunts reported by Allen are the same sounds because recordings made in a tank can distort sounds (Parmentier et al. 2014; Akamatsu et al. 2002). In *A. clarkii* and *A. frenatus*, Chen and Mok (1988) noted shaking sounds corresponded to submissive displays with the belly facing the dominant recipient. In a detailed study on *A. frenatus*, submissive sounds were produced when subordinates displayed submissive posture as a reaction to charge and chase by dominants, which means that these sounds were never recorded from dominant females (Colleye and Parmentier 2012). From all these studies, there are two types of sounds produced by the anemonefishes. Threatening and fighting sounds form the first group. The sound-producing mechanism of these sounds is most probably the same way because the major difference between them is only the number of pulses. In different pomacentrid species, a lower number of pulses is usually found during fighting than during threatening behaviors (Parmentier et al. 2010; Mann and Lobel 1998; Parmentier et al. 2021). The shaking sounds constitute the second group of sounds and correspond to a submissive behavior. In *A. frenatus*, a comparative study between threatening and shaking sounds allows to better distinguish them (Colleye and Parmentier 2012). In comparison to aggressive sounds, shaking sounds are always composed of several pulses forming a unit that can be produced alone or in series, whereas aggressive sounds are composed of a single pulse that can be emitted alone or in series (Figure 10.1). Consequently, aggressive sounds can be made of a single pulse which is not the case for submissive sounds (Figure 10.1). In *A. frenatus*, shaking sounds also exhibit significantly shorter pulse periods (12 ms versus 106 ms) and shorter pulse durations (8 ms versus 14 ms) than aggressive sounds (Colleye and Parmentier 2012). However, the frequency range of both kinds of sounds is equivalent. Although we do not have data from calibrated hydrophones to support our claims, shaking sounds clearly possess a lower amplitude than aggressive sounds. Aggressive sounds can be easily recorded in the field but shaking sounds are quite hard to distinguish from background noise and have never been recorded in situ. These sounds (and corresponding submissive behavior) were recorded only in tanks with low background noise conditions.

10.3 RELATIONSHIPS BETWEEN SOUNDS AND BEHAVIORS

The implication of acoustic signals in agonistic interactions may be a simple strategy to avoid conflicts, which otherwise might escalate to a severe outcome (Colleye and Parmentier 2012). In fact, in tank experiments, where specimens cannot escape, confrontations always start with sound production and charges before physical attacks that can lead to death. Between conspecifics, aggressive sounds are mainly produced by dominant individuals during charges, chases, and threat displays during agonistic interactions (Takemura 1983; Colleye et al. 2009). They can be produced by individuals of different sexual statuses (females, males, and non-breeders) during charge-and-chase displays when another con- or heterospecific (including humans!) approaches the sea anemone in which they dwell (Colleye et al. 2009). These sounds were first attributed to territory defence against hetero- or conspecifics (Schneider 1964; Allen 1972). The reason sounds are produced could however be more complex. In the fight against heterospecifics, the intruder can be deterred. However, in the fight between conspecifics, the confrontation does not always end with the departure of one antagonist since specimens can share the same host. In this case, the aggressive behavior does not correspond to physical territory defence.

Anemonefishes live in social groups composed of a breeding pair accompanied by no or several non-breeders. Group members are not related and non-breeders do not provide alloparental care (Buston 2004a; 2004b; Buston et al. 2007). Within each group, there is a size hierarchy: the female is the largest individual, the male is the second largest, and non-breeders get progressively smaller (Buston 2003). The size hierarchy represents the queue to become a breeding member: if the female of the group dies, then the male changes sex (Casadevall et al. 2009; Casas et al. 2016) and becomes the new female; the largest non-breeder becomes the new male (Buston 2004). As all individuals grow, the smallest individual is always the last recruit. In this system, without predation, individuals are thought to wait peacefully to inherit breeding positions following the death of the breeders (Branconi et al. 2020). In *A. percula*, individuals adjacent in rank are separated by body size ratios: the growth of individuals is regulated so that each dominant ends up being about 1.26 times the size of its immediate subordinate (Buston and Cant 2006). The same kind of ratio (< 1.30) is observed in the different groups of *A. frenatus* (Colleye and Parmentier 2012). In this small society, numerous agonistic interactions occur and appear to play an important role in maintaining these observed size differences between individuals that are adjacent in rank (Fricke 1979; Buston 2003). Larger fishes chase smaller ones, which means that the smallest one is the recipient of numerous charges (Fricke 1979). These chasing behaviors are accompanied by sound production. In anemonefish species, the size hierarchy is perfectly mirrored in the acoustic features. In aggressive sounds, pulse duration and dominant frequency are highly correlated with standard length ($r = 0.97$): smaller individuals produce higher frequency and shorter duration pulses than larger individuals (Figure 10.2), irrespective of the sexual status (Colleye et al. 2009). Consequently, these sonic features might be useful

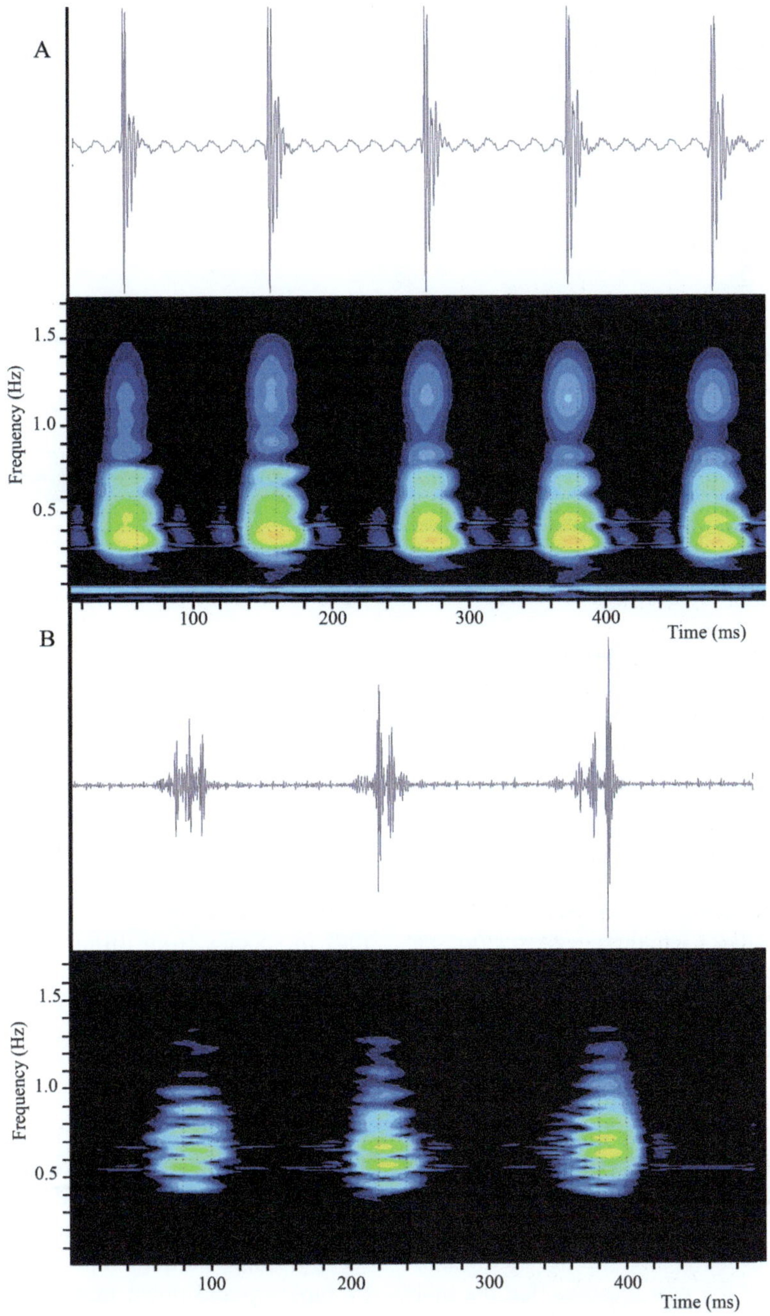

FIGURE 10.1 Oscillograms and spectrograms illustrating agonistic (A) and submissive (B) sounds produced by *Amphiprion frenatus* during interactions. Panels were placed at the same time scale to allow comparison. Note submissive sounds show a group of pulses with short periods whereas agonistic sounds possess longer periods between pulses. Redrawn from Colleye and Parmentier (2012).

cues for individual recognition within a group and may convey information on the social rank of the emitter within the group (Colleye et al. 2009; Colleye and Parmentier 2012). Additional studies should be conducted to determine experimentally whether a fish can use sounds to infer the size and establish the social hierarchy of conspecifics. Interestingly, the relationship between peak frequency and size is equivalent across the different clownfish species (Figure 10.3), supporting that the size-related vocal message should be the same within the taxa. A recent experimental study has used different sensory cues (mechanosensory [pressure and/or touch], auditory, chemosensory, and/or visual) to show that juvenile anemonefish likely require the use of mechanosensory (pressure and/or touch) cues to assess the size of conspecifics (Desrochers et al. 2020). However, the experimental design of these authors did not indicate if sounds were produced during the duration of the experiment. The function of the sound can thus be hardly assessed.

Sound production abilities are not restricted to anemonefish since they can also be found in many different pomacentrid taxa (Parmentier et al. 2016). In *Dascyllus* species, up to six different types of sounds have been described

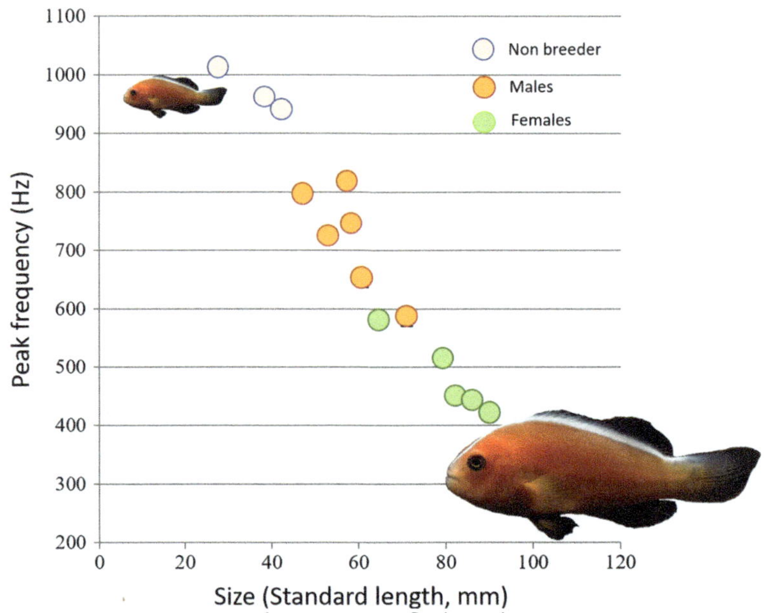

FIGURE 10.2 Relationships between standard length and peak frequency in *Amphiprion akallopisos*. The size ratio between both shown fish is respected. Redrawn from Colleye et al. (2009).

(Lobel and Mann 1995; Mann and Lobel 1998; Parmentier et al. 2010, 2021). As in anemonefish, other pomacentrid species also produce aggressive sounds during territory defence against conspecifics and heterospecifics, but also fighting sounds and submissive sounds. Moreover, they produce sounds during courtship. Many pomacentrid males produce signal jumps to attract females. In *Dascyllus* species, sounds are also emitted when females visit male nests. To date, sound production during reproductive periods has been reported in *A. ocellaris*, *A. frenatus*, and *A. sandaracinos* (Takemura 1983). However, these observations need to be carefully considered since, according to the author, these species are supposed to emit weak sounds with high-frequency components of more than 2 kHz during reproduction. Anemonefish are not able to hear these frequencies (Parmentier et al. 2009). Therefore, high frequencies (>2 kHz) have never been reported in other studies on the same species. The preparation of the nest requires that the male and female peck up the surface of the rock to clean it for correct egg adhesion. These high-pitched sounds could be a by-product of the nest cleaning activities, corresponding to gratings of teeth on the rocks. Cleaning rocks for spawning could originate different incidental sounds but it is not communication. Moreover, Takemura (1983) noted also in his description "sounds were not so closely connected with spawning, because these sounds were not always heard and were the same as usual sounds". In other words, sounds produced during the reproductive period are probably not related to reproduction or spawning. In addition, spawning events were also observed and audio-recorded in *A. akindynos*, *A. clarkii*, *A. perideraion*, *A. melanopus*, and *A. percula*. No sound has been recorded during a total of 13 complete spawning events (Colleye and Parmentier 2012). The lack of sounds during spawning events is probably an evolutionary outcome related to their peculiar way of life. In their restricted territory, partners do not need to make sounds to attract females. Within pomacentrids It is however very interesting to note that the same kind of sounds (=groups of pulses), produced with the same mechanism (Parmentier et al. 2016), can have different meanings in species from different genera (e.g., *Dascyllus* and *Amphiprion*). Additional studies using play-back experiments are however required to better understand how these sounds are used by these species.

10.4 MECHANISM

The mechanism of sound production has been discussed since the first report on sound production in Pomacentridae. The hypotheses were sprawling and sometimes quite contradictory. Some authors claimed that sounds were produced by rapid up-and-down movements of the opercula and by movements of the mouth bones (Verwey 1930). In *Abudefduf luridus*, Santiago and Castro (1997) hypothesized that sound production involves a swim bladder mechanism, but extrinsic muscles attached to the swim bladder were never found in Pomacentridae species. Schneider (1964) noted "neither the gill-teeth nor the teeth on the upper and lower jaw are engaged in sound production" but other authors reported later that sounds could be produced by grating pharyngeal teeth and then be amplified by the swim bladder (Luh and Mok 1986; Rice and Lobel 2003). In all recorded anemonefishes, call duration is related to the number of pulses, suggesting there is a fixed mechanism with a motor pattern. Moreover, the peak frequency (between 350 and 1,100 Hz, according to the fish size) is too low for typical stridulatory mechanisms and too high for swim bladder sounds driven as a forced response to sonic muscle contraction.

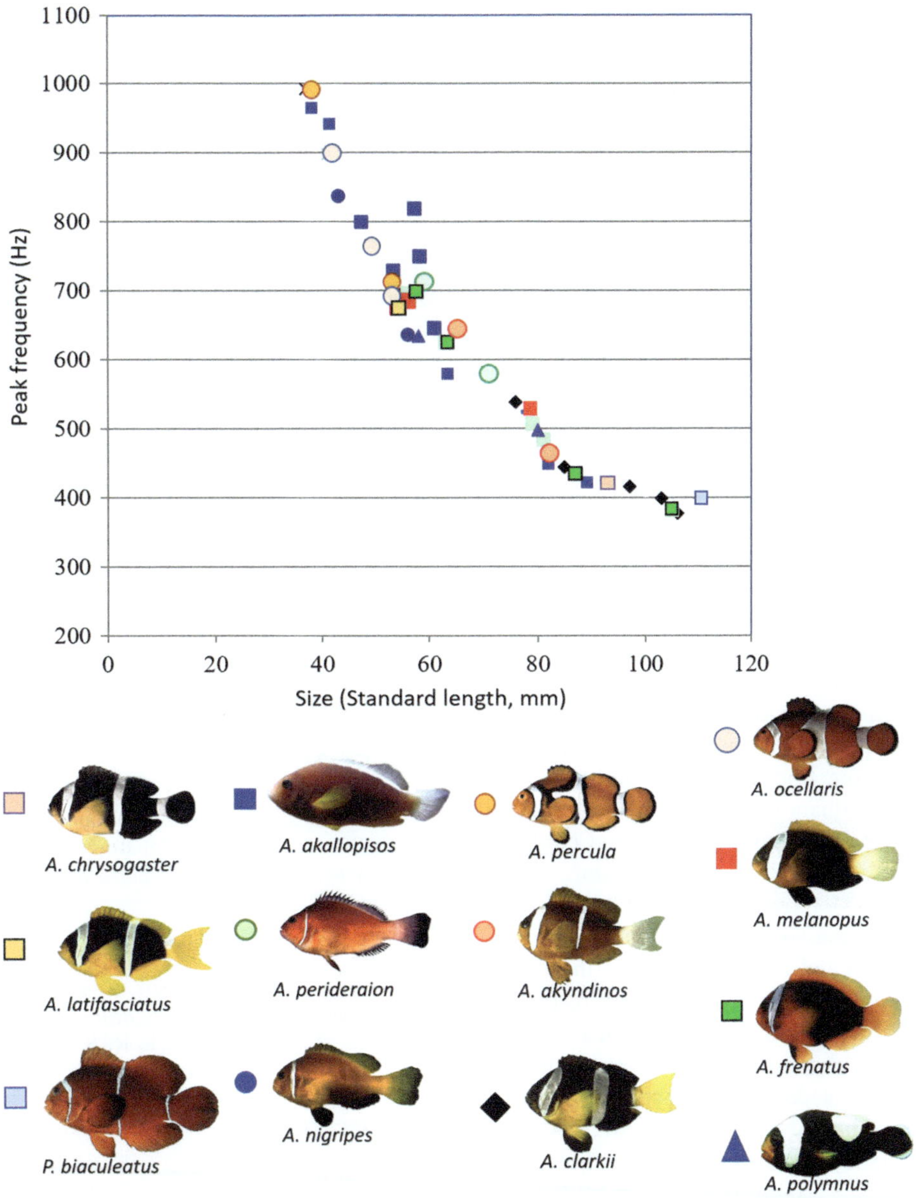

FIGURE 10.3 Correlation of peak frequency against fish size (SL) on acoustic variables in 13 clownfish species. Fishes ranged from 37 to 110 mm (n=43). Results are expressed as mean values of 50 recorded pulses for each individual.

Using manipulations of freshly euthanized fish and high-speed videos coupled (or not) with X-rays and synchronized with sound recordings, it has been shown that aggressive sounds emitted by the yellowtail anemonefish *A. clarkii* result from the teeth collision induced by a fast jaw slam (Parmentier et al. 2007; Damien Olivier et al. 2015). This rapid mouth closing movement is caused by the cerato-mandibular (c-md) ligaments (right and left) joining the hyoid bars (at the level of the ceratohyal) to the medial sides of the mandibles (at the level of the coronoid process of the angular), a synapomorphic trait of the Pomacentridae (Stiassny 1981; Olivier et al. 2014). Consequently, Pomacentridae are the only known teleosts that use that kind of mechanism to close their mouth. According to the review of Olivier et al. (2016), the kinematic pattern during sound production in *A. clarkii* can be divided into three phases: initial, mouth-opening, and mouth-closing. (1) During the initial phase, the mouth is closed, the neurocranium is held at rest, and the hyoid apparatus is not depressed. At this moment, the cerato-mandibular ligament is loose and does not transmit any tension to the lower jaw. (2) During the mouth-opening phase, the neurocranium is elevated and the hyoid apparatus is depressed causing the downward rotation of the lower jaw. As a result, the insertion points of the c-md ligament are moved away from one another, causing tension in the ligament. With accentuated depression of the hyoid apparatus, the c-md ligament acts as a cord, forcing the lower jaw to rotate clockwise around its quadrate articulation. It induces the mouth to close within a few ms. Comparisons with available data in the literature show that the mechanism of the c-md ligament provides the damselfish with one of the fastest mouth-closing mechanisms in teleosts (Olivier et al. 2015; Olivier et al. 2016). The transection of the c-md ligaments prevents sound production. This suite of events only explains the onset of the sounds. Using a combination of different approaches, further studies provided complementing explanations (Colleye and Parmentier 2012). In anemonefish, sound duration and frequency are known to be morphologically determined signals strongly related to fish size (Colleye et al. 2009; Parmentier et al. 2009; Colleye et al. 2011). This suggests that these acoustic features are subject to a morphological constraint. Considering the positive relationship existing between fish size and swim bladder volume, this organ could be the structure responsible for the size-related variations in acoustic features (Colleye et al. 2012). However, the swim bladder itself is too inefficient to act as a resonator. The walls of the swim bladder are rigidly attached to the articulated (and thus movable) ribs. The combination of ribs and swim bladder wall probably forms a structure analogous to a loudspeaker membrane. Vibrations of the rib cage could be the driver because they would provoke movements of this membrane, thus initiating sound production. The swim bladder wall is driven by bone movements. Experimental manipulation of the swim bladder confirmed its function related to sound production. Pulse duration and dominant frequency changed when filling the swim bladder with physiological liquid confirmed its function since it changes both the pulse duration and dominant frequency. Moreover, strikes of the rib cage with a hammer generated sounds with size-related variations in sound duration and frequency, suggesting that the vibrating properties of the rib cage might be responsible for the size-related variations observed in acoustic features (Colleye et al. 2012). Results of this kind were not found when striking the swim bladder wall, probably because this structure has high intrinsic damping (Fine 2012). All these experiments were mainly conducted using the anemonefish *A. clarkii*. However, since dominant frequency and pulse duration were strongly predicted by body size among 14 different anemonefish species, this highlights that all species use the same vocalization mechanism (Colleye et al. 2011). In conclusion to this part, we hypothesize that the pulse within a sound is initiated by a slam of the jaws. It provokes skeleton vibration at the origin of rib cage shaking. Close association of the rib cage with the swim bladder wall could constitute a structure analogous to a loudspeaker membrane causing the second part of the sound. The frequency and pulse duration are constrained by the size of the fish. Therefore, the only parameter that can be used to produce species-specific sounds has to be related to the motor pattern responsible for the fast mouth closing. It can vary at two levels: the number of slams and the speed (rhythm) between two slams, corresponding to the pulse period.

Lastly, because anemonefishes are confined to an anemone for habitat, and anemones are very sensitive to climate change (bleaching of its tissue due to thermal stress), the sound of anemonefish could be an indicator of coral reef health. Biodiversity assessment remains one of the most difficult challenges encountered by ecologists and conservation biologists, especially in hyper-diverse ecosystems such as coral reefs (Barnosky et al. 2011; Wilkinson et al. 2013). Biological sounds have been suggested as a means to quantify ecosystem health and biodiversity (Bertucci et al. 2015, 2016; Di Iorio et al. 2021; Mooney et al. 2021). By taking advantage of the sounds produced by clown species in healthy or bleached anemones, coral reef health could be monitored and surveyed.

REFERENCES

Akamatsu, T., T. Okumura, N. Novarini, and H. Y. Yan. 2002. Empirical Refinements Applicable to the Recording of Fish Sounds in Small Tanks. *The Journal of the Acoustical Society of America* 112: 3073–82.

Allen, G. R. 1972. *The Anemonefishes: Their Classification and Biology*. Neptune City: T.F.H. Publications.

Amorim, M. C. P., C. Conti, T. Modesto, A. Gonçalves, and P. J. Fonseca. 2015. Agonistic sounds signal male quality in the Lusitanian toadfish. *Physiology & Behavior* 149: 192–198.

Barnosky, A. D, N. Matzke, S. Tomiya, G. O. U. Wogan, B. Swartz, T. B. Quental, C. Marshall, et al. 2011. Has the Earth's Sixth Mass Extinction Already Arrived? *Nature* 471 (7336): 51–57.

Bertucci, F., E. Parmentier, L. Berten, R. M. Brooker, and D. Lecchini. 2015. Temporal and Spatial Comparisons of Underwater Sound Signatures of Different Reef Habitats in Moorea Island, French Polynesia. *PLoS ONE* 10 (9): e0135733.

Bertucci, F., E. Parmentier, G. Lecellier, A. D. Hawkins, and D. Lecchini. 2016. Acoustic Indices Provide Information on the Status of Coral Reefs: An Example from Moorea Island in the South Pacific. *Scientific Reports* 6: 33326.

Branconi, R., T. A. Barbasch, R. K. Francis, M. Srinivasan, G. P. Jones, and P. M. Buston. 2020. Ecological and Social Constraints Combine to Promote Evolution of Non-Breeding Strategies in Clownfish. *Communications Biology* 3 (1): 649.

Buston, P. 2004a. Does the Presence of Non-Breeders Enhance the Fitness of Breeders? An Experimental Analysis in the Clown Anemonefish *Amphiprion percula. Behavioral Ecology and Sociobiology* 57: 23–31.

Buston, P. M. 2003. Size and Growth Modification in Clownfish. *Nature* 424: 145–46.

Buston, P. M. 2004b. Territory Inheritance in Clownfish. *Proceedings. Biological Sciences/The Royal Society* 271 Suppl: S252–54.

Buston, P. M., S. M. Bogdanowicz, A. Wong, and R. G. Harrisson. 2007. Are Clownfish Groups Composed of Close Relatives? An Analysis of Microsatellite DNA Variation in *Amphiprion percula. Molecular Ecology* 16 (17): 3671–78.

Buston, P. M., and M. A. Cant. 2006. A New Perspective on Size Hierarchies in Nature: Patterns, Causes, and Consequences. *Oecologia* 149: 362–72.

Casadevall, M., E. Delgado, O. Colleye, S. B. Monserrat, and E. Parmentier. 2009. Histological Study of the Sex-Change in the Skunk Clownfish *Amphiprion akallopisos. The Open Fish Science Journal* 2: 55–58.

Casas, L., F. Saborido-Rey, T. Ryu, C. Michell, T. Ravasi, and X. Irigoien. 2016. Sex Change in Clownfish: Molecular Insights from Transcriptome Analysis. *Scientific Reports* 6 (1): 35461.

Chen, K.-C., and H.-K. Mok. 1988. Sound Production in the Anemonfishes, *Amphiprion clarkii* and *A. frenatus* (Pomcentridae), in Captivity. *Japanese Journal of Ichthyology* 35: 90–97.

Colleye, O., B. Frederich, P. Vandewalle, M. Casadevall, and E. Parmentier. 2009. Agonistic Sounds in the Skunk Clownfish *Amphiprion akallopisos*: Size-Related Variation in Acoustic Features. *Journal of Fish Biology* 75: 908–16.

Colleye, O., M. Nakamura, B. Frederich, and E. Parmentier. 2012. Further Insight into the Sound-Producing Mechanism of Clownfishes: What Structure Is Involved in Sound Radiation? *Journal of Experimental Biology* 215: 2192–202.

Colleye, O., and E. Parmentier. 2012. Overview on the Diversity of Sounds Produced by Clownfishes (Pomacentridae): Importance of Acoustic Signals in Their Peculiar Way of Life. *PLoS ONE* 7 (11): e49179.

Colleye, O., P. Vandewalle, D. Lanterbecq, D. Lecchini, and E. Parmentier. 2011. Interspecific Variation of Calls in Clownfishes: Degree of Similarity in Closely Related Species. *BMC Evolutionary Biology* 11 (1): 365.

Collingwood, C. 1868. Note on the Existence of Gigantic Sea-Anemones in the China Sea, Containing within Them Quasi-Parisitc Fish. *Annals and Magazine of Natural History* 4: 33.

Desrochers, L., R. Branconi, E. Schlatter, B. Dent, and P. Buston. 2020. Sensory Cues Underlying Competitive Growth in the Clown Anemonefish (*Amphiprion percula*). *Behavioural Processes* 181: 104276.

Emery, A. R. 1973. Comparative ecology and functional osteology of fourteen species of damselfish (Pisces: Pomacentridae) at Alligator Reef, Florida Keys. *Bulletin of Marine Science* 23: 649–770.

Fine, M. L. 2012. Swimbladder Sound Production: The Forced Response versus the Resonant Bubble. *Bioacoustics* 21 (1): 5–7.

Fish, M. P., and W. H. Mowbray. 1970. *Sounds of the Western North Atlantic Fishes*. Baltimore: Johns Hopkins University Press.

Frédérich, B., and E. Parmentier. 2016. *Biology of Damselfishes*. Boca Raton: CRC Press, Taylor & Francis.

Fricke, H. W. 1979. Mating System, Resource Defense and Sex Change in the Anemonefish *Amphiprion akallopisos*. *Zeitschrift Für Tierpsychologie* 50: 313–26.

Graham, R. 1992. Sounds fishy. *Australia's Geographic Magazine* 14: 76–83.

Iorio, L. D., M. Audax, J. Deter, F. Holon, J. Lossent, C. Gervaise, and P. Boissery. 2021. Biogeography of Acoustic Biodiversity of NW Mediterranean Coralligenous Reefs. *Scientific Reports* 11 (1): 16991.

Limbaugh, C. 1964. Notes on the life history of two California pomacentrids: Garibaldis, Hypsypops rubicunda (Girard), and blacksmiths, Chromis punctipinnis (Cooper). *Pacific Science* 28: 41–50.

Lobel, P. S., and D. A. Mann. 1995. Spawning Sounds of the Damselfish, *Dascyllus albisella* (Pomacentridae), and Relationship to Male Size. *Bioacoustics* 6: 187–98.

Lobel, P. S., I. M. Kaatz, and A. N. Rice. 2010. Acoustical Behavior of Coral Reef Fishes. In *Reproduction and Sexuality in Marine Fishes: Patterns and Processes*, Ed. K. S. Cole, 307–348. Berkely: Uuniversity of California Press.

Luh, H. K., and H. K. Mok. 1986. Sound Production in the Domino Damselfish, *Dascyllus trimaculatus* (Pomacentridae) under Laboratory Conditions. *Japanese Journal of Ichthyology* 33: 70–74.

Mann, D., and P. S. Lobel. 1998. Acoustic Behaviour of the Damselfish *Dascyllus albisella*: Behavioural and Geographic Variation. *Environmental Biology of Fishes* 51: 421–28.

Mann, D. A., and P. S. Lobel. 1997. Propogation of damselfish (Pomacentridae) courtship sounds. *Journal of the Acoustic Society of America* 101: 3783–3791.

Mooney, T. A., L. D. Iorio, M. Lammers, T-H. Lin, S. L. Nedelec, M. Parsons, C. Radford, et al. 2021. Listening Forward: Approaching Marine Biodiversity Assessments Using Acoustic Methods. *Royal Society Open Science* 7 (8): 201287.

Moyer, J. T., and C. E. Sawyers. 1973. Territorial Behavior of the Anemonefish *Amphiprion xanthurus* with Notes on the Life History. *Japanese Journal of Ichthyology* 20: 85–93.

Myrberg, A. A. Jr, E. Spanier, and S. J. Ha. 1978. Temporal patterning in acoustical communication. In *Contrasts in Behavior*. Ed. E. S. Reese and F. J. Lighter, 137–179. New York: Wiley Interscience.

Myrberg, A. A. Jr, S. J. Ha, and M. J. Shamblott. 1993. The sounds of bicolor damselfish (Pomacentrus partitus): predictors of body size and a spectral basis for individual recognition and assessment. *Journal of the Acoustical Society of America* 94: 3067–3070.

Olivier, D., B. Frédérich, A. Herrel, and E. Parmentier. 2015. A Morphological Novelty for Feeding and Sound Production in the Yellowtail Clownfish. *Journal of Experimental Zoology Part A: Ecological Genetics and Physiology* 323 (4): 227–38.

Olivier, D., B. Frédérich, and E. Parmentier. 2016. Cerato-Mandibular Ligament: A Key Trait in Damselfishes? In *Biology of Damselfishes*, eds B. Frédérich and E. Parmentier, 204–28. Boca Raton: CRC Press, Taylor & Francis.

Olivier, D., B. Frédérich, M. Spanopoulos-Zarco, E. F. Balart, and E. Parmentier. 2014. The Cerato-Mandibular Ligament: A Key Functional Trait for Grazing in Damselfishes (Pomacentridae). *Frontiers in Zoology* 11: 63.

Parmentier, E., O. Colleye, M. L. Fine, B. Frédérich, P. Vandewalle, and A. Herrel. 2007. Sound Production in the Clownfish *Amphiprion clarkii*. *Science* 316 (May 2007): 1006.

Parmentier, E., O. Colleye, and D. Mann. 2009. Hearing Ability in Three Clownfish Species. *The Journal of Experimental Biology* 212: 2023–26.

Parmentier, E., L. Kéver, M. Casadevall, and D. Lecchini. 2010. Diversity and Complexity in the Acoustic Behaviour of *Dacyllus flavicaudus* (Pomacentridae). *Marine Biology* 157: 2317–27.

Parmentier, E., D. Lecchini, and D. A. Mann. 2016. Sound Production in Damselfishes. In *Biology of Damselfishes*, eds B. Frédérich and E. Parmentier, 204–28. Boca Raton: CRC Press, Taylor & Francis.

Parmentier, E., R. Scalbert, X. Raick, C. Gache, B. Frédérich, F. Bertucci, and D. Lecchini. 2021. First Use of Acoustic Calls to Distinguish Cryptic Members of a Fish Species Complex. *Zoological Journal of the Linnean Society*, 195(3): 964–975.

Parmentier, E., J. Tock, J. C. Falguière, and M. Beauchaud. 2014. Sound Production in *Sciaenops ocellatus*: Preliminary Study for the Development of Acoustic Cues in Aquaculture. *Aquaculture* 432: 204–11.

Rice, A. N., and P. S. Lobel. 2003. The Pharyngeal Jaw Apparatus of the Cichlidae and Pomacentridae: Function in Feeding and Sound Production. *Reviews in Fish Biology and Fisheries* 13 (4): 433–44.

Santiago, J. A, and J. J. Castro. 1997. Acoustic Behaviour of Abudefduf Luridus. *Journal of Fish Biology* 51 (5): 952–59.
Schneider, H. 1964. Bioakustische Untersuchungen an Anemonenfischen Der Gattung Amphiprion (Pisces). *Zeitschrift für Morphologie und Ökologie der Tiere* 53: 453–74.
Spanier, E. 1979. Aspects of species recognition by sound in four species of damselfishes, genus Eupomacentrus (Pisces: Pomacentridae). *Zeitschrift für Tierpsychologie* 51: 301–316.
Stiassny, M. L. J. 1981. The Phyletic Status of the Family Cichlidae (Pisces, Perciformes): A Comparative Anatomical Investigation. *Netherlands Journal of Zoology* 31: 275–314.
Takemura, A. 1983. Studies on the Underwater Sound - VIII. Acoustical Behavior of Clownfishes (Amphiprion spp.). *Bulletin of the Faculty of Fisheries Nagasaki University* 54: 21–27.
Verwey, J. 1930. Coral Reef Studies. The Symbiosis between Damselfishes and Sea Anemones in Batavia Bay. *Treubia* 12: 305–55.
Wilkinson, C., M. Nowak, I. Miller, and V. Baker. 2013. Status of Caribbean Coral Reefs in Seven Countries in 1986. *Marine Pollution Bulletin* 70 (1): 7–9.

11 Neuroendocrinology of Life History and Stress in Anemonefishes

Mélanie Dussenne, Alexander Goikoetxea*, Benjamin Geffroy, and Laurence Besseau*

CONTENTS

11.1 INTRODUCTION

Environmental changes affect the evolution of complex life-history stages, notably phases of development, growth, reproduction, and senescence, each of them characterized by distinctive phenotypes (Wingfield 2008). These life-history transitions are under the control of the neuroendocrine system, which integrates both environmental cues and information provided by internal biological timers. The neuroendocrine system leads to the efficient synchronization of vital physiological functions with environmental cues (Wingfield 2015; Hazlerigg and Wagner 2006), allowing phenotypic plasticity in accordance with environmental changes.

In vertebrates, environmental cues (e.g., temperature, photoperiod, chemicals) as well as internal information (e.g., nutrition, metabolic signals) are integrated by the central nervous system (CNS), which results in the release of neuroendocrine signals notably from the hypothalamus. This triggers the downstream release of hormones by the anterior pituitary gland (i.e., the adenohypophysis) and, in turn, peripheral endocrine secretions that regulate morphological, physiological, and behavioral responses (Falcón et al. 2007). So far, most of the existing data regarding the impact of the environment on the fish life cycle derives from canonical models, such as zebrafish (*Danio rerio*) or medaka (*Oryzias latipes*) (both freshwater species), or aquaculture breeding species such as salmonids or the European sea bass (*Dicentrarchus labrax*) (Jonsson and Jonsson 2014). Hence, there is now a need to study alternative models such as anemonefish, in which very little data is available to date.

This chapter summarizes recent findings with regard to how the environment can trigger morphological, physiological, and behavioral responses that orchestrate the anemonefish life cycle at two main life-history steps: i) the post-embryonic phase transforming larvae into juveniles, called metamorphosis, and ii) the adult phase regarding sexual development, including protandrous (or male-to-female) sex change. We also link these two fascinating processes with the stress pathway. Finally, we give insights into how modifications associated with climate change could affect the ecology, physiology, and, eventually, survival of these fishes.

11.2 THE NEUROENDOCRINE SYSTEM IN FISH: HOW TO LIVE IN A FLUCTUATING ENVIRONMENT

In fish, as in all vertebrates, the brain constitutes the main interface between an individual's physiology and its environment. This allows organisms to maintain their homeostasis in a given environment and to face environmental fluctuations through acclimation. Following the pioneering work of Scharrer (1928), suggesting the presence of secretory cells in the brain of teleost fish, neuroendocrinology emerged as a new discipline in neurosciences (Cerdá-Reverter and Canosa 2009). Subsequently, the existence of the hypothalamo-pituitary system was established in fish, thus linking the CNS and the endocrine system which ensure the regulation of a variety of vital functions and processes. Both tightly linked systems contribute to adapting the response of the organism to changes in the environment and internal milieu, to ultimately control body homeostasis (Zohar et al. 2021; Vissio et al. 2021).

The hypothalamo-pituitary (HP) complex in teleosts is divided into three main areas: the hypothalamus (which

* Equal co-authors.

DOI: 10.1201/9781003125365-13

is part of the diencephalon), the neurohypophysis (which derives from the ventral diencephalon and represents the neural compartment of the pituitary), and the adenohypophysis (which is the secretory part of the gland). The environmental information received by the hypothalamus is transferred through axonic projections to the neurohypophysis where neurosecretory fibres release various peptides (e.g., corticotropin-releasing hormone [CRH], thyrotropin-releasing hormone [TRH], or gonadotropin-releasing hormone [GnRH]) in the vicinity of the cells of the adenohypophysis (Figure 11.1) (Zohar et al. 2010). According to their localization within the pituitary, these cells synthesize at least eight different hormones, belonging to different families depending on their action. Four hypothalamo-pituitary (HP) axes can be thus defined: i) the hypothalamo-pituitary somatotropic (HPS) axis which consists in the secretion of growth hormone (GH), prolactin (PRL), and somatolactin (SL) controlling growth, but also osmoregulation; ii) the hypothalamo-pituitary gonadal (HPG) axis characterized by two gonadotropins, the follicle-stimulating hormone (FSH), and the luteinizing hormone (LH) which are involved in the regulation of sexual reproduction; iii) the hypothalamo-pituitary interrenal (HPI) axis consisting of peptides derived from a common precursor, the proopiomelanocortin (POMC), leading to the melanocyte-stimulating hormone (MSH), the adrenocorticotropic hormone (ACTH), and the β-endorphin, both involved in the control of stress and immunity in fish; and iv) the hypothalamo-pituitary thyroid (HPT) axis, characterized by the secretion of thyroid-stimulating hormone (TSH) mainly involved in teleost fish development and metamorphosis, but also in metabolism, reproduction, growth, and osmoregulation (Deal and Volkoff 2020; Zohar et al., 2021).

In this chapter, we review the available data concerning the involvement of the aforementioned neuroendocrine systems in two main transition phases shaping the life cycle of anemonefishes: metamorphosis and sex change.

11.3 ANEMONEFISHES AND THEIR ENVIRONMENT

Like most coral reef fishes, anemonefish exhibit a biphasic life cycle that includes a dispersive planktonic larval phase in the open ocean followed, after the recruitment phase, by a sedentary reef phase for the juveniles and adults (Roux et al. 2020). From an ecological point of view, one of the most pressing questions regarding the life history of anemonefish has been to decipher how anemonefish larvae manage to navigate within the pelagic environment and then orient themselves to a suitable benthic recruitment site. Although the exact mechanisms used by anemonefish larvae at this crucial step of their life cycle still remain unclear (Dixson et al. 2011), the behavioral capabilities of these and other reef fish larvae suggest they exhibit a strong determinism concerning their final settlement destination (Leis 2007) thanks to extraordinary swimming capabilities, endurance, orientation skills, and acute sensory abilities (Dixson et al. 2011).

During life transition periods, the neuroendocrine control system plays a crucial role: the fish brain (i.e., the hypothalamus) receives and integrates environmental information as well as tactile, visual, and chemical communication signals transmitted from the central and peripheral nervous systems. The detection of environmental signals stimulates the hypothalamic neurons to secrete peptides such as CRH. Nerve endings terminate in close association with corticotropic cells (cells secreting POMC) and thyrotropic cells (cells secreting TSH) of the adenohypophysis (Figure 11.1). Thus, environmental information is transmitted from the brain to peripheral organs thanks to a particular temporal pattern of secretion of several hormones that in turn control various physiological, developmental, and behavioral processes that lead to the final phenotype (Suzuki et al. 2020).

One of the pivotal components of the neuroendocrine system in teleosts is the hypothalamo-pituitary interrenal (HPI) axis. Once triggered by environmental stressors, the HPI axis releases glucocorticoid hormones (GCs) that have multiple regulatory effects on the organism's biology. GCs can interact with other hormonal pathways, among them: i) the HPT axis, leading to the release of thyroid hormones (TH) which control larval development and metamorphosis (Blanton and Specker 2007; Laudet 2011; Denver 2017); and ii) the HPG axis, which regulates the release of sex steroid hormones. Therefore, GCs play a crucial role in enabling vertebrates to cope with and respond to environmental factors in the wild. However, despite evidence that environmental variations activate the HPI axis, commonly referred to as the stress axis, direct links have rarely been established in anemonefish (Mills et al. 2020).

The settlement habitat of anemonefish is complex, it is composed of a sea anemone, a colony of congeners and a cortege of other reef species of invertebrates and vertebrates, among which some predators (Fautin and Allen 1992). Each part of this complex ecosystem constitutes a source of stress (positive or negative) either by direct visual detection (Casas and Saborido-Rey 2021), acoustic communication (Colleye and Parmentier 2012), or olfactive detection of the chemical cues produced (Barth et al. 2015). Like in many other coral reef fish (Atema et al. 2002; Lecchini et al. 2005; Gerlach et al. 2007; Leis et al. 2011), olfaction plays a critical role in locating settlement habitat after the larval dispersal stage in the orange anemonefish (*Amphiprion percula*) (Dixson et al. 2008). While the exact mechanisms (sensory receptors, communication molecules, etc.) that enable *A. percula* larvae to locate suitable settlement sites are still unknown, Dixson et al. (2008) suggest that an olfactory response to multiple chemical cues (e.g., those emanating from island vegetation, host sea anemones, and/or congeners) may play a role in the final choice of settlement. Therefore, a specific "olfactive landscape" might assist larvae in navigating to reefs where suitable settlement habitats can be found (Veilleux et al. 2013), afterwards triggering the required neuroendocrine signalling for subsequent morphological, physiological, or behavioral events during this life history transition. However, recent studies

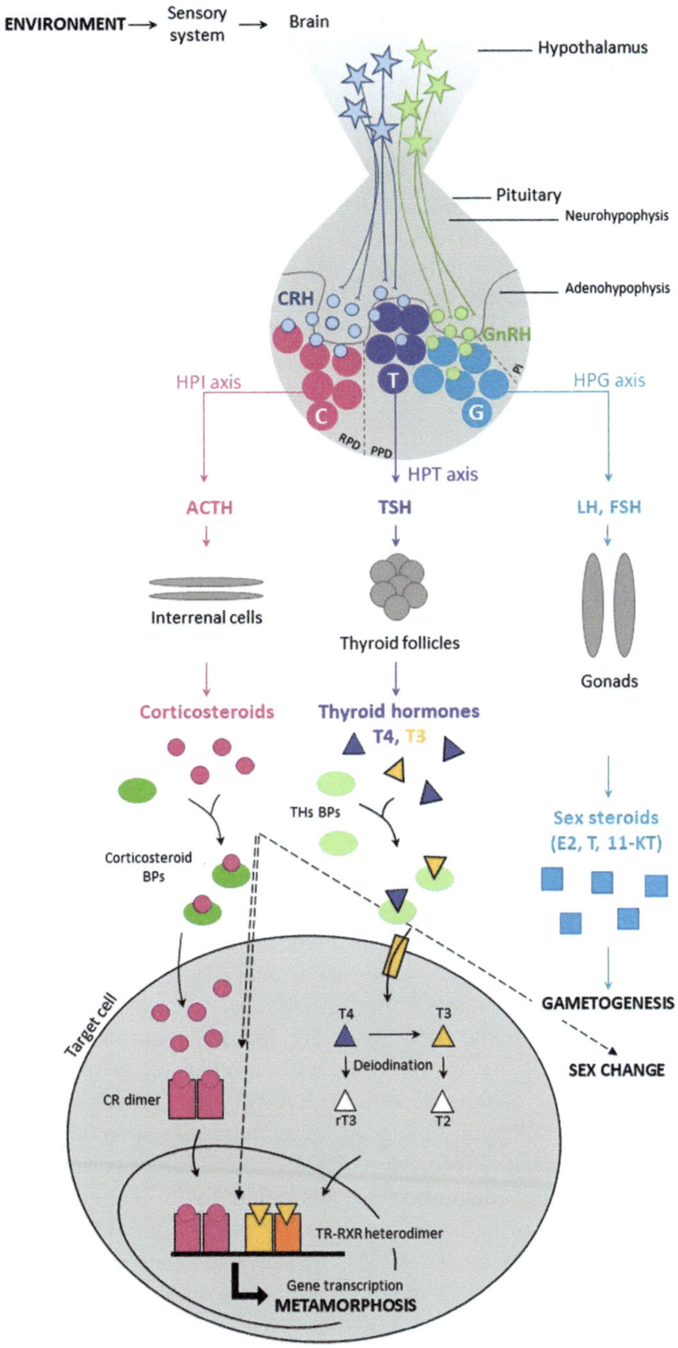

FIGURE 11.1 Illustration representing the HPI, HPT, and HPG axes in non-mammalian species (pink, purple, and blue, respectively). Environmental cues are perceived by sensory organs and processed in the brain, where corticotropin-releasing hormone (CRH) and gonadotropin-releasing hormone (GnRH) are secreted. In non-mammalian species, CRH is thought to control both the HPI and HPT axes. CRH acts on the pituitary cells to stimulate the synthesis and release of i) adrenocorticotropic hormone (ACTH) from corticotropic cells (C, pink circles) and ii) thyroid-stimulating hormone (TSH) from thyrotropic cells (T, purple circles). HPI axis: ACTH stimulates the secretion of glucocorticoids (GCs) by the interrenal cells. Once secreted, GCs are transported in the blood by corticosteroid binding proteins (BPs). GCs enter target cells where they bind to corticosteroid receptor dimers which then translocate into the nucleus where they bind to specific DNA regions to induce gene expression. HPT axis: TSH stimulates the production of thyroid hormones (THs, T4, and T3) by thyroid follicles. THs are transported to target cells *via* their specific BPs. Inside target cells, THs are substrates of deiodinase enzymes which can both activate (i.e., forming T3) or inactivate (i.e., rT3 and T2) THs. THs are transported to the cell nucleus where they bind to TR-RXR heterodimers, which activate gene transcription, *in fine* resulting in metamorphosis. HPG axis: GnRH reaches gonadotropic cells to induce synthesis and release of gonadotropins (LH and FSH). LH and FSH both reach the gonads where they stimulate gonadal maturation and function (i.e., synthesis of sex steroids: E2, T, and 11-KT). The balance between these sex steroids, in turn, regulates gametogenesis and controls sex change in hermaphroditic species. Finally, corticosteroids can influence both metamorphosis and sex change processes (dotted black lines). RPD: rostral *pars distalis*, PPD: proximal *pars distalis*, PI: *pars intermedia* of the adenohypophysis. Adapted from Denver (2017).

indicate that this landscape could be modified with global change, as indicated by i) behavioral experiments showing that anemonefish juveniles can discriminate between olfactive cues emanating from bleached and unbleached host anemones (Scott and Dixson 2016), and ii) the fact that some chemical cues degrade more quickly in conditions simulating global change (UV radiation and ocean acidification) compared to control conditions (Chivers et al. 2014). Therefore, modifications of the olfactive landscape could lead to misinterpretations by anemonefish larvae of the location of their settlement site or their ability to escape predation (Munday et al. 2009; Dixson et al. 2010; Biswal et al. 2021).

11.4 NEUROENDOCRINE CONTROL OF METAMORPHOSIS

As previously stated, very few species are used as models to investigate developmental biology, with anemonefish just emerging as a new model organism (Roux et al. 2020). As the signalling pathway underlying metamorphosis is strongly conserved among vertebrates, we will focus in this section on anemonefish whenever possible, and will otherwise refer more generally to other teleost fish and amphibians in which the mechanisms leading to metamorphosis have been more extensively described.

Metamorphosis is classically defined as an abrupt and irreversible life-history transition in which a larva transforms into a juvenile (Laudet 2011). In most marine teleost species, the life cycle is biphasic: pelagic larvae disperse into the ocean, then undergo metamorphosis that is concomitant with the recruitment to the adult habitat. During metamorphosis, larval features are lost and adult features develop, enabling the colonization of the new environment. This morphological transformation is thus accompanied by profound physiological and ecological changes (McMenamin and Parichy 2013; Bishop et al. 2006). More or less important alterations occur during metamorphosis, from a total remodelling of the body in flatfishes and amphibians to more subtle modifications in most teleost fishes, such as the development of fins, formation of scales, ossification, maturation of organs, as well as the acquisition of the adult color pattern (Laudet 2011). For example, in the anemonefish *A. ocellaris*, first the notochord bends, then soft rays appear in the anal and dorsal fins that later acquire their spines, pelvic fins grow, and finally, the color pattern develops, with the white bars appearing in a rostro-caudal gradient (Roux et al. 2019b).

In amphibians and in the teleost species studied so far, metamorphosis is triggered by the action of THs on thyroid receptors (TRs), which has been confirmed *in vivo* using either THs treatments, pharmacological treatments (using chemical compounds called goitrogens that block TH synthesis), or transgenic animals expressing a dominant-negative form of TRs (Inui and Miwa 1985; Schreiber et al. 2001; Buchholz et al. 2003; Salis et al. 2021; see McMenamin and Parichy 2013 for review). Recently, the importance of THs in metamorphosis has been extended to several coral reef fishes including *A. ocellaris* (Holzer et al. 2017; Salis et al. 2021). THs are the final product of the HPT axis, beginning in the brain. In mammals, thyrotropin-releasing hormone (TRH) stands at the top of this HPT axis. In non-mammalian species, however, it seems that CRH has taken up this role (Laudet 2011). CRH stimulates pituitary cells (i.e., thyrotropic cells) that consequently synthesize and release thyroid-stimulating hormone (TSH) into the global circulation. TSH reaches thyroid follicles that mainly produce thyroxine (T4, main form) and 3,5,3'-triiodothyronine (T3, biologically active form) (Laudet 2011; Denver 2017). T4 and T3 can both reversibly bind to circulating binding proteins (BPs) to be transported from the site of production (thyroid follicles) to target cells in peripheral tissues (Power et al. 2000). Thanks to transmembrane transporters, THs enter target cells in which they can be enzymatically activated or inactivated under the action of deiodinases (DIO), therefore regulating their intracellular availability in target tissues. This way, T4 can be activated into T3; T4 and T3 can be inactivated into rT3 and T2, respectively (Blanton and Specker 2007; Denver 2017). THs are then transported from the cytoplasm to the cell nucleus where they bind to thyroid hormone receptors (TRs). Teleost fish possess at least two TR genes named TRα and TRβ, with some species possessing additional copies of these genes (McMenamin and Parichy 2013). In teleosts, two distinct genes encoding TRα have been identified (TrαA and TRαB). Additionally, isoforms of TRβ have been described, resulting either from alternative splicing or transcription of two different genes (i.e., such as in the Japanese conger eel *Conger myriaster*) (Marchand et al. 2001; Kawakami et al. 2003a, 2003b; Galay-Burgos et al. 2008). TRs form dimers either with another TR or most often with a retinoid X receptor (RXR). These TR-RXR heterodimers act as transcriptional activators only upon THs binding which induces the transcription of target genes, giving rise to the transformations associated with metamorphosis (Sachs et al. 2002; Denver 2017) (see Figure 11.1). Generally speaking, whole-body TH levels, together with the expression of TSH and TRβ genes, increase at the onset of metamorphosis, peak during climax, and then decrease when the juvenile stage is reached (see Campinho 2019 for review). This general pattern has been confirmed in the coral reef fish *Acanthurus triostegus*, in which individuals being recruited show a rise in THs level and TRs expression, both dropping right after recruitment (Holzer et al. 2017). In other coral reef species (e.g., *Rhinecanthus aculeatus*, *Chromis viridis*, *Chaetodon lunula*, and *Ostorhinchus angustatus*) recruiting larvae also experience the same TH levels fluctuations, suggesting that TH-mediated metamorphosis could be generalized to all coral reef fishes (Holzer et al. 2017). Similar expression profiles of THs and TRs genes have been observed during the metamorphosis of *A. ocellaris* (Roux et al. 2019a).

Originally, CRH was known for its role in stimulating the HPI axis. By the end of the 20th century, although there was no doubt THs were essential for tadpole metamorphosis, it was unclear which neurotransmitter could stimulate the HPT axis, as all attempts to induce TSH secretion with TRH injection were unsuccessful (reviewed in Ball 1981). Later on, *in vivo* experiments showed that in tadpoles experiencing environmental stress (pond drying, overcrowding, etc.) the time to complete metamorphosis was shortened, which is thought to maximize the probability of survival. Such metamorphosis acceleration could be reversed if the larvae were treated with a CRH antagonist (Newman 1992; Denver 1997). Concomitantly, *in vitro* and *in vivo* experiments showed that CRH could elevate TSH, THs, and stress-related hormone (glucocorticoids, GCs) levels while accelerating tadpole metamorphosis (Denver and Licht 1989; Denver 1993; Denver 1997). Therefore, CRH, stimulating both the HPI and HPT axes, cross-links environmental stress and TH homeostasis, procuring some developmental plasticity to larvae living in a fluctuating environment. Whether TRH and/or CRH stimulate the HPT axis in teleost fishes remains an open question; the regulation of the HPT axis might be species-specific (Larsen et al. 1998; Eales and Himick 1988; Geven et al. 2009; Galas et al. 2009; Rousseau et al. 2021).

Briefly, as soon as the HPI axis is stimulated, CRH is synthesized by the brain and reaches the pituitary to stimulate the synthesis and release of adrenocorticotropic hormone (ACTH) (derived from proopiomelanocortin, POMC) to the global circulation. ACTH then reaches its target cells in the interrenal gland to induce the synthesis of GCs, mainly corticosterone and cortisol, the latter being the main stress hormone in fish (Wendelaar-Bonga 1997). GCs are then transported to the blood bound to corticosteroid BPs and finally enter target cells where they bind to corticosteroid receptor (CR) dimers (glucocorticoid [GRs] or mineralocorticoid [MRs] receptors) that then translocate into the cell nucleus to induce gene transcription (Bury and Sturm 2007; Denver 2017). GCs can modulate various functions in fish, such as glycogen metabolism (energy is mobilized in a stressful situation) but also osmoregulation or blood pressure (see Mommsen et al. 1999 for a review). Additionally, GCs seem to have an important role during metamorphosis. Indeed, *in vivo* and *in vitro* experiments in frogs showed that GCs administered alone had no effect on the HPT axis but when administered together with T3, the interaction of both hormones strongly stimulated TRs and deiodinases gene expression, suggesting that tissue sensitivity to THs was increased (Bonett et al. 2010). Other reports showed that deiodinases activity could be enhanced thanks to this simultaneous treatment, increasing T3 bioavailability in target tissues, resulting in a rapid metamorphosis (Galton 1990; Kühn et al. 2005). Thus, stimulation of the HPI axis may accelerate metamorphosis *via* two different modes of action: i) centrally, stimulating the synthesis of CRH in the brain, therefore increasing the synthesis of TSH and stimulating the whole HPT axis; and ii) peripherally, increasing tissue sensitivity to THs at the peripheral level.

11.5 NEURAL REGULATION OF THE STRESS RESPONSE AND SEXUAL DEVELOPMENT

11.5.1 Neuroendocrinology of Plasticity in Teleost Fishes

Environmental factors (e.g., social cues) are known to be transduced into physiological changes that can promote alterations in the nervous system, notably affecting brain structure and functionality (Ebbesson and Braithwaite 2012). Neural plasticity is more pronounced in fishes compared to mammalian systems, and some brain processes such as neurogenesis can happen throughout the whole life (Zupanc et al. 2005). Although the remodelling of the fish brain following a stressful situation is not yet fully understood (reviewed by Sørensen et al. 2013), it has been hypothesized that cell proliferation in the brain might be, in part, influenced by the levels of GCs and their effect mediated by the corresponding receptors (i.e., GRs and MRs) (Montaron et al. 2003; Wong and Herbert 2005; reviewed by Sørensen et al. 2013). Such a link between the stress axis and neurogenesis is further reflected by the decrease in the proliferation of neural cells observed in fish exposed to chronic stress (i.e., social subordination) (Sørensen et al. 2013). This underscores a tight relationship between adaptation to social changes and neural plasticity, which results in a form of behavioral plasticity (Sørensen et al. 2011). Such malleability translates into an extraordinary capacity to adapt to stressful events of diverse magnitude, type, and/or duration. In fish, the stress response is regulated by the HPI axis. Under a situation of stress, activation of the HPI axis begins in the preoptic area of the hypothalamus, setting off a series of chain reactions, as described in the previous section, that ultimately promote the release of cortisol by the interrenal cells of the head kidney (Mommsen et al. 1999; Wendelaar Bonga 1997). The mechanisms underlying the stress response in fish have been extensively reviewed, including a few great examples such as Wendelaar Bonga (1997), Barton (2002), Flik et al. (2006), and Gorissen and Flik (2016), which should be referred to for further detail. Although not much is known yet regarding the molecular mechanisms involved in stress regulation in anemonefish in particular, they likely involve the same classical pathways. Like many other teleost fish, the yellowtail anemonefish *Amphiprion clarkii* expresses two glucocorticoid receptors, Gr1 and Gr2 (Zhang et al. 2020). Relatively high levels of *gr1* transcripts are found in the hypothalamus, cerebellum, and pituitary independently of the sex, whereas *gr2* peaks in the medulla oblongata of males only (Zhang et al. 2020). There are, however, no differences in *gr1* and *gr2* expression according to the sexual status of individuals (Zhang et al. 2020). Undoubtedly, further research into the relationship between environmental stressors and the elicited downstream reactions in the neuroendocrine system of anemonefishes will help improve our understanding of neural plasticity in these and other species. Moreover, addressing this lack of knowledge will shed light on how fish and vertebrates in general may adapt to their changing environment.

Cortisol is released into the blood between 30 and 120 minutes following a stressful event (e.g., capture and/or restraint) in *Amphiprion chrysopterus* and *Amphiprion akallopisos*, levels similar to those observed in other fish (Mills et al. 2015, 2020). Interestingly, the maximum temperature encountered by *A. akallopisos* in their lifetime appears to modulate cortisol production in relation to stress in both male and female individuals, meaning that a correlation exists between increased cortisol release and previous exposure to high temperatures independent of sex (Mills et al. 2015). This exemplifies the tight link that exists between metabolism and stressful events in regulating cortisol production (Bessa et al. 2021). Basal cortisol levels reported in anemonefishes (Godwin 1994; Godwin and Thomas 1993a; Iwata et al. 2008, 2012; Mills et al. 2015, 2020) are also in line with what is generally found in other fish species. Such levels seem, nevertheless, modulated by the social (dominant versus subordinate) and reproductive (mature versus immature) status, although no clear and significant patterns have been observed (Godwin 1994; Godwin and Thomas 1993a; Iwata et al. 2008, 2012).

11.5.2 Social Control of Sex Change – A Superb Example of Sexual Plasticity

Social and sexual plasticity constitute some of the most extreme examples of phenotypic plasticity, and they become especially apparent in hermaphroditic fish (fish with the ability to change sex naturally during their life cycle, whose sexual fate is not fixed at birth). In some species, such as those belonging to the anemonefish subfamily, this incredible transformation is regulated by modifications in the social structure (reviewed in Olivotto and Geffroy 2017). Socially regulated sex change is a spectacular illustration of reproductive plasticity which has now been reported in multiple species (Choat and Robertson 1975; Jones and Thompson 1980; Liu et al. 2017; Thomas et al. 2019; Todd et al. 2019; Warner and Swearer 1991). While the gonadal and morphological alterations underlying sex change have now been well described in anemonefishes, the molecular cascade controlling this process remains to be fully understood, particularly at the level of the brain. However, it is in the brain where we may find the key to fully comprehending the triggering mechanisms underlying sex change, as visual cues the fish perceive (i.e., removal of the dominant female in the case of anemonefishes) set off a series of interactions, first promoting behavioral sex change (Fricke and Fricke 1977) and then followed by transformation of the gonads (Casas et al. 2016). For this reason, fulfilling this gap in knowledge will be essential to understanding the global picture of this astonishing process. While pieces of the puzzle are still missing, a number of signalling molecules associated with neural activity following the trigger of sex change in fish has been identified, including monoamine neurotransmitters (e.g., dopamine, norepinephrine, serotonin) and neuropeptides (e.g., gonadotropin-releasing hormone, arginine vasotocin, isotocin) (Godwin 2010; Lamm et al. 2015; Thomas et al. 2019). For example, arginine vasotocin (AVT) signalling seems of primary importance in the establishment of dominance in false clown anemonefish, *A. ocellaris* (Yaeger et al. 2014). Inhibiting this signalling pathway significantly reduces aggressive behaviors (bites and charges) as well as the number of *c-fos* positive cells (indicative of high neuroplasticity) in the preoptic area of the hypothalamus (Yaeger et al. 2014). Anemonefish also produce a variety of specific agonistic sounds that are likely involved in the establishment and maintenance of the social hierarchy (Colleye et al. 2011; Colleye and Parmentier 2012). Together, these agonistic sounds and acts have led to the straightforward hypothesis that stress, and more specifically cortisol, would be involved in socially controlled sex change. Recent results indeed pinpointed the essential role of this hormone in natural hermaphroditism, though mainly in protogynous fish (Goikoetxea et al. 2017). For instance, cortisol administration in *Halichoeres trimaculatus* successfully triggered sex change from female to male (Nozu and Nakamura 2015). Nevertheless, regarding protandrous sex change in anemonefishes, results have been inconclusive until now. For example, while serum cortisol levels were observed to increase during sex change in the cinnamon anemonefish *A. melanopus*, this rise took place in the late stages of the transformation (Godwin and Thomas 1993b), suggesting such fluctuation may have been a consequence rather than a cause of the redirection of sexual fate. In *A. ocellaris*, dominant (presumably future female) individuals tended to have less basal cortisol than subordinates after 180 days of inter-individual relationships, though the sex was not clearly established (Iwata et al. 2008). In another experiment involving groups of three immature false clown anemonefish, the dominant individual presented the highest level of cortisol after ten days of interaction, but this was unrelated to sex (Iwata et al. 2012). The fact that no clear patterns are observable could be due to the difficulty in obtaining reliable basal cortisol levels without disturbing the fish, but we believe that new matrices to measure cortisol (i.e., water, mucus, scales, faeces) will ultimately help in this direction for future studies (Sadoul and Geffroy 2019). It is also possible that other mechanisms (e.g., epigenetic) would be key transducers of socially induced sex change (Piferrer 2013; Todd et al. 2019).

11.6 ANEMONEFISH AS A MODEL TO UNDERSTAND MARINE FISH STRESS IN A CHANGING WORLD

The increasing use of anemonefishes as field and laboratory study species suggests their potential as a new biological marine model in multiple areas of research. For example, the recent review by Roux et al. (2020) presented the false anemonefish *A. ocellaris* as a promising fish model for developmental, ecological, and evolutionary biology. Furthermore, in the current landscape of climate change, fish and other aquatic organisms are bound to suffer the intensifying consequences of thermal stress, as the temperature of water bodies on the planet increases (Laffoley and Baxter 2016). Therefore, understanding how such dramatic environmental changes will affect the stress physiology of fishes is of the utmost importance (Alfonso et al. 2021; Brierley and Kingsford 2009; Geffroy and Wedekind 2020).

Since the pre-industrial era, the release of greenhouse gases (carbon dioxide [CO_2], methane, etc.) into the atmosphere has drastically increased and has reached levels unprecedented during the last two million years (IPCC, 2021). As such, atmospheric CO_2 levels have increased from 280 ppm in 1800 to 410 ppm nowadays and could further increase to reach 900–1,000 ppm by 2100 (Meinshausen et al. 2011; IPCC 2021). Part of this atmospheric CO_2 is being absorbed by the oceans. This has two major consequences: i) the sea surface temperature (SST) rises and ii) the pH of the water decreases, a process known as "ocean acidification" (OA).

OA directly affects calcifying organisms (e.g., corals) because the availability of calcium carbonate ions is reduced. OA, together with other factors (pollution, overfishing, etc.), threatens the existence of corals and accelerates their global decline by creating a vicious circle. The decline of corals makes room for seaweed growth, which then becomes dominant in the reef and chemically repulses coral and fish larvae, limiting their recruitment (Dixson et al. 2014). Certainly, corals are essential for reef fish communities and their decline may seriously threaten the sustainability of coral reef fish communities.

Moreover, OA also directly affects the physiology and behavior of coral reef fishes, and anemonefishes are a valuable model to estimate its extent. As previously mentioned in this chapter, anemonefish larvae rely on their sensory capabilities (i.e., smell, hearing, and sight) to discover an adequate reef in which to settle. However, when larvae are raised at high CO_2 levels (i.e., end of the century CO_2 levels, simulating OA), their capabilities to discriminate between chemical cues are disrupted. Larvae become more attracted to chemical cues they usually avoid and lose the ability to discriminate between different types of chemical cues (parents vs non-parents; predators vs non-predators) (Munday et al. 2009; Dixson et al. 2010; Munday et al. 2010; Jarrold et al. 2017). Surprisingly, young coral reef fish can even become attracted to their predators' smell and reduce their antipredatory response when exposed to danger, showing a riskier behavior that could lead to death (Munday et al. 2010; Ferrari et al. 2011). Preference for specific acoustic environments is also thought to be altered in high CO_2 conditions, with larvae becoming deterred by the soundscape of settling habitats but attracted to the soundscapes of unfamiliar or dangerous environments (Simpson et al. 2011; Rossi et al. 2015, 2018). It is still unclear whether sound reception or sound processing by the brain could cause this phenomenon (Holmberg et al. 2019). Studies show that these alterations in sensory functions and behavior might be caused by impaired $GABA_A$ receptors function (Nilsson et al. 2012; Chivers et al. 2014; see Schunter et al. 2019 for more details). Importantly, it must be pointed out that some of the previously cited results are currently under debate (Clark et al. 2020; Clements et al. 2022).

Finally, coral reef fishes have to face an increased SST concomitant with OA. Even though the effects of high temperatures on coral reef fish have seldom been assessed, preliminary results seem to indicate an altered escape response of larvae exposed to a predation simulation, questioning their potential survival in the wild (Warren et al. 2017). Moreover, *in situ* experiments show that sea warming events cause stress in anemonefish (i.e., elevated cortisol circulating levels) (Mills et al. 2015; Beldade et al. 2017). At the genetic and physiological levels, it seems clear that a rise in temperatures creates stress in fish, with the expression of heat shock protein genes drastically increasing, which is associated with an increased CRH gene expression in the brain and increased cortisol circulating concentration (Currie et al. 2000; Madeira et al. 2017; Liu et al. 2018; Uchimura et al. 2019; Goikoetxea et al. 2021). As previously suggested in this chapter, the hypothalamic factor CRH may control both the HPI and HPT axes, themselves regulating the onset and continuity of metamorphosis. Consequently, one wonders whether larvae evolving at high temperatures would experience a precocious metamorphosis or not. Nonetheless, the combined effects of OA and increased SST on coral fish larvae have rarely been investigated to date and surely deserve more attention.

11.7 FUTURE DIRECTIONS

Throughout this chapter, we reviewed the current knowledge about the neuroendocrine control of metamorphosis and protandrous sex change in anemonefish and highlighted how stress could influence these processes. The use of anemonefish as model species is very recent and many questions regarding the biology and life history remain unanswered to date. Indeed, there is a need for basic characterization of the metamorphosis process in its globality, from the perception of environmental cues to the hormones secreted downstream the neuroendocrine cascades, leading to the physiological, morphological, and phenotypical changes transforming a larva into a juvenile. In regards to the stunning process of male-to-female sex change observed in multiple species of the anemonefish subfamily, further research using these remarkable fish as study models will shed light on the underlying mechanisms governing this transformation. Future studies may be able to finally discern the potential role of stress and epigenetic pathways involved in this process. The full characterization of the processes of metamorphosis and protandrous sex change will help us understand how anemonefish communities will be able to respond to future environmental challenges during key life-history steps.

REFERENCES

Alfonso, S., M. Gesto, and B. Sadoul. 2021. Temperature increase and its effects on fish stress physiology in the context of global warming. *Journal of Fish Biology* 98: 1496–1508.

Atema, J., M. J. Kingsford, and G. Gerlach. 2002. Larval reef fish could use odour for detection, retention and orientation to reefs. *Marine Ecology Progress Series* 241: 151–160.

Ball, J. N. 1981. Hypothalamic control of the pars distalis in fishes, amphibians, and reptiles. *General and Comparative Endocrinology* 44: 135–170.

Barth, P., I. Berenshtein, M. Besson, N. Roux, E. Parmentier, B. Banaigs, and D. Lecchini. 2015. From the ocean to a reef habitat: How do the larvae of coral reef fishes find their way home? A state of art on the latest advances. *Vie et milieu* 65: 91–100.

Barton, B. A. 2002. Stress in fishes: A diversity of responses with particular reference to changes in circulating corticosteroids. *Integrative and Comparative Biology* 42: 517–525.

Beldade, R., A. Blandin, R. O'Donnell, and S. C. Mills. 2017. Cascading effects of thermally induced anemone bleaching on associated anemonefish hormonal stress response and reproduction. *Nature Communications* 8: 716.

Bessa, E., B. Sadoul, D. J. Mckenzie, and B. Geffroy. 2021. Group size, temperature and body size modulate the effects of social hierarchy on basal cortisol levels in fishes. *Hormones and Behavior* 136: 105077.

Bishop, C. D., D. F. Erezyilmaz, T. Flatt, C. D. Georgiou, M. G. Hadfield, A. Heyland, J. Hodin, et al. 2006. What is metamorphosis? *Integrative and Comparative Biology* 46(6): 655–661.

Biswal, A., P. P. Srivastava, and T. Paul. 2021. Effect of climate change on endocrine regulation of fish reproduction. In *Recent Updates in Molecular Endocrinology and Reproductive Physiology of Fish*, eds Sundaray, J. K., et al., 335–349. Singapore: Springer.

Blanton, M. L., and J. L. Specker. 2007. The hypothalamic-pituitary-thyroid (HPT) axis in fish and its role in fish development and reproduction. *Critical Reviews in Toxicology* 37: 97–115.

Bonett, R. M., E. D. Hoopfer, and R. J. Denver. 2010. Molecular mechanisms of corticosteroid synergy with thyroid hormone during tadpole metamorphosis. *General and Comparative Endocrinology* 168: 209–219.

Brierley, A. S., and M. J. Kingsford. 2009. Impacts of climate change on marine organisms and ecosystems. *Current Biology* 19: 602–614.

Buchholz, D. R., S. C. V. Hsia, L. Fu, and Y. B. Shi. 2003. A dominant-negative thyroid hormone receptor blocks amphibian metamorphosis by retaining corepressors at target genes. *Molecular and Cellular Biology* 23(19): 6750–6758.

Bury, N. R., and A. Sturm. 2007. Evolution of the corticosteroid receptor signalling pathway in fish. *General and Comparative Endocrinology* 153: 47–56.

Campinho, M. A. 2019. Teleost metamorphosis: The role of thyroid hormone. *Frontiers in Neuroendocrinology* 10: 383.

Casas, L., and F. Saborido-Rey. 2021. Environmental cues and mechanisms underpinning sex change in fish. *Sexual Development* 15: 108–121.

Casas, L., F. Saborido-Rey, T. Ryu, C. Michell, T. Ravasi, and X. Irigoien. 2016. Sex change in clownfish: Molecular insights from transcriptome analysis. *Scientific Reports* 6: 1–19.

Cerdá-Reverter, J. M. and L. F. Canosa. 2009. Neuroendocrine systems of the fish brain. *Fish Physiology* 28: 3–74.

Chivers, D. P., M. I. Mccormick, G. E. Nilsson, P. L. Munday, S. A. Watson, M. G. Meekan, M. G. Mitchell, et al. 2014. Impaired learning of predators and lower prey survival under elevated CO_2: A consequence of neurotransmitter interference. *Global Change Biology* 20: 515–522.

Chivers, D. P., R. A. Ramasamy, M. I. McCormick, S. A. Watson, U. E. Siebeck, and M. C. O. Ferrari. 2014. Temporal constraints on predation risk assessment in a changing world. *Science of the Total Environment* 500–501: 332–338.

Choat, J. H., and D. R. Robertson. 1975. Protogynous hermaphroditism in fishes of the family Scaridae. In *Intersexuality in the Animal Kingdom*, ed. Reinboth, R., 263–283. Berlin, Heidelberg: Springer.

Clark, T. D., G. D. Raby, D. G. Roche, S. A. Binning, B. Speers-Roesch, F. Jutfelt, and J. Sundin. 2020. Ocean acidification does not impair the behaviour of coral reef fishes. *Nature* 577: 370–375.

Clements, J. C., J. Sundin, T. D. Clark, and F. Jutfelt. 2022. Meta-analysis reveals an extreme "decline effect" in the impacts of ocean acidification on fish behavior. *PLOS Biology* 20: e3001511.

Colleye, O., and E. Parmentier. 2012. Overview on the diversity of sounds produced by clownfishes (Pomacentridae): Importance of acoustic signals in their peculiar way of life. *PLoS ONE* 7(11): e49179.

Colleye, O., P. Vandewalle, D. Lanterbecq, D. Lecchini, and E. Parmentier. 2011. Interspecific variation of calls in clownfishes: Degree of similarity in closely related species. *BMC Evolutionary Biology* 11: 365.

Currie, S., C. D. Moyes, and B. L. Tufts. 2000. The effects of heat shock and acclimation temperature on hsp70 and hsp30 mRNA expression in rainbow trout: *In vivo* and *in vitro* comparisons. *Journal of Fish Biology* 56: 398–408.

Deal, C. K., and H. Volkoff. 2020. The role of the thyroid axis in fish. *Frontiers in Endocrinology* 11: 1–25.

Denver, R. J. 1993. Acceleration of anuran amphibian metamorphosis by corticotropin-releasing hormone-like peptides. *General and Comparative Endocrinology* 91: 38–51.

Denver, R. J. 1997. Environmental stress as a developmental cue: Corticotropin-releasing hormone is a proximate mediator of adaptive phenotypic plasticity in amphibian metamorphosis. *Hormones and Behavior* 31: 169–179.

Denver, R. J. 2017. Endocrinology of complex life cycles: Amphibians. In *Hormones, Brain and Behavior*, eds Pfaff, D. W., et al., 145–168. San Diego, CA: Academic Press.

Denver, R. J., and P. Licht. 1989. Neuropeptide stimulation of thyrotropin secretion in the larval bullfrog: Evidence for a common neuroregulator of thyroid and interrenal activity in metamorphosis. *Journal of Experimental Zoology* 252: 101–104.

Dixson, D. L., D. Abrego, and M. E. Hay. 2014. Chemically mediated behavior of recruiting corals and fishes: A tipping point that may limit reef recovery. *Science* 345(6199): 892–897.

Dixson, D. L., G. P. Jones, P. L. Munday, S. Planes, M. S. Pratchett, M. Srinivasan, C. Syms, et al. 2008. Coral reef fish smell leaves to find island homes. *Proceedings of the Royal Society B: Biological Sciences* 275: 2831–2839.

Dixson, D. L., P. L. Munday, and G. P. Jones. 2010. Ocean acidification disrupts the innate ability of fish to detect predator olfactory cues. *Ecology Letters* 13: 68–75.

Dixson, D. L., P. L. Munday, M. Pratchett, and G. P. Jones. 2011. Ontogenetic changes in responses to settlement cues by Anemonefish. *Coral Reefs* 30: 903.

Eales, J. G., and B. A. Himick. 1988. The effects of TRH on plasma thyroid hormone levels of rainbow trout (*Salmo gairdneri*) and arctic charr (*Salvelinus alpinus*). *General and Comparative Endocrinology* 72: 333–339.

Ebbesson, L. O. E., and V. A. Braithwaite. 2012. Environmental effects on fish neural plasticity and cognition. *Journal of Fish Biology* 81: 2151–2174.

Falcón, J., L. Besseau, S. Sauzet, and G. Bœuf. 2007. Melatonin effects on the hypothalamo-pituitary axis in fish. *Trends in Endocrinology & Metabolism* 18(2): 81–8.

Fautin, D. G., and G. R. Allen. 1992. *Field Guide to Anemonefishes and Their Host Sea Anemones*. Perth, WA: Western Australian Museum.

Ferrari, M. C. O., M. I. McCormick, P. L. Munday, M. G. Meekan, D. L. Dixson, Ö. Lonnstedt, and D. P. Chivers. 2011. Putting prey and predator into the CO_2 equation: Qualitative and quantitative effects of ocean acidification on predator-prey interactions. *Ecology Letters* 14: 1143–1148.

Flik, G., P. H. M. Klaren, E. H. Van den Burg, J. R. Metz, and M. O. Huising. 2006. CRF and stress in fish. *General and Comparative Endocrinology* 146: 36–44.

Fricke, H., and S. Fricke. 1977. Monogamy and sex change by aggressive dominance in coral reef fish. *Nature* 266: 830–832.

Galas, L., E. Raoult, M. C. Tonon, R. Okada, B. G. Jenks, J. P. Castaño, S. Kikuyama, et al. 2009. TRH acts as a multifunctional hypophysiotropic factor in vertebrates. *General and Comparative Endocrinology* 164: 40–50.

Galay-Burgos, M., D. M. Power, L. Llewellyn, and G. E. Sweeney. 2008. Thyroid hormone receptor expression during metamorphosis of Atlantic halibut (*Hippoglossus hippoglossus*). *Molecular and Cellular Endocrinology* 281: 56–63.

Galton, A. V. 1990. Mechanisms underlying the acceleration of thyroid hormone-induced tadpole metamorphosis by corticosterone. *Endocrinology* 127(5): 2997–3002.

Geffroy, B., and C. Wedekind. 2020. Effects of global warming on sex ratios in fishes. *Journal of Fish Biology* 97: 596–606.

Gerlach, G., J. Atema, M. J. Kingsford, K. P. Black, and V. Miller-Sims. 2007. Smelling home can prevent dispersal of reef fish larvae. *Proceedings of the National Academy of Sciences of the United States of America*, 104: 858–863.

Geven, E. J. W., G. Flik, and P. H. M. Klaren. 2009. Central and peripheral integration of interrenal and thyroid axes signals in common carp (*Cyprinus carpio* L.). *Journal of Endocrinology* 200: 117–123.

Godwin, J. R. 1994. Behavioural aspects of protandrous sex change in the anemonefish, *Amphiprion melanopus*, and endocrine correlates. *Animal Behaviour* 48: 551–567.

Godwin, J. R. 2010. Neuroendocrinology of sexual plasticity in teleost fishes. *Frontiers in Neuroendocrinology* 31: 203–216.

Godwin, J. R., and P. Thomas. 1993. Sex change and steroid profiles in the protandrous anemonefish *Amphiprion melanopus* (Pomacentridae, Teleostei). *General and Comparative Endocrinology* 91: 144–157.

Goikoetxea, A., B. Sadoul, E. Blondeau-Bidet, J. Aerts, M. O. Blanc, H. Parrinello, C. Barrachina, et al. 2021. Genetic pathways underpinning hormonal stress responses in fish exposed to short-and long-term warm ocean temperatures. *Ecological Indicators* 120: 106937.

Goikoetxea, A., E. V. Todd, and N. J. Gemmell. 2017. Stress and sex: Does cortisol mediate sex change in fish? *Reproduction* 154: 149–160.

Gorissen, M., and G. Flik. 2016. The endocrinology of the stress response in fish. In *Biology of Stress in Fish, Fish Physiology*, eds Schreck, C. B., et al., 75–111. Amsterdam, The Netherlands: Academic Press.

Hazlerigg, D. G., and G. C. Wagner. 2006. Seasonal photoperiodism in vertebrates: From coincidence to amplitude. *Trends in Endocrinology and Metabolism* 17: 83–91.

Holmberg, R. J., E. Wilcox-Freeburg, A. L. Rhyne, M. F. Tlusty, A. Stebbins, S. W. Nye, A. Honig, et al. 2019. Ocean acidification alters morphology of all otolith types in Clark's anemonefish (*Amphiprion clarkii*). *PeerJ* 7: e6152.

Holzer, G., M. Besson, A. Lambert, L. François, P. Barth, B. Gillet, S. Hughes, et al. 2017. Fish larval recruitment to reefs is a thyroid hormone-mediated metamorphosis sensitive to the pesticide chlorpyrifos. *eLife* 6: e27595.

Inui, Y., and S. Miwa. 1985. Thyroid hormone induces metamorphosis in flounder larvae. *General and Comparative Endocrinology* 60: 450–454.

IPCC. 2021. Climate change 2021: The physical science basis. Contribution of working group I to the sixth assessment report of the intergovernmental panel on climate change, eds Masson-Delmotte, V., et al. Cambridge, UK and New York, NY: Cambridge University Press.

Iwata, E., K. Mikami, J. Manbo, K. Moriya-Ito, and H. Sasaki. 2012. Social interaction influences blood cortisol values and brain aromatase genes in the protandrous false clown anemonefish, *Amphiprion ocellaris*. *Zoological Science* 29: 849–855.

Iwata, E., Y. Nagai, M. Hyoudou, and H. Sasaki. 2008. Social environment and sex differentiation in the false clown anemonefish, *Amphiprion ocellaris*. *Zoological Science* 25: 123–128.

Jarrold, M. D., C. Humphrey, M. I. McCormick, and P. L. Munday. 2017. Diel CO_2 cycles reduce severity of behavioural abnormalities in coral reef fish under ocean acidification. *Scientific Reports* 7: 10153.

Jones, G. P., and S. M. Thompson. 1980. Social inhibition of maturation in females of the temperate wrasse *Pseudolabrus celidotus* and a comparison with the blennioid *Tripterygion varium*. *Marine Biology* 59: 247–256.

Jonsson, B., and N. Jonsson. 2014. Early environment influences later performance in fishes. *Journal of Fish Biology* 85: 151–188.

Kawakami, Y., M. Tanda, S. Adachi, and K. Yamauchi. 2003a. Characterization of thyroid hormone receptor α and β in the metamorphosing Japanese conger eel, *Conger myriaster*. *General and Comparative Endocrinology* 132: 321–332.

Kawakami, Y., M. Tanda, S. Adachi, and K. Yamauchi. 2003b. cDNA cloning of thyroid hormone receptor βs from the conger eel, Conger myriaster. *General and Comparative Endocrinology* 131: 232–240.

Kühn, E. R., B. De Groef, S. Van Der Geyten, and V. M. Darras. 2005. Corticotropin-releasing hormone-mediated metamorphosis in the neotenic axolotl *Ambystoma mexicanum*: Synergistic involvement of thyroxine and corticoids on brain type II deiodinase. *General and Comparative Endocrinology* 143(1): 75–81.

Laffoley, D., and J. M. Baxter. 2016. *Explaining Ocean Warming: Causes, Scale, Effects and Consequences*. Gland, Switzerland: IUCN.

Lamm, M. S., H. Liu, N. J. Gemmell, and J. R. Godwin. 2015. The need for speed: Neuroendocrine regulation of socially-controlled sex change. *Integrative and Comparative Biology* 55: 307–322.

Larsen, D. A., P. Swanson, J. T. Dickey, J. Rivier, and W. W. Dickhoff. 1998. *In vitro* thyrotropin-releasing activity of corticotropin-releasing hormone-family peptides in coho salmon, *Oncorhynchus kisutch*. *General and Comparative Endocrinology* 109: 276–285.

Laudet, V. 2011. The origins and evolution of vertebrate metamorphosis. *Current Biology* 21: 726–737.

Lecchini, D., J. Shima, B. Banaigs, and R. Galzin. 2005. Larval sensory abilities and mechanisms of habitat selection of a coral reef fish during settlement. *Oecologia* 143: 326–334.

Leis, J. M. 2007. Behaviour as input for modelling dispersal of fish larvae: Behaviour, biogeography, hydrodynamics, ontogeny, physiology and phylogeny meet hydrography. *Marine Ecology Progress Series* 347: 185–193.

Leis, J. M., U. Siebeck, and D. L. Dixson. 2011. How Nemo finds home: The neuroecology of dispersal and of population connectivity in larvae of marine fishes. *Integrative and Comparative Biology* 51: 826–843.

Liu, H., E. V. Todd, P. M. Lokman, M. S. Lamm, J. R. Godwin, and N. J. Gemmell. 2017. Sexual plasticity: A fishy tale. *Molecular Reproduction and Development* 84: 171–194.

Liu, X., H. Shi, Z. Liu, Y. Kang, J. Wang, and J. Huang. 2018. Effect of heat stress on heat shock protein 30 (Hsp30) mRNA expression in rainbow trout (*Oncorhynchus mykiss*). *Turkish Journal of Fisheries and Aquatic Sciences* 19(8): 681–688.

Madeira, C., D. Madeira, M. S. Diniz, H. N. Cabral, and C. Vinagre. 2017. Comparing biomarker responses during thermal acclimation: A lethal vs non-lethal approach in a tropical reef clownfish. *Comparative Biochemistry and Physiology -Part A: Molecular and Integrative Physiology* 204: 104–112.

Marchand, O., R. Safi, H. Escriva, E. Van Rompaey, P. Prunet, and V. Laudet. 2001. Molecular cloning and characterization of thyroid hormone receptors in teleost fish. *Journal of Molecular Endocrinology* 26: 51–65.

McMenamin, S. K., and D. M. Parichy. 2013. Metamorphosis in teleosts. *Current Topics in Developmental Biology* 103: 127–165.

Meinshausen, M., S. J. Smith, K. Calvin, J. S. Daniel, M. L. T. Kainuma, J. Lamarque, K. Matsumoto, et al. 2011. The RCP greenhouse gas concentrations and their extensions from 1765 to 2300. *Climatic Change* 109: 213–241.

Mills, S. C., R. Beldade, P. Chabanet, L. Bigot, J. L. O'Donnell, and G. Bernardi. 2015. Ghosts of thermal past: Reef fish exposed to historic high temperatures have heightened stress response to further stressors. *Coral Reefs* 34: 1255–1260.

Mills, S. C., R. Beldade, L. Henry, D. Laverty, S. L. Nedelec, S. D. Simpson, and A. N. Radford. 2020. Hormonal and behavioural effects of motorboat noise on wild coral reef fish. *Environmental Pollution* 262: 114250.

Mommsen, T. P., M. M. Vijayan, and T. W. Moon. 1999. Cortisol in teleosts: Dynamics, mechanisms of action, and metabolic regulation. *Reviews in Fish Biology and Fisheries* 9: 211–268.

Montaron, M. F., P. V. Piazza, C. Aurousseau, A. Urani, M. Le Moal, and D. N. Abrous. 2003. Implication of corticosteroid receptors in the regulation of hippocampal structural plasticity. *European Journal of Neuroscience* 18: 3105–3111.

Munday, P. L., D. L. Dixson, J. M. Donelson, G. P. Jones, M. S. Pratchett, G. V. Devitsina, and K. B. Døving. 2009. Ocean acidification impairs olfactory discrimination and homing ability of a marine fish. *Proceedings of the National Academy of Sciences of the United States of America* 106: 1848–1852.

Munday, P. L., D. L. Dixson, M. I. McCormick, M. Meekan, M. C. O. Ferrari, and D. P. Chivers. 2010. Replenishment of fish populations is threatened by ocean acidification. *Proceedings of the National Academy of Sciences of the United States of America* 107(29): 12930–12934.

Newman, R. A. 1992. Adaptive plasticity in amphibian metamorphosis. *BioScience* 42(9): 671–678.

Nilsson, G. E., D. L. Dixson, P. Domenici, M. I. McCormick, C. Sørensen, S. A. Watson, and P. L. Munday. 2012. Near-future carbon dioxide levels alter fish behaviour by interfering with neurotransmitter function. *Nature Climate Change* 2: 201–204.

Nozu, R., and M. Nakamura. 2015. Cortisol administration induces sex change from ovary to testis in the protogynous wrasse, *Halichoeres trimaculatus*. *Sexual Development* 9: 118–124.

Olivotto, I., and B. Geffroy. 2017. Clownfish. In *Marine Ornamental Species Aquaculture*, eds Calado, R., et al., 177–199. Hoboken, NJ: John Wiley & Sons, Ltd.

Piferrer, F. 2013. Epigenetics of sex determination and gonadogenesis. *Developmental Dynamics* 242: 360–370.

Power, D. M., N. P. Elias, S. J. Richardson, J. Mendes, C. M. Soares, and C. R. A. Santos. 2000. Evolution of the thyroid hormone-binding protein, transthyretin. *General and Comparative Endocrinology* 119: 241–255.

Rossi, T., I. Nagelkerken, S. D. Simpson, J. C. A. Pistevos, S. A. Watson, L. Merillet, P. Fraser, et al. 2015. Ocean acidification boosts larval fish development but reduces the window of opportunity for successful settlement. *Proceedings of the Royal Society B: Biological Sciences* 282: 20151954.

Rossi, T., J. C. A. Pistevos, S. D. Connell, and I. Nagelkerken. 2018. On the wrong track: Ocean acidification attracts larval fish to irrelevant environmental cues. *Scientific Reports* 8: 5840.

Rousseau, K., S. Dufour, and L. Sachs. 2021. Interdependence of thyroid and corticosteroid signaling in vertebrate developmental transitions. *Frontiers in Ecology and Evolution* 9: 735487.

Roux, N., P. Salis, A. Lambert, V. Logeux, O. Soulat, P. Romans, B. Frédérich, et al. 2019b. Staging and normal table of post-embryonic development of the clownfish (*Amphiprion ocellaris*). *Developmental Dynamics* 248: 545–568.

Roux, N., P. Salis, and V. Laudet. 2019a. Les larves de poissons coralliens: Un nouveau modèle d'étude de la métamorphose et des hormones thyroïdiennes. *Biologie aujourd'hui* 213: 27–33.

Roux, N., P. Salis, S. H. Lee, L. Besseau, and V. Laudet. 2020. Anemonefish, a model for Eco-Evo-Devo. *EvoDevo* 11: 20.

Sachs, L. M., P. L. Jones, E. Havis, N. Rouse, B. A. Demeneix, and Y. B. Shi. 2002. Nuclear receptor corepressor recruitment by unliganded thyroid hormone receptor in gene repression during Xenopus laevis development. *Molecular and Cellular Biology* 22(24): 8527–8538.

Sadoul, B., and B. Geffroy. 2019. Measuring cortisol, the major stress hormone in fishes. *Journal of Fish Biology* 94: 540–555.

Salis, P., N. Roux, D. Huang, A. Marcionetti, P. Mouginot, M. Reynaud, O. Salles, et al. 2021. Thyroid hormones regulate the formation and environmental plasticity of white bars in clownfishes. *Proceedings of the National Academy of Sciences of the United States of America* 118(23): e2101634118.

Scharrer, E. 1928. Die Lichtempfindlichkeit blinder Elritzen (Untersuchungen über das Zwischenhirn der fische). *Zeitschrift für Vergleichende Physiologie Berlin* 7: 1–36.

Schreiber, A. M., B. Das, H. Huang, N. Marsh-Armstrong, and D. D. Brown. 2001. Diverse developmental programs of *Xenopus laevis* metamorphosis are inhibited by a dominant negative thyroid hormone receptor. *Proceedings of the National Academy of Sciences of the United States of America* 98(19): 10739–10744.

Schunter, C., T. Ravasi, P. L. Munday, and G. E. Nilsson. 2019. Neural effects of elevated CO_2 in fish may be amplified by a vicious cycle. *Conservation Physiology* 7: 1–8.

Scott, A., and D. L. Dixson. 2016. Reef fishes can recognize bleached habitat during settlement: Sea anemone bleaching alters anemonefish host selection. *Proceedings of the Royal Society B: Biological Sciences* 283: 20152694.

Simpson, S. D., P. L. Munday, M. L. Wittenrich, R. Manassa, D. L. Dixson, M. Gagliano, H. Y. Yan. 2011. Ocean acidification erodes crucial auditory behaviour in a marine fish. *Biology Letters* 7: 917–920.

Sørensen, C., L. C. Bohlin, Ø. Øverli, and G. E. Nilsson. 2011. Cortisol reduces cell proliferation in the telencephalon of rainbow trout (*Oncorhynchus mykiss*). *Physiology and Behavior* 102: 518–523.

Sørensen, C., I. B. Johansen, and Ø. Øverli. 2013. Neural plasticity and stress coping in teleost fishes. *General and Comparative Endocrinology* 181: 25–34.

Suzuki, Y., K. Z. McKenna, and H. F. Nijhout. 2020. Regulation of phenotypic plasticity from the perspective of evolutionary developmental biology. In *Phenotypic Switching, Implications in Biology and Medicine*, eds Levine, H., et al., 403–442. Academic Press.

Thomas, J. T., E. V. Todd, S. Muncaster, P. M. Lokman, E. L. Damsteegt, H. Liu, K. Soyano, F. Gleonnec, et al. 2019. Conservation and diversity in expression of candidate genes regulating socially-induced female-male sex change in wrasses. *PeerJ* 7: e7032.

Todd, E. V., O. Ortega-Recalde, H. Liu, M. S. Lamm, K. M. Rutherford, H. Cross, M. A. Black, et al. 2019. Stress, novel sex genes, and epigenetic reprogramming orchestrate socially controlled sex change. *Science Advances* 5: eaaw7006.

Uchimura, T., S. Hara, T. Yazawa, Y. Kamei, and T. Kitano. 2019. Involvement of heat shock proteins on the transcriptional regulation of corticotropin-releasing hormone in medaka. *Frontiers in Endocrinology* 10: 1–9.

Veilleux, H. D., L. Van Herwerden, N. J. Cole, E. K. Don, C. De Santis, D. L. Dixson, A. S. Wenger, and P. L. Munday. 2013. Otx2 expression and implications for olfactory imprinting in the anemonefish, *Amphiprion percula*. *Biology Open* 2: 907–915.

Vissio, P. G., M. P. Di Yorio, D. I. Pérez-Sirkin, G. M. Somoza, K. Tsutsui, and J. E. Sallemi. 2021. Developmental aspects of the hypothalamic-pituitary network related to reproduction in teleost fish. *Frontiers in Neuroendocrinology* 63: 100948.

Warner, R. R., and S. E. Swearer. 1991. Social control of sex change in the bluehead wrasse, *Thalassoma bifasciatum* (Pisces: Labridae). *The Biological Bulletin* 181: 199–204.

Warren, D. T., J. M. Donelson, and M. I. McCormick. 2017. Extended exposure to elevated temperature affects escape response behaviour in coral reef fishes. *PeerJ* 5: e3652.

Wendelaar Bonga, S. E. 1997. The stress response in fish. *Physiological Reviews* 77: 591–625.

Wingfield, J. C. 2008. Comparative endocrinology, environment and global change. *General and Comparative Endocrinology* 157: 207–216.

Wingfield, J. C. 2015. Coping with change: A framework for environmental signals and how neuroendocrine pathways might respond. *Frontiers in Neuroendocrinology* 37: 89–96.

Wong, E. Y. H., and J. Herbert. 2005. Roles of mineralocorticoid and glucocorticoid receptors in the regulation of progenitor proliferation in the adult hippocampus. *European Journal of Neuroscience* 22: 785–792.

Yaeger, C., A. M. Ros, V. Cross, R. S. DeAngelis, D. J. Stobaugh, and J. S. Rhodes. 2014. Blockade of arginine vasotocin signaling reduces aggressive behavior and c-Fos expression in the preoptic area and periventricular nucleus of the posterior tuberculum in male *Amphiprion ocellaris*. *Neuroscience* 267: 205–218.

Zhang, Y., H. Zhang, J. Wang, X. Zhang, S. Bu, X. Liu, Q. Wang, et al. 2020. Molecular characterization and expression patterns of glucocorticoid receptor (GR) genes in protandrous hermaphroditic yellowtail clownfish, *Amphiprion clarkii*. *Gene* 745: 144651.

Zohar, Y., J. A. Muñoz-Cueto, A. Elizur, and O. Kah. 2010. Neuroendocrinology of reproduction in teleost fish. *General and Comparative Endocrinology* 165(3): 438–455.

Zohar, Y. 2021. Fish reproductive biology – Reflecting on five decades of fundamental and translational research. *General and Comparative Endocrinology* 300: 113544.

Zupanc, G. K. H., K. Hinsch, and F. H. Gage. 2005. Proliferation, migration, neuronal differentiation, and long-term survival of new cells in the adult zebrafish brain. *The Journal of Comparative Neurology* 488: 290–319.

Part III

Reproduction and Social Behavior

12 Sex Change from Male to Female
Active Feminization of the Brain, Behavior, and Gonads in Anemonefish

Laura Casas, Coltan Gable Parker, and Justin S. Rhodes

CONTENTS

12.1 INTRODUCTION

In the majority of vertebrates, including fishes, sex is determined during early development and remains unchanged throughout life. Nonetheless, a small number of teleosts have the capacity to change functional sex in adulthood, exhibiting environmental sex determination. This unique sexual strategy is called sequential hermaphroditism and occurs sporadically among fish. It has been described in 462 species from 41 families and 17 orders, accounting for about 1.5% of teleosts (Nelson et al. 2016; Kuwamura et al. 2020). The majority of these species occur in the marine realm, especially on coral reefs, as only 2% of fish with the capacity to change sex inhabit freshwaters (Pandian 2010; Kuwamura et al. 2020). Three different forms of sex change have been described among sequentially hermaphroditic fishes: protandrous (fish mature first as males and then change to females), protogynous (fish mature first as females and then change to males), and bidirectional (fish can change back and forth between male and female phases). The most common type of sequential hermaphroditism in teleosts is protogyny (female first) with a prevalence five times higher than the two other strategies (Kuwamura et al. 2020).

12.1.1 Who, When, and Why?

The size-advantage model (SAM) is the most accepted theory to explain the occurrence of sex change and predict the timing and direction of the switch (Ghiselin 1969; Warner 1975, 1988; Munday et al. 2006). According to SAM, sex change should be favoured when the reproductive success of the opposite sex is enhanced at larger body sizes than the initial sex (Figure 12.1). Thus, the time at which sex change occurs is determined to maximize the lifetime reproductive output of an individual.

Factors triggering sex change in teleosts differ among species but are generally either size (and/or age) dependent or socially regulated and are closely related to the mating system and social structure of the species (Munday et al. 2006; Godwin 2009; Kobayashi et al. 2013). In most protogynous species, sex change is mediated by variations in the social context as they display mainly polygynous mating systems (Warner 1984; Kuwamura and Nakashima 1998; Munday et al. 2006). Most protandrous species need to attain a threshold age or size to change sex. One interesting exception are the anemonefishes (subfamily Amphiprioninae) since sex change is regulated socially in this group despite being protandrous hermaphrodites.

DOI: 10.1201/9781003125365-15

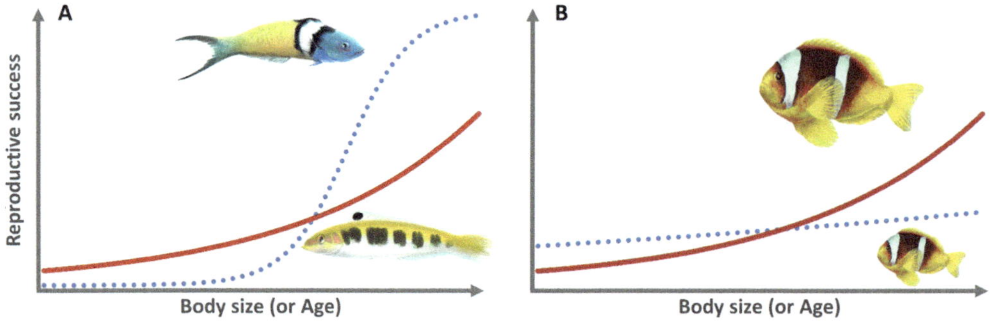

FIGURE 12.1 Graphical representation of the size-advantage model of sex change. A) Protogyny is characterized by a greater increase of reproductive success with age/size in males than females. The bluehead wrasse is among the best-studied female-to-male sex changers and has proven to be a useful model for understanding protogyny. In B) protandry, reproductive success increases faster with age/size in females than in males. Anemonefishes are iconic representatives of protandrous sex change and have been used as models to investigate male-to-female sex change. Red (solid line) and blue (dotted line) indicate female and male growth curves, respectively. Modified from Warner (1975).

12.1.2 The Particular Case of Anemonefishes

Anemonefishes display a distinctive breeding strategy among teleosts. All studied species are monogamous protandrous hermaphrodites with strictly structured societies where sex change is controlled socially (Fricke and Fricke 1977; Moyer and Nakazono 1978; Ross 1978a; Fricke 1983; Godwin and Thomas 1993; Godwin 1994b). Anemonefishes live in pairs or groups organized by a strong size hierarchy that functions as a breeding queue. Within each group, the largest individual is always the female, surrounded by a male and a variable number of immature juvenile non-breeders. If the dominant female disappears, all her subordinates take the opportunity to ascend in rank and grow. The mature male that assumes the dominant position changes sex, while the largest immature fish differentiates into a mature male to complete the breeding pair. It is believed that ecological and social constraints lie at the heart of this particular breeding strategy (Rueger et al. 2021). Anemonefishes live in close association with cnidarian hosts that provide shelter from predators, food, and a refuge to lay their eggs (Fautin 1986, 1992). The patchy distribution of anemone hosts makes travel across the reef dangerous for a widowed fish in search of a new mate. Sex change allows anemonefishes to circumvent this danger (Elliott et al. 1995).

Similarly, non-breeders choose to forgo reproduction and avoid the risk of moving between hosts since inheritance of the territory provides the potential for future reproduction (Buston 2004; Rueger et al. 2021). Field observations in reefs of Japan have reported the presence of an individual of *A. clarkii* in the same anemone for 14+ years, going from subordinate immature to dominant male and subsequently changing sex to female (Moyer, 1986). The social hierarchy regulates the growth of subordinates to maintain defined size differences between individuals adjacent in rank, diminishing the risk of eviction due to direct competition for breeding positions (Buston 2003; Buston and Cant 2006; Branconi et al. 2020).

This breeding strategy is costly, not only energetically but also in terms of time. Thus, in more favourable environments where the danger of movement between host anemones is reduced, social assemblages and pair bonds are not as stable, at least in some anemonefish species. In environments with a high density of host anemones, migration of mature males between territories to establish new breeding pairs is more frequent as it saves the time and energy required to change sex, a period during which the fish are not able to reproduce (Ochi 1989a). Moreover, when a breeding space is vacated, large non-breeders in the vicinity, migrate to fulfil the vacated position (Ochi 1989b). Movement of non-breeders among social groups is promoted as it favours rank improvement faster than waiting for a vacancy in its own group.

Sex change involves coordinated changes along multiple body axes. Once the dominance hierarchy is altered, rapid neurochemical changes in the brain reflect the upgraded status of social sex changers by adjusting the behavior within minutes to hours. The male, previously dominated by the female, starts displaying aggression and dominance. Subsequently, changes in the brain are transmitted to the gonads triggering a complete reorganization of the gonadal tissue. Together, a coordinated cascade of behavioral, physiological, and morphological changes results in a functional male becoming a functional female.

12.2 ACTIVE FEMINIZATION OF THE BRAIN AND BEHAVIOR

The great majority of what is known about sexual differentiation of the brain and periphery comes from mammalian models, where feminization occurs "by default". Whereas masculinization is driven by gonadal hormones which shift the trajectory of development from female to male (McCarthy 2008; McCarthy et al. 2017; Tsukahara and Morishita 2020), feminization occurs in absence of gonadal steroids. However, just because a process occurs in the absence of gonadal hormones, does not mean it is a passive process. Evidence in mice establishes active proliferation and survival of cells in the brain and active suppression of masculinizing genes (Ahmed et al. 2008; Nugent et al. 2015; Mohr et al. 2016). Hence, at least in mammals, feminization is not a "passive" process so much as it is a developmental process without a specific identifiable trigger like gonadal hormones, and thus the mechanisms have been more difficult to identify and sort out. The anemonefish represent an outstanding model to study the process of active feminization. Feminization is "active" in anemonefishes in the sense that female sexual differentiation follows from a social trigger and involves complete transformation from male to female, and so the active mechanisms can be tracked in real time as the fish transforms. This is less tractable in a mammalian model or other vertebrate models that display genetic sex determination or sex change from female to male in which there is no clear signal initiating the cascade of feminization.

There are three conceptually distinct stages of the brain's involvement in the process of sex change in anemonefish. 1) *Initiation of sex change*: the brain senses and perceives ascent to dominant status and initiates sex change. 2) *Behavioral sex change*: male fish behave differently from female fish, and these behavioral differences must result from changes in physiology, connectivity, and/or activation of certain brain circuits. 3) *Hypothalamic sex change*: major changes in the numbers and types of neurons and glia in the preoptic area (POA) of the hypothalamus (and likely other hypothalamic sources of gonadotroph innervation, including the ventral tuberal and mediobasal hypothalamus) are needed to appropriately regulate the male versus female gonad via the hypothalamic-pituitary-gonad axis (Elofsson et al. 1997; Zohar et al. 2010; Trudeau 2018; Dodd et al. 2019).

The time course and process that ensues for producing female behavior and a female POA have not been fully worked out for anemonefishes (Dodd et al. 2019). Nevertheless, the collective evidence suggests the process is likely very different from protogynous (female-to-male) sex change (Figure 12.2).

The beginning is the same: the fish perceive a change in the social hierarchy. However, in protogynous sex change,

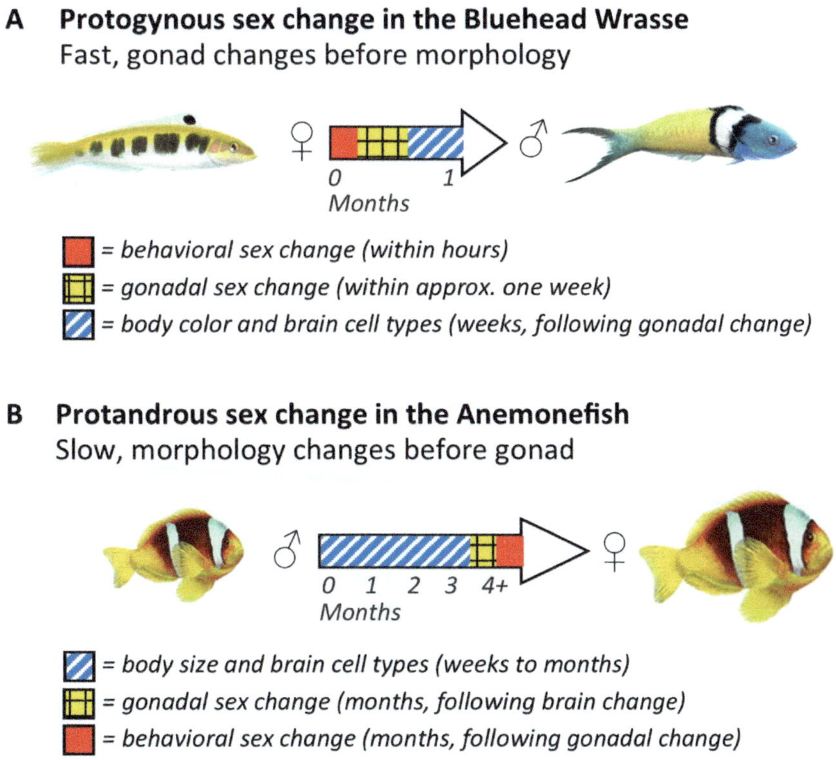

FIGURE 12.2 Time-course of sex change from female to male in the bluehead wrasse as compared to male to female sex change in anemonefish. A) Female-to-male sex change in the bluehead wrasse is well established to occur in less than one month (Warner and Swearer 1991). Behavioral sex change occurs first, before gonadal and hormonal sex change. Masculinization of the gonads, which occurs within a week, precedes and orchestrates morphological changes in the POA and periphery via male sex hormones (Grober et al. 1991). B) In contrast, male-to-female sex change in anemonefish is less well understood. The process occurs over a wider time frame. In some species, gonadal sex change can be completed relatively rapidly after female removal (e.g., 26 days in *A. bicintus* or 45 days in *A. melanopus* [Fricke and Fricke 1977; Godwin 1994a] while for *A. ocellaris* and other species [Moyer and Nakazono 1978; Hattori 1991; Godwin 1994a; Dodd et al. 2019] it takes four months or longer). In anemonefishes, while aggressive and dominance-associated behavior changes nearly immediately in response to ascension in the dominance hierarchy, the full behavioral phenotype is not intermediate between male and female and is more male-like than female-like. Behavioral sex change in anemonefishes occurs after gonadal sex change and after the sex steroids have completely changed. Hence, the process is nearly opposite to that described for the bluehead wrasse and other protogynous species.

behavior changes to that of the opposite sex within hours to days, well before the gonads change. Within a few days to weeks the brain signals to the gonads to change, and testicular tissue develops (Warner and Swearer 1991; Todd et al. 2019). Then, androgens are released from the gonads to complete the morphological sex change, including alterations in body size, changes in colouration and changes in numbers of GnRH neurons in the anterior POA that are involved in regulating gonadal function (Grober et al. 1991; Warner and Swearer 1991; Semsar and Godwin 2004; Todd et al. 2019). In anemonefishes, gonadal sex change (as defined by both the presence of vitellogenic oocytes in the gonad and the presence of a female-typical sex hormone profile) appears to be highly variable and can take a long time, up to years (described in more detail in the following section, "Gonadal Sex Change"). However, changes in the numbers of cells in the anterior POA appear to follow a predictable time-course and are completed in several months (Dodd et al. 2019). The precise time-course of cell addition to the POA has yet to be evaluated. The gonadal hormones (androgen and estrogen) of sex-changing fish are also in between male and female levels, but are overall more male-like than female-like (Dodd et al. 2019). The implication is that in anemonefishes, the parts of the brain that control the gonads (e.g., POA) change sex first, while the gonads and behavior revert to a reproductively dormant state not unlike that seen in juvenile non-breeder fish. When the fish is ready (and it is not clear exactly what constitutes "readiness", perhaps a certain threshold of body growth or brain change), these brain parts signal to the gonads to feminize, and the sex hormone profile follows after vitellogenic eggs develop (Dodd et al, 2019).

12.2.1 Initiation of Sex Change

The sensory stimuli that inform the male of his new dominant status remain to be determined. These stimuli likely include visual interpretation of behavioral displays and interactions with the other fish in the group, possible detection of pheromones, changes in the perception of psychological stress exerted by the other fish, or perception of sounds emitted by the other fish (Colleye and Parmentier 2012; Johnston and Dixson 2017; Desrochers et al. 2020). How such social signals are processed in the brain to initiate sex change remains completely unknown in any species of sex-changing fish to our knowledge.

12.2.2 POA Sex Change

In anemonefishes, the anatomy of the POA changes to the female phenotype before gonadal sex change and before behavioral sex change occurs. A cell population has been identified in the anemonefish anterior POA that contains roughly twice as many cells in females compared to males, controlling for body size differences (Dodd et al. 2019; Figure 12.3A–D). This cell population is feminized by six months after the initiation of sex change, before the sex-changing fish displays vitellogenic oocytes and female-typical gonadal sex hormones. A recent analysis of single nuclei RNA sequencing (snRNA-seq) of the POA region from six female and six male *A. ocellaris* revealed a number of different neuron populations that are sexually dimorphic in cell number, including several candidates displaying an increased number of inhibitory neurons in females (Figure 12.3E–F). The anterior POA population is of particular interest as a putative homologue or phenologue of the rodent anteroventral periventricular nucleus (AVPV). The AVPV contains more inhibitory neurons in females compared to males and is crucial in regulating ovarian hormone cycling and ovulation. If functional or developmental similarities can be established between the anemonefish anterior POA and the rodent AVPV, anemonefishes could prove indispensable to the study of vertebrate brain feminization.

12.2.3 Behavioral Sex Change

Some of the earliest published work on anemonefishes described their remarkable sex differences in aggressive and parental behavior (Ross 1978a, 1978b). Females are much more aggressive towards other females as compared to males, defending their territory vigorously from female intruders and reinforcing the social hierarchy within their group using a combination of aggressive displays, charging, and biting. Anemonefish males provide the majority of parental care to developing eggs, tending to them with frequent fanning (to keep water circulating over the eggs and remove debris drifting nearby) and mouthing (to keep nearby algae and other threats to the eggs in check) (see Chapter 15, "Parental Care: Patterns, Proximate, and Ultimate Causes and Consequences").

12.2.3.1 Sex Change in Aggression

Sexually differentiated aggression in anemonefishes has been well described, but only one published study has actually assessed aggression *during* sex change. Field work with *A. melanopus* assessed aggressive behavior of dominant sex-changing fish towards conspecific female intruders (Godwin 1994b). Fish were tested 15 days after the resident female was removed, and were found to display approximately two-fold greater aggression than a resident male. While an increase in aggression does represent a shift towards more "female-like" behavior, it is important to compare this with the patterns of aggression expected from true females. Earlier fieldwork with *A. melanopus* assessed aggression from resident males and resident females towards a conspecific female intruder and found that females displayed between two- and 12-fold greater aggression than their male partners, depending on the pair and the distance of the intruder from the centre of the territory (Ross 1978a). More recent laboratory work with *A. ocellaris* has found that female residents display approximately five-fold greater aggression, on average, than male residents towards a female intruder (Iwata and Manbo

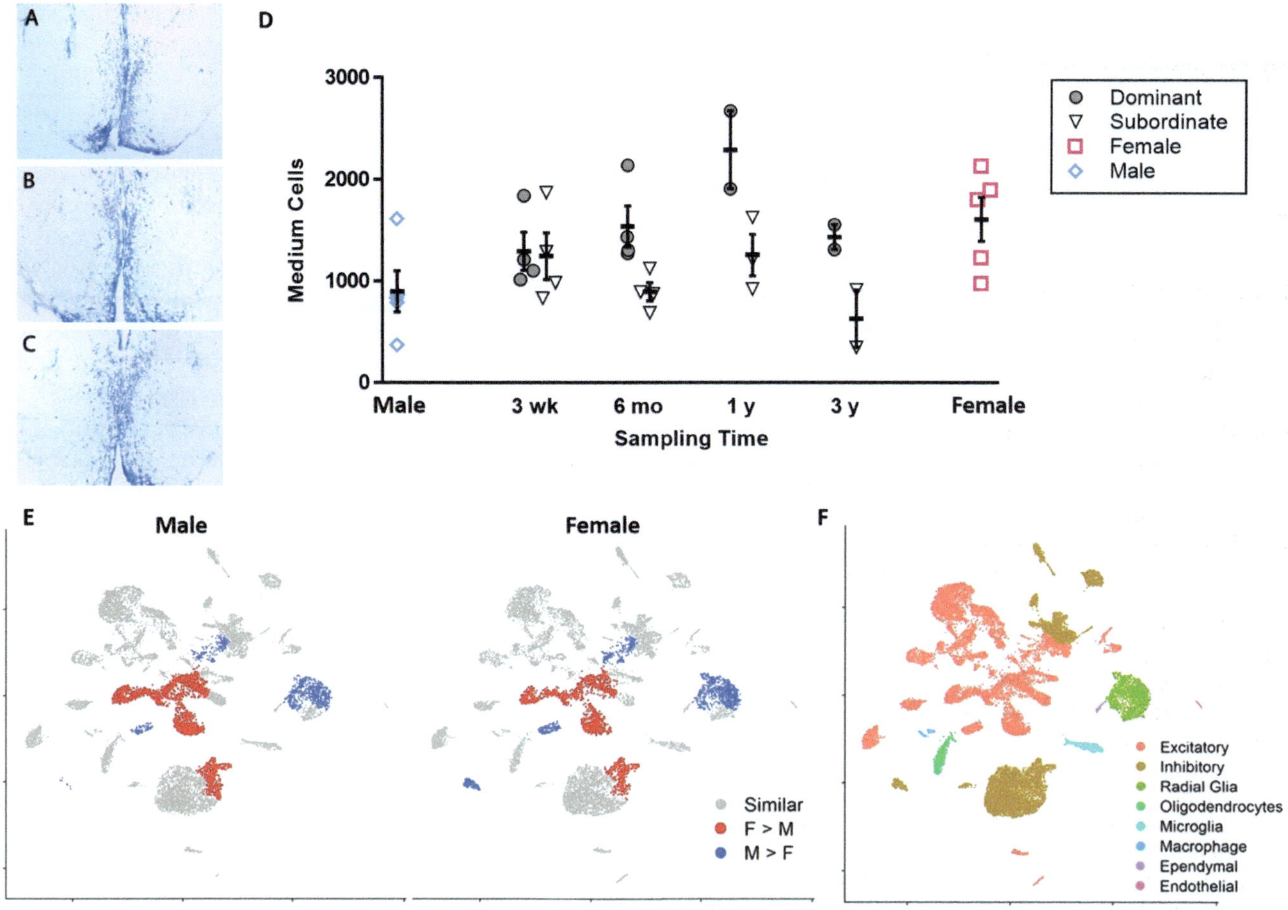

FIGURE 12.3 Sexually dimorphic cell populations in *Amphiprion ocellaris* POA region. A–C) Representative sections through the anterior POA from a reproductive male, a sex-changing fish six months after induction, and a reproductive female, respectively. Photos were taken at 50X magnification. D) Total number of medium sized cells in the anterior POA plotted against time-point following induction of sex change with reproductive males (M) on the left and reproductive females (F) on the right. Means are shown as horizontal bars within the scatter plot with standard errors bars in the vertical position. Significant differences were observed between dominant (sex-changing) and subordinate (not sex-changing) at the six-month and one-year time points. None of the subordinate groups differed from males, whereas dominant members from the six-month and one-year time-points displayed significantly greater numbers of medium cells than males. Females were significantly different from six-month subordinate and three-year subordinate groups. E) Approximately 24,000 cellular nuclei from the POA (and surrounding region) of six male and six female *A. ocellaris* were analyzed by single nuclei RNA-sequencing. Each cell is plotted as a point on two multi-dimensional scaling axes such that cells that are farther apart are more distinct in their nuclear transcript profiles. The first two plots show males and females separately and identify the clusters of cells that are sexually dimorphic in number. Red clusters represent cell types that are more numerous in females than males (i.e., had a greater proportion of cells of that type), whereas blue clusters were more numerous in males than females. F) The third plot shows all the nuclei (males and females together) and identifies the major cell types based on multiple canonical cell-type specific markers.

2013). This study also assessed aggression towards male intruders, finding that resident males and females display comparable levels of aggression towards a male intruder, and at a level much lower than the level of aggression seen by females towards females.

Altogether, with regard to conspecific aggression, females are not simply more aggressive. Rather, they are selectively more aggressive towards female conspecifics, and to a degree that is unmatched by males in any context. Recent data suggest that while sex-changing fish display slightly elevated aggression towards both males and females, the levels of aggression displayed towards females are nowhere near those displayed by typical females towards females. Thus, the aggression phenotype of a sex-changing *A. ocellaris* is neither male nor female. It is unique in that it constitutes moderately elevated indiscriminate aggression towards both sexes. Moreover, as described below, the phenotype of a sex-changing *A. ocellaris* is decidedly male-like when it comes to parental care.

12.2.3.2 Sex Change in Parental Care

While female anemonefish spend their time patrolling their territory and preserving its integrity, male anemonefish contribute to the success of the mating pair by picking up parenting duty (Ross, 1978b; DeAngelis et al. 2017, 2018). Unlike aggressive and defensive behavior, which are just as necessary during sex change as any other time, parental behavior is not expressed under natural conditions for a sex-changing fish, as they will not have offspring. However, recent research shows that even lone non-breeding fish with no parenting experience indiscriminately care for unrelated conspecific eggs in their territory (Phillips et al. 2020). A recent experiment in which sex-changing fish and their partner were given eggs from another spawning pair in the colony, established the sex-changing fish as the primary caregiver of the foster eggs. Sex-changing fish displayed similar levels of foster care as mature males did. The subordinate member of the pair spent significantly less time in the nest and less time caring for the eggs. These data reinforce the notion that sex-changing anemonefish behave more like males than females. Further, even in those pairs in which the dominant sex-changing fish had developed female gonads with vitellogenic oocytes by the time of behavioral assessment, these fish provided male-like parental care. Taken together the data suggest that behavioral sex change occurs after gonadal sex change in *A. ocellaris*.

12.2.3.3 Neural Mechanisms of Behavioral Sex Change

General neuroendocrine mechanisms underlying aggression and parental care in anemonefish have been explored, but how such mechanisms are modified during protandrous sex change has not been investigated yet to our knowledge. Isotocin and arginine-vasotocin both play a role in regulating parental care and aggression in anemonefish (DeAngelis et al. 2017, 2020), and estradiol administration increases aggression at least in undifferentiated non-breeder fish (Iwata and Suzuki, 2020). Conspecific aggression (which is highly sexually differentiated, as described earlier) is positively regulated by arginine-vasotocin signalling and is associated with cell activation in the POA and the periventricular nucleus of the posterior tuberculum (Yaeger et al. 2014). Female anemonefish have higher whole-brain aromatase (estradiol synthase) gene expression, and males have higher whole-brain isotocin receptor gene expression (Casas et al. 2016; DeAngelis et al. 2018). These differences in gene expression may play a role in mediating the sexual differentiation of aggressive and parental behavior. Current theories about the regulation of social behaviors emphasize the role of neuropeptides like isotocin and arginine-vasotocin, and possibly also neurosteroids like estradiol, in modulating patterns of correlated cell activation across a network of brain regions involved in social decision making (Johnson and Young 2017). Sex differences in aromatase and isotocin receptor expression may be confined to one or many nodes of the network, and changes in expression levels would be a convenient mechanism by which gonad-independent behavioral sex change may be accomplished.

12.3 GONADAL SEX CHANGE

The complex genomic response of the brain associated with sex change is subsequently transmitted to the gonads along the hypothalamic-pituitary-gonadal axis. Receptors on the gonadal tissue receive the hormonal signals that stimulate the corresponding resorption or extension, completing gonadal reorganization (Kobayashi et al. 2009). This process involves a complete restructuring of the gonadal tissue, which can be accomplished in a few weeks or take years, depending on both the social environment and the anemonefish species. The fastest completion of sex change reported in the genus corresponded to *A. bicintus*, with the sex-changing mate laying eggs 26 days after the disappearance of the dominant female (Fricke and Fricke 1977). In several other species, including *A. clarkii*, *A. akallopisos*, and *A. melanopus*, the process lasts several months while in *A. frenatus* it can take up to several years (Fricke and Fricke 1977; Moyer and Nakazono 1978; Hattori 1991; Godwin 1994a). Moreover, in captive conditions, reorganization of the gonads might be delayed for long periods after brain feminization (Dodd et al. 2019). It has been suggested that the timing of feminization might be size-dependent and that sex change is contingent upon the presence of a smaller conspecific to complete the breeding pair (Fricke 1983; Hattori 1991; Casas et al. 2016). Nonetheless, other factors potentially explaining differences within and among anemonefish species have not yet been described.

12.3.1 Histological Changes across Sex Change in *Amphiprion*

As protandrous species, the primary development in anemonefishes is female while male development is a temporary phase (Figure 12.4). During early gonadal development, immature individuals show mainly ovarian tissue with primary perinucleolus oocytes and no distinguishable spermatogenic cells. Later, in the juvenile phase, ovotestis show primary growth phase oocytes and male germinal cells at different stages of development, including some spermatozoa (Casadevall et al. 2009). As the fish matures into a breeding male, the testicular tissue expands gradually occupying most of the gonadal cavity while the female region is restricted to the periphery. The male stage is characterized by seminiferous tubules with cells at different stages of development, organized in cysts that open up upon completion of spermiogenesis to discharge the spermatozoa into the lobular lumen and, subsequently, the sperm duct. Males only possess oocytes in the primary growth phase (i.e., oogonia, chromatin-nucleolar, and perinucleolar stages), peripherally located (Figure 12.4A). The transition from male to female starts with the formation of an ovarian cavity or lumen and the displacement of the spermatogenic tissue towards the periphery of the gonad. The ovarian tissue is still mainly composed of oocytes in the primary growth phase but cells in the stage of cortical alveoli start to form, indicating the readiness to start vitellogenesis. In the testicular tissue, all the spermatogenic stages (from spermatogonia to spermatozoa) are still present (Figure 12.4B). As sex change progresses, the delimitation of the lumen completes, and the testicular tissue starts to degenerate while the number of oocytes in the stage of cortical alveoli increases significantly

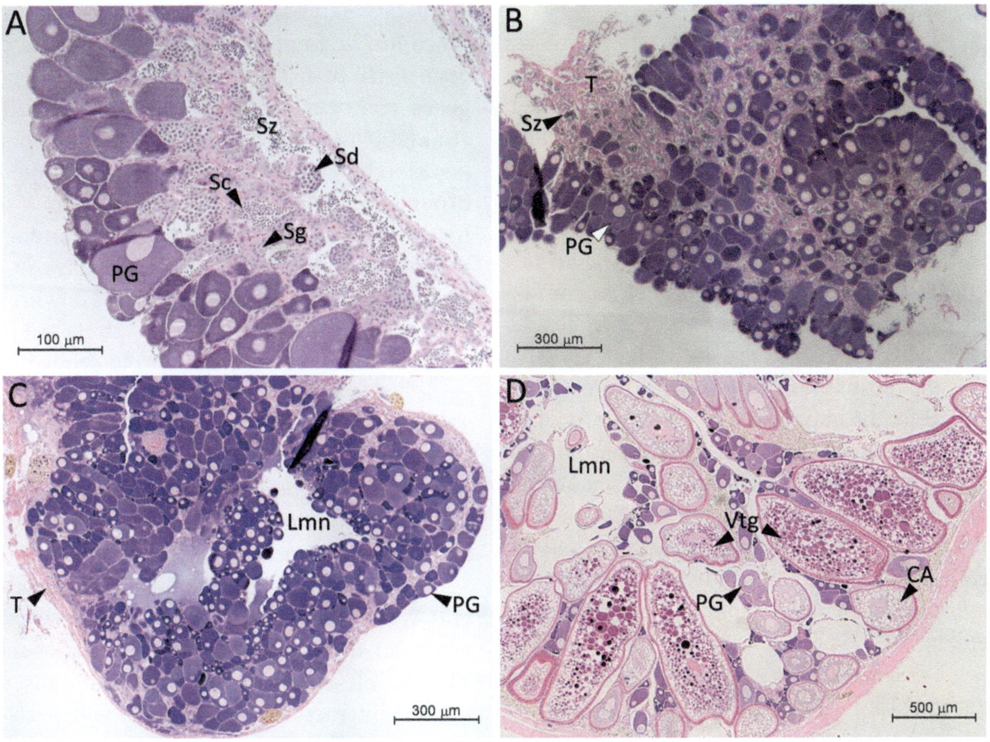

FIGURE 12.4 Histological sections of the Red Sea anemonefish (*Amphiprion bicinctus*) ovotestis across sex change displaying four different gonadal stages from testis to functional ovary. A) Longitudinal section of a functional male. Ovary and testes occupy half of the gonad each, oocytes are in primary growth stage (PG), while various stages of development can be observed in male germ cells (Sg: Spermatogonia; Sd: Spermatida; Sc Spermatocyte; Sz: Spermatozoid). B) Transverse section of a specimen at an early transitional stage of sex change. Testicular tissue (T) has reduced and mostly spermatozoids are visible, while other male germ cells are degenerated. Ovary tissue has expanded and only PG oocytes are present. C) Transverse section of an individual at a late transitional stage of sex change (immature female). Testicular tissue (T) is restricted to a small portion in the periphery of the gonad and highly degenerated while no germ cells are detected. The lumen of the ovary (Lmn) is visible, but still only PG oocytes are present. D) Transverse section of a mature female. Oocytes in secondary growth stages are visible (CA: cortical alveoli; Vtg: vitellogenic oocyte). Figure originally published in Casas et al. (2016).

(Figure 12.4C). The transition from male to female is completed when oocytes begin to mature and the testicular tissue of the ovotestis is resorbed. The full degeneration of the testicular tissue indicates that the sex change is irreversible (Casas et al. 2016). The mature female gonad is characterized by well-developed ovarian tissue with oocytes of all stages (chromatin-nucleolar, perinucleolar, and cortical alveoli, vitellogenic stages, mature and atretic oocytes) and asynchronous ovarian development (Figure 12.4D).

12.3.2 Endocrine Regulation of Sex Change: Sex Steroids

Sex change was first described in anemonefishes in the late 1970s (Fricke and Fricke 1977) and 16 more years were necessary for the characterization of the first key players involved in the process (Godwin and Thomas 1993). Significant progress has been made since, but it was not until recently that a pioneering wide-genome study of sex change in the Red Sea anemonefish provided insights into the global molecular mechanism orchestrating social sex change and gonadal restructuring (Casas et al. 2016; Casas et al. 2018).

The sexual identity depends on the balance between gonadal estrogen and androgen production, which promote ovarian and testicular function, respectively, in vertebrates. In teleosts, the major relevant steroids are 17β-estradiol (E2) and 11-ketotestosterone (11-KT). Their production depends on the bioconversion of testosterone (T), which acts as a prohormone in fish, via two opposing pathways (Frisch 2004; Guiguen et al. 2010). In all sex-changing fishes, pronounced shifts in the balance of sex steroids occur across sex change and marked sex-biased levels characterize males and females. In anemonefishes specifically, a sharp drop in 11-KT levels precedes testis degeneration and is followed by a gradual increase in E2 production accompanying the progression of sex change (Godwin and Thomas 1993). High 11-KT levels and low E2 levels are present in functional males, whereas the opposite trends characterize functional females (Nakamura et al. 1994; Kobayashi et al. 2010; Mills et al. 2018). Moreover, treatment of differentiating juveniles with exogenous E2 prevents the differentiation of testicular cells forming exclusively ovarian tissues (Miura et al. 2013).

12.3.3 Molecular Pathways Underlying Gonadal Sex Change

At the molecular level, the maintenance of the sexual identity in anemonefishes is dependent on male and female pathways, well conserved in all vertebrates, acting antagonistically (Figure 12.5). Thus, the simultaneous activation of the appropriate sex-specific network and the suppression of the sex-opposing network actively regulate gonadal fate (Capel 2017). One of the best-known players in the feminizing gene network is the *cyp19a1a* gene, which encodes the gonadal aromatase and controls the sex steroid balance. This enzyme catalyzes the biosynthesis of estrogens from androgens and is essential for the maintenance of ovarian function in vertebrates (Guiguen et al. 2010). The ovarian aromatase gene has been proposed to play a central role in sex change in anemonefishes acting as the potential switch of the female pathway by causing the rise of estrogen production and the concomitant collapse of the male network. The expression of this gene shows a sharp upregulation in transitional females and remains overexpressed until the completion of sex change in *A. bicinctus* (Casas et al. 2016). The administration of an aromatase inhibitor to functional females of A. clarkii causes their masculinization, resulting in active spermatogenic germ cells in the ovarian tissue (Nakamura et al. 2015). Another key member of the feminizing network in vertebrates, including mammals and fish, is the forkhead transcriptional factor L2 (*foxl2*) (Uhlenhaut et al. 2009; Siegfried 2010; Georges et al. 2014). It is essential for ovarian differentiation and maintenance in fish since it regulates aromatase expression and hence, the synthesis of estrogen. Since *foxl2* can upregulate aromatase and estrogens upregulate *foxl2* in fish, a positive feedback loop between both genes has been proposed (Wang et al. 2007; Yamaguchi et al. 2007; Guiguen et al. 2010). In anemonefishes, is proposed to be pivotal for the activation of the female pathway driving the gonadal transformation from testis to ovary during sex change. Accordingly, the spatio-temporal expression profiles of both genes across sex change in *Amphiprion* are highly correlated showing a marked downregulation in male stages (Casas and Saborido-Rey 2021). Additionally, several genes required for ovarian steroid production (the steroidogenic acute regulatory protein [*Star*], the enzyme estradiol 17β-dehydrogenase 1 [*hsd17b1*]), have also been proposed as relevant members of the feminizing network in anemonefishes. Their expression is upregulated in mature females and correlates significantly with the regression of the testis and the development of the ovary.

The opposing-male promoting network consists of several widely conserved components of sex determination, testicular differentiation, and spermatogenesis in teleosts. The doublesex and mab-3 related transcription factor 1 (*dmrt1*) interacts antagonistically with *foxl2* to regulate aromatase expression and thus, estrogen production (Matson et al. 2011; Li et al. 2013). Gene *dmrt1* is not only fundamental in the differentiation and maintenance of the gonadal male identity across vertebrates but also for the inhibition of the female pathway (Matson and Zarkower 2012). The expression profile of *dmrt1* in anemonefishes and other protandrous fish shows a steady decline across

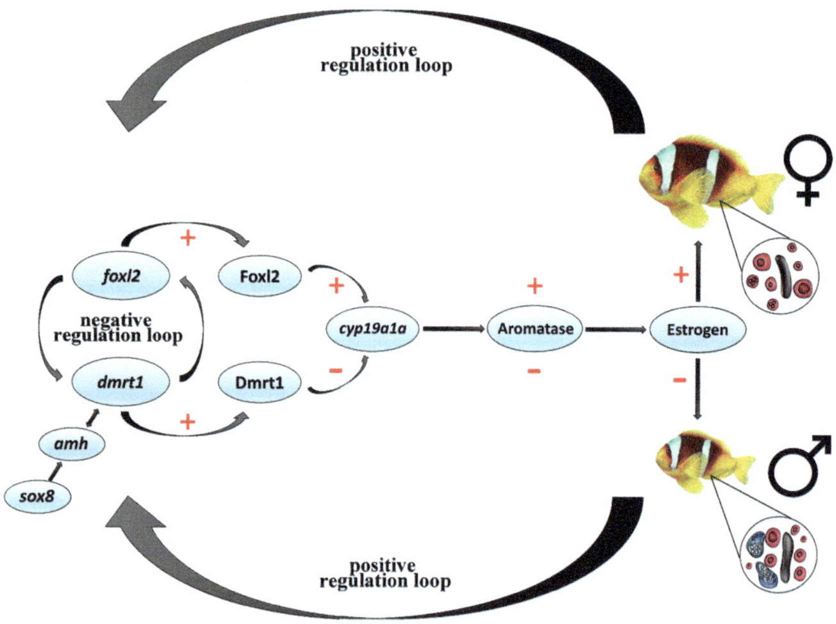

FIGURE 12.5 Model of the molecular mechanism underlying gonadal sex change in anemonefish. When sex change is triggered, the hypothalamic-pituitary-gonadal axis communicates the brain response to the gonad, which induces a rapid spike of aromatase production. This produces the upregulation of *foxl2* and the synthesis of steroid hormones tipping the sex steroid balance towards the estrogen. At the same time, a negative feedback loop between *foxl2* and *dmrt1* downregulates the male pathway genes shutting down the male network, allowing gonad remodelling. Gene names are written with lowercase italics while, for protein names, non-italic and the first letter in uppercase are used, following the zebrafish nomenclature convention. Figure modified from Casas et al. (2016).

sex change, paralleling the regression of the testis (He et al. 2003; Casas et al. 2016; Zhang et al. 2019). A second essential member of the male pathway in fish is the anti-Müllerian hormone (*amh*) gene (Pfennig et al. 2015; Adolfi et al. 2019). The exact function of *amh* in fish remains unclear but it has been proposed as a candidate to downregulate aromatase (Wang and Orban 2007). In anemonefish and the protogynous black porgy, a steady decrease in *amh* expression values is observed throughout sex change, following closely the expression of *dmrt1* (Wu et al. 2010; Zhang et al. 2019; Casas and Saborido-Rey 2021). The activity of *amh* is regulated by *sox* genes, members of a large family of transcription factors that encode key mediators of testis determination and male fertility maintenance in mammals (Jiang et al. 2013). The most prominent member of this family is *sox9* as it is necessary and sufficient to induce testis differentiation in various vertebrates (Vining et al. 2021). Although its function is poorly understood in fish, *sox9* has been reported as either male-restricted or strongly male-biased in at least ten hermaphroditic species, including protogynous and protandrous fish (Manousaki et al. 2014; Liu et al. 2015; Tsakogiannis et al. 2018, 2019). However, studies in anemonefish do not support an important role of this gene in sex change. An alternative family member, *sox8*, has been proposed as an important determinant for the maintenance of testis cell identity (Casas et al. 2016). This gene is known to be essential for male fertility maintenance in mammals and functions redundantly with *sox9* in the maintenance of spermatogenesis (Barrionuevo et al. 2016). In anemonefish, expression of *sox8* is upregulated in males and steadily decreases as sex change progresses (Casas et al. 2016).

The molecular mechanism described here involves a complex regulatory loop combining steroid hormonal activity with transcriptional regulation of well-conserved genes among vertebrates and teleosts. It has been proposed to be common across sequential hermaphrodites, with protandrous species sharing a mirrored mechanism with protogynous sex-changers (Casas and Saborido-Rey 2021). However, the precise organization of the gene network directing sex reprogramming has not been elucidated in anemonefishes or any other hermaphrodite fish species yet. Similarly, few studies have addressed the epigenetic processes mediating sex change in fish and a detailed mechanism is still lacking.

12.4 CONCLUSION

Anemonefishes are now established as the best-studied group of protandrous fishes with socially controlled sex change, as evidenced by the rich literature surveyed in this book. They are well suited to the study of protandrous sex change, in both its molecular mechanisms and the peculiar ecological factors that define the anemonefish subfamily and drove the emergence of their particular mode of sex change. There is no doubt that future research in this model will continue to surprise us and enrich our understanding of sex change and vertebrate sexual development more generally.

REFERENCES

Adolfi, M. C., R. T. Nakajima, R. H. Nóbrega, and M. Schartl. 2019. Intersex, hermaphroditism, and gonadal plasticity in vertebrates: Evolution of the Müllerian Duct and Amh/Amhr2 signaling. *Annual Review of Animal Biosciences* 7: 149–172.

Ahmed, E. I., J. L. Zehr, K. M. Schulz, B. H. Lorenz, L. L. DonCarlos, and C. L. Sisk. 2008. Pubertal hormones modulate the addition of new cells to sexually dimorphic brain regions. *Nature Neuroscience* 11: 995–997.

Barrionuevo, F. J., A. Hurtado, G. J. Kim, F. M. Real, M. Bakkali, J. L. Kopp, M. Sander, et al. 2016. Sox9 and Sox8 protect the adult testis from male-to-female genetic reprogramming and complete degeneration. *Elife* 5: e15635.

Branconi, R., T. A. Barbasch, R. K. Francis, M. Srinivasan, G. P. Jones, and P. M. Buston. 2020. Ecological and social constraints combine to promote evolution of non-breeding strategies in clownfish. *Communications Biology* 3: 649.

Buston, P. M. 2003. Size and growth modification in clownfish. *Nature* 424: 145–146.

Buston, P. 2004. Does the presence of non-breeders enhance the fitness of breeders? An experimental analysis in the clown anemonefish *Amphiprion percula*. *Behavioral Ecology and Sociobiology* 57: 23–31.

Buston, P. M., and M. A. Cant. 2006. A new perspective on size hierarchies in nature: Patterns, causes, and consequences. *Oecologia* 149: 362–372.

Capel, B. 2017. Vertebrate sex determination: Evolutionary plasticity of a fundamental switch. *Nature Reviews Genetics* 18: 675–689.

Casadevall, M., E. Delgado, O. Colleye, S. B. Monserrat, and E. Parmentier. 2009. Histological study of the sex-change in the skunk clownfish *Amphiprion akallopisos*. *Open Fish Science Journal* 2: 55–58.

Casas, L., F. Saborido-Rey, T. Ryu, C. Michell, T. Ravasi, and X. Irigoien. 2016. Sex change in clownfish: Molecular insights from transcriptome analysis. *Scientific Reports* 6: 35461.

Casas, L., P. Saenz-Agudelo, and X. Irigoien. 2018. High-throughput sequencing and linkage mapping of a clownfish genome provide insights on the distribution of molecular players involved in sex change. *Scientific Reports* 8: 4073.

Casas, L., and F. Saborido-Rey. 2021. Environmental cues and mechanisms underpinning sex change in fish. *Sexual Development* 15: 108–121.

Colleye, O., and E. Parmentier. 2012. Overview on the diversity of sounds produced by clownfishes (Pomacentridae): Importance of acoustic signals in their peculiar way of life. *PLoS One* 7: e49179.

DeAngelis, R., J. Gogola, L. Dodd, and J. S. Rhodes. 2017. Opposite effects of nonapeptide antagonists on paternal behavior in the teleost fish *Amphiprion ocellaris*. *Hormones and Behavior* 90: 113–119.

DeAngelis, R., L. Dodd, A. Snyder, and J. S. Rhodes. 2018. Dynamic regulation of brain aromatase and isotocin receptor gene expression depends on parenting status. *Hormones and Behavior* 103: 62–70.

DeAngelis, R., L. Dodd, and J. Rhodes. 2020. Nonapeptides mediate trade-offs in parental care strategy. *Hormones and Behavior* 121: 104717.

Desrochers, L., R. Branconi, E. Schlatter, B. Dent, and P. Buston. 2020. Sensory cues underlying competitive growth in the clown anemonefish (*Amphiprion percula*). *Behavioural Processes* 181: 104276.

Dodd, L. D., E. Nowak, D. Lange, C. G. Parker, R. DeAngelis, J. A. Gonzalez, and J. S. Rhodes. 2019. Active feminization of the preoptic area occurs independently of the gonads in *Amphiprion ocellaris*. *Hormones and Behavior* 112: 65–76.

Elliott, J. K., J. M. Elliott, and R. N. Mariscal. 1995. Host selection, location, and association behaviors of anemonefishes in field settlement experiments. *Marine Biology* 122: 377–389.

Elofsson, U., S. Winberg, and R. C. Francis. 1997. Number of preoptic GnRH-immunoreactive cells correlates with sexual phase in a protandrously hermaphroditic fish, the dusky anemonefish (Amphiprionmelanopus). *Journal of Comparative Physiology A* 181: 484–492.

Fautin, D. G. 1986. Why do anemonefishes inhabit only some host actinians? *Environmental Biology of Fishes* 15: 171–180.

Fautin, D. G. 1992. Anemonefish recruitment: The roles of order and chance. *Symbiosis* 14: 143–160.

Fricke, H., and S. Fricke. 1977. Monogamy and sex change by aggressive dominance in coral reef fish. *Nature* 266: 830–832.

Fricke, H. W. 1983. Social control of sex: Field experiments with the anemonefish *Amphiprion bicinctus*. *Zeitschrift für Tierpsychologie* 61: 71–77.

Frisch, A. 2004. Sex-change and gonadal steroids in sequentially-hermaphroditic teleost fish. *Reviews in Fish Biology and Fisheries* 14: 481–499.

Georges, A., L'Hote, D., A. L. Todeschini, A Auguste, B. Legois, A. Zider, and R. A. Veitia. 2014. The transcription factor FOXL2 mobilizes estrogen signaling to maintain the identity of ovarian granulosa cells. *Elife* 3: 1–19.

Ghiselin, M. T. 1969. The evolution of hermaphroditism among animals. *The Quarterly Review of Biology* 44: 189–208.

Godwin, J., and P. Thomas. 1993. Sex change and steroid profiles in the protandrous anemonefish *Amphiprion melanopus* (Pomacentridae, Teleostei). *General and Comparative Endocrinology* 91: 144–157.

Godwin, J. 1994a. Histological aspects of protandrous sex change in the anemonefish *Amphiprion melanopus* (Pomacentridae, Teleostei). *Journal of Zoology* 232: 199–213.

Godwin, J. R. 1994b. Behavioural aspects of protandrous sex change in the anemonefish, *Amphiprion melanopus*, and endocrine correlates. *Animal Behaviour* 48: 551–567.

Godwin, J. 2009. Social determination of sex in reef fishes. *Seminars In Cell and Developmental Biology* 20: 264–270.

Grober, M. S., I. M. Jackson, and A. H. Bass. 1991. Gonadal steroids affect LHRH preoptic cell number in a sex/role changing fish. *Journal of Neurobiology* 22: 734–741.

Guiguen, Y., A. Fostier, F. Piferrer, and C. F. Chang. 2010. Ovarian aromatase and estrogens: A pivotal role for gonadal sex differentiation and sex change in fish. *General and Comparative Endocrinology* 165: 352–366.

Hattori, A. 1991. Socially controlled growth and size-dependent sex change in the Anemonefish *Amphiprion frenatus* in Okinawa, Japan. *Japanese Journal of Ichthyology* 38: 165–177.

He, C. L., J. L. Du, G. C. Wu, Y. H. Lee, L. T. Sun, and C. F. Chang. 2003. Differential Dmrt1 transcripts in gonads of the protandrous black porgy, *Acanthopagrus schlegeli*. *Cytogenetic and Genome Research* 101: 309–313.

Iwata, E., and J. Manbo. 2013. Territorial behaviour reflects sexual status in groups of false clown anemonefish (*Amphiprion ocellaris*) under laboratory conditions. *Acta Ethologica* 16: 97–103.

Iwata, E., and N. Suzuki. 2020. Steroidal regulation of the aromatase gene and dominant behavior in the false clown anemonefish *Amphiprion ocellaris*. *Fisheries Science* 86: 457–463.

Jiang, T., C. C. Hou, Z. Y. She, and W. X. Yang. 2013. The *SOX* gene family: Function and regulation in testis determination and male fertility maintenance. *Molecular Biology Reports* 40: 2187–2194.

Johnson, Z. V., and L. J. Young. 2017. Oxytocin and vasopressin neural networks: Implications for social behavioral diversity and translational neuroscience. *Neuroscience & Biobehavioral Reviews* 76: 87–98.

Johnston, N. K., and D. L. Dixson. 2017. Anemonefishes rely on visual and chemical cues to correctly identify conspecifics. *Coral Reefs* 36: 903–912.

Kobayashi, Y., M. Nakamura, T. Sunobe, T. Usami, T. Kobayashi, H. Manabe, B. Paul-Prasanth, et al. 2009. Sex change in the gobiid fish is mediated through rapid switching of gonadotropin receptors from ovarian to testicular portion or vice versa. *Endocrinology* 150: 1503–1511.

Kobayashi, Y., R. Horiguchi, S. Miura, and M. Nakamura. 2010. Sex- and tissue-specific expression of P450 aromatase (*cyp19a1a*) in the yellowtail clownfish, *Amphiprion clarkii*. *Comparative Biochemistry and Physiology Part A: Molecular and Integrative Physiology* 155: 237–244.

Kobayashi, Y., Y. Nagahama, and M. Nakamura. 2013. Diversity and plasticity of sex determination and differentiation in fishes. *Sexual Development* 7: 115–125.

Kuwamura, T., and Y. Nakashima. 1998. New aspects of sex change among reef fishes: Recent studies in Japan. *Environmental Biology of Fishes* 52: 125–135.

Kuwamura, T., T. Sunobe, Y. Sakai, T. Kadota, and K. Sawada. 2020. Hermaphroditism in fishes: An annotated list of species, phylogeny, and mating system. *Ichthyological Research* 67: 341–360.

Li, M. H., H. H. Yang, M. R. Li, Y. L. Sun, X. L. Jiang, Q. P. Xie, T. R. Wang, et al. 2013. Antagonistic roles of Dmrt1 and Foxl2 in sex differentiation via estrogen production in Tilapia as demonstrated by TALENs. *Endocrinology* 154: 4814–4825.

Liu, H., M. S. Lamm, K. Rutherford, M. A. Black, J. R. Godwin, and N. J. Gemmell. 2015. Large-scale transcriptome sequencing reveals novel expression patterns for key sex-related genes in a sex-changing fish. *Biology of Sex Differences* 6: 26.

Manousaki, T., A. Tsakogiannis, J. Lagnel, E. Sarropoulou, J. Z. Xiang, N. Papandroulakis, C. C. Mylonas, et al. 2014. The sex-specific transcriptome of the hermaphrodite sparid sharpsnout seabream (*Diplodus puntazzo*). *BMC Genomics* 15: 655.

Matson, C. K., M. W. Murphy, A. L. Sarver, M. D. Griswold, V. J. Bardwell, and D. Zarkower. 2011. DMRT1 prevents female reprogramming in the postnatal mammalian testis. *Nature* 476: 101–104.

Matson, C. K., and D. Zarkower. 2012. Sex and the singular DM domain: Insights into sexual regulation, evolution and plasticity. *Nature Reviews Genetics* 13: 163–174.

McCarthy, M. M. 2008. Estradiol and the developing brain. *Physiological Reviews* 88: 91–134.

McCarthy, M. M., B. M. Nugent, and K. M. Lenz. 2017. Neuroimmunology and neuroepigenetics in the establishment of sex differences in the brain. *Nature Reviews Neuroscience* 18: 471–484.

Mills, S. C., J. L. O'Donnell, G. Bernardi, and R. Beldade. 2018. Natural endocrine profiles of the group-living skunk anemonefish *Amphiprion akallopisos* in relation to their size-based dominance hierarchy. *Journal of Fish Biology* 92: 773–789.

Miura, S., Y. Kobayashi, R. K. Bhandari, and M. Nakamura. 2013. Estrogen favors the differentiation of ovarian tissues in the ambisexual gonads of anemonefish *Amphiprion clarkii. The Journal of Experimental Zoology Part A* 319: 560–568.

Mohr, M. A., F. L. Garcia, L. L. DonCarlos, and C. L. Sisk. 2016. Neurons and glial cells are added to the female rat anteroventral periventricular nucleus during puberty. *Endocrinology* 157: 2393–2402.

Moyer, J. T., and A. Nakazono. 1978. Protandrous hermaphroditism in six species of the anemonefish genus *Amphiprion* in Japan. *Japanese Journal of Ichthyology* 25: 101–106.

Moyer, J. T. 1986. Longevity of the anemonefish *Amphiprion clarkii* at Miyake-jima, Japan with notes on four other species (1986). *Copeia* 1: 135–139.

Munday, P. L., P. M. Buston, and R. R. Warner. 2006. Diversity and flexibility of sex-change strategies in animals. *Trends in Ecology & Evolution* 21: 89–95.

Nakamura, M., T. Mariko, and Y. Nagahama. 1994. Ultrastructure and in vitro steroidogenesis of the gonads in the protandrous anemonefish *Amphiprion frenatus. Japanese Journal of Ichthyology* 41: 47–56.

Nakamura, M., S. Miura, and R. Nozu. 2015. Opposite-directional sex change in functional female protandrous anemonefish, *Amphiprion clarkii*: Effect of aromatase inhibitor on the ovarian tissue. *Zoological Letters* 1: 30.

Nelson, J. S., T. C. Grande, and M. V. H. Wilson. 2016. *Fishes of the World*. 5th Edition. Hoboken: John Wiley & Sons.

Nugent, B. M., C. L. Wright, A. C. Shetty, G. E. Hodes, K. M. Lenz, A. Mahurkar, S. J. Russo, et al. 2015. Brain feminization requires active repression of masculinization via DNA methylation. *Nature Neuroscience* 18: 690–697.

Ochi, H. 1989a. Mating behavior and sex change of the anemonefish, *Amphiprion clarkii*, in the temperate waters of southern Japan. *Environmental Biology of Fishes* 26: 257–275.

Ochi, H. 1989b. Acquisition of breeding space by nonbreeders in the anemonefish *Amphiprion clarkii* in temperate waters of southern Japan. *Ethology* 83: 279–294.

Pandian, T. J. 2010. *Sexuality in Fishes*. Boca Raton: CRC Press/Taylor & Francis.

Pfennig, F., A. Standke, and H. O. Gutzeit. 2015. The role of Amh signaling in teleost fish – Multiple functions not restricted to the gonads. *General and Comparative Endocrinology* 223: 87–107.

Phillips, E., R. DeAngelis, J. V. Gogola, and J. S. Rhodes. 2020. Spontaneous alloparental care of unrelated offspring by non-breeding *Amphiprion ocellaris* in absence of the biological parents. *Scientific Reports* 10: 1–11.

Ross, R. M. 1978a. Territorial behavior and ecology of the anemonefish *Amphiprion melanopus* on Guam. *Zeitschrift für Tierpsychologie* 46: 71–83.

Ross, R. M. 1978b. Reproductive behavior of the anemonefish *Amphiprion melanopus* on Guam. *Copeia* 1978: 103–107.

Rueger, T., R. Branconi, C. Y. M. Froehlich, S. J. Heatwole, M. L. Wong, and P. M. Buston. 2021. The next frontier in understanding the evolution of coral reef fish societies. *Frontiers in Marine Science* 8: 437.

Semsar, K., and J. Godwin. 2004. Multiple mechanisms of phenotype development in the bluehead wrasse. *Hormones and Behavior* 45: 345–353.

Siegfried, K. R. 2010. In search of determinants: Gene expression during gonadal sex differentiation. *Journal of Fish Biology* 76: 1879–1902.

Todd, E. V., O. Ortega-Recalde, H. Liu, M. S. Lamm, K. M. Rutherford, H. Cross, M. A. Black, et al. 2019. Stress, novel sex genes, and epigenetic reprogramming orchestrate socially controlled sex change. *Science Advances* 5(7): eaaw7006.

Trudeau, V. L. 2018. Facing the challenges of neuropeptide gene knockouts: Why do they not inhibit reproduction in adult teleost fish? *Frontiers in Neuroscience* 12: 302.

Tsakogiannis, A., T. Manousaki, J. Lagnel, A. Sterioti, M. Pavlidis, N. Papandroulakis, C. C. Mylonas, et al. 2018. The transcriptomic signature of different sexes in two protogynous hermaphrodites: Insights into the molecular network underlying sex phenotype in fish. *Scientific Reports* 8: 3564.

Tsakogiannis, A., T. Manousaki, J. Lagnel, N. Papanikolaou, N. Papandroulakis, C. C. Mylonas, and C. S. Tsigenopoulos. 2019. The gene toolkit implicated in functional sex in Sparidae hermaphrodites: Inferences from comparative transcriptomics. *Frontiers in Genetics* 9: 749.

Tsukahara, S., and M. Morishita. 2020. Sexually dimorphic formation of the preoptic area and the bed nucleus of the stria terminalis by neuroestrogens. *Frontiers in Neuroscience* 14: 797.

Uhlenhaut, N. H., S. Jakob, K. Anlag, T. Eisenberger, R. Sekido, J. Kress, A. C. Treier, et al. 2009. Somatic sex reprogramming of adult ovaries to testes by FOXL2 ablation. *Cell* 139(6): 1130–1142.

Vining, B., Z. Ming, S. Bagheri-Fam, and V. Harley. 2021. Diverse regulation but conserved function: SOX9 in vertebrate sex determination. *Genes* 12: 486.

Wang, D. S., T. Kobayashi, L. Y. Zhou, B. Paul-Prasanth, S. Ijiri, F. Sakai, K. Okubo, et al. 2007. Foxl2 up-regulates aromatase gene transcription in a female-specific manner by binding to the promoter as well as interacting with Ad4 binding protein/steroidogenic factor 1. *Molecular Endocrinology* 21: 712–725.

Wang, X. G., and L. Orban. 2007. Anti-Müllerian hormone and 11 β-hydroxylase show reciprocal expression to that of aromatase in the transforming gonad of zebrafish males. *Developmental Dynamics* 236: 1329–1338.

Warner, R. R. 1975. The adaptive significance of sequential hermaphroditism in animals. *The American Naturalist* 109: 61–82.

Warner, R. R. 1984. Mating behavior and hermaphroditism in coral reef fishes: The diverse forms of sexuality found among tropical marine fishes can be viewed as adaptations to their equally diverse mating systems. *American Scientist* 72: 128–136.

Warner, R. R. 1988. Sex change and the size-advantage model. *Trends in Ecology and Evolution* 3: 334–336.

Warner, R. R., and S. E. Swearer. 1991. Social control of sex change in the bluehead wrasse, *Thalassoma bifasciatum* (Pisces: Labridae). *The Biological Bulletin* 181: 199–204.

Wu, G. C., P. C. Chiu, Y. S. Lyu, and C. F. Chang. 2010. The expression of *amh* and *amhr2* is associated with the development of gonadal tissue and sex change in the protandrous black porgy, *Acanthopagrus schlegeli*. *Biology of Reproduction* 83: 443–453.

Yaeger, C., A. M. Ros, V. Cross, R. S. DeAngelis, D. J. Stobaugh, and J. S. Rhodes. 2014. Blockade of arginine vasotocin signaling reduces aggressive behavior and c-Fos expression in the preoptic area and periventricular nucleus of the posterior tuberculum in male *Amphiprion ocellaris*. *Neuroscience* 267: 205–218.

Yamaguchi, T., S. Yamaguchi, T. Hirai, and T. Kitano. 2007. Follicle-stimulating hormone signaling and Foxl2 are involved in transcriptional regulation of aromatase gene during gonadal sex differentiation in Japanese flounder, *Paralichthys olivaceus*. *Biochemical and Biophysical Research Communications* 359: 935–940.

Zhang, K., J. Xu, Z. Zhang, Y. Huang, Z. Ruan, S. Chen, F. Zhu, et al. 2019. A comparative transcriptomic study on developmental gonads provides novel insights into sex change in the protandrous black porgy (*Acanthopagrus schlegelii*). *Genomics* 111: 277–283.

Zohar, Y., J. A. Muñoz-Cueto, A. Elizur, and O. Kah. 2010. Neuroendocrinology of reproduction in teleost fish. *General and Comparative Endocrinology* 165: 438–455.

13 Anemonefish Behavior and Reproduction

Ricardo Beldade, Giacomo Bernardi, and Suzanne C. Mills

CONTENTS

13.1 INTRODUCTION

Anemonefishes have increasingly been used as model systems to study a wide variety of subjects which include: mutualism ecology and evolution (Litsios et al. 2012; Ollerton et al. 2007), social dynamics (Hattori 2002, 2004; Hattori and Yamamura 1995), population and meta-population dynamics (Almany et al. 2007; Pinsky et al. 2010; Beldade et al. 2012; Catalano et al. 2021; Dedrick et al. 2021; Harrison et al. 2012), developmental biology (Salis et al. 2021), physiological plasticity (Mills et al. 2015; Norin et al. 2018; Mills et al. 2020; Cortese et al. 2021), phylogenetics (Elliott et al. 1999; Litsios et al. 2014), and genomics (Marcionetti et al. 2018; Lehmann et al. 2019). Anemonefish have also been used as model organisms for conservation biology (Frisch et al. 2016; Haguenauer et al. 2021). Yet, they are most attractive as models for their ease in measuring fitness changes. Beyond measuring survival, sublethal effects may manifest on reproduction due to social context (Rueger et al. 2018), natural environmental variation (Barbasch et al. 2020; Buston 2004; Buston and Elith 2011), such as climate change (Beldade et al. 2017; Cortese et al. 2021; Miller et al. 2015; Saenz-Agudelo et al. 2011), as well as other anthropogenic stressors (Fobert et al. 2021; Frisch et al. 2016; Mills et al. 2020; Schligler et al. 2021). In this chapter, we will broadly cover reproduction, including reproductive behavior (courtship and nest preparation) and endocrine control to answer fundamental questions in reproduction such as *Who? How? Where? When? How many?* and *Why does it vary?*

DOI: 10.1201/9781003125365-16

13.2 *WHO* REPRODUCES?

13.2.1 Group Structure and Hormonal Profiles

Social hierarchies are widespread in many group-living species across the animal kingdom (Drews 1993) and anemonefishes are no exception (Fricke 1979; Buston 2003; Buston and Cant 2006). Groups of anemonefish inhabiting anemones consist of a mated adult pair (female- and male-functioning individuals) and typically a variable number of immature and non-reproductive individuals (Fricke 1979; Fautin and Allen 1992; Godwin and Thomas 1993; Ross 1978a, 1978b). Breeding status in groups of anemonefish of more than two individuals is restricted to the two largest fish, according to a size-based dominance hierarchy; the female is largest, the male is second largest, and the non-breeders get progressively smaller as the hierarchy is descended (Fricke 1979; Buston 2003). Sex, as well as breeding and social status, are linked with the functioning of the hypothalamus-pituitary-gonadal (HPG) axis (e.g., Cardwell et al. 1996) with different behavioral and hormonal profiles for individual anemonefish in relation to their sex and position in the social hierarchy.

Blood plasma levels of the fish-specific androgen, 11-ketotestosterone (11-KT), rather than testosterone (T), provide the best androgen correlate of male sex in anemonefish, with males of the cinnamon anemonefish, *Amphiprion melanopus* (Godwin and Thomas 1993), the false clown anemonefish, *Amphiprion ocellaris* (Iwata et al. 2012), and the skunk anemonefish *Amphiprion akallopisos* (Mills et al. 2010; Mills et al. 2018) showing three- to eight-fold higher levels of 11-KT compared to females. The primary functions of 11-KT are spermatogenic and spermiogenic as well as inducing secondary sex characteristics (Borg 1994). Levels of 17ß-estradiol are the main predictor of female sex in anemonefish, with females showing seven-fold higher 17ß-estradiol levels than males (Godwin and Thomas 1993; DeAngelis and Rhodes 2016; Mills et al. 2018). 17ß-estradiol plays a major role in teleost reproduction, particularly sexual maturation and differentiation, including oogenesis and vitellogenesis (Ng and Idler 1983; Fostier et al. 1983; Lazier et al. 1987). In addition, breeding females and males, as well as the largest non-breeding male skunk anemonefish, all have similar levels of T which may be linked to their aggressive behavior when defending their host anemone against heterospecific intruders or predators (e.g., Section 13.3.3), rather than breeding status (Mills et al. 2018).

Aggressive and dominant behaviors are also determined by 11-KT (Wingfield et al. 1990; Borg 1994). In skunk anemonefish males, 11-KT levels decline as the hierarchy is descended (Mills et al. 2018), a common finding in dominant compared to subordinate teleost males (Brantley et al. 1993). The highest-ranked male monopolizes the anemone, displays the most aggression, especially towards the largest non-breeding male, receives the least amount of aggression but has the highest cortisol level (Iwata et al. 2012). The largest non-breeding male directs its aggressive acts towards the next smallest male, but they have similar cortisol levels (Iwata et al. 2012). As social stimulation via male–male interactions leads to higher levels of 11-KT (Oliveira et al. 1996), descending levels of 11-KT down the hierarchy may be a consequence, rather than a predictor, of different social stimuli experienced by different conspecific males (Cardwell and Liley 1991; Oliveira et al. 1996). Evidence that non-breeding male skunk anemonefish more similar in size to breeding males have higher 11-KT levels relative to less similar-sized non-breeding males supports this hypothesis (Mills et al. 2018). As such, the breeding male may be suppressing the reproductive function of smaller non-breeding males through aggressive social interactions, which, via actions on the HPG axis, and in particular on 11-KT, ultimately represses fertility.

13.2.2 Protandrous Sequential Hermaphroditism

Anemonefishes are protandrous sequential hermaphrodite fish which change sex from male to female. Sex change does not occur when the male attains a certain size, but rather after a social trigger, the disappearance of the female and all the subordinates ascend in rank, becoming sexually mature or changing sex (Moyer and Nakazono 1978; Fricke 1983). This strict control of sexual maturity by social rank in anemonefish is triggered when the breeding female or male of the group dies. When the female rank is vacant, the breeding male rapidly changes sex to become female and the largest immature, non-reproductive individual turns into a mature male completing the new breeding pair (Moyer and Nakazono 1978; Fricke and Fricke 1977). Sexual maturity, either from non-breeder to male, non-breeder to female or male to female following the "opening" of one or more positions can take as little as 26 days and up to two years (Fricke and Fricke 1977; Hattori 1991; Ochi 1989). The speed of sex change is facilitated by the fact that the gonad of functional males of *Amphiprion* species is an ovotestis, with the presence of both testicular and ovarian tissues (Brusle-Sicard et al. 1994; Godwin and Thomas 1993; Shapiro 1992; Casadevall et al. 2009). In functional males the testes are mature, but 55% of the gonad is female tissue, the ovary, in an immature phase (with only oogonia and primary growth oocytes). When the male begins to change sex the ovotestis enters a transition phase (Sabordio-Rey 2016), characterized by the progressive degeneration of the testes and proliferation of ovarian tissue. At the end of the transitional phase the testicular tissue is resorbed and 40 days after the female rank became vacant the amount of ovarian tissue reaches 87%. After 40–50 days, large oocytes are seen but are not yet ripe as vitellogenesis has not yet commenced, but these females are now capable of spawning.

The physiology, gonadal structure, importance of gonadal sex steroid hormones, especially estrogens, and changes that accompany sex change are well described (Godwin and Thomas 1993; Kobayashi et al. 2010; Khoo et al. 2018) (more details in Chapter 12; Casas et al. 2022). However, the upstream mechanisms controlling the production and activity of gonadal steroid hormones during sex change in clownfish remain largely unknown (Kobayashi et al. 2013). A recent study has found that *cyp19a1b* (also known as aromatase or *P450aromA*) plays a central role in the mechanism of sex change in the brain of *Amphiprion bicinctus* by modulating the balance between estrogen and androgen signalling (Casas et al. 2016). Both *sox6* and *foxp4* also may play a role in regulating the expression of aromatase and/or other genes involved in the sex steroidogenic pathway in *A. bicinctus* at the brain level. At the gonadal level, a feedback loop between *dmrt1* and *foxl2* regulates the estrogen/androgen balance that drives sex change in *A. bicinctus* (Casas et al. 2016).

In small groups, with no breeding adult pair, the earliest reported age that males become sexually mature is at four months for *Amphiprion polymnus* and between six and 14 months for females (Rattanayuvakorn et al. 2006), as well as just under 12 months for *Amphiprion chrysopterus* (Schliger et al. 2021).

13.3 *HOW* DOES REPRODUCTION OCCUR?

13.3.1 Spermatogenesis

Functional males have an ovotestis, which is composed of active spermatogenic tissue as well as previtellogenic oocytes (Brusle-Sicard et al. 1994; Godwin and Thomas 1993; Shapiro 1992). Male gametogenesis, spermatogenesis, entails the development of a small number of diploid spermatogonial stem cells into many highly differentiated spermatozoa with a haploid, recombined genome over four phases (Shulz et al. 2010). The *mitotic* or spermatogonial phase describes the mitotic divisions of spermatogonia, spermatocytogenesis, resulting in the first spermatocytes. Follicule stimulating hormone (FSH) plays a major regulatory role during these early stages and FSH is the key gonadotropin for the initiation of spermatogenesis (Shulz et al. 2010). The *meiotic* phase involves primary spermatocytes dividing meiotically into secondary spermatocytes. *Spermiogenesis* entails secondary spermatocytes dividing into two haploid spermatids and their development into motile, flagellated genome vectors, spermatozoa. Luteinizing hormone (LH) mainly regulates processes during this spermiogenic phase. The concentration of androgens (T and 11-KT) increases gradually as spermatogenesis proceeds and decrease at the final phase, *spermiation*, which is the release of mature spermatids from Sertoli cells into the seminiferous tubule lumen prior to their passage to the epididymis (Shulz et al. 2010). Fish spermatozoa can be classified into two forms, aquasperm and introsperm, according to the external or internal mode of fertilization, respectively (Jamieson 1991). Spermatozoa of teleost fish generally have no acrosome, and the impenetrable egg chorion is pierced by a micropyle providing access to the membrane of the oocyte.

13.3.2 Oogenesis

After sex change and once ovarian tissues have taken over the ovotestis female gametogenesis may begin and female anemonefishes are oviparous, producing yolk-containing eggs. Female gametogenesis (oogenesis) encompasses the morphological and functional changes by which an oogonium develops into an oocyte with the potential to be fertilized and begins once females are sexually mature (Grier et al. 2009). Oogenesis is comprehensively described (e.g., Lubzens et al. 2010; Lubzens et al. 2017), but here we outline the main stages and highlight those stages during which maternal reserves and hormones are accumulated, which is interesting for intergenerational mechanisms of acclimation and plasticity (e.g., Section 13.7.2). Oogenesis begins with the *previtellogenic* stage, that of primary oocyte growth and their accumulation of protein-rich vesicles (cortical alveolus) (Lubzens et al. 2010). During the *vitellogenic* stage, the oocyte begins to incorporate the egg precursor protein, vitellogenin (Vtg) and oocytes are maturationally competent, responding to maturation-inducing hormones (MIH) (Lubzens et al. 2017). Oocytes grow considerably during vitellogenesis, accumulating nutritional reserves needed for the development of the embryo, as well as maternal mRNAs, proteins, lipids, carbohydrates, vitamins E and A (recruited from peripheral tissues and muscles in the adult female), and hormones (Lubzens et al. 2010). During vitellogenesis the number and/or size of developing oocytes are adjusted to match available energy resources (Luckenbach et al. 2008) and the majority of maternal cortisol is incorporated into the yolk (Faught and Vijayan 2018). Furthermore, follicle loss may also occur because of environmental stressors (e.g., Section 13.7.1). Finally, during the *oocyte maturation* stage, meiosis is resumed and the germinal vesicle migrates from the centre of the oocyte to the periphery and the nuclear membrane breaks down, culminating in a mature egg ready for ovulation and spawning. During this stage, cortisol may have a regulatory role in promoting hydration, the action of MIH, and ovulation (Faught and Vijayan 2018). At this stage, the egg is fully formed and contains all the molecules and nutritive reserves needed for embryonic development after fertilization (Brooks et al. 1997). Oogenesis is enhanced with increasing plasma levels of 17ß-estradiol whose maximal levels are reached just prior to spawning triggered by high levels of T (Shin et al. 2013).

Three main types of oogenesis have been described for fish (Wallace and Selman 1981). First, is synchronous, where all oocytes develop and ovulate at the same time. Second is group-synchronous, where at least two populations of oocytes can be recognized in the ovary throughout the reproductive season (i.e., vitellogenic and maturing oocytes). Third is asynchronous oogenesis, in which ovaries contain a random mixture of oocytes at every stage (Wallace and Selman 1981). The type of oogenesis will directly impact a species' spawning pattern (e.g., Section 13.5). Individuals with synchronous oogenesis will shed all of their eggs in a single episode or over a short period of time – synchronous ovulation. On the other hand, individuals with asynchronous development recruit eggs from the heterogeneous mix of developing oocytes into maturation and are subsequently ovulated in several batches over the spawning season – asynchronous ovulation or batch spawning (Murua and Saborido-Rey 2003). Full-grown oocytes within the ovary undergo maturation and ovulation at each spawning event.

Oogenesis in anemonefish is rarely described outside of sex inversion. However, ovaries from female *Amphiprion frenatus* revealed the presence of yolky oocytes at the same time as young oocytes (i.e., containing previtellogenic as well as vitellogenic oocytes) (Nakamura et al. 1994). Similarly, *Amphiprion clarkii* ovaries contained various developmental stages of oocytes (perinucleolus, cortical alveolus, and vitellogenesis stages) (Hattori 1991). Ovarian histological evidence therefore suggests that anemonefish show asynchronous oogenesis, which agrees with their pattern of batch spawning (e.g., Section 13.5).

13.3.3 Behavior Prior to Spawning

Anemonefish reproductive behaviors have been described since the early papers dedicated to this group of fishes. The behaviors described fall into three main categories: (i) pre-spawning, (ii) during spawning, and (iii) during egg development. In this chapter, we will not cover behaviors during development, including parental care, which is already the subject of Chapter 15 (Barbasch et al. 2022), except to emphasize the aggressiveness of anemonefish which increases when they are guarding eggs. Anemonefish are aggressive towards predators of their host anemone, for example, butterflyfish (Chaetodontidae) and turtles (Godwin and Fautin 1992), as well as towards their own potential predators, including sharks, that pass within their territory (Schligler et al. 2022).

Pre-spawning behaviors associated with reproduction comprise courtship and nest preparation (specifically anemone biting or tentacle nibbling and substrate biting). Courtship in anemonefish is ritualized as in most Pomacentrids. During the nuptial period, the male chases and nips at the female (Fautin and Allen 1997). Nest-preparation behaviors can be performed by both males and females, by clearing algae and debris from the oviposition site on a rock surface adjacent to the anemone up to a day prior to egg-laying (e.g., in *A. melanopus* [Ross 1978a] and in *A. ocellaris* [Raheem et al. 2021]). Spawning behaviors, which can last from 20 to 120 min (Fautin and Allen 1997; Ross 1978a; Anil et al. 2012; Raheem et al. 2021), include "skimming" or "brushing" over the oviposition site, firstly by the female who is closely followed by the male. During these passes, the female lays rows of conical-shaped eggs, extruded from a ventral conical urogenital ovipositor, that are promptly fertilized by the male.

13.3.4 Egg Development and Hatching

Eggs are deposited from the centre to the edges of the nest. While releasing their gametes, both parents avidly flutter their pectoral fins (Ross 1978a), potentially in order to limit the dispersal of sperm and insure maximum fertilization rate. Newly laid eggs are identified by their vivid orange color (Figure 13.1A) and they are elliptical, about 3–4 mm in length, and adhere to the nest surface by a stalk of short filaments (Madhu et al. 2012a). Egg characteristics, mainly length, width, and area, are related to parental size (Green and McCormick 2005). Egg size also varies with species, for example from 1.5–3.0 × 0.8–1.84 mm in *A. ocellaris* (Madhu et al. 2006a), 2.0–2.3 × 1.0–1.2 mm in *A. nigriceps* (Dhaneesh et al. 2009), 2.4 × 0.9 mm in *A. chrysopterus* (Allen 1980) and 2.7 × 0.9 mm in *A. clarkii* (Ghosh et al. 2012) to 2.8–3.5 × 1.1–1.7 mm in *Premnas biaculeatus* (Madhu et al. 2006b). Egg size also varies according to their position within the nest, with larger length and volume in the centre, but this was not attributable to the quantity of yolk in eggs (Green et al. 2006; Kunzmann and Diemel 2020). The amount of yolk in *A. melanopus* eggs does however vary across individuals (Green et al. 2006). Egg size increases slightly throughout development, for example, Green (2004) found that *A. melanopus* eggs ranged from 1.76–2.51 mm at spawning to 2.24–2.67 mm at hatching. Developmental time depends on species and season or temperature, for example, the development in *A. chrysopterus* takes six days from January to July (austral summer), but seven days from August to December (austral winter) (Cortese 2021). During development, eggs transition in color with the yolk sac responsible for the general color of the entire egg mass (Figure 13.1B–C). The bright orange colouration of newly laid eggs darkens over time and the embryos develop readily identifiable features such as large, pigmented eyes with silvery irises (Figure 13.1C). Unfertilized eggs are also easily identifiable after the first day of incubation (Figure 13.1C).

Hatching takes place at night, in the first hours after sunset, enabled by vigorous movements of the larva inside the capsule (e.g., Ross 1978a). Hatching is controlled by a hatching enzyme regulated by photoreceptors (Helvik and Walther 1992; McAlary and McFarland 1993) and stimulated by prolactin (Schoots et al. 1982). Hatching is mostly synchronous (i.e., all larvae in a nest will hatch on the same night with few exceptions). On rare occasions, unhatched eggs have been observed to hatch one day later.

13.4 *WHERE* DOES REPRODUCTION OCCUR?

13.4.1 Geographically

Anemonefishes are distributed throughout the Indo-Pacific, from the Red Sea to French Polynesia, and from Japan to southern Australia, but are notably absent from the Hawaiian Archipelago (Fautin and Allen 1997). Ten anemone species host anemonefishes (Titus et al. 2020; Emms et al. 2020), and most of them are attached to rocks or other hard surfaces (Fautin and Allen 1997). Anemonefish populations are known to mostly live in shallow waters between 1 m and 40 m depth, however, recently with the advent of deep scientific diving techniques and especially closed-circuit

FIGURE 13.1 (A) A breeding pair of orange-fin anemonefish, *Amphiprion chrysopterus*, at the end of spawning in Moorea, French Polynesia (Cécile Berthe); (B) an *A. chrysopterus* male guarding his nest at the end of development at a depth of 55 m in Tikehau, French Polynesia (Frédéric Zuberer); (C) a panel showing *A. chrysopterus* embryonic development over seven days in Moorea during the Austral winter (Suzanne C. Mills).

diving (Pyle 1996), the bathymetric limit of anemonefish has been extended from 40 m down to 60 m in *A. bicinctus* (Brokovich et al. 2008), *A. akindynos*, *A. perideraion* (Bridge et al. 2012), and *A. chrysopterus* (Haguenauer et al. 2021) and even down to 70 m in *A. clarkii* (Coleman et al. 2018). Reproduction has also been reported at these mesophotic depths (Haguenauer et al. 2021) (Figure 13.1B).

13.4.2 Oviposition Site

Anemonefish choose an oviposition site under the tentacles of the anemones, likely to protect the eggs from predation (Fautin and Allen 1992) as well as to imprint a potential olfactory settlement cue on unhatched larvae (Arvedlund et al. 2000). There are anemone species that occur in seagrass patches or in sandy bottoms, such as *Heteractis crispa* and *Stichodactyla haddoni*, around which an oviposition site may not be readily available. To circumvent this, *A. polymnus* hosted by these anemones pushes and drags suitable oviposition surfaces (bivalve shells, marine litter, palm fronds) to its host anemone (Moyer and Steene 1979). Some sea anemones hosting anemonefish, for example *Macrodactyla doreensis*, can fully retract, leaving the nest and parents exposed to potential predators (Fautin and Allen 1997). Anemones that host anemonefish can change their position, sometimes even detaching and moving away from their initial location, they might also disappear, either due to predation or to adverse environmental stresses. Long-term monitoring of wild breeding pairs has also revealed that oviposition sites can change through time, perhaps as a response to environmental cues.

13.5 *WHEN* DOES REPRODUCTION OCCUR?

Spawning rhythms can be described following a progressive temporal scale from the time of day, lunar month, season, as well as year, and even as lifetime reproductive output in anemonefish. While reports of spawning events have been described in early classical papers from the smallest up to yearly temporal scales, variation in individual lifetime reproductive estimates have only recently received attention.

13.5.1 Daily (Light–Dark) and Tidal Spawning Rhythms

Ross (1978a) reports spawning observations within 2 h after sunrise, which has been confirmed for *Amphiprion chrysogaster* (Gopakumar et al. 1994), *A. sebae* (Ignatius et al. 2001), *A. nigripes* (Anil et al. 2012), and *A. ephippium* (Rohini et al. 2018). *P. biaculeatus* is also reported to spawn in the late afternoon (Madhu et al. 2012a). In Moorea, French Polynesia, *A. chrysopterus* shows variable egg-laying timings ranging from early morning to late afternoon (pers. Obs. RB & SCM). Hatching occurs after sunset, during high tides (Ross 1978a), a strategy presumed to increase larval dispersal away from the nest, and also to avoid potential predation.

13.5.2 Lunar Spawning Rhythms

Anemonefish, depending on the individual and the region they inhabit, have variable lunar spawning patterns. In tropical regions, lunar spawning periodicity patterns cover all the possibilities including no pattern, lunar (spawning once per lunar month), semi-lunar (spawning twice per lunar month), and trient-lunar (spawning three times per lunar month) (Richardson 1999; Holtswarth et al. 2017; Seymour et al. 2018). In addition to the tropics, this range of patterns is observed across subtropical and temperate regions (Richardson et al. 1997). Studies have found that *A. clarkii* spawned an average of 0.5 nests month^{-1} (Ochi 1985, 1989), *A. perideraion* spawned 0.7 nests month^{-1} (Allen 1975), *A. melanopus* 1.6 nests month^{-1} (Ross 1978a), and *A. percula* and *A. chrysopterus* between 1, 2, and 3 nests month^{-1} (Seymour et al. 2018; Cortese 2021). Distinct periodicities can be found across individuals in the same population, for example, Seymour et al. (2018) report distinct patterns in individual *A. percula* and highlight that conservation strategies based on a one single lunar spawning strategy (for example the most frequent), will likely be disregarding (or selecting against) alternative reproductive strategies.

However, plasticity in periodicities can also be found (Seymour et al. 2018), for example, *A. ephippium* shows semi- and trient-lunar spawning patterns except in the cooler months when spawning frequency is reduced to lunar (Rohini et al. 2018). Similarly, *A. akallopisos* shows an increased spawning rate (2.67 times month^{-1}) in the warmer months with a decrease in winter cooler months (2 times month^{-1}) (Dhaneesh et al. 2012). The plasticity in spawning cycles is understudied, as is the relative importance of other parameters that explain the variability in reproduction of anemonefish (see the last section of this chapter). Finally, a higher spawning frequency was observed in individuals held in captivity (average 2.2 ± 0.8 times month^{-1}) as compared with wild ones (Gordon and Bok 2001).

13.5.3 Seasonal Spawning Rhythms

The high temperatures and thermally stable characteristics of tropical regions have for a long time been used to justify a seemingly persistent reproductive season for most tropical *Amphiprion*. For example, Allen observed a lack of seasonality for several *Amphiprion* species in Eniwetok atoll (Allen 1975). However, other tropical species show seasonal variation in which water temperature influences spawning (i.e., *A. sebae*), with decreased spawning frequencies in cooler months (Madhu and Madhu 2007; Dhaneesh et al. 2012). Breeding and reproductive output of *A. clarkii* which occurs in both tropical and subtropical or temperate regions showed a similarly clear seasonal trend (Bell 1976), where reproductive output also decreased in the colder months. Recent studies have, however, shed more light on these contrasting patterns. Holtswarth et al. (2017) found the opposite seasonal pattern in a tropical population of *A. clarkii* from the Philippines, where fewer spawning events and smaller clutch sizes occurred during the warmer months. Another tropical anemonefish, *A. chrysopterus*, also displayed decreases in reproduction associated with seasonally elevated temperatures (Beldade et al. 2017). Reproductive output is therefore tracking seasonal temperature changes, either warm or cold, across their thermal range with decreasing reproductive output at both of the thermal extremes.

13.5.4 Yearly Spawning Rhythms

Few papers have reported spawning frequencies over the course of at least one year. Spawning frequency not only changes as a function of the species but also as a function of female size. Ochi (1989) found a decrease in the spawning frequency with size of *A. clarkii* in the temperate waters of Okinawa, southern Japan. However, the differences in fecundity due to female size guarantee that the largest females have a greater number of eggs per year than the smaller females.

13.6 *HOW* MANY EGGS?

13.6.1 Fecundity

Reproductive effort relates to the number of eggs laid, fecundity, and accurately measuring fecundity is of great importance for studies estimating fitness (Figure 13.2A). These, in turn, provide some direct measurement that allows for comparisons between pairs that have different characteristics such as age, size, location, and habitat or even within the same pair before and after exposure to natural or manipulated environmental stressors (e.g., Beldade et al 2017). While there is a growing number of studies reporting fecundity measurements, mostly described as the number of eggs laid in a nest, few have quantified parameters such as egg mortality through development or egg number immediately prior to hatching. Egg mortality from day one to hatching in *A. chrysopterus* in Moorea, French Polynesia, ranges from 20% to 80% (Cortese 2021). These distinct fecundity measures likely relate to different processes: fecundity at day one relates more closely to female investment, whereas egg mortality and fecundity measured

FIGURE 13.2 (A) *Amphiprion chrysopterus* egg clutch (Anne Haguenauer); (B) measurement of *A. chrysopterus* juvenile metabolic rate in bleached anemones (Suzanne C. Mills); (C) measurement of *A. chrysopterus* larval metabolic rate (Shaun Killen); and (D) measurement of *A. chrysopterus* embryonic heart beat rate (Jules Schligler).

just prior to hatching include fertilisation success and also the quality of parental (male) care during development.

Fecundity in many *Amphiprion* species shows a positive allometric relation to female size (Khoo et al. 2018; Hattori 2012). There are also multiple measurements of the number of eggs per clutch for several *Amphiprion* species, for example, *A. melanopus* 200–400 eggs; *A. ocellaris* at Pulau Tioman, Malaysia 582 ± 478 SD 23–1,518 (Khoo et al. 2018), and *A. chrysopterus* in Moorea, French Polynesia 322–4,601 eggs (Cortese 2021).

13.6.2 Annual Fecundity

Annual fecundity per breeding pair of *A. clarkii* has been reported from 8,000 to 17,500 (Bell 1976; Ochi 1989), 7,200 eggs for *A. melanopus* (Ross 1978a), 1,752–11,832 eggs for *P. biaculeatus* (Madhu and Madhu 2012a) but up to 53,229 eggs in the larger *A. chrysopterus* from Moorea, French Polynesia (Cortese 2021).

13.6.3 Lifetime Estimates of Reproductive Success

Lifetime estimates of reproductive success have recently received much attention given that this is a fundamental parameter in population dynamics and may vary due to genetic, maternal, and environmental components (Salles et al. 2020; Dedrick et al. 2021). Determining the number of successful offspring an individual contributes to the next generation over its lifetime is very difficult to calculate empirically. On the one hand, sporadic measurements of fecundity are likely not good representations of individual lifetime reproductive success as there is a considerable variation even in yearly fecundity estimates (see earlier). On the other hand, given the potential life expectancy of some species, empirical measurements of reproductive success are nearly impossible to obtain. Furthermore, to reach a good estimate of fecundity over the lifetime of a given individual, one would need an individual's size at sexual maturity, their longevity and naturally their average reproductive output throughout the whole period as well as over the whole area into which their larvae could potentially settle.

The size of males when they acquire breeding status can vary in a population. In some anemonefish species, movement between groups can shorten such periods and new males were shown to even step-father an egg clutch in case it is present (Hattori 1994, 2004; Hattori and Yamamura 1995; Kobayashi and Hattori 2006; Yanagisawa and Ochit 1986). By being sequential protandrous hermaphrodites, anemonefishes have a reproductive output that will generally entail both the male phase, as well as the female phase. Reproductive senescence has also been seen in captive populations (Madhu et al. 2012b), but no information is known for wild populations. Given the potential longevity of anemonefish (14 years for *A. clarkii*, Moyer 1986; 18 years for *A. frenatus* and *A. perideraion*, Fautin and Allen 1997; five years for *A. chrysopterus*, Allen 1975; seven years for *A. melanopus*, Ross 1978a; 30 years for *A. percula*, Buston and García 2007) and with breeding status beginning early, at three years for *A. clarkii* (Moyer 1986), and under one year for *A. chrysopterus* (Schligler et al. 2021) and *A. polymnus* (Rattanayuvakorn et al. 2006), the time fish can be reproductive varies between four and 15 years, which translates into a reproductive output of approximately 40,000 to 800,000 eggs over a lifetime of a given species.

13.7 *WHY* DOES REPRODUCTION VARY?

13.7.1 Natural Environmental and Individual Factors

Reproductive output in anemonefishes is known to vary according to multiple parameters that can be grouped into environmental (anemone species, habitat quality, food availability, temperature), individual (female and male size – length or mass, growth, experience, behavior), and ultimately endocrine control (influence or are influenced by all of the above).

As mentioned earlier, the attraction of using anemonefish as a model system is the ease of manipulation, including a proper estimate of fecundity. It is possible to study all the factors described, in isolation, to determine their relative effects on fitness. In the case of environmental factors, both depth and anemone diameter are good predictors of the number of eggs at hatching in *A. percula* (Buston and Elith 2011) and artificially increasing the total surface area of anemones was found to increase fecundity (Cortese 2021). Beldade et al (2016) found that certain areas around the island of Moorea where anemonefish were found were more likely to harbour anemonefish that produce self-recruiting larvae. In another study, anemone species and geographic location were found to be major determinants of lifetime reproductive success in *A. percula* (Salles et al. 2020). Change in male size (length), female size (mass), female size (length), and *experience* (as indicated by the increase in number of eggs laid in consecutive spawning events) are all phenotypic characteristics that may be used to predict fecundity (Buston and Elith 2011). Finally, positive correlations have been observed between circulating 17ß-estradiol levels and fecundity *A. chrysopterus* (Cortese 2021), therefore any environmental, social, or individual impacts on fecundity likely occur via their impacts on the HPG axis and circulating hormones.

13.7.2 Environmental Stressors

Stress is closely associated with reproductive function in most species including anemonefish. Environmental or social reproductive suppression may occur in two ways. Firstly, physical (abiotic) and psychological (e.g., perceived threat of predation) stressors can suppress the activity of the HPG axis, usually via the hypothalamic-pituitary-interrenal (HPI) axis (in fish), compromising reproduction (Geraghty and Kaufer 2015). Low levels of sex steroids suppress oogenesis such that it does not surpass the early vitellogenic stage, resulting in reproductive arrest (Varela et al. 2017). Once the stressors have passed, reproduction is reactivated, but the endocrine and molecular processes involved in the dynamics of oogenesis reactivation are currently not known. Secondly, environmental or social reproductive suppression may occur via ovarian atresia, in which certain ovarian follicles recruited into the vitellogenesis pool fail to complete maturation and ovulation (Saidapur 1978). Ovarian atresia is essential for the maintenance of ovarian homeostasis and is usually seen at the end of each reproductive cycle (Krysko et al. 2008). However, several factors cause increased follicular atresia in fish, such as starvation, temperature changes, and stress (Guraya 1986) and ovarian atresia is common in fish ovaries under natural, experimental, and anthropogenic conditions.

Our increasing human population imposes considerable local environmental disturbances, for example, fisheries, habitat loss, chemical, noise, and light pollution, which are further exacerbated by global phenomena such as climate change. Such disturbances have resulted in the degradation of habitats worldwide, in particular coral reef ecosystems, and have contributed to biodiversity loss and extinction. Many of these anthropogenic stressors have impacted reproduction, behavior, and endocrine control in multiple ways, and as anemonefish are site-attached, living in tight association with anemones, they are unable to move away from localized stressors and are therefore especially vulnerable to environmental perturbations.

13.7.2.1 Climate Change

The impacts of climate change on anemonefish reproduction were determined during the 2016 large-scale sea warming event that caused global bleaching on coral reefs. A strong correlation was found between anemone bleaching (zooxanthellae loss), anemonefish stress response (increased levels of cortisol), and reproductive hormones (decreased levels of 17ß-estradiol and 11-KT) that decreased fecundity by 73% (Beldade et al. 2017). Anemone bleaching, independent of elevated temperature, also reduced orange-fin anemonefish activity and lowered metabolic rate and reduced growth (Norin et al. 2018; Cortese et al. 2021) (Figure 13.2B–C). Reproductive failure, lower growth, and energetic disadvantages lasted for the period during which the anemones remained bleached, and as the duration of bleaching periods are increasing, in the absence of acclimation to climate change this is a worrying finding for anemonefish population demography.

13.7.2.2 Motorboat Noise

Human activities that generate sound have increased since the Industrial Revolution (Hildebrand 2009) and anthropogenic noise pollution has negative effects on wildlife, altering individual behavior as well as conspecific and heterospecific interactions (Nedelec et al. 2017), translating into consequences for individual fitness, and potentially populations and communities (Simpson et al. 2016; Nedelec et al. 2017; Nedelec et al. 2016). Motorboat noise changes anemonefish endocrine control and behaviors. Underwater motorboat noise playback increases aggression in free-living orange-fin anemonefish (correlated with levels of 11-KT and T), potentially increasing their metabolism, but motorboat noise also increased hiding (correlated with cortisol) and decreased the distance moved out of the anemone, potentially reducing feeding (Mills et al. 2020). Furthermore, longer-term noise exposure led to higher levels of cortisol, and stress responses to an additional environmental challenge in both sexes were impaired. Embryonic fish also respond to sound, with increases seen in the heart rate of three-day-old anemonefish eggs (Simpson et al. 2005). Considering the functional links between androgens and reproduction, and direct impacts on embryos, it is likely that motorboat noise also decreases reproduction with repercussions on fitness in anemonefish.

13.7.2.3 Artificial Light a Night

Artificial light at night (ALAN) is another globally widespread environmental pollutant with direct ecological impacts on multiple terrestrial and aquatic ecosystems (Sanders and Gaston 2018). Recent studies have highlighted impacts on anemonefish reproduction. Exposure of *A. ocellaris* parents to ALAN in a short-term laboratory study reduced egg hatching and reproductive success (Fobert et al. 2019; Fobert et al. 2021) and the yolk sacs of *A. ocellaris* larvae were smaller when parents and eggs were exposed to ALAN in aquaria (Fobert et al. 2021). A longer-term manipulative experiment in the wild showed that in addition to reducing fecundity, ALAN also lowered the heart rate of *A. chrysopterus* eggs on the day of hatching from 168 to 153 beats per minute (bpm) (Jules Schligler pers. comm.) (Figure 13.2D). A similar embryonic heart rate has been reported for the tomato anemonefish, *A. frenatus*, at 134 bpm on the day of hatching eggs (Ha 2002) and is also strongly correlated with offspring health (Perrichon et al. 2017) with links between embryonic cardiac rhythmicity and larval swimming performance in mahi-mahi (Mager et al. 2014). Therefore, a decrease in embryonic heart rate after exposure to ALAN may have consequences across life stages. In addition, not only are there immediate impacts of ALAN on juvenile *A. chrysopterus* survival, but the decreased growth of surviving individuals may also have considerable fitness consequences later in life, as a poor start in life might compromise adult reproduction (Schligler et al. 2021).

A glimmer of hope has been demonstrated in orange-fin anemonefish wherein intergenerational plasticity in fecundity, egg, and larval traits associated with dispersal was found after artifical manipulation of parental habitat quality (Cortese 2021). This is a rare example in the marine realm in which the parental environment influences reproduction and offspring dispersal-related traits across the life history, mediated by parents, and may prove to be an important acclimation mechanism in response to our changing world.

13.8 FINAL REMARKS

Anemonefish include approximately 28 species of obligate residents of ten species of anemones. Some anemonefishes are exclusively found in a single anemone species, while others, such as *A. clarkii*, are generalists. In the most recent comprehensive phylogeny of damselfishes, Tang et al. (2021) show that the basal group of anemonefishes includes the clownfishes *A. percula* and *A. ocellaris*, together with the spinecheek anemonefish *Premnas/Amphiprion biaculeatus*. This phylogenetic arrangement may give us an idea of what reproductive mode may have looked like in early anemonefishes. Interestingly, these anemonefish species are not generalists, but rather are specialists of a single or a few anemone species.

Reproductive characteristics of anemonefishes are likely to have played an important role in shaping speciation events. As described earlier, reproduction characteristics are very variable in anemonefish species. Timing, area, tidal and lunar phases, frequency, and phenotypic factors all play a role in shaping the outcome of a reproductive event. Slight variation in such variables is likely to have generated genetic divergences that ultimately resulted in speciation events.

In this chapter, we have tried to give an overview of the reproductive patterns observed in anemonefishes. Anthropogenic factors, such as ALAN and global climate change have a direct effect on both fish (stress hormones) and anemones (bleaching). Anemonefish are particularly vulnerable to anthropogenic effects because they directly impact reproductive output, living habitat, and the protection of clutches. The fragile symbiosis between fish and anemone has been the engine for the remarkable diversification of this unique group of fish, but is also its Achilles' heel in the era of anthropogenic change.

Much work remains to be done. It is likely that other, unexplored factors, may play a very important role in the complex dynamics of the relationship between anemones and anemonefishes. For example, microbiome studies, as well as epigenetic interactions between host anemones and fish are a priority to uncover new insights into the underpinnings of the relationship. In recent years, new developments in husbandry techniques and fieldwork have allowed us to study the reproductive behavior of anemonefishes at a different level, and it is likely that new genomic approaches will open the doors to new fields of investigation.

REFERENCES

Allen, G. R. 1975. *Anemonefishes: Their Classification and Biology*. Neptune City, NJ: T. F. H. Publications.

Allen, G. R. 1980. *Anemone Fishes of the World: Species, Care and Breeding*. Mentor, OH: Aquarium Systems.

Almany, G. R., M. L. Berumen, S. R. Thorrold, S. Planes, and G. P. Jones. 2007. Local replenishment of coral reef fish populations in a marine reserve. *Science* 316 (5825): 742–44.

Anil, M. K., B. Santhosh, B. O. Prasad, and M. G. Rani. 2012. Broodstock development and breeding of black-finned anemone fish *Amphiprion nigripes* Regan, 1908 under captive conditions. *Indian Journal of Fisheries* 59 (1): 77–82.

Arvedlund, M., I. Bundgaard, and L. E. Nielsen. 2000. Host imprinting in anemonefishes (Pisces : Pomacentridae): Does it dictate spawning site preferences? *Environmental Biology of Fishes* 58: 203–13.

Barbasch, T.A., DeAngelis, R., Rhodes, J., and Buston, P.M. 2022. Parental care: Patterns, proximate and ultimate causes, and consequences. In *Evolution, Development and Ecology of Anemonefishes: Model Organisms for Marine Science*, edited by V. Laudet and T. Ravasi, 159–166. Boca Raton, FL: CRC Press.

Barbasch, T. A., T. Rueger, M. Srinivasan, M. Y. L. Wong, G. P. Jones, and P. M. Buston. 2020. Substantial plasticity of reproduction and parental care in response to local resource availability in a wild clownfish population. *Oikos* 129 (12): 1844–55.

Beldade, R., A. Blandin, R. O'Donnell, and S. C. Mills. 2017. Cascading effects of thermally-induced anemone bleaching on associated anemonefish hormonal stress response and reproduction. *Nature Communications* 8 (1): 716.

Beldade, R., S. J. Holbrook, R. J. Schmitt, S. Planes, and G. Bernardi. 2016. Spatial patterns of self-recruitment of a coral reef fish in relation to island-scale retention mechanisms. *Molecular Ecology* 25: 5203–11.

Beldade, R., S. J. Holbrook, R. J. Schmitt, S. Planes, D. Malone, and G. Bernardi. 2012. Larger female fish contribute disproportionately more to self-replenishment. *Proceedings of the Royal Society B: Biological Sciences* 279 (1736): 2116–21.

Bell, L. J. 1976. Notes on the nesting success and fecundity of the anemonefish *Amphiprion clarkii* at Miyake-Jima, Japan. *Japanese Journal of Ichthyology* 22 (4): 207–11.

Borg, B. 1994. Androgens in teleost fishes. *Comparative Biochemistry and Physiology Part C: Pharmacology, Toxicology and Endocrinology* 109 (3): 219–45.

Brantley, R. K., J. C. Wingfield, and A. H. Bass. 1993. Sex steroid levels in *Porichthys notatus*, a fish with alternative reproductive tactics, and a review of the hormonal bases of male dimorphism among teleost fishes. *Hormones and Behaviour* 27: 332–47.

Bridge, T., A. Scott, and D. Steinberg. 2012. Abundance and diversity of anemonefishes and their host sea anemones at two mesophotic sites on the Great Barrier Reef, Australia. *Coral Reefs* 31 (4): 1057–62.

Brokovich, E., S. Einbinder, N. Shashar, M. Kiflawi, and S. Kark. 2008. Descending to the twilight-zone: Changes in coral reef fish assemblages along a depth gradient down to 65 m. *Marine Ecology Progress Series* 371: 253–62.

Brooks, S., C. R. Tyler, and J. P. Sumpter. 1997. Egg quality in fish: What makes a good egg? *Reviews in Fish Biology and Fisheries* 7 (4): 387–416.

Brusle-Sicard, S., R. Reinboth, and B. Fourcault. 1994. Germinal potentialities during sexual state changes in a protandric hermaphrodite, *Amphiprion frenatus* (Teleostei, Pomacentridae). *Journal of Fish Biology* 45: 597–611.

Buston, P. M. 2003. Mortality is associated with social rank in the clown anemonefish (*Amphiprion percula*). *Marine Biology* 143 (4): 811–15.

Buston, P. M. 2004. Does the presence of non-breeders enhance the fitness of breeders? An experimental analysis in the Clown anemonefish *Amphiprion percula*. *Behavioral Ecology and Sociobiology* 57: 23–31.

Buston, P. M., and M. A. Cant. 2006. A new perspective on size hierarchies in nature: Patterns, causes, and consequences. *Oecologia* 149 (2): 362–372.

Buston, P. M., and J. Elith. 2011. Determinants of reproductive success in dominant pairs of clownfish: A boosted regression tree analysis. *Journal of Animal Ecology* 80: 528–38.

Buston, P. M., and M. B. García. 2007. An extraordinary life span estimate for the clown anemonefish *Amphiprion percula*. *Journal of Fish Biology* 70 (6): 1710–19.

Cardwell, J. R., and N. R. Liley. 1991. Androgen control of social status in males of a wild population of stoplight parrotfish, *Sparisoma viride* (Scaridae). *Hormones and Behaviour* 25: 1–18.

Cardwell, J. R., P. W. Sorensen, and G. J. van der Kraak. 1996. Effect of dominance status on sex hormone levels in laboratory and wild-spawning male trout. *General and Comparative Endocrinology* 101: 333–41.

Casadevall, M., E. Delgado, O. Colleye, S. B. Monserrat, and E. Parmentier. 2009. Histological study of the sex-change in the skunk clownfish *Amphiprion akallopisos*. *The Open Fish Science Journal* 2: 55–58.

Casas, L., Parker, C.G., and Rhodes, J.S. 2022. Sex change from male to female: Active feminization of the brain, behaviour, and gonads in anemonefish. In *Evolution, Development and Ecology of Anemonefishes: Model Organisms for Marine Science*, edited by V. Laudet and T. Ravasi, 117–128. Boca Raton, FL: CRC Press.

Casas, L., F. Saborido-Rey, T. Ryu, C. Michell, T. Ravasi, and X. Irigoien. 2016. Sex change in clownfish: Molecular insights from transcriptome analysis. *Scientific Reports* 6: 35461.

Catalano, K. A., A. G. Dedrick, M. R. Stuart, J. B. Puritz, H. R. Montes, and M. L. Pinsky. 2021. Quantifying dispersal variability among nearshore marine populations. *Molecular Ecology* 30 (10): 2366–77.

Coleman, R. R., J. M. Copus, D. M. Coffey, R. K. Whitton, and B. W. Bowen. 2018. Shifting reef fish assemblages along a depth gradient in Pohnpei, Micronesia. *PeerJ* 6: e4650.

Cortese, D. 2021. *Environmental and Parental Determinants on Larval Dispersal and Post-Settlement Physiology*. PhD diss., Ecole Pratique des Hautes Etudes.

Cortese, D., T. Norin, R. Beldade, A. Crespel, S. S. Killen, and S. C. Mills. 2021. Physiological and behavioural effects of anemone bleaching on symbiont anemonefish in the wild. *Functional Ecology* 35 (3): 663–74.

DeAngelis, R. S, and J. S. Rhodes. 2016. Sex differences in steroid hormones and parental effort across the breeding cycle in *Amphiprion ocellaris*. *Copeia* 104 (2): 586–93.

Dedrick, A. G., K. A. Catalano, M. R. Stuart, J. W. White, H. R. Montes, and M. L. Pinsky. 2021. Persistence of a reef fish metapopulation via network connectivity: Theory and data. *Ecology Letters* 24 (6): 1121–32.

Dhaneesh, K. V., T. T. Ajith Kumar, and T. Shunmugaraj. 2009. Embryonic development of percula clownfish *Amphiprion percula* (Lacepede 1802). *Middle-East Journal of Scientific Research* 4 (2): 84–89.

Dhaneesh, K. V., K. N. Devi, T. T. A. Kumar, T. Balasubramanian, and K. Tissera. 2012. Breeding, embryonic development and salinity tolerance of skunk clownfish *Amphiprion akallopisos*. *Journal of King Saud University – Science* 24 (3): 201–209.

Drews, C. 1993. The concept and definition of dominance in animal behaviour. *Behaviour* 125 (3–4): 283–313.

Elliott, J. K., S. C. Lougheed, B. Bateman, L. K. McPhee, and P. T. Boag. 1999. Molecular phylogenetic evidence for the evolution of specialization in anemonefishes. *Proceedings of the Royal Society of London, Series B: Biological Sciences* 266 (1420): 677–85.

Emms, M. A., P. Saenz-Agudelo, E. C. Giles, R. Gatins, G. B. Nanninga, A. Scott, J-P. A. Hobbs, et al. 2020. Comparative phylogeography of three host sea anemones in the Indo-Pacific. *Journal of Biogeography* 47: 487–500.

Faught, E., and M. M. Vijayan. 2018. Maternal stress and fish reproduction: The role of cortisol revisited. *Fish and Fisheries* 19: 1016–1030.

Fautin, D. J., and G. R. Allen. 1992. *Field Guide to Anemone Fishes and Their Host Sea Anemones*. Perth: Western Australian Museum.

Fautin, D. J., and G. R. Allen. 1997. *Anemone Fishes and Their Host Sea Anemones*. Perth, WA: Western Australian Museum.

Fobert, E. K., K. Burke da Silva, and S. E. Swearer. 2019. Artificial light at night causes reproductive failure in clownfish. *Biology Letters* 15 (7): 20190272.

Fobert, E. K., K. P. Schubert, and K. Burke da Silva. 2021. The influence of spectral composition of artificial light at night on clownfish reproductive success. *Journal of Experimental Marine Biology and Ecology* 540: 151559.

Fostier, A., B. Jalabert, R. Billard, B. Breton, and Y. Zohar. 1983. The gonadal steroids. In *Fish Physiology*, edited by W. S. Hoar, D. J. Randall, and E. M. Donaldson, 277–372. New York: Academic Press.

Fricke, H., and S. Fricke. 1977. Monogamy and sex change by aggressive dominance in anemonefish. *Nature* 266 (5605): 830–32.

Fricke, H. W. 1979. Mating system, resource defence and sex change in the anemonefish *Amphiprion akallopisos. Zeitschrift Für Tierpsychologie* 50 (3): 313–26.

Fricke, H. W. 1983. Social control of sex: Field experiments with the anemonefish *Amphiprion bicinctus. Zeitschrift Für Tierpsychologie* 61 (1): 71–77.

Frisch, A. J., J. R. Rizzari, K. P. Munkres, and J-P. A. Hobbs. 2016. Anemonefish depletion reduces survival, growth, reproduction and fishery productivity of mutualistic anemone–anemonefish colonies. *Coral Reefs* 35 (2): 375–86.

Geraghty, A. C., and D. Kaufer. 2015. Glucocorticoid regulation of reproduction. *Glucocorticoid Signalling* 872: 253–78.

Ghosh, S., T. Thankappanpillai, and A. Kumar. 2012. Embryology of Maldives clownfish, *Amphiprion nigripes* (Amphiprioninae). *Oceanic and Coastal Sea Research* 11: 174–80.

Godwin, J. R., and P. Thomas. 1993. Sex change and steroid profiles in the protandrous anemonefish *Amphiprion melanopus* (Pomacentridae, Teleostei). *General and Comparative Endocrinology* 91: 144–57.

Godwin, J. and D. Fautin. 1992. Defense of host actinians by anemonefishes. *Copeia* 3: 902–908

Gopakumar, G., R. M. George, and S. Jasmin. 1994. Breeding and larval rearing of the clownfish *Amphiprion chrysogaster. Marine Fisheries Information Service* 161: 8–11.

Gordon, A. K., and A. W. Bok. 2001. Frequency and periodicity of spawning in the clownfish *Amphiprion akallopisos* under aquarium conditions. *Aquarium Sciences and Conservation* 3 (4): 307–13.

Green, B. S. 2004. Embryogenesis and oxygen consumption in benthic egg clutches of a tropical clownfish, *Amphiprion melanopus* (Pomacentridae). *Comparative Biochemistry and Physiology - A Molecular and Integrative Physiology* 138 (1): 33–38.

Green, B. S., K. R. N. Anthony, and M. I. McCormick. 2006. Position of egg within a clutch is linked to size at hatching in a demersal tropical fish. *Journal of Experimental Marine Biology and Ecology* 329 (1): 144–52.

Green, B. S., and M. I. McCormick. 2005. Maternal and paternal effects determine size, growth and performance in larvae of a tropical reef fish. *Marine Ecology Progress Series* 289: 263–72.

Grier, H. J., M. C. Uribe-Aranzábal, and R. Patino 2009. The ovary, folliculogenesis, and oogenesis in teleosts. In *Reproductive Biology and Phylogeny of Fishes (Agnathans and Bony Fishes)*, edited by B. G. M. Jamieson, 25–84. Enfield, NH: Science Publishers.

Guraya, S. S. 1986. The cell and molecular biology of fish oogenesis. *Monographs in Developmental Biology* 18: 1–233.

Ha, L. T. L. 2002. The embryonic and larvae developments of tomato anemonefish (*Amphiprion frenatus* Brevoort, 1856) in Khanh Hoa coast. *Collection of Marine Research Works* XII: 233–42.

Haguenauer, A., F. Zuberer, G. Siu, D. Cortese, R. Beldade, and S. C. Mills. 2021. Deep heat: A comparison of water temperature, anemone bleaching, anemonefish density and reproduction between shallow and mesophotic Reefs. *Fishes* 6 (3): 37.

Harrison, H. B., D. H. Williamson, R. D. Evans, G. R. Almany, S. R. Thorrold, G. R. Russ, K. A. Feldheim, et al. 2012. Larval export from marine reserves and the recruitment benefit for fish and fisheries. *Current Biology* 22 (11): 1023–28.

Hattori, A. 1991. Socially controlled growth and size-dependant sex change in the anemonefish *Amphiprion frenatus* in Okinawa, Japan. *Japanese Journal of Ichthyology* 38 (2): 165–77.

Hattori, A. 1994. Inter-group movement and mate acquisition tactics of the protandrous anemonefish, *Amphiprion clarkii*, on a coral reef, Okinawa. *Japanese Journal of Ichthyology* 41: 159–65.

Hattori, A. 2002. Small and large anemonefishes can coexist using the same patchy resources on a coral reef, before habitat destruction. *Journal of Animal Ecology*. 71 (5): 824–31.

Hattori, A. 2004. High mobility of the protandrous anemonefish *Amphiprion frenatus*: Nonrandom pair formation in limited shelter space. *Animal Behaviour*, 52 (1): 57–63.

Hattori, A. 2012. Determinants of body size composition in limited shelter space: Why are anemonefishes protandrous? *Behavioral Ecology* 23 (3): 512–20.

Hattori, A., and N. Yamamura. 1995. Co-existence of subadult males and females as alternative tactics of breeding post acquisition in a monogamous and protandrous anemonefish. *Evolutionary Ecology* 9: 292–303.

Helvik, J.V., and Walther, B.T. 1992. Photo-regulation of the hatching process of halibut (*Hippoglossus hippoglossus*) eggs. *Journal of Experimental Zoology* 263: 204–209.

Hildebrand, J. A. 2009. Anthropogenic and natural sources of ambient noise in the ocean. *Marine Ecology Progress Series* 395: 5–20.

Holtswarth, J., S. S. Jose, H. Montes Jr, J. W. Morley, and M. Pinsky. 2017. The reproductive seasonality and fecundity of yellowtail clownfish (*Amphiprion clarkii*) in the Philippines. *Bulletin of Marine Science* 93 (4): 997–1007.

Ignatius, B., G. Rathore, I. Jagadis, D. Kandasami, and A. C. C. Victor. 2001. Spawning and larval rearing technique for tropical clown fish *Amphiprion sebae* under captive conditions. *Journal of Aquaculture in the Tropics* 16 (3): 241–49.

Iwata, E., K. Mikami, J. Manbo, K. Moriya-Ito, and H. Sasaki. 2012. Social interaction influences blood cortisol values and brain aromatase genes in the protandrous false clown anemonefish, *Amphiprion ocellaris. Zoological Science* 29 (12): 849–55.

Jamieson, B. G. M. 1991. *Fish Evolution and Systematics: Evidence from Spermatozoa*. Cambridge: Cambridge University Press.

Khoo, M. L., S. K. Das, and M. A. Ghaffar. 2018. Growth pattern, diet and reproductive biology of the clownfish *Amphiprion ocellaris* in waters of Pulau Tioman, Malaysia. *Egyptian Journal of Aquatic Research* 44 (3): 233–39.

Kobayashi, M., and A. Hattori. 2006. Spacing pattern and body size composition of the protandrous anemonefish *Amphiprion frenatus* inhabiting colonial host anemones. *Ichthyological Research* 53 (1): 1–6.

Kobayashi, Y., R. Horiguchi, S. Miura, and M. Nakamura. 2010. Sex-and tissue-specific expression of P450 aromatase (cyp19a1a) in the yellowtail clownfish, *Amphiprion clarkii. Comparative Biochemistry and Physiology Part A: Molecular & Integrative Physiology* 155: 237–44.

Kobayashi, Y., Y. Nagahama, and M. Nakamura. 2013. Diversity and plasticity of sex determination and differentiation in fishes. *Sexual Development* 7: 115–125.

Krysko, D. V., A. Diez-Fraile, G. Criel, A. A. Svistunov, P. Vandenabeele, and K. D'Herde. 2008. Life and death of female gametes during oogenesis and folliculogenesis. *Apoptosis* 13: 1065–1087.

Kunzmann, A., and V. C. Diemel. 2020. The effect of spatial position and age within an egg-clutch on embryonic development and key metabolic enzymes in two clownfish species, *Amphiprion ocellaris* and *Amphiprion frenatus*. *PLoS ONE* 15 (1): 20226600.

Lazier, C. B., M. Mann, and T. P. Mommsen. 1987. Estrogen receptors and regulation of vitellogenesis. In *Third International Symposium on Reproductive Physiology of Fish*, edited by Idler, D. R., L. W. Grim, and J. M. Walsh, 178–82. Newfoundland, St. John's: Memorial University.

Lehmann, R., D. J. Lightfoot, C. Schunter, C. T. Michell, H. Ohyanagi, K. Mineta, S. Foret, et al. 2019. Finding nemo's genes: A chromosome-scale reference assembly of the genome of the orange clownfish *Amphiprion percula*. *Molecular Ecology Resources* 19 (3): 570–85.

Litsios, G., P. B. Pearman, D. Lanterbecq, N. Tolou, and N. Salamin. 2014. The radiation of the clownfishes has two geographical replicates. *Journal of Biogeography* 41: 2140–49.

Litsios, G., C. A. Sims, R. O. Wüest, P. B. Pearman, N. E. Zimmermann, and N. Salamin. 2012. Mutualism with sea anemones triggered the adaptive radiation of clownfishes. *BMC Evolutionary Biology* 12 (1): 1–15.

Lubzens, E., J. Bobe, G. Young, and C. V. Sullivan. 2017. Maternal investment in fish oocytes and eggs: The molecular cargo and its contributions to fertility and early development. *Aquaculture* 472: 107–43.

Lubzens, E., G. Young, J. Bobe, and J. Cerdà. 2010. Oogenesis in teleosts: How fish eggs are formed. *General and Comparative Endocrinology* 165 (3): 367–89.

Luckenbach, J. A., D. B. Iliev, F. W. Goetz, and P. Swanson. 2008. Identification of differentially expressed ovarian genes during primary and secondary oocyte growth in coho salmon, *Oncorhynchus kisutch*. *Reproductive Biology and Endocrinology* 6: 2–16.

Madhu, K., and R. Madhu. 2007. Influence of lunar rhythm on spawning of clown anemonefish *Amphiprion percula* under captive condition in Andaman and Nicobar Islands. *Journal of the Marine Biological Association of India* 49 (1): 58–64.

Madhu, K., R. Madhu, G. Gopakumar, and C. S. Sasidharan. 2006b. Breeding larval rearing and seed production of maroon clown *Premnas biaculeatus* under captive conditions. *Marine Fisheries Information Service, Technical and Extension Series* 190: 1–5.

Madhu, K., R. Madhu, L. Krishnan, C. S. Sasidharan, and K. M. Venugopalan. 2006a. Spawning and larval rearing of *Amphiprion ocellaris* under captive conditions. *Marine Fisheries Information Service, Technical and Extension Series* 188: 1–5.

Madhu, K., R. Madhu, and T. Retheesh. 2012a. Broodstock development, breeding, embryonic development and larviculture of spine-cheek anemonefish, *Premnas biaculeatus* (Bloch, 1790). *Indian Journal of Fisheries* 59 (1): 65–75.

Madhu, R., K. Madhu, and T. Retheesh. 2012b. Life history pathways in false clown *Amphiprion ocellaris* Cuvier, 1830: A journey from egg to adult under captive condition. *Journal of the Marine Biological Association of India* 54 (1): 77–90.

Mager, E. M., A. J. Esbaugh, J. D. Stieglitz, R. Hoenig, C. Bodinier, J. P. Incardona, N. L. Scholz, et al. 2014. Acute embryonic or juvenile exposure to deepwater horizon crude oil impairs the swimming performance of Mahi-Mahi (*Coryphaena hippurus*). *Environmental Science and Technology* 48 (12): 7053–61.

Marcionetti, A., V. Rossier, J. A. M. Bertrand, G. Litsios, and N. Salamin. 2018. First draft genome of an iconic clownfish species (*Amphiprion frenatus*). *Molecular Ecology Resources* 18 (5): 1092–101.

McAlary, F.A., and McFarland, W.N. 1993. The Effect of light and darkness on hatching in the pomacentrid *Abudefduf saxatilis*. *Environmental Biology of Fishes* 37: 237–244.

Miller, G. M., F. J. Kroon, S. Metcalfe, and P. L. Munday. 2015. Temperature is the evil twin: Effects of increased temperature and ocean acidification on reproduction in a reef fish. *Ecological Applications* 25 (3): 603–20.

Mills, S. C., R. Beldade, P. Chabanet, L. Bigot, J. L. O'Donnell, and G. Bernard. 2015. Ghosts of thermal past: Reef fish exposed to historic high temperatures have heightened stress response to further stressors. *Coral Reefs* 34: 1255–60.

Mills, S. C., R. Beldade, L. Henry, D. Laverty, S. L. Nedelec, S. D. Simpson, and A. N. Radford. 2020. Hormonal and behavioural effects of motorboat noise on wild coral reef fish. *Environmental Pollution* 262: 114250.

Mills, S. C., J. Mourier, and R. Galzin. 2010. Plasma cortisol and 11-ketotestosterone enzyme immunoassay (EIA) kit validation for three fish species: The orange clownfish *Amphiprion percula*, the orangefin anemonefish *Amphiprion chrysopterus* and the blacktip reef shark *Carcharhinus melanopterus*. *Journal of Fish Biology* 77: 769–77.

Mills, S. C., J. L. O'Donnell, G. Bernardi, and R. Beldade. 2018. Natural endocrine profiles of the group-living skunk anemonefish *Amphiprion akallopisos* in relation to their size-based dominance hierarchy. *Journal of Fish Biology* 92 (3): 773–89.

Moyer, J. T. 1986. Longevity of the anemonefish *Amphiprion clarkii* at Miyake-Jima, Japan with notes on four other species. *Copeia* 1: 135–39.

Moyer, J. T., and A. Nakazono. 1978. Protandous hermaphroditism in six species of the anemonefish genus Amphiprion in Japan. *Japanese Journal of Ichthyology* 25: 101–106.

Moyer, J. T., and R. C. Steene. 1979. Nesting behaviour of the anemonefish *Amphiprion polymnus*. *Japanese Journal of Ichthyology* 26 (2): 209–14.

Murua, H., and F. Saborido-Rey. 2003. Female reproductive strategies of marine fish species of the North Atlantic. *Journal of Northwest Atlantic Fishery Science* 33: 23–31.

Nakamura, M., T. Mariko, and Y. Nagahama. 1994. Ultrastructure and in vitro steroidogenesis of the gonads in the protandrous anemonefish *Amphiprion frenatus*. *Japanese Journal of Ichthyology* 41 (1): 47–56.

Nedelec, S. L., S. C. Mills, D. Lecchini, B. Nedelec, S. D. Simpson, and A. N. Radford. 2016. Repeated exposure to noise increases tolerance in a coral reef fish. *Environmental Pollution* 216: 428–36.

Nedelec, S. L., S. C. Mills, A. N. Radford, R. Beldade, S. D. Simpson, B. Nedelec, and I. M. Côté. 2017. Motorboat noise disrupts co-operative interspecific interactions. *Scientific Reports* 7 (1): 1–8.

Ng, T. B., and D. R. Idler. 1983. Yolk formation and differentiation in teleost fishes. In *Fish Physiology*, edited by W. S. Hoar, D. J. Randall, and E. M. Donaldson, 373–404. New York: Academic Press.

Norin, T., S. C. Mills, A. Crespel, D. Cortese, S. S. Killen, and R. Beldade. 2018. Anemone bleaching increases the metabolic demands of symbiont anemonefish. *Proceedings of the Royal Society of London, Series B: Biological Sciences* 285: 20180282.

Ochi, H. 1985. Temporal patterns of breeding and larval settlement in a temperate population of the tropical anemonefish, *Amphiprion clarkii*. *Japanese Journal of Ichthyology* 32 (2): 248–57.

Ochi, H. 1989. Mating behavior and sex change of the anemonefish, *Amphiprion clarkii*, in the temperate waters of Southern Japan. *Environmental Biology of Fishes* 26: 257–75.

Oliveira, R. F., V. C. Almada, and A. V. M. Canario. 1996. Social modulation of sex steroid concentrations in the urine of male cichlid fish *Oreochromis mossambicus*. *Hormones and Behaviour* 12 (2): 2–12.

Ollerton, J., D. McCollin, D. G. Fautin, and G. R. Allen. 2007. Finding NEMO: Nestedness engendered by mutualistic organization in anemonefish and their hosts. *Proceedings of the Royal Society B: Biological Sciences* 274 (1609): 591–98.

Perrichon, P., M. Grosell, and W. W. Burggren. 2017. Heart performance determination by visualization in larval fishes: Influence of alternative models for heart shape and volume. *Frontiers in Physiology* 8: 1–10.

Pinsky, M. L., H. R. Jr. Montes, and S. R. Palumbi. 2010. Using isolation by distance and effective density to estimate dispersal scales in anemonefish. *Evolution* 64 (9): 2688–700.

Pyle, R. L. 1996. Exploring deep coral reefs: How much biodiversity are we missing? *Global Biodiversity* 6: 3–7.

Rattanayuvakorn, S., P. Mungkornkarn, A. Thongpan, and K. Chatchavalvanich. 2006. Gonadal development and sex inversion in saddleback anemonefish *Amphiprion polymnus* Linnaeus (1758). *Kasetsart Journal: Natural Science* 40 (1): 196–203.

Richardson, D. L. 1999. Correlates of environmental variables with patterns in the distribution and abundance of two anemonefishes (Pomacentridae: Amphiprion) on an eastern Australian sub-tropical reef system. *Environmental Biology of Fishes* 55 (3): 255–63.

Richardson, D. L., P. L. Harrison, and V. J. Harriott. 1997. Timing of spawning and fecundity of a tropical and subtropical anemonefish (Pomacentridae: Amphiprion) on the east coast of Australia. *Marine Ecology Progress Series* 156: 175–81.

Raheem, P. K., M. V. R. Krishna, S. Surya, P. Gomathi, A. P. Gop, B. Santhosh, B. Raju,et al. 2021. Breeding, larval rearing and growth of black *Amphiprion ocellaris* (Cuvier, 1830) under captivity. *Indian Journal of Fisheries* 68 (2): 60–69.

Rohini, K. M. V., M. K. Anil, and R. P. Neethu. 2018. Larval development and growth of Red Saddleback Anemonefish, *Amphiprion ephippium* (Bloch, 1790) under captive conditions. *Indian Journal of Geo-Marine Sciences* 47: 2421–28.

Ross, R. M. 1978a. Reproductive behavior of the anemonefish *Amphiprion melanopus* on Guam. *Copeia* 1: 103–107.

Ross, R. M. 1978b. Territorial behavior and ecology of the anemonefish *Amphiprion melanopus* on Guam. *Zeitschrift Für Tierpsychologie* 46 (1): 71–83.

Rueger, T., T. A. Barbasch, M. Y. L. Wong, M. Srinivasan, G. P. Jones, and P. M. Buston. 2018. Reproductive control via the threat of eviction in the Clown Anemonefish. *Proceedings of the Royal Society B: Biological Sciences* 285 (1891): 20181295.

Saborido-Rey, F. 2016. Fish reproduction. *Reference Module in Earth Systems and Environmental Sciences* 1: 232–45.

Saenz-Agudelo, P., G. P. Jones, S. R. Thorrold, and S. Planes. 2011. Detrimental effects of host anemone bleaching on anemonefish populations. *Coral Reefs* 30 (2): 497–506.

Saidapur, S. K. 1978. Follicular atresia in the ovaries of nonmammalian vertebrates. *International Review of Cytology* 54: 225–44.

Salis, P., N. Roux, D. Huang, A. Marcionetti, P. Mouginot, M. Reynaud, O. Salles, et al. 2021. Thyroid hormones regulate the formation and environmental plasticity of white bars in clownfishes. *PNAS*. 118 (23) e2101634118.

Salles, O. C., G. R. Almany, M. L. Berumen, G. P. Jones, P. Saenz-Agudelo, M. Srinivasan, S. R. Thorrold, et al. 2020. Strong habitat and weak genetic effects shape the lifetime reproductive success in a wild clownfish population. *Ecology Letters* 23 (2): 265–73.

Sanders, D., and K. J. Gaston. 2018. How ecological communities respond to artificial light at night. *Journal of Experimental Zoology Part A: Ecological and Integrative Physiology* 329 (8–9): 394–400.

Schligler, J., A. Blandin, R. Beldade, and S. C. Mills. 2022. Aggression of an orange-fin anemonefish to a blacktip reef shark: A potential example of fish mobbing? *Marine Biodiversity*, 52: 17.

Schligler, J., D. Cortese, R. Beldade, S. E. Swearer, and S. C. Mills. 2021. Long-term exposure to artificial light at night in the wild decreases survival and growth of a coral reef fish. *Proceedings of the Royal Society B: Biological Sciences* 288 (1952): 20210454.

Schoots, A.F.M., de Bont R.G., van Eys, G.J.J.M., and Denucé, J.M. 1982. Evidence for a stimulating effect of prolactin on teleostean hatching enzyme secretion. *Journal of Experimental Zoology* 219: 129–132.

Schulz, R. W., L. R. De Franca, J. J. Lareyre, F. Legac, H. Chiarini-Garcia, R. H. Nobrega, and T. Miura. 2010. Spermatogenesis in fish. *General and Comparative Endocrinology* 165: 390–411.

Seymour, J. R., T. A. Barbasch, and P. M. Buston. 2018. Lunar cycles of reproduction in the clown anemonefish *Amphiprion percula*: Individual-level strategies and population-level patterns. *Marine Ecology Progress Series* 594: 193–201.

Shapiro, D. Y. 1992. Plasticity of gonadal development and protandry in Fishes. *Journal of Experimental Zoology* 261: 194–203.

Shin, H. S., N. N. Kim, Y. J. Choi, H. R. Habibi, J. W. Kim, and C. Y. Choi. 2013. Light-emitting diode spectral sensitivity relationship with reproductive parameters and ovarian maturation in yellowtail damselfish, *Chrysiptera parasema*. *Journal of Photochemistry and Photobiology B: Biology* 127: 108–13.

Simpson, S. D., A. N. Radford, S. L. Nedelec, M. C. O. Ferrari, D. P. Chivers, M. I. McCormick, and M. G. Meekan. 2016. Anthropogenic noise increases fish mortality by predation. *Nature Communications* 7 (1): 1–7.

Simpson, S. D., H. Y. Yan, M. L. Wittenrich, and M. G. Meekan. 2005. Response of embryonic coral reef fishes (Pomacentridae: Amphiprion spp.) to noise. *Marine Ecology Progress Series* 287: 201–208.

Tang, K. L., M. L. J. Stiassny, R. L. Mayden, and R. DeSalle. 2021. Systematics of damselfishes. *Ichthyology and Herpetology* 109 (1): 258–318.

Titus, B. M., R. Laroche, E. Rodríguez, H. Wirshing, and C. P. Meyer. 2020. Host identity and symbiotic association affects the taxonomic and functional diversity of the clownfish-hosting sea anemone microbiome. *Biology Letters* 16: 20190738.

Varela, M. L., M. F. Ferreira, R. H. da Cuña, F. L. lo Nostro, G. Genovese, and F. J. Meijide. 2017. Dynamics of ovarian maturation throughout the reproductive cycle of the neotropical cichlid fish *Cichlasoma dimerus* (Teleostei, Cichliformes). *Canadian Journal of Zoology* 95 (7): 485–98.

Wallace, R. A., and K. Selman. 1981. Cellular and dynamic aspects of oocyte growth in teleosts. *American Zoologist* 21: 325–43.

Wingfield, J. C., R. E. Hegner, A. M. Dufty, Jr., and G. F. Ball. 1990. The "challenge hypothesis": Theoretical implications for patterns of testosterone secretion, mating systems, and breeding strategies. *American Naturalist* 136 (6): 829–46.

Yanagisawa, Y., and H. Ochit. 1986. Step-fathering in the anemonefish *Amphiprion clarkii*: A removal study. *Animal Behaviour* 34 (6): 1769–80.

14 Social Evolution in Anemonefishes
Formation, Maintenance, and Transformation of Social Groups

Peter M. Buston, Rebecca Branconi, and Theresa Rueger

CONTENTS

14.1 INTRODUCTION

Animal societies in which some individuals forgo their own reproduction and help others to reproduce are one of the most remarkable products of evolution (Emlen 1991,1997; Seger 1991; Smith and Szathmary 1995; Bourke 1997). They are found in a wide variety of taxa, including ants, bees, wasps, termites, aphids, spiders, shrimp, fishes, lizards, birds, and mammals (Rubenstein and Abbott 2017). Such societies can be considered to lie on a continuum from the simpler social groups or cooperative breeding systems that are common in social vertebrates, in which some members are non-breeding helpers but are not irreversibly committed to that role, to the more complex social groups or eusocial systems that are common in social insects, in which some individuals are irreversibly committed to a sterile worker role (Sherman et al. 1995; Bourke 2011). These social systems raise a number of intriguing questions. Why do non-breeders forgo their own reproduction? Why do breeders tolerate non-breeders? How are conflicts of interest among group members resolved? How do simple social groups transition to become more complex social groups? Answering these questions has been a major goal of behavioral ecologists and evolutionary biologists.

The evolution of animal societies can be thought of in three stages: social group formation; social group maintenance; and, social group transformation (Bourke 2011). Social group formation refers to the processes that bring

DOI: 10.1201/9781003125365-17

individuals together (i.e., favour the genes underlying cooperative behaviors). Here, concepts such as kin selection and ecological constraints have played a major role in our understanding (Hamilton 1964; Emlen 1982; Keller and Reeve 1994). Social group maintenance refers to the processes that keep individuals together and confer group stability (i.e., that favour the genes underlying conflict resolution. Here, concepts such as dominance, punishment, and hidden threats have played a major role in our understanding) (Ratneiks and Wenseleers 2005; Cant 2011; Raihani et al. 2012). Social group transformation refers to the processes that convert simple social groups into complex social groups (i.e., that favour the genes underlying novel social behaviors and increased division of labour). Here, the size-complexity hypothesis has played a major role in our understanding (Alexander 1991; Bonner 2004; Bourke 2011). While the lines between these three stages of social evolution are blurred, and the processes involved in all stages will occur both sequentially and simultaneously, the framework is a useful one for organizing our thinking.

Anemonefishes present exciting opportunities to test the robustness of current theories of and generate new insights into social evolution (Allen 1972; Fricke 1979; Krebs and Davies 1993). While there is some interspecific variation in anemonefish societies, the generic system can be described as follows (Figure 14.1). Anemonefish are found in close association with sea anemones that provide protection, food, and oviposition sites (Mariscal 1970; Fautin 1992; Verde et al. 2015; see Chapter 19). Each anemone (or cluster of anemones) hosts a single group of anemonefish composed of a breeding pair and a small number of non-breeders (Fricke 1979; Mitchell 2003a; Buston 2004a). Within each group there is a size-based dominance hierarchy: the female is largest, the male is second largest, and the non-breeders get progressively smaller (Fricke 1979; Buston 2003a). Anemonefishes are protandrous hermaphrodites (Fricke and Fricke 1977; Moyer and Nakazono 1978; see Chapter 12): if the female of the group dies, the male changes sex and takes the position vacated by the female, and the largest non-breeder from the anemone (or a nearby anemone) inherits the position vacated by the male (Fricke 1979; Buston 2004b; Mitchell 2005). Reproduction occurs on lunar cycles (Ross 1978; Seymour et al. 2018; see Chapter 13); each lunar month the female lays several hundred eggs, which the male takes care of until they hatch (Buston and Elith 2011; Barbasch et al. 2020; see Chapter 15). Once they hatch, the larvae disperse (Jones et al. 2005; Catalano et al. 2021; see Chapter 20) before settling into an anemone and joining a group (Elliott et al. 1995; Buston 2003b; see Chapter 16).

Here, we review what is, and what is not, known about social evolution in anemonefishes. We use the society on which the most work has been conducted to date – the clown anemonefish or orange clownfish (*Amphiprion percula*) – to shed light on the processes involved in social group formation, social group maintenance, and social group transformation. We then consider how subtle changes in these processes can cause interspecific variation in anemonefish societies. Finally, we highlight several avenues of research that we feel are particularly exciting. This review complements a recent review of social group formation in coral reef fishes more generally (Rueger et al. 2021a), going into more detail and considering social group maintenance and

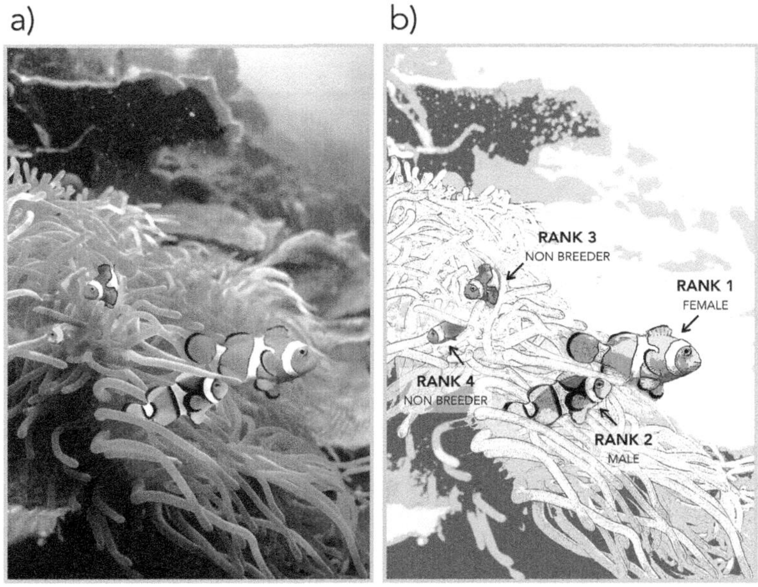

FIGURE 14.1 The generic anemonefish social system: a. composite photograph of a social group of the clown anemonefish *Amphiprion percula* in Kimbe Bay, Papua New Guinea; b. each anemonefish social group is composed of a breeding pair and a small number of non-breeders; the female is largest, the male is second largest, and the non-breeders get progressively smaller; if the female of the group dies, the male changes sex and takes the position vacated by the female. (Photo and illustration by R. B.)

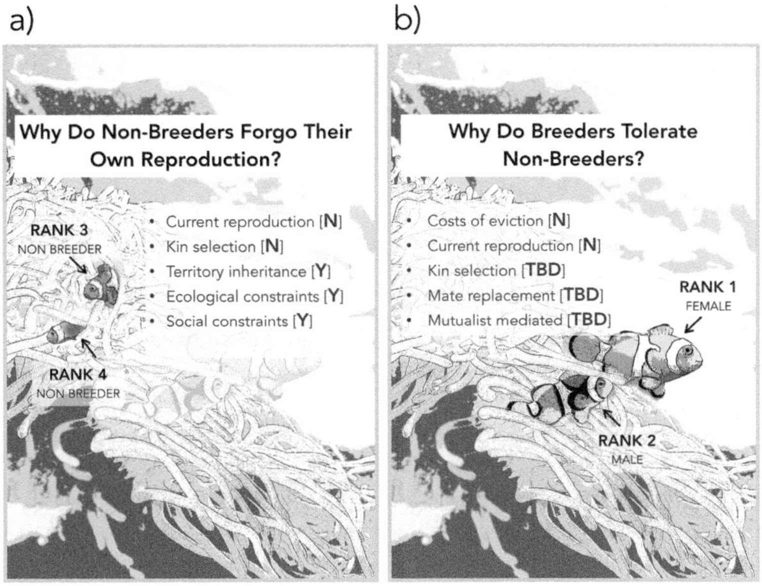

FIGURE 14.2 Social group formation in the clown anemonefish *Amphiprion percula*. The evolution of social behavior can be understood using Hamilton's inequality: a. natural selection favours individuals that adopt subordinate non-breeding positions, because they stand to inherit the territory in the future, and because their alternative options of moving to breed elsewhere or contesting to breed at home are poor due to harsh ecological and social constraints; b. natural selection favours dominant breeders that tolerate subordinate non-breeders because their alternative option of evicting the subordinates and living alone is relatively poor, likely because non-breeders confer some combination of weak kin-selected benefits, mate-replacement benefits, and mutualist-mediated benefits. Text refers to hypotheses and symbols indicate that hypothesis was tested and rejected [N], hypothesis was tested and supported [Y], hypothesis is plausible but critical tests are yet to be done [TBD]. (Illustrations by R. B.)

social group transformation in anemonefishes. Our goal is to provide the reader with a framework for thinking about anemonefish social evolution as they become a model system for marine science.

14.2 SOCIAL GROUP FORMATION

14.2.1 Why Do Non-Breeders Forgo Their Own Reproduction?

The evolution of non-breeding and helping strategies depends on the costs and benefits of cooperative options and alternative options for the non-breeder and its relatives (Figure 14.2). The behavior favoured by selection can be determined using Hamilton's Rule (Hamilton 1964). On the one hand, the cooperative option can be favoured if it enables the individual to reproduce in the present (direct genetic benefits), has beneficial effects on kin (indirect genetic benefits), or has beneficial effects in the future (future genetic benefits). On the other hand, the cooperative option can be favoured if the outside option is poor due to ecological constraints and if the inside option is poor due to social constraints. In the case of the clown anemonefish the cooperative option (waiting peacefully to breed) will be favoured over the outside option (dispersing to breed elsewhere) or the inside option (contesting to breed at home) if

$$X_i + rY_i + fZ_i > X_j + rY_j + fZ_j \qquad (14.1)$$

where *Xi* or *Xj* is the non-breeder's fitness associated with its actions in the present; *Yi* or *Yj* is the breeder's fitness associated with the non-breeder's actions, and *r* is the probability that the two individuals share a copy of a particular gene identical by descent; *Zi* or *Zj* is the non-breeder's fitness associated with its actions in the future, and *f* is the probability that benefits will be realized in the future; *i* terms represent payoffs associated with cooperative option and *j* terms represent payoffs associated with alternative options. In the following, we describe how each of the direct genetic benefits, indirect genetic benefits, future genetic benefits, ecological constraints, and social constraints hypotheses have been tested in the clown anemonefish *Amphiprion percula*.

14.2.1.1 Direct Genetic Benefits: Current Reproduction

First, we need to address whether subordinates in groups of *A. percula* are truly forgoing their own reproduction. Subordinate *A. percula* do not have functional gonads, and they avoid developing functional gonads due to the threat of eviction (Rueger et al. 2018). Subordinate *A. percula* are old enough to have functional gonads, and they can develop functional gonads within months when breeding opportunities arise (Buston 2004b). This confirms that subordinates are indeed non-breeders ($X_i = 0$ in Inequality 14.1) and it leaves us with the question of why they choose the cooperative option within social groups.

14.2.1.2 Indirect Genetic Benefits: Kin Selection

The kin selection hypothesis emphasizes that there are two ways for individuals to get their genes into the next generation – either directly, by producing their own offspring, or indirectly, by enhancing the offspring production of their relatives (Hamilton 1964). Helping relatives can be favoured by kin selection because the relatives' offspring share copies of the helpers' genes. The kin selection hypothesis makes two critical predictions: first, non-breeders enhance the fitness of breeders; and, second, non-breeders are closely related to breeders (Hamilton 1964; Emlen and Wrege 1988; Griffin and West 2003). We have tested both of these predictions using a population of *Amphiprion percula* in Madang Lagoon, Papua New Guinea (PNG). We monitored survival, growth and reproduction of 71 groups over 12 months (Buston 2002). We also removed all non-breeders from 14 of the 71 groups and examined the effect of the removal on the fitness of the breeders (Buston 2004a). Further, we investigated the genetic relatedness of 35 individuals from nine groups using seven microsatellite loci (Buston et al. 2007). The non-breeders had no effect on the survival, growth or reproduction of the breeders and they were not closely related to the breeders (Buston 2004a; Buston et al. 2007; $rY_i = 0$ in Inequality 14.1). The first prediction was also tested and refuted in *A. percula*'s sister taxon *A. ocellaris* (Mitchell 2003a). These results suggest that kin selection, a founding concept of social evolution, does not play an important role in the evolution of non-breeding strategies in anemonefishes (Buston 2004a; Buston et al. 2007).

14.2.1.3 Future Genetic Benefits: Territory Inheritance

The territory inheritance hypothesis emphasizes that there is yet another way for individuals to get their genes into the next generation – directly, by producing their own offspring in the future. Forgoing reproduction in the present can be favoured by selection if it maximizes the chances of reproducing successfully in the future. The territory inheritance hypothesis makes two critical predictions: first, non-breeders have the capacity to reproduce in the future; second, the probability of territory inheritance is high (Kokko and Johnstone 1999). We have tested both predictions by monitoring 57 of the 71 groups of *A. percula* in Madang Lagoon described earlier and recording which individuals filled breeding vacancies when they arose (Buston 2002). We also removed a breeder from 16 groups and determined which individuals filled the breeding vacancy (Buston 2004b). In all cases the largest non-breeder from the anemone inherited the breeding vacancy and began reproducing successfully; in no case did a smaller non-breeder from the anemone or a non-breeder from elsewhere usurp the vacancy ($fZ_i >> 0$ in Inequality 14.1). Territory inheritance via queuing was also documented in *A. percula's* sister taxon *A. ocellaris* (Mitchell 2005). These results suggest that territory inheritance is a driving force behind the evolution of non-breeding strategies in anemonefishes (Buston 2004b).

14.2.1.4 Poor Outside Options: Ecological Constraints

The ecological constraints hypothesis emphasizes that there are two options available to individuals – either they can disperse to breed elsewhere or they can stay on their current territory as non-breeders and wait to inherit the territory (Emlen 1982). Staying on the current territory can be favoured by selection if the habitat is saturated, if it's dangerous to move between patches of habitat, or if it's difficult to establish new territories. The ecological constraints hypothesis makes two predictions: first, individuals will engage in cooperative actions when there is some ecological constraint; second, critically, the likelihood of individuals leaving to breed will increase when the ecological constraint is relaxed (Emlen 1982; Cant and Johnstone 2009). We have tested the first prediction in *A. percula* monitoring anemone occupancy and movement among 97 anemones on three reefs in Madang Lagoon (Buston 2002). All but one anemone were continuously occupied by two or more *A. percula* for the entire year, indicating that the breeding habitat was fully saturated (Buston 2003a, c). The anemones were tens of metres apart, and clownfish are poor swimmers and can be rapidly preyed upon (Mariscal 1970; Elliott et al. 1995), indicating that it is risky to move between anemones. Taken together, the habitat saturation and risks of movement are suggestive of strong ecological constraints ($X_j + rY_j + fZ_j \approx 0$ in Inequality 14.1).

To test the critical prediction, we experimentally relaxed habitat saturation (*sensu* Pruett-Jones and Lewis 1990; Komdeur 1992) by removing non-breeders from 14 groups and removing males from 16 groups (Buston 2003a, 2004a, 2004b). In no case did a non-breeder disperse to take advantage of these habitat vacancies. These results suggest that strong ecological constraints in the form of risks of movement play a role in the evolution of non-breeding strategies in *A. percula*, but they can also be interpreted as evidence for two alternative hypotheses: i) non-breeders do not disperse because their home anemone confers higher expected reproductive success than alternatives; ii) non-breeders do not disperse because there is limited plasticity of movement in clownfish.

To tease apart these alternatives we conducted a series of experimental manipulations using a population of *A. percula* in Kimbe Bay, PNG (Branconi et al. 2020). We manipulated risks of movement by altering the distance between anemones using 32 anemone pairs and examined the effect of the risks of movement on the likelihood of subordinate dispersal (Branconi et al. 2020). The results were clear: when anemones were 0.5 m apart, non-breeders dispersed often; when anemones were 5.0 m apart, non-breeders dispersed rarely. Indeed, the likelihood of non-breeders moving between anemones was effectively zero when anemones were five meters apart, and anemones are mostly much further apart than this in nature. Taken together, these results demonstrate that individuals will adopt non-breeding positions because of the risks of movement, indicating that strong ecological constraints are a driving force behind the evolution of non-breeding strategies in anemonefishes (Branconi et al. 2020).

14.2.1.5 Poor Inside Options: Social Constraints

The social constraints hypothesis emphasizes that there are two options available to individuals – either they can contest to breed or they can wait peacefully to inherit breeding positions. Waiting peacefully can be favoured by selection if contesting would result in unproductive inbreeding or if contests are dangerous and difficult to win. The social constraints hypothesis makes two predictions: first, individuals will engage in cooperative actions when there is some social constraint; second, critically, the likelihood of individuals contesting to breed will increase when the social constraint is relaxed (Muthoo 2000; Buston and Zink 2009). Within groups of *Amphiprion sp.* there is a size-based dominance hierarchy, where the female is largest, the male is second largest, and the non-breeders get progressively smaller (Fricke and Fricke 1977). In *A. percula*, this is not simply a random collection of individuals of different sizes: well-defined size ratios are maintained between individuals adjacent in rank, by precise regulation of subordinate growth (Buston 2003a; Buston and Cant 2006). Further, dominants occasionally evict or kill subordinates that are similar in size to themselves (Allen 1972; Buston 2003b). Taken together, the well-defined size hierarchy and the threat of eviction are suggestive of a strong social constraint, that would reduce the payoff associated with contesting for breeding positions ($X_j + rY_j + fZ_j \approx 0$ in Inequality 14.1).

To test the critical prediction, we manipulated the likelihood of contesting using 16 focal groups of *A. percula* in Kimbe Bay (Branconi et al. 2020). The likelihood of the subordinate contesting and winning increases as it becomes more similar in size to its dominant (Wong et al. 2016). We staged contests in which we manipulated the size ratio between individuals adjacent in rank and examined the effect of the size ratio on the likelihood of a contest and the outcome of a contest (Branconi et al. 2020). The results were striking: when the introduced rank 3 was less than 80% of the size of the breeding male it contested rarely, and was tolerated; when the introduced rank 3 was more similar in size to the breeding male it contested often, but ended up being evicted in the majority of cases. Taken together, these results demonstrate that individuals will adopt non-breeding positions because of the risks of eviction and subsequent depredation, indicating strong social constraints are a driving force behind the evolution of non-breeding strategies in anemonefishes (Branconi et al. 2020).

14.2.2 Why Do Breeders Tolerate Non-Breeders?

The evolution of cooperative behavior in breeders (i.e., tolerating non-breeders) depends on the costs and benefits of cooperative options and alternative options for the breeder and its relatives) (Figure 14.2). The behavior favoured by selection can be determined using Hamilton's rule (Hamilton 1964). In the case of the clown anemonefish, the cooperative option (toleration of the non-breeder) will be favoured over the alternative option (eviction of the non-breeder) if

$$X_i + rY_i + fZ_i > X_j + rY_j + fZ_j \qquad (14.2)$$

where *Xi* or *Xj* is the breeder's fitness associated with its actions in the present; *Yi* or *Yj* is the non-breeder's fitness associated with the breeder's actions, and r is the probability that the two individuals share a copy of a particular gene identical by descent; *Zi* or *Zj* is the breeder's fitness associated with its actions in the future, and *f* is the probability that benefits will be realized in the future; *i* terms represent payoffs associated with cooperative option, and *j* terms represent payoffs associated with alternative options. Here, we describe where we are in the process of testing each of the eviction constraints, direct genetic benefits, indirect genetic benefits, and future genetic benefits (mate-replacement benefits and mutualist-mediated benefits) hypotheses in the clown anemonefish *Amphiprion percula*.

14.2.2.1 Poor Outside Options: Eviction Constraints

The eviction constraints hypothesis emphasizes that there are two options available to dominant individuals – either they can evict subordinates or they can tolerate them. Toleration can be favoured by selection if it is difficult or dangerous to evict, and if the benefits of toleration outweigh the costs. The eviction constraints hypothesis predicts, critically, that non-breeders will be evicted when their cost increases. We have tested this critical prediction in two ways. First, we manipulated the reproductive state of *A. percula* subordinates (rank 3) in 13 groups, by removing the resident rank 3 (non-breeder) and replacing it with a similarly sized breeding male or non-breeder from another group (Rueger et al. 2018). In this experiment, males were more likely to be evicted than non-breeders (Rueger et al. 2018). Second, we manipulated the size of subordinates (rank 3) in 16 groups, by removing the resident rank 3 and replacing it with a non-breeder a few millimetres larger or a few millimetres smaller (Branconi et al. 2020). In this experiment, the larger non-breeders were more likely to be evicted than the smaller non-breeders (Branconi et al. 2020). Taken together, the results of these experiments demonstrated that breeders can evict non-breeders and do so when the potential costs of tolerating them are increased. This suggests that in anemonefishes under natural conditions breeders tolerate non-breeders either because they are neutral or because they provide some benefit, ($X_i + rY_i + fZ_i \geq X_j + rY_j + fZ_j$ in Inequality 14.2).

14.2.2.2 Direct Genetic Benefits: Current Reproduction

This hypothesis focuses on the immediate fitness gains that breeders might accrue from tolerating non-breeders if, for example, the non-breeders assist with defending against predators, finding good food patches, or providing care to the young. This hypothesis predicts that the breeders' present direct fitness is enhanced by tolerating non-breeders (Woolfenden and Fitzpatrick 1984; Emlen and Wrege 1988). We have tested this prediction by monitoring the survival, growth, and reproduction of 71 groups of *A. percula* in Madang Lagoon over 12 months (Buston 2002). We also removed all non-breeders from 14 of the 71 groups and examined the effect of the removal on the fitness of the breeders (Buston 2004a). The non-breeders did not assist the breeders in any obvious way, e.g., by providing conspicuous alloparental care. The non-breeders had no effect on the survival, growth, or reproduction of the breeders (Buston 2004a; Buston and Elith 2011). These results support the null hypothesis that the breeders' present direct fitness is neither enhanced (nor reduced) by tolerating non-breeders ($X_i \approx X_j$, Inequality 14.2), and indicate that if there are benefits to tolerating non-breeders they must be accrued either indirectly or in the future.

14.2.2.3 Indirect Genetic Benefits: Kin Selection

The kin selection hypothesis, framed from the dominants' perspectives, focuses on the indirect genetic benefits that breeders might accrue from tolerating non-breeders if the non-breeders are relatives that have little chance of surviving to breed elsewhere. The hypothesis predicts that breeders will tolerate non-breeders because they are close relatives who go on to inherit the territory (Kokko and Johnstone 1999). We have tested this using naturally occurring (n = 9) and experimentally created (n = 16) breeding vacancies in *A. percula* (Buston 2004). The results were unambiguous: resident non-breeders did go on to inherit the territory (Buston 2004). We have conducted a preliminary test of the second prediction that non-breeders are closely related to the breeders, by genotyping 35 individuals from nine groups using seven microsatellite loci (Buston et al. 2007). That study suggested that non-breeders were not closely related to the breeders (Buston et al. 2007). However, more recent evidence suggests that non-breeders can sometimes be related to the breeders in *A. percula* (e.g., rare occurrences of full-sibs, uncles/aunts, nieces/nephews; Salles et al. 2016), creating the potential for weak kin selection that might favour breeders tolerating rather than evicting non-breeders as seen in other reef fishes (Rueger et al. 2020, 2021b). This weak kin selection hypothesis is yet to be rigorously tested in *A. percula* or any other anemonefish, but dominants might accrue a 1–2% increase in relative fitness ($rY_i > rY_j$, Inequality 14.2) tipping the balance in favour of tolerating rather than evicting subordinates.

14.2.2.4 Future Genetic Benefits I: Mate-Replacement Benefits

The mate-replacement hypothesis focuses on the future genetic benefits that the breeders might accrue from tolerating non-breeders if the non-breeders serve as rapid mate replacements following the death of one of the breeders. The original mate-replacement benefit hypothesis predicts that widowed breeders will take longer to recommence breeding in groups from which non-breeders are removed than in control groups in which non-breeders are present (Fricke 1979; Buston 2004). We have tested this hypothesis in *A. percula* from the female's perspective, using two experiments: i) the removal of non-breeders (n = 13) to estimate the time for recruitment to occur in the absence of non-breeders; and ii) a male removal (n=16) to estimate the time for a non-breeder to start functioning as a male (Buston 2004). Results suggest that females that tolerate non-breeders might have a 2% higher relative fitness than those that do not (Buston 2004). While this effect is small, it is another slight gain in relative fitness ($fZ_i > fZ_j$, Inequality 14.2) that might tip the balance in favour of tolerating rather than evicting subordinates. While the original mate-replacement hypothesis focuses on the difference in the *mean* fitness associated with the alternative options, it is possible that the main benefit of tolerating non-breeders comes from the reduction in the *variance* in time to recommence breeding that might be caused by spatial and temporal stochasticity in time for recruitment to occur. This new bet-hedging hypothesis (Rubenstein 2011) is yet to be rigorously tested in *A. percula* or any other anemonefish, but dominants might accrue significant increases in relative fitness ($fZ_i >> fZ_j$, Inequality 14.2) by avoiding rare, long wait times to recommence breeding that impact their fitness.

14.2.2.5 Future Genetic Benefits II: Mutualist-Mediated Benefits

The mutualist-mediated hypothesis focuses on the future genetic benefits that the breeders might accrue from tolerating non-breeders if the non-breeders enhance the fitness of the anemone on which the breeders depend. The mutualist-mediated benefits hypothesis makes two critical predictions: first, that non-breeders enhance the survival, growth, and size of the anemone; and second, that large anemones enhance the survival, growth, size, and reproduction of the breeders. These two predictions have not been tested simultaneously in *A. percula* or any anemonefish species, but various lines of evidence suggest that the hypothesis is plausible.

Considering the first prediction, three experimental studies in three different anemone-anemonefish pairings have shown that the number of fish present in the anemone positively impacts anemone size, due to their effects on anemone growth and expansion behavior (*A. bicinctus* and *Entacmaea quadricolor*, Porat and Chadwick-Furman 2004; *A. chrysopterus and Heteracis magnifica*, Holbrook and Schmitt 2005; *A. melanopus and E. quadricolor*, Frisch et al. 2016). These effects might be caused by multiple mechanisms: the fish defend the anemone against predators (e.g., butterflyfishes and turtles), clean the anemone of debris, provide nutrients to the anemone via their waste products, and oxygenate the anemone via their movements (Mariscal 1966; Ross 1978; Fricke 1979; Cleveland et al. 2011; Iwata and Manbo 2013; Szczebak et al., 2013).

Considering the second prediction, observational studies of *A. percula* and *H. magnifica* interaction have indicated that anemone size is positively correlated with the growth of the female (Buston 2002), explaining why large anemones are associated with large females (Fautin 1992; Chausson et al. 2018), and large females lay more eggs, resulting in more paternal care, and higher embryo survival (Barbasch et al. 2020). These effects might be caused by two mechanisms: larger anemones might provide greater foraging area (Barbasch et al. 2020) or more nutritious egesta (Verde et al. 2015).

This hypothesis has yet to have both predictions tested simultaneously and experimentally in *A. percula* or any other anemonefish species, but dominants might accrue significant increases in relative fitness ($fZ_i >> fZ_j$, Inequality 14.2) by tolerating subordinates because of mutualist-mediated benefits: a 1 cm^2 increase in anemone area is predicted to cause approximately 0.07 more eggs to be laid per clutch, which sounds trivial until one realizes that *H. magnifica* vary over one order of magnitude in area from $\approx$400 cm^2 to $\approx$3,400 cm^2 in Kimbe Bay, so individuals might produce $\approx$200 more eggs per clutch in large anemones relative to small anemones, and relative fitness in large anemones may be nearly double that in small anemones (data from Barbasch et al. 2020). Of course, what matters here is the gain in anemone size that comes from tolerating non-breeders, i.e., the difference between fZ_i and fZ_j in Inequality 14.2. That is to say that some fraction of the fitness benefits of being in a large anemone may be due to the presence of non-breeders, but some fraction will be due to sheer luck of where individuals settled (Buston and Elith 2011; Salles et al. 2016; Barbasch et al. 2020).

BOX 14.1: THE ROLE OF PERSONALITIES IN SOCIAL GROUP FORMATION

While we have focused on the breeders' net benefit of tolerating versus evicting the average non-breeder, not all non-breeders are created equal. In many animal populations, including anemonefishes, there is variation in behavior among individuals and individuals exhibit consistent behavioral traits, or personalities, whereby some individuals are more active, more social, or more caring than others (Dall et al. 2004; Reale et al. 2007; Dingemanse and Wolf 2010; Wong et al. 2013; Barbasch and Buston 2018). These individual personality traits, or the combination of personality traits within a group, can influence the costs and benefits of group living. For example, Schmiege et al. (2017) demonstrated that boldness/shyness of *Amphiprion percula* influences the growth of their surrogate hosts in an aquarium setting, likely because shier individuals spend more time among the anemone's tentacles where they might provide nutrients or oxygenation (Cleveland et al. 2011; Szczebak et al. 2013). To date, little work has been done on anemonefish personality traits in the wild (but see Wong et al. 2017) but there's lots of exciting work to be done investigating how variation in personalities influences the functioning of social groups and why variation in personalities is maintained by selection (see Chapter 15 for more on personalities).

14.3 SOCIAL GROUP MAINTENANCE

14.3.1 How Are Conflicts of Interest among Group Members Resolved?

Within any social group, there will be conflicts of interest between genetically selfish individuals over access to reproduction, and these conflicts must be resolved for the group to remain stable. Nowhere is this clearer than in anemonefishes where potential conflict over access to reproduction can result in actual conflict, intense contests, forcible eviction, and the subsequent demise of the losing party (Buston 2003a; Wong et al. 2016; Rueger et al. 2018; Branconi et al. 2020). The conflict arises because subordinates benefit from settling in an anemone and queuing for breeding position, but they are always potential challengers for a dominant's social rank and the access to reproduction that it confers (Buston 2004a, 2004b). The question is, how are these potential conflicts resolved so that anemonefish societies are not in a constant state of turmoil, fighting, and evictions. The answer depends on the context in which conflict occurs and one of the most remarkable anemonefish adaptations: their ability to modify their growth and size in response to the growth and size of those around them.

BOX 14.2: LIMITATION OF EXPLOITATION BY OUTSIDERS

A key component of social group maintenance, in addition to the resolution of conflict among insiders (see main text), is the prevention of the exploitation of the group's resources by outsiders of the same species and other species (Bourke 2011). In general, defence against exploitation by outsiders relies on recognition of outsiders, followed by their exclusion (Sherman et al. 1997). Anemonefishes defend their anemone against members of the same species, but they seem more interested in excluding uncooperative individuals than in excluding outsiders per se (perhaps because groups are composed of non-relatives, so insiders and outsiders are more interchangeable). For example, in the clown anemonefish *Amphiprion percula*, outsiders will be tolerated if they are behaving cooperatively, i.e., not inflicting costs: insiders will tolerate outsiders of a given size, as long as they are not reproductively active (Rueger et al. 2018); insiders will tolerate outsiders that are non-reproductive, as long as they are not too large (Branconi et al. 2020). Anemonefishes also defend their host anemones against members of other species (Fautin 1986; Srinivasan 1999), but they do sometimes coexist (see Chapter 18 for more on competition and coexistence). Such coexistence likely indicates a failure of exclusion (or a success of intrusion). Curiously, the two species that commonly coexist as subordinates, *Amphiprion perideraion* and *A. sandaracinos*, have colouration that is equivalent to recent settlers (pinkish with one white bar) which might dupe the dominant individuals. Investigating the production, perception, and action components of recognition systems that prevent exploitation by outsiders (Sherman et al. 1997) is a promising area of future work.

14.3.2 Non-Lethal Resolution of Conflict in Stable Groups

In the clown anemonefish *Amphiprion percula* the size-based dominance hierarchy is particularly well-defined (Buston 2003b): a non-random distribution of size ratios is found between individuals adjacent in rank, with subordinates being about 80% of the size of their immediate dominant (Buston and Cant 2006). These size ratios are maintained, with millimetric precision, by the subordinate regulating its growth (Buston 2003b; Buston and Cant 2006). At the time this was described, it was hypothesized that this regulation of growth resolves the potential evolutionary conflict over social rank and access to reproduction: subordinates would benefit from regulating their growth because, in doing so, they would avoid becoming a threat to their immediate dominant and thereby avoid being punished by eviction; dominants would benefit from using the threat of eviction to force their immediate subordinates to regulate their growth because, in doing so, they would prevent challenges to their social rank (Buston 2003a, b). Recently, the critical test of this hypothesis was conducted, removing individuals from stable size hierarchies and introducing individuals a few millimetres larger or smaller and determining their fate: the smaller introducees were tolerated whereas the larger introducees were evicted from their social group (Branconi et al. 2020). In stable social groups, subordinate clownfish can reduce their growth to resolve potentially lethal conflicts over social rank (Figure 14.3).

BOX 14.3: THE ROLE OF VISUAL SIGNALS IN SOCIAL GROUP MAINTENANCE

The use of threats in conflict resolution relies on some individuals being able to accurately target others, which in turn requires a system to recognize individuals. Selection can favour individuals that signal their identity if it confers some benefit, e.g., if by signalling identity and behaving cooperatively they avoid eviction (Dale et al. 2001; Tibbetts and Dale 2007). The color and pattern of *Amphiprion percula* and its sister taxon *A. ocellaris* – three white bars, bordered with black, on an orange background – are highly

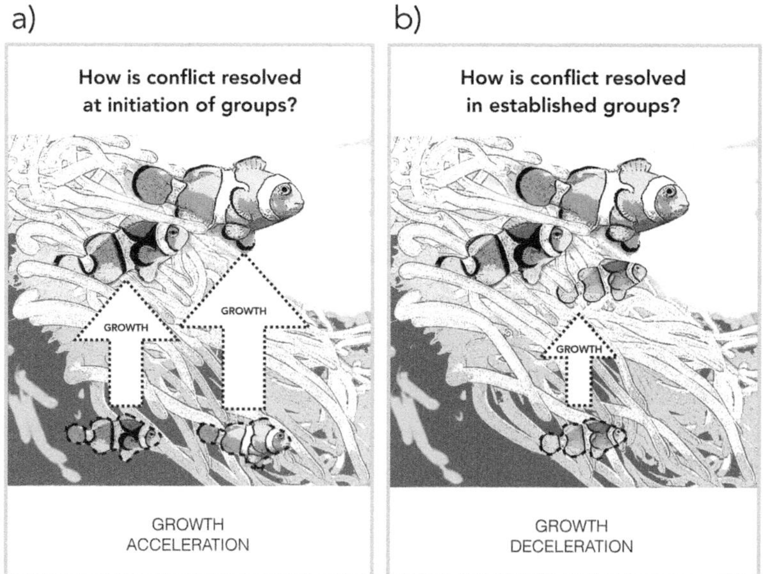

FIGURE 14.3 Social group maintenance in the clown anemonefish *Amphiprion percula*. Both breeders and non-breeders benefit from forming social groups, but there will always be potential conflict over social rank that must be resolved for groups to be stable: a. at the initiation of groups, size-matched rivals increase their growth in an attempt to outgrow each other and attain dominance; b. in established groups, subordinates decrease their growth to avoid coming into conflict with their immediate dominant and avoid eviction. Fish with dashed outlines represent initial sizes, fish with solid outlines represent final sizes, and arrows represent relative growth within and across panels. (Illustrations by R. B.)

variable which is a common characteristic of individual recognition signals (Dale et al. 2001; Maytin et al. 2018). Indeed, these markings are so variable that the investigators can use them to tell individuals apart in the field (Nelson et al. 1994; Buston 2003b; see also Fricke 1973). Interestingly, not all members of the genus seem to have these highly variable markings, though color pattern variation has not been quantified. This difference might be anticipated if some social contexts select for signalling of individual identity while others do not (see Section 14.5), as is the case in *Polistes* wasps (Tibbetts 2002, Tibbetts and Dale 2004). Investigating the function of anemonefish color patterns in light of their social evolution is a promising avenue of future research (see Chapter 7 for more on the structure and function of color patterns).

14.3.3 Non-Lethal Resolution of Conflict at the Initiation of Groups

Following the discovery that clownfish could resolve potential conflict by reducing their growth in response to those around them (Buston 2003b), investigators of other social vertebrates began to look at growth and size in a new light (Heg et al. 2004; Wong et al. 2007; Ang 2010; Dantzer et al. 2013). In 2016, Huchard and others demonstrated that in Kalahari meerkats, where social rank and breeding status are also dependent on size, subordinates of both sexes responded to experimentally induced increases in growth of size-matched rivals by raising their own growth rate and food intake (Huchard et al. 2016). A similar, singular observation had been made in the clown anemonefish *Amphiprion percula* in the field: at the initiation of social groups, individuals living with similarly sized rivals grew more than individuals living alone (Buston 2002). Recently, we tested the hypothesis that *A. percula* individuals paired with size-matched rivals would increase their growth. We conducted an experiment housing size-matched individuals in pairs or solitarily and measuring their growth (Reed et al. 2019). As predicted, individuals living in pairs grew faster than individuals living solitarily, despite being on the same food ration (Reed et al. 2019). At the initiation of social groups, clownfish can increase their growth to resolve potentially lethal conflicts over social rank (Figure 14.3).

BOX 14.4: PROXIMATE MECHANISMS FACILITATING SOCIAL GROUP MAINTENANCE

The proximate mechanisms by which anemonefishes regulate their growth in response to those around them remain a mystery. The problem is complex: individuals must be able to assess their own size and the size of others, process that information, and change their behavior, hormones, or gene expression to achieve an appropriate growth response. One preliminary study of *A. percula* suggests that information on size might be conveyed by mechanosensory cues mediated via the lateral line, or require unobstructed interactions (Desrochers et al. 2020). Studies of other species suggest that subordinates can modify their foraging rates to decrease (Wong et al. 2008; Ang et al. 2010) or increase growth (Huchard et al. 2016) as needed. Depending on the species, the threat of eviction by the dominant may suffice or aggression from the dominant may be needed to motivate the subordinate to modify its growth (Buston and Cant 2006; Ang et al. 2010). To date, little work has been done on the hormonal mechanisms (but see Mills et al. 2018) and no work has been done on the genetic mechanisms underlying strategic growth, but this represents a fascinating avenue for future research (see Chapter 11 for more on the neuroendocrinology of stress and growth).

14.3.4 Adaptive Size Modification on Acquisition of Breeding Status

A third context in which individuals adjust their growth and size has less to do with resolving conflict and more to do with maximizing reproductive potential. In cooperative breeders such as Kalahari meerkats (Russell et al. 2004; Huchard et al. 2016), Damaraland mole-rats (Young and Bennett 2010; Thorley et al. 2018), and naked mole-rats (O'Riain et al. 2000; Dengler-Crish and Catania 2007), where a single female monopolizes reproduction and breeding competition among females is intense, females that acquire the breeding position show a rapid increase in body size and changes in body shape that enhance their fecundity (and may also help them meet challenges by competitors). In the clown anemonefish *Amphiprion percula*, individuals that have just acquired female breeding positions show an increase in length and mass relative to size-matched individuals who remain male and relative to size-matched individuals who have been female for some time (data from Buston 2002). Interestingly, there is anecdotal evidence of a similar phenomenon in the pink anemonefish *A. perideraion* (Allen 1972) and the tomato clownfish *A. frenatus* (Hattori 1991). In addition, in anemonefishes, as in cooperatively breeding mammals, increases in female size are associated with increases in clutch size (Buston and Elith 2011; Saenz-Agudelo et al. 2015). How big the dominant female grows seems to be dependent on the size of the anemone, with dominant females growing larger in larger anemones (Fautin 1991). The size modification by clownfish on acquisition of breeding status likely maximizes their reproductive value given the constraints imposed by their anemone.

14.4 SOCIAL GROUP TRANSFORMATION

14.4.1 How Do Small, Simple Groups Become Large, Complex Groups?

Social group transformation is the process that turns stable social groups into more cohesive, integrated entities. It's the process that transforms simple groups, like those of cooperatively breeding vertebrates, into complex groups, like those of naked mole-rats, and many ants and termites. The primary hypothesis for social group transformation is the size-complexity hypothesis – the idea that an increase in the size of social groups causes an increase in social complexity (Alexander et al. 1991; Bourke 1999; Jeon and Choe 2003; Bonner 2004; Bourke 2011). The hypothesis makes three predictions: i) some ecological or evolutionary driver favours increases in social group size; ii) increases in social group size favour the expression of novel traits or behaviors and a greater division of labour that result in increases in social complexity; and iii) increases in social complexity favour further increases in social group size via positive feedback. The idea has received much less theoretical or empirical study than social group formation or social group maintenance, perhaps due to its experimental intractability. While anemonefishes would be squarely categorized as simple groups, like other cooperatively breeding vertebrates, they also provide insights into the incipient stages of social group transformation. What follows is speculative, but worthy of being critically tested.

14.4.2 The Size-Complexity Hypothesis

In the clown anemonefish *Amphiprion percula* the size of the social group is dependent on the size of the dominant female and the rules of the size-based dominance hierarchy (Buston 2003b; Buston and Cant 2006). When the dominant female is large, it is possible to fit more fish in the anemone while maintaining the required size ratio between individuals adjacent in rank (Buston and Cant 2006). The addition of each new rank creates the potential for new individuals to engage in new behaviors that maximize their effect on the survival and growth of their anemone, given their size and given what others are doing. Low-rank individuals are likely to perform different tasks from high-rank individuals because the benefits of performing each task are likely to show diminishing returns and the costs of performing each task are likely to depend on the size of the individual. So, for example, larger, higher rank individuals may engage in more defence of the anemone, while smaller, lower rank individuals may engage in more cleaning of the anemone (Rueger et al. 2022). These behaviors can enhance the size of the anemone which can further enhance the size of the female, creating a positive feedback loop whereby increases in social group size beget increases in social complexity beget increases in social group size (Figure 14.4).

BOX 14.5: EXTRAORDINARY LIFE SPANS OF ANEMONEFISHES

Many eusocial organisms have evolved extraordinary life spans, including ants (Keller and Genoud 1997),

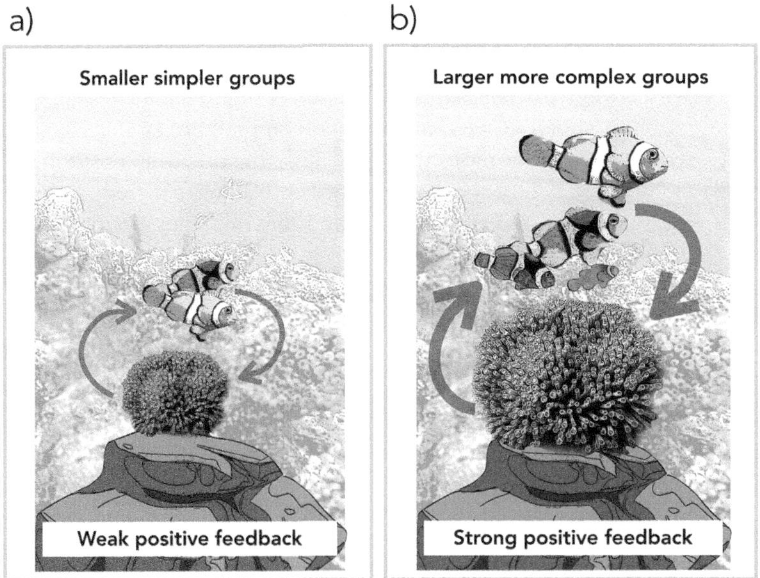

FIGURE 14.4 Social group transformation in the clown anemonefish *Amphiprion percula*. While clownfish would be considered simple social groups, the discrete variation in group size creates an opportunity for investigating social group transformation by testing the predictions of the size-complexity hypothesis: a. small females only facilitate small groups that are predicted to have a lesser division of labour and diversity of social behaviors, that are in turn predicted to only weakly enhance anemone size, creating only a weak positive feedback loop; b. large females facilitate large groups that are predicted to have a greater division of labour and diversity of social behaviors, which are in turn predicted to strongly enhance anemone size, creating a strong positive feedback loop. (Illustrations by R. B.)

naked mole-rats (Sherman and Jarvis 2002), and anemonefishes (Buston and García 2007). The first evidence for long life spans in anemonefishes came from observations suggesting that *A. clarkii* lived 12 years in the wild (Moyer 1986) and *A. frenatus* and *A. perideraion* lived 18+ years in captivity (Fautin and Allen 1992). More recently, Buston and García (2007) used field data and a matrix model to estimate the life expectancy of female *A. percula* to be 31 years (95% confidence interval = 22 years–90 years), an extraordinary life expectancy relative to other coral reef damselfishes of the same size. Evolutionary hypotheses of ageing suggest senescence is the result of selection favouring alleles that increase reproductive success early in life even if they have negative effects later in life, or the result of weak selection against alleles that have deleterious effects later in life due to the reduced proportion of each cohort subject to selection (Williams 1957; Rose 1991; Hamilton 1966). These hypotheses predict that in safe habitats where extrinsic sources of mortality are rare (e.g., when protected by sea anemones, selection will favour individuals that show delayed senescence and increased longevity) (Keller and Genoud 1997; Blanco and Sherman 2005; Buston and García 2007). Regardless of the precise evolutionary cause, the socio-ecological context of anemonefishes seems to have resulted in the evolution of extraordinary lifespans, creating the potential for them to be used as a model for understanding proximate causes of ageing and senescence (see Chapter 8 for more on age and longevity).

14.5 INTERSPECIFIC VARIATION IN ANEMONEFISH SOCIETIES

14.5.1 What Causes Interspecific Variation in Anemonefish Societies?

Anybody who has observed anemonefishes in the wild will know that their societies vary from one species to the next (Allen 1972; Elliott and Mariscal 2001; Hattori 2002; Mitchell 2005; Cleveland et al. 2011), just as societies of cooperatively breeding birds, mammals, and insects vary among species within a genus. Making sense of this variation among animal societies is a major goal of evolutionary biology. The aforementioned framework provides a good starting point for thinking about the causes of this variation, considering how differences in the terms on both sides of Hamilton's inequality might cause differences in the social structure of groups within a population, populations within a species, or species within the genus. When comparing just two species it will always be hard to isolate what causes the differences between them. Proper phylogenetically controlled comparisons, using the most recent *Amphiprion* phylogeny (see Chapter 2), will be required to test hypotheses concerning species differences. Here, we propose plausible but untested hypotheses for the causes of variation in anemonefish societies.

14.5.2 Variation in Strength of Indirect Genetic Benefits and Ecological Constraints

Considering the major hypotheses in turn, variation in the strength of kin selection seems less likely and variation in the strength of ecological constraints seems more likely to cause variation in the costs and benefits of social group formation. Variation in the strength of kin selection is an unlikely cause because, in all species studied to date, the larvae disperse from tens of meters to tens of kilometres from their natal anemone (*A. polymnus*, Jones et al. 2005; Saenz-Agudelo et al. 2011; *A. percula*, Almany et al. 2007; Planes et al. 2009; Buston et al. 2012; Almany et al. 2017; *A. omanensis*, Simpson et al. 2014; *A. bicinctus*, Nanninga et al. 2015; *A. clarkii*, Pinsky et al. 2010; Catalano et al. 2021). In contrast, variation in the strength of ecological constraints is a likely cause because they clearly vary within and among species. Ecological constraints can vary among groups within a population because their proximity to other groups varies (e.g., some anemones may be within centimetres of each other while others may be hundreds of meters from their nearest neighbour). Ecological constraints can vary among populations within a species because the ecology varies (e.g., in subtropical populations of *A. clarkii*, habitat saturation and risks of predation seem lower than in tropical populations, resulting in more movement of *A. clarkii* in subtropical populations) (Hattori 1994). Ecological constraints can vary among species within the genus because their traits vary (e.g., *A. clarkii* has a larger body size and presumably lower risk or predation than *A. perideraion*, resulting in more movement of *A. clarkii* on the same reefs) (Figure 14.5; Hattori 1995; Elliott and Mariscal 2001; Hattori 2002; Cleveland et al. 2011).

14.5.3 Variation in Strength of Future Genetic Benefits and Social Constraints

A reduction in the strength of ecological constraints seems likely to have knock-on effects on the future genetic benefits of forming social groups, from the perspective of both breeders and non-breeders: it creates the potential for non-breeders to have their inheritance usurped by outsiders; it means that breeders don't need resident non-breeders to serve as rapid mate-replacements. This, in turn, will have knock-on effects on conflict resolution and the size hierarchy, as there will be an incentive for subordinates to grow larger to prevent usurpation and dominants will have to use consistent aggression (not just the threat of eviction) to get subordinates to regulate their growth (Figure 14.5). A reduction in ecological constraints and increased movement seems likely to have knock-on effects on mutualist-mediated benefits also: if individuals are more active outside

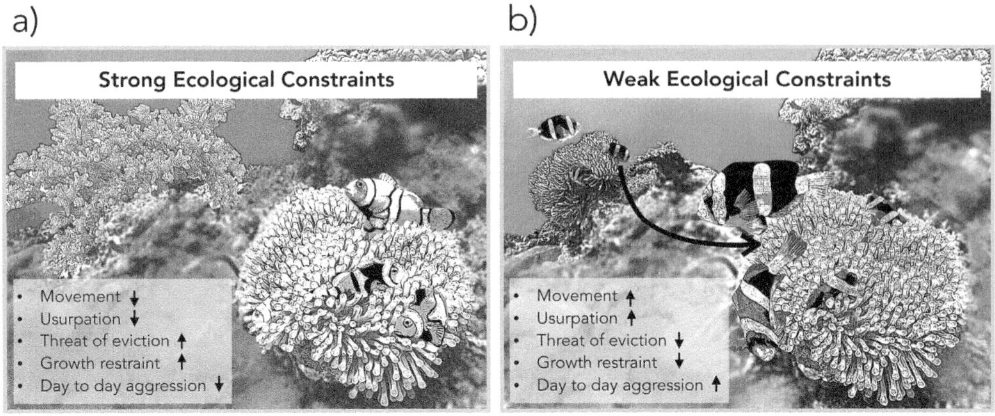

FIGURE 14.5 Interspecific variation in anemonefish societies. A key driver of variation in anemonefish societies seems to be variation in the strength of ecological constraints, which creates continuous variation with two types of extremes: a. strong ecological constraints, less opportunity for movement, lower probability of social rank being usurped by outsiders, incentive for subordinates to remain small and avoid eviction, dominants maintain control via hidden threat of eviction (e.g., *Amphiprion percula*); b. weak ecological constraints, more opportunity for movement, higher probability of social rank being usurped by outsiders, incentive for subordinates to grow large and avoid usurpation, dominants maintain control with constant aggression (e.g., *Amphiprion clarkii*). (Illustrations by R. B.)

of their anemones then they have less of a positive impact on their anemones (Cleveland et al. 2011; Schmiege et al. 2017), and the anemone has less of a positive impact on them (Verde et al. 2015). As noted above, proper phylogenetically controlled comparisons will be required to rigorously test these hypotheses, but in-depth studies of other species in the genus will undoubtedly generate new insights into social evolution in their own right.

14.6 CONCLUDING COMMENTS AND FUTURE DIRECTIONS

Animal societies in which some individuals forgo their own reproduction and help others to reproduce are one of the most remarkable products of evolution, and they have been the focal point for tests of social evolution theory. Anemonefish societies present opportunities to test the robustness of current theories, in a relatively understudied taxon and environment, and generate new insights. The majority of experimental work to date has been conducted on the clown anemonefish or orange clownfish *Amphiprion percula*. Here, we place that body of work in the context of Bourke's (2011) principles of social evolution, providing a conceptual framework for understanding the evolution of anemonefish societies and foundations for future research.

The ecology of anemonefishes, like the ecology of many cooperative breeders, sets the stage for their social group formation: there are a finite number of anemones, they are fully occupied, and it is dangerous to move between them. In this context, natural selection favours settlers that adopt non-breeding positions and queue to inherit the territory in the future, rather than moving to breed elsewhere or contesting to breed at home. Selection will favour breeders that tolerate non-breeders, rather than evicting them, if they are distant relatives, if they reduce the variance in time to recommence breeding following mate loss, and if they enhance anemone size which enhances breeder size and reproductive success. We say "if" because, while there's plenty of supporting evidence, the critical experiments remain to be done from the breeder's perspective.

The observation that both breeders and non-breeders benefit from forming social groups, but non-breeders are always potential challengers to breeders, sets the stage for social group maintenance. Conflicts are resolved non-lethally by the regulation of growth – one of the most remarkable of anemonefish adaptations. At the initiation of groups, size-matched rivals increase their growth in an attempt to outgrow each other and attain dominance. In stable groups, subordinates decrease their growth to avoid coming into conflict with their immediate dominant and avoid eliciting eviction. This strategic growth and its ultimate causation have now been demonstrated in a wide variety of social vertebrates (Buston and Clutton-Brock 2022), but to date, little work has been done on the proximate mechanisms that underpin it (Box 14.3).

The maintenance of stable social groups sets the stage for social group transformation – the emergence of larger, more socially complex groups. While anemonefish would be squarely considered simple social groups, the discrete variation in group size makes them tractable for testing the size-complexity hypothesis. We already know that large anemones are associated with large females and large females facilitate large groups. The outstanding questions are i) are large groups more socially complex than small groups, i.e., do they have a greater division of labour and novel social behaviors, and ii) do large, socially complex groups enhance anemone size, creating a positive feedback loop. As in all taxa, more work has been done on social group formation and maintenance than has been done on social group transformation.

The diversity of social systems found among anemonefishes also demands explanation. A key driver of that diversity seems to be variation in ecological constraints among species, caused by variation in ecology or variation in traits. Relaxed ecological constraints will change the breeder and non-breeder inequalities: non-breeders gain less from queuing for breeding positions because of the risk of being queue jumped; breeders gain less from tolerating non-breeders because mate replacements can come from elsewhere, and non-breeders will have less impact on their anemones. The knock-on effect of this is that non-breeders will be less inclined to remain small and wait peacefully, generating conflict between breeders and non-breeders. The variation in movement and aggression among anemonefish societies is particularly interesting.

Finally, there are many traits that evolve in the context of the social lives of anemonefish. Just as in naked mole-rats and other highly social animals, their intense social lives have selected for some remarkable adaptations. Their personalities (Box 14.1; Chapter 15), strategic growth (Section 14.3), social signals (Box 14.2; Chapter 7), parental care (Chapter 15), sex change (Chapter 12), and extraordinary life spans (Box 14.5; Chapter 8) have all evolved in the context of the social group and are inextricably intertwined with their social evolution. Studying these topics in an integrated fashion, from proximate and ultimate perspectives, will undoubtedly yield new and fascinating insights.

REFERENCES

Alexander, R. D., K. M. Noonan, and B. J. Crespi. 1991. The evolution of eusociality. In *The biology of the naked mole-rat*, eds P. W. Sherman, J. U. M. Jarvis, and R. D. Alexander, 3–44. Princeton: Princeton University Press.

Allen, G. R. 1972. *The anemonefishes: Their classification and biology*. Neptune City, NJ: TFH Publications.

Almany, G. R., M. L. Berumen, S. R. Thorrold, S. Planes, and G. P. Jones. 2007. Local replenishment of coral reef fish populations in a marine reserve. *Science* 316: 742–744.

Almany, G. R., S. Planes, S. R. Thorrold, M. L. Berumen, M. Bode, P. Saenz-Agudelo, M. C. Bonin, A. J. Frisch, H. B. Harrison, V. Messmer, G. B. Nanninga, and G. P. Jones. 2017. Larval fish dispersal in a coral-reef seascape. *Nature Ecology & Evolution* 1: 1–7.

Ang, T. Z., and A. Manica. 2010. Unavoidable limits to group size in a body size-based linear hierarchy. *Behavioral Ecology* 21: 819–25.

Barbasch, T. A., and P. M. Buston. 2018. Plasticity and personality of parental care in the clown anemonefish. *Animal Behaviour* 136: 65–73.

Barbasch, T. A., T. Rueger, M. Srinivasan, M. Y. Wong, G. P. Jones, and P. M. Buston. 2020. Substantial plasticity of reproduction and parental care in response to local resource availability in a wild clownfish population. *Oikos* 129: 1844–55.

Blanco, M. A., and P. W. Sherman. 2005. Maximum longevities of chemically protected and non-protected fishes, reptiles, and amphibians support evolutionary hypotheses of aging. *Mechanisms of ageing and development* 126: 794–803.

Bonner, J. T. 2004. Perspective: The size-complexity rule. *Evolution* 58: 1883–90.

Bourke, A. F. G. 1997. Sociality and kin selection in insects. In *Behavioural ecology: An evolutionary approach*, 4th ed., eds J. R. Krebs and N. B. Davies, 205–27 Oxford: Blackwell.

Bourke, A. F. G. 1999. Colony size, social complexity and reproductive conflict in social insects. *Journal of Evolutionary Biology* 12: 245–57.

Bourke, A. F. G. 2011. *Principles of social evolution*. Oxford: Oxford University Press.

Branconi, R., T. A. Barbasch, R. K. Francis, M. Srinivasan, G. P. Jones, and P. M. Buston. 2020. Ecological and social constraints combine to promote evolution of non-breeding strategies in clownfish. *Communications Biology* 3: 1–7.

Buston, P. M. 2002. *Group structure of the clown anemonefish, Amphiprion percula*. PhD Diss, Cornell University, Ithaca, NY.

Buston, P. M. 2003a. Forcible eviction and prevention of recruitment in the clown anemonefish. *Behavioral Ecology* 14: 576–82.

Buston, P. M. 2003b. Size and growth modification in clownfish. *Nature* 424: 145–46.

Buston, P. M. 2003c. Mortality is associated with social rank in the clown anemonefish (*Amphiprion percula*). *Marine Biology* 143: 811–15.

Buston, P. M. 2004a. Territory inheritance in the clown anemonefish. *Proceedings of the Royal Society B: Biological Sciences* 271: S252–54.

Buston, P. M. 2004b. Does the presence of non-breeders enhance the fitness of breeders? An experimental analysis in the clown anemonefish *Amphiprion percula*. *Behavioral Ecology and Sociobiology* 57: 23–31.

Buston, P. M., S. W. Bogdanowicz, M. Y. Wong, and R. G. Harrison. 2007. Are clownfish groups composed of close relatives? An analysis of microsatellite DNA variation in *Amphiprion percula*. *Molecular Ecology* 16: 3671–78.

Buston, P. M., and M. A. Cant. 2006. A new perspective on size hierarchies in nature: Patterns, causes, and consequences. *Oecologia* 149: 362–72.

Buston, P. M., G. P. Jones, S. Planes, S. R. Thorrold. 2012. Probability of successful larval dispersal declines fivefold over 1 km in a coral reef fish. *Proceedings of the Royal Society B: Biological Sciences* 279: 1883–1888.

Buston, P. M., and T. Clutton-Brock. 2022. Strategic growth in social vertebrates. *Trends in Ecology and Evolution* in press. https://doi.org/10.1016/j.tree.2022.03.010

Buston, P. M., and J. Elith. 2011. Determinants of reproductive success in dominant pairs of clownfish: A boosted regression tree analysis. *Journal of Animal Ecology* 80: 528–38.

Buston, P. M., and M. B. García. 2007. An extraordinary life span estimate for the clown anemonefish Amphiprion percula. Journal of Fish Biology, 70(6), 1710–1719.

Buston, P. M., and A. G. Zink. 2009. Reproductive skew and the evolution of conflict resolution: A synthesis of transactional and tug-of-war models. *Behavioral Ecology* 20: 672–84.

Cant, M. A., and R. A. Johnstone. 2009. How threats influence the evolutionary resolution of within-group conflict. *American Naturalist* 173: 759–71.

Cant, M. A. 2011. The role of threats in animal cooperation. *Proceedings of the Royal Society B: Biological Sciences* 278: 170–178.

Catalano, K. A., A. G. Dedrick, M. R. Stuart, J. B. Puritz, H. R. Montes Jr., and M. L. Pinsky. 2021. Quantifying dispersal variability among nearshore marine populations. *Molecular Ecology* 30: 2366–77.

Chausson, J., M. Srinivasan, and G. P. Jones. 2018. Host anemone size as a determinant of social group size and structure in the orange clownfish (*Amphiprion percula*). *PeerJ* 6: e5841.

Cleveland, A., E. A. Verde, and R. W. Lee. 2011. Nutritional exchange in a tropical tripartite symbiosis: Direct evidence for the transfer of nutrients from anemonefish to host anemone and zooxanthellae. *Marine Biology* 158: 589–602.

Dale, J., D. B. Lank, and H. K. Reeve. 2001. Signaling individual identity versus quality: A model and case studies with ruffs, queleas, and house finches. *The American Naturalist* 158: 75–86.

Dall, S. R., A. I. Houston, and J. M. McNamara. 2004. The behavioural ecology of personality: Consistent individual differences from an adaptive perspective. *Ecology Letters* 7: 734–39.

Dantzer, B., A. E. Newman, R. Boonstra, R. Palme, S. Boutin, M. M. Humphries, and A. G. McAdam. 2013. Density triggers maternal hormones that increase adaptive offspring growth in a wild mammal. *Science* 340: 1215–17.

Dengler-Crish, C. M., and K. C. Catania. 2007. Phenotypic plasticity in female naked mole-rats after removal from reproductive suppression. *Journal of Experimental Biology* 210: 4351–58.

Desrochers, L., R. Branconi, E. Schlatter, B. Dent, and P. M. Buston. 2020. Sensory cues underlying competitive growth in the clown anemonefish (*Amphiprion percula*). *Behavioral Processes* 181: 104276.

Dingemanse, N. J., and M. Wolf. 2010. Recent models for adaptive personality differences: A review. *Philosophical Transactions of the Royal Society B: Biological Sciences* 365: 3947–58.

Elliott, J. K., J. M. Elliott, and R. N. Mariscal. 1995. Host selection, location, and association behaviors of anemonefishes in field settlement experiments. *Marine Biology* 122: 377–89.

Elliott, J. K., and R. N. Mariscal. 2001. Coexistence of nine anemonefish species: differential host and habitat utilization, size and recruitment. *Marine Biology* 138: 23–36.

Emlen, S. T. 1982. The evolution of helping. I. An ecological constraints model. *American Naturalist* 119: 29–39.

Emlen, S. T. 1991. Evolution of cooperative breeding in birds and mammals. In *Behavioral ecology*, 2nd ed., eds J. R. Krebs and N. B. Davies, 305–35. Oxford: Blackwell.

Emlen, S. T. 1997. Predicting family dynamics in social vertebrates. In *Behavioural ecology*, 4th ed., eds J. R. Krebs and N. B. Davies, 228–53. Oxford: Blackwell.

Emlen, S. T., and P. H. Wrege. 1988. The role of kinship in helping decisions among white-fronted bee-eaters. *Behavioral Ecology and Sociobiology* 23: 305–15.

Fautin, D. G. 1986. Why do anemonefishes inhabit only some host actinians? *Environmental Biology of Fishes* 15: 171–180

Fautin, D. G. 1991. The anemonefish symbiosis: What is known and what is not. *Symbiosis* 10: 23–46.

Fautin, D. G. 1992. Anemonefish recruitment: The roles of order and chance. *Symbiosis* 14: 143–60.

Fautin, D. G., and G. R. Allen. 1992. *Field guide to anemonefishes and their host sea anemones.* Perth, WA: Western Australian Museum.

Fricke, H., and S. Fricke. 1977. Monogamy and sex change by aggressive dominance in coral reef fish. *Nature* 26: 830–32.

Fricke, H. W. 1973. Individual partner recognition in fish: Field studies on *Amphiprion bicinctus. Naturwissenschaften* 60: 204–205.

Fricke, H. W. 1979. Mating system, resource defence and sex change in the anemonefish *Amphiprion akallopisos. Zeitschrift für Tierpsychology* 50: 313–26.

Frisch, A. J., J. R. Rizzari, K. P. Munkres, and J. P. A. Hobbs. 2016. Anemonefish depletion reduces survival, growth, reproduction and fishery productivity of mutualistic anemone–anemonefish colonies. *Coral Reefs* 35: 375–86.

Griffin, A. S., and S. A. West. 2003. Kin discrimination and the benefit of helping in cooperatively breeding vertebrates. *Science* 302: 634–36.

Hamilton, W. D. 1964. The genetical evolution of social behaviour. II. *Journal of Theoretical Biology* 7: 17–52.

Hamilton, W. D. 1966. The moulding of senescence by natural selection. *Journal of theoretical biology* 12: 12–45.

Hattori, A. 1991. Socially controlled growth and size-dependent sex change in the anemonefish *Amphiprion frenatus* in Okinawa, Japan. *Japanese Journal of Ichthyology* 38: 165–77.

Hattori, A. 1994. Inter-group movement and mate acquisition tactics of the protandrous anemonefish *Amphiprion clarkii*, on a coral reef, Okinawa. *Japanese Journal of Ichthyology* 41: 159–65.

Hattori, A. 1995. Coexistence of two anemonefishes, *Amphiprion clarkii* and *A. perideraion*, which utilize the same host sea anemone. *Environmental Biology of Fishes* 42: 345–53.

Hattori, A. 2002. Small and large anemonefishes can coexist using the same patchy resources on a coral reef, before habitat destruction. *Journal of Animal Ecology* 71: 824–31.

Heg, D., Z. Bachar, L. Brouwer, and M. Taborsky. 2004. Predation risk is an ecological constraint for helper dispersal in a cooperatively breeding cichlid. *Proceedings of the Royal Society B: Biological Sciences* 271: 2367–74.

Holbrook, S. J., and R. J. Schmitt. 2005. Growth, reproduction and survival of a tropical sea anemone (Actiniaria): Benefits of hosting anemonefish. *Coral Reefs* 24: 67–73.

Huchard, E., S. English, M. B. Bell, N. Thavarajah, and T. Clutton-Brock. 2016. Competitive growth in a cooperative mammal. *Nature* 533: 532–34.

Iwata, E., and J. Manbo. 2013. Territorial behaviour reflects sexual status in groups of false clown anemonefish (*Amphiprion ocellaris*) under laboratory conditions. *Acta Ethol* 16: 97–103.

Jeon, J., and J. C. Choe. 2003. Reproductive skew and the origin of sterile castes. *American Naturalist* 161: 206–24.

Jones, G. P., S. Planes, and S. R. Thorrold. 2005. Coral reef fish larvae settle close to home. *Current Biology* 15: 1314–18.

Keller, L., and M. Genoud. 1997. Extraordinary lifespans in ants: a test of evolutionary theories of ageing. *Nature* 389: 958–960.

Keller, L., and H. K. Reeve. 1994. Partitioning of reproduction in animal societies. *Trends in Ecology and Evolution* 9: 98–102.

Kokko, H., and R. A. Johnstone. 1999. Social queuing in animal societies: A dynamic model of reproductive skew. *Proceedings of the Royal Society B: Biological Sciences* 266: 571–78.

Komdeur, J. 1992. Importance of habitat saturation and territory quality for evolution of cooperative breeding in the Seychelles warbler. *Nature* 358: 493–95.

Krebs, J. R., and N. B. Davies. 1993. *An introduction to behavioural ecology.* Oxford: Blackwell.

Mariscal, R. N. 1966. The symbiosis between tropical sea anemones and fishes: A review. In *The Galapagos: Proceedings of the Symposia of the Galapagos International Scientific Project*, ed. R. I. Bowman, 157–71. Berkeley and Los Angeles: University of California Press.

Mariscal, R. N. 1970. The nature of the symbiosis between Indo-Pacific anemone fishes and sea anemones. *Marine Biology* 6: 58–65.

Maytin, A. K., S. W. Davies, G. E. Smith, S. P. Mullen, and P. M. Buston. 2018. De novo transcriptome assembly of the clown anemonefish (*Amphiprion percula*): A new resource to study the evolution of fish color. *Frontiers in Marine Science* 5: 284.

Mills, S. C., J. L. O'Donnell, G. Bernardi, and R. Beldade. 2018. Natural endocrine profiles of the group-living skunk anemonefish Amphiprion akallopisos in relation to their size-based dominance hierarchy. *Journal of Fish Biology* 92: 773–89.

Mitchell, J. S. 2003. Social correlates of reproductive success in false clown anemonefish: Subordinate group members do not pay-to-stay. *Evolutionary Ecology Research* 5: 89–104.

Mitchell, J. S. 2005. Queue selection and switching by false clown anemonefish, *Amphiprion ocellaris. Animal Behaviour* 69: 643–52.

Moyer, J. T., and A. Nakazono. 1978. Protandrous hermaphroditism in six species of the anemonefish genus Amphiprion in Japan. *Japanese Journal of Ichthyology* 25: 101–106.

Moyer, J. T. 1986. Longevity of the anemonefish Amphiprion clarkii at Miyake-jima, Japan with notes on four other species. *Copeia* 1: 135–139.

Muthoo, A. 2000. A non-technical introduction to bargaining theory. *World Economics* 1: 145–66.

Nanninga, G. B., P. Saenz-Agudelo, P. Zhan, I. Hoteit, and M. L. Berumen. 2015. Not finding Nemo: limited reef-scale retention in a coral reef fish. *Coral Reefs* 34: 383–392.

Nelson, J. S., L. M. Chou, and V. P. E. Phang. 1994. Pigmentation variation in the anemonefish *Amphiprion ocellaris* (teleostei, pomacentridae)-type, stability and its usefulness for individual identification. *Raffles Bulletin of Zoology* 42: 927–30.

O'Riain, M. J., J. U. M. Jarvis, R. Alexander, R. Buffenstein, and C. Peeters. 2000. Morphological castes in a vertebrate. *Proceedings of the National Academy of Sciences* 97: 13194–97.

Pinsky, M. L., H. R. Montes, Jr., and S. R. Palumbi. 2010. Using isolation by distance and effective density to estimate dispersal scales in anemonefish. *Evolution: International Journal of Organic Evolution* 64: 2688–700.

Planes, S., G. P. Jones, and S. R. Thorrold. 2009. Larval dispersal connects fish populations in a network of marine protected areas. *Proceedings of the National Academy of Sciences* 106: 5693–5697.

Porat, D., and N. E. Chadwick-Furman. 2004. Effects of anemonefish on giant sea anemones: Expansion behavior, growth, and survival. *Hydrobiologia* 530/531: 513–20.

Pruett-Jones, S. G., and M. J. Lewis. 1990. Sex ratio and habitat limitation promote delayed dispersal in superb fairy-wrens. *Nature* 348: 541–42.

Raihani, N. J., A. Thornton, and R. Bshary. 2012. Punishment and cooperation in nature. *Trends in Ecology and Evolution* 27: 288–95.

Ratnieks, F. L., and T. Wenseleers. 2005. Policing insect societies. *Science* 307: 54–56.

Réale, D., S. M. Reader, D. Sol, P. T. McDougall, and N. J. Dingemanse. 2007. Integrating animal temperament within ecology and evolution. *Biological Reviews* 82: 291–318.

Reed, C., R. Branconi, J. Majoris, C. Johnson, and P. M. Buston. 2019. Competitive growth in a social fish. *Biology Letters* 15: 20180737.

Rose, J. H. 1991. A life course perspective on health threats in aging. *Journal of Gerontological Social Work* 17: 85–97.

Ross, R. M. 1978. Reproductive behavior of the anemonefish *Amphiprion melanopus* on Guam. *Copeia* 1: 103–107.

Rubenstein, D. R. 2011. Spatiotemporal environmental variation, risk aversion, and the evolution of cooperative breeding as a bet-hedging strategy. *Proceedings of the National Academy of Sciences* 108 Supplement 2: 10816–22.

Rubenstein, D. R., and P. Abbot. 2017. *Comparative social evolution.* Cambridge: Cambridge University Press.

Rueger, T., T. A. Barbasch, M. Y. Wong, M. Srinivasan, G. P. Jones, and P. M. Buston. 2018. Reproductive control via the threat of eviction in the clown anemonefish. *Proceedings of the Royal Society B: Biological Sciences* 285: 20181295.

Rueger, T., H. B. Harrison, P. M. Buston, N. M. Gardiner, M. L. Berumen, and G. P. Jones. 2020. Natal philopatry increases relatedness within groups of coral reef cardinalfish. *Proceedings of the Royal Society B: Biological Sciences* 287: 20201133.

Rueger, T., R. Branconi, C. Y. Froehlich, S. J. Heatwole, M. Y. Wong, and P. M. Buston. 2021a. The next frontier in understanding the evolution of coral reef fish societies. *Frontiers in Marine Science* 8: 665780.

Rueger, T., P. M. Buston, S. M. Bogdanowicz, and M. Y. Wong. 2021b. Genetic relatedness in social groups of the emerald coral goby *Paragobiodon xanthosoma* creates potential for weak kin selection. *Molecular Ecology* 30: 1311–1321.

Rueger, T., H. B. Harrison, P. M. Buston, N. M. Gardiner, M. L. Berumen, and G. P. Jones. 2020. Natal philopatry increases relatedness within groups of coral reef cardinalfish. *Proceedings of the Royal Society B: Biological Sciences* 287: 20201133.

Rueger, T., S. J. Heatwole, and M. Y. L. Wong. 2022. Social rank and species-specific differences influence cooperation and conflict in anemonefish societies. *Animal Behaviour* 187: 105–116.

Russell, A. F., A. A. Carlson, G. M. McIlrath, N. R. Jordan, and T. Clutton-Brock. 2004. Adaptive size modification by dominant female meerkats. *Evolution* 58: 1600–607.

Saenz-Agudelo, P., G. P. Jones, S. R. Thorrold, and S. Planes. 2011. Connectivity dominates larval replenishment in a coastal reef fish metapopulation. *Proceedings of the Royal Society B: Biological Sciences* 278: 2954–2961.

Saenz-Agudelo, P., G. P. Jones, S. R. Thorrold, and S. Planes. 2015. Mothers matter: Contribution to local replenishment is linked to female size, mate replacement and fecundity in a fish metapopulation. *Marine Biology* 162: 3–14.

Salles, O. C., B. Pujol, J. A. Maynard, G. R. Almany, M. L. Berumen, G. P. Jones, and S. Planes. 2016. First genealogy for a wild marine fish population reveals multigenerational philopatry. *Proceedings of the National Academy of Sciences* 113: 13245–50.

Seger, J. 1991. Cooperation and conflict in social insects. In: *Behavioural ecology: An evolutionary approach,* 3rd ed., eds J. R. Krebs and N. B. Davies, 338–73. Oxford: Blackwell.

Seymour, J. R., T. A. Barbasch, and P. M. Buston. 2018. Lunar cycles of reproduction in the clown anemonefish *Amphiprion percula*: Individual-level strategies and population-level patterns. *Marine Ecology Progress Series* 594: 193–201.

Schmiege, P. F., C. C. D'Aloia, and P. M. Buston. 2017. Anemonefish personalities influence the strength of mutualistic interactions with host sea anemones. *Marine Biology* 164: 1 9.

Sherman, P. W., E. A. Lacey, H. K. Reeve, and L. Keller. 1995. Forum: The eusociality continuum. *Behavioral Ecology* 6: 102–108.

Sherman, P. W., and J. U. Jarvis. 2002. Extraordinary life spans of naked mole-rats (*Heterocephalus glaber*). *Journal of Zoology* 258: 307–311.

Simpson, S. D., H. B. Harrison, M. R. Claereboudt, S. Planes. 2014. Long-distance dispersal via ocean currents connects Omani clownfish populations throughout entire species range. *PLoS One* 9: e107610.

Sherman, P. W., H. K. Reeve, and D. W. Pfennig. 1997. Recognition systems. In *Behavioural ecology: An evolutionary approach*, 4th ed., eds J. R. Krebs and N. B. Davies, 69-96, Oxford: Blackwell.

Smith, J. M., and E. Szathmáry. 1995. *The major transitions in evolution*. Oxford: Oxford University Press.

Srinivasan, M., G. P. Jones, and M. J. Caley. 1999. Experimental evaluation of the roles of habitat selection and interspecific competition in determining patterns of host use by two anemonefishes. *Marine Ecology Progress Series* 186: 283–92.

Szczebak, J. T., R. P. Henry, F. A. Al-Horani, and N. E. Chadwick. 2013. Anemonefish oxygenate their anemone hosts at night. *Journal of Experimental Biology* 216: 970–76.

Thorley, J., N. Katlein, K. Goddard, M. Zöttl, and T. Clutton-Brock. 2018. Reproduction triggers adaptive increases in body size in female mole-rats. *Proceedings of the Royal Society B: Biological Sciences* 285: 20180897.

Tibbetts, E. A. 2002. Visual signals of individual identity in the wasp *Polistes fuscatus*. *Proceedings of the Royal Society of London. Series B: Biological Sciences* 269: 1423–28.

Tibbetts, E. A., and J. Dale. 2004. A socially enforced signal of quality in a paper wasp. *Nature* 432: 218–222.

Tibbetts, E. A., and J. Dale. 2007. Individual recognition: It is good to be different. *Trends in Ecology and Evolution* 22: 529–37.

Verde, E. A., A. Cleveland, and R. W. Lee. 2015. Nutritional exchange in a tropical tripartite symbiosis II: Direct evidence for the transfer of nutrients from host anemone and zooxanthellae to anemonefish. *Marine Biology* 162: 2409–29.

West-Eberhard, M. J. 1975. The evolution of social behavior by kin selection. *Quarterly Review of Biology* 50: 1–33.

Williams, E. C. 1957. The commoner urologic problems of the aging. *Southern medical journal* 50: 235–236.

Woolfenden, G. E., and J. W. Fitzpatrick. 1984. *The Florida scrub jay: Demography of a cooperative-breeding bird*. Princeton: Princeton University Press.

Wong, M. Y., A. L. Beasley, T. Douglass, S. Whalan, and A. Scott. 2017. Some anemonefish lack personality: A comparative assessment of behavioral variation and repeatability in relation to environmental and social factors. *Coral Reefs* 36: 1307–16.

Wong, M. Y., P. M. Buston, P. L. Munday, and G. P. Jones. 2007. The threat of punishment enforces peaceful cooperation and stabilizes queues in a coral-reef fish. *Proceedings of the Royal Society B: Biological Sciences* 274: 1093–99.

Wong, M. Y., A. Medina, C. Uppaluri, S. Arnold, J. R. Seymour, and P. M. Buston. 2013. Consistent behavioural traits and behavioural syndromes in pairs of the false clown anemonefish *Amphiprion ocellaris*. *Journal of Fish Biology* 83: 207–13.

Wong, M. Y., P. L. Munday, P. M. Buston, and G. P. Jones. 2008. Fasting or feasting in a fish social hierarchy. *Current Biology* 18: R372–73.

Wong, M. Y., C. Uppaluri, A. Medina, J. Seymour, and P. M. Buston. 2016. The four elements of within-group conflict in animal societies: An experimental test using the clown anemonefish, *Amphiprion percula*. *Behavioral Ecology and Sociobiology* 70: 1467–75.

Young, A. J., and N. C. Bennett. 2010. Morphological divergence of breeders and helpers in wild damaraland mole-rat societies. *Evolution: International Journal of Organic Evolution* 64: 3190–97.

15 Parental Care
Patterns, Proximate and Ultimate Causes, and Consequences

Tina A. Barbasch, Ross DeAngelis, Justin Rhodes, and Peter M. Buston

CONTENTS

15.1 INTRODUCTION

Parents can go to incredible lengths to improve the survival of their offspring (Royle et al. 2014). Yet, care is often costly and given at the expense of other activities, such as foraging or territory defence, resulting in stark tradeoffs (Stearns 1989). The tradeoffs involved in parenting are particularly interesting in species with biparental care, as parents must not only assess their own condition, but also the motivations of their partner. Moreover, the proximate mechanisms underlying the expression of parental care can influence whether and how parents respond to the demands of parenting: selection for parental care can result in the evolution of mechanisms and ontogenies that facilitate plasticity in parental care; however, both mechanistic and ontogenetic causes can also impose constraints on how parents to respond to their environments (Sinervo and Svensson 1998). Thus, there are several important questions concerning parental care that arise: how do parents successfully raise offspring despite the demands, tradeoffs, and constraints involved in parenting? How do parents adjust their behavior in response to environmental changes? How do parents adjust their behavior in response to the behavior of their partner? And finally, how does the parental brain manage the multiple, often competing demands involved with parenting? Answering these questions is critical to understanding how much care offspring receive from their parents, which influences offspring fate and has ecological and evolutionary consequences for future generations.

Anemonefishes are an attractive system for studying parental care in part because their tractability allows for comprehensive studies of both proximate and ultimate causes in the lab and the field. While there is interspecific variation in anemonefish parental care (Allen 1972; Ghosh et al. 2012), the general pattern of care can be described as follows. Anemonefish live in social groups composed of a breeding pair and a small number of non-breeding subordinates (see Chapter 14). The breeding pair lay eggs together up to three times per lunar month for many years (Buston and Elith 2011; Thomas and Prakash 2015; Seymour et al. 2018; see Chapter 13). A few days before breeding, parents first clear a nest site on a hard substrate near the base of the anemone (Moyer and Bell 1976; Green and McCormick 2005; Ghosh et al. 2012). Once a clutch is laid, both parents care for the eggs for six to nine days, during which time the male provides the majority of care (Green and McCormick 2005; Ghosh et al. 2012). Direct care takes the form of mouthing or nipping eggs, fanning the clutch, and defending against intruders and egg predators (Moyer and Bell 1976; Moyer 1980; Green and McCormick 2005), all of which collectively enhance embryo survival (Moyer and Sawyers 1973; Ghosh et al. 2012). As embryos develop, parental care increases, which may reflect responsiveness to increased metabolic needs or increased reproductive value of older clutches (Green and McCormick 2005; Ghosh et al. 2012). On the night of hatching, parents provide additional care, which may serve to synchronize hatching and facilitate the transition of the benthic embryos to pelagic larvae (Moyer and Bell 1976; Ross 1978; Pacaro et al. 2022). Parental care ceases with hatching and the larvae disperse (Jones et al. 2005; see Chapter 20) before setting into another anemone (Elliott et al. 1995; see Chapter 16).

Here, we review the most recent work on parental care in anemonefishes. We focus on the two species whose parental care has been studied most extensively in the last five years: the clown anemonefish *Amphiprion percula* and its

DOI: 10.1201/9781003125365-18

sister taxon, the false clown anemonefish *Amphiprion ocellaris*. First, tapping into the rich literature on phenotypic plasticity and animal personalities, we consider how parents respond to changes in resource availability and social roles. Second, linking to recent theoretical advances in the study of parental care, we consider how parents respond to changes in each other's behavior (i.e., how parents negotiate care). Third, we discuss a rare and interesting form of parental care that occurs in anemonefishes as a byproduct of their social system: step-fathering. Fourth, we dive into the proximate mechanisms underlying parental care. This chapter complements chapters on reproduction (Chapter 13) and dispersal (Chapter 20), revealing the central role that parents play in the life cycle and population dynamics of anemonefishes. Our goal is to provide the reader with a review of the most recent advances in anemonefish parental care research and highlight promising future directions.

15.2 PLASTICITY AND PERSONALITY OF PARENTAL CARE

15.2.1 Plasticity and Personality of Parental Care in Response to Changes in Resources

Plasticity, the capacity of individuals to respond to environmental changes by modifying traits or behaviors, is critical to our understanding of whether and how populations can persist under environmental change (West-Eberhard 2003). In the clown anemonefish *Amphiprion percula*, parental care is plastic in response to changes in resource availability, reflecting the vital role of habitat quality on reproductive success. A field study revealed that in groups that occupied large anemones and in groups supplemented with food, females laid more eggs and both males and females increased time spent tending to those eggs (Barbasch et al. 2020). Additionally, there was support for a causal pathway linking anemone size and reproductive success through effects on egg-laying by females and parental care by males (Figure 15.1; Barbasch et al. 2020). Combined these results indicate that plasticity in reproduction and parental care in response to short- and long-term changes in resource availability can generate among-group variation in embryo survival and larval production (Box 15.1; Barbasch et al. 2020). Troublingly, *A. percula* may be particularly susceptible to environmental degradation due to the strong effects of anemone quality on reproductive success (Salles et al. 2016). Plasticity in reproduction and parental care, if adaptive, may allow parents to optimize reproductive output to take advantage of short-term increases in resource availability, while limiting investment in costly reproduction when resources are scarce. Therefore, plasticity provides some hope for the capacity for population persistence in the face of their rapidly changing environment.

BOX 15.1: CONSEQUENCES OF VARIATION IN PARENTAL CARE

Parents play a pivotal role in offspring development, but not all parents are good parents. Variation in parental care, within and among individuals, may have cross-generational consequences by influencing the *number* and *quality* of offspring that survive and reproduce (Mousseau and Fox 1998). In anemonefishes, parental care is positively related to the number of offspring produced (larval number, Figure 15.1; Barbasch et al. 2020; also Ghosh et al. 2012), and the number of offspring produced is related to recruitment success (Figure 15.1; Saenz-Agudelo et al. 2015). Whether parental care is also related to the quality of offspring produced is an outstanding question (larval quality, Figure 15.1). In anemonefishes, larvae undergo a dispersal phase after which they attempt to recruit to an anemone – this is a phase during which the chances of success may be 1 in 10,000. The distance a larva disperses and whether it can recruit are major determinants of population connectivity and structure (see Chapters 17). Dispersal distance can be related to larval size and swimming performance (Leis 2007; Nanninga and Manica 2018; Majoris et al. 2019), yet little is known about the role of parental care in generating variation in these traits (Figure 15.1). Experimental tests of parental effects on larval size, swimming ability, and local recruitment, will provide new insights into the consequences of variation in parental care for population dynamics.

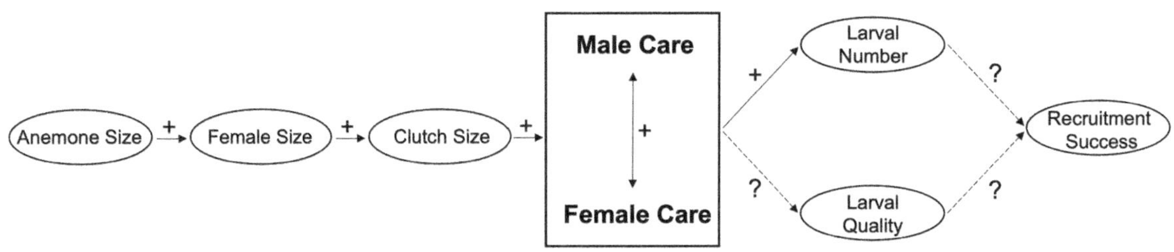

FIGURE 15.1 Parental care plays a central role in the population dynamics of anemonefishes. Hypothesized causal pathway showing the central role that parental care plays in linking anemone size to recruitment success in the clown anemonefish *Amphiprion percula* (adapted from Barbasch et al. 2020). Solid arrows represent relationships between habitat traits, fish traits, and reproduction which have empirical support in *A. percula* (see text); double-headed arrow represents a relationship between male and female care with no hypothesized causal direction, and dashed arrows represent hypothesized relationships that remain untested in *A. percula*. The (+) indicates a positive association, the (–) a negative association, and the (?) an untested, hypothesized relationship.

In addition to plasticity, individuals can exhibit "personalities", consistent behavioral traits maintained across contexts. Plasticity might be favoured by selection if it allows for adaptive responses to environmental changes, but it is not immediately apparent why personality variation would be maintained by selection (Dingemanse and Wolf 2013; Alonzo 2015). In the clownfish *Amphiprion percula* and the false clownfish *A. ocellaris*, individuals differ from each other and are repeatable through time and across contexts in parental behaviors (DeAngelis et al. 2017; Barbasch and Buston 2018), raising the question of why this variation exists. One adaptive explanation is that personalities can reflect stable among-individual differences in state (e.g., size, age, or physiological condition) (Dingemanse and Wolf 2010), which might exist in anemonefishes due to variation among groups in habitat quality. Larger or higher-quality anemones result in resident fish that grow larger, reproduce more, and provide more care than residents of smaller or lower-quality anemones (Chausson et al. 2018; Salles et al. 2020; Barbasch et al. 2020). Additionally, personality may be due to mechanistic constraints, for example, due to pleiotropy or limited ability to express the optimal phenotype across all environments (Alonzo 2015). One important future step in understanding why personalities exist is to determine to what extent they are heritable, and if so, whether they are adaptive.

Plasticity and personality are not mutually exclusive, reflected in individual-by-environment interactions, IxE (Royle et al. 2014). It is not intuitive why this IxE variation exists – why would individuals respond differently to the same environmental changes? In *A. percula*, the average level of parental care increases in response to an increase in food availability, but individuals vary significantly in the magnitude and direction of this response (Barbasch and Buston 2018). One plausible adaptive explanation for an IxE interaction in anemonefish parental care centres around social contexts. Models reveal that consistent behavioral types may favour individuals that modify their behavior in response to their social partner, which selects for individuals that are consistent in their responses, resulting in co-existence of responsive and non-responsive individuals (Wolf et al. 2011). This model may help explain IxE variation in anemonefish, as parents must interact repeatedly to coordinate offspring care. Even if IxE variation is not heritable or adaptive, its existence can have consequences for populations, as IxE can theoretically stabilize population-level responses to environmental changes simply due to the diversity of existing responses (West-Eberhard 2003; Dingemanse and Wolf 2013).

15.2.2 Plasticity and Personality of Parental Care in Response to Sex Change

In addition to exhibiting plasticity of care in response to changes in ecological context (resource availability), anemonefishes also exhibit plasticity of care in response to changes in social roles (across sex change). Anemonefishes are protandrous hermaphrodites and individuals have the capacity to change sex from male to female (Chapter 12). Therefore, any average differences between males and females in parental care reflect plasticity across sex change. Although *A. percula* and *A. ocellaris* males and females differ in their average level of care, representing within-individual plasticity (DeAngelis and Rhodes 2016; Barbasch and Buston 2018), parental care may also be correlated across sex change, such that males that provide a relatively high level of care also provide a relatively high level of care as a female, representing personality across sex change.

If male and female care optima differ, cross-sex correlations, as might occur in *Amphiprion*, suggest that there may be some mechanistic constraint on the independent evolution of male and female parenting behavior (Box 15.2). One hypothesis is that constraints on plasticity in the expression of isotocin and arginine vasotocin, which have antagonistic effects on parental care and territory defence (see "Mechanisms Underlying Parental Care"), could help explain variation among individuals in parental care. Future studies exploring the behavioral and molecular mechanisms underlying how a single individual can rapidly and dramatically shift parental roles, as well as the constraints involved, will ultimately help us understand why individual variation in parental care exists.

BOX 15.2: FUNCTION OF FEMALE CARE

In *A. percula*, male care is a strong predictor of embryo survival (Figure 15.1; Barbasch and Buston 2018; Barbasch et al. 2020); however, the function of female care is less clear. During the day, *A. percula* females spend time in proximity to the clutch (referred to as tending) but do very little mouthing and next to no fanning (Barbasch and Buston 2018). The amount of time a female spends tending is sensitive to food availability, suggesting that females may face a tradeoff between tending and other activities like foraging (Barbasch and Buston 2018). If female tending does not enhance offspring survival, why do females engage in tending at all? One hypothesis is simply that we have not had the statistical power to detect an effect of female tending on embryo survival. An alternative hypothesis is that female tending has more to do with monitoring clutch development, in preparation for their involvement on the night of hatching (Pacaro 2022). Another alternative is that female tending is indicative of her monitoring her partner's efforts, as part of a negotiation over levels of care (Barbasch et al. 2021). Finally, it's also plausible that selection for male care is strong and selection against low levels of female care is weak, meaning that female care may be a case of intergender hitchhiking (Clint et al. 2012) in these sex-changing fish. Novel experiments manipulating levels of female tending are needed to understand the function of female care.

15.3 NEGOTIATIONS OVER CARE

In the previous section, we demonstrated that parents exhibit plasticity in response to changing ecological conditions and social roles. However, anemonefishes are biparental, and thus *A. percula* parents are also faced with variation in the social environment created by interactions with their partner. In species with biparental care, conflict arises between parents because each parent benefits from shifting the burden of care to their partner. Theory predicts that parents should respond to each other to reach a negotiated settlement over how much care each should provide (McNamara et al. 1999).

While the theory is sound, a test of a series of negotiation models in *A. percula* revealed that no current models fully explain whether and how anemonefish parents negotiate (Barbasch et al. 2021). When one parent was experimentally handicapped via fin clipping, the other parent seemingly did not respond: males tended just as much when the female was handicapped and females tended just as much when the male was handicapped. However, when one parent was handicapped, pairs did not face any consequences in terms of larval production, suggesting that parents are fully compensating in some way for changes in their partner's effort.

Apparent full compensation, as seen in *A. percula*, presents an evolutionary conundrum, because if parents are completely making up for reductions in their partner's effort, one parent should cease care altogether, resulting in uniparental care as the evolutionary outcome (McNamara et al. 1999). One explanation is that, in *A. percula*, there may not be sufficient conflict to be resolved by negotiation. Actual conflict might be low because ecological constraints and the social hierarchy within groups enforce cooperation (Chapter 14; Rueger et al. 2018; Branconi et al. 2020). However, potential conflict still exists because the non-breeding group members can rapidly replace either member of the breeding pair, thus serving as future reproductive opportunities (Buston 2004a).

The role of conflict in negotiations could be tested by manipulating the presence/absence of non-breeders. Conflict should theoretically be greater in groups with non-breeders, which represent future mates for the breeding pair, than in groups without non-breeders (Buston 2004a; Chapter 14). If non-breeders were experimentally removed, parental interests would become strongly (if not fully) aligned, and thus full compensation for a reduction in partner effort makes sense. However, when non-breeders are present and one parent is experimentally handicapped, such generosity is not predicted. Both parents have other options, in the form of non-breeders, if their partner is unable to provide sufficient care, therefore the predictions of negotiation models are expected to hold. Another potential explanation for full compensation in negotiations is that parental care was not sufficiently reduced to incite conflict between parents. This hypothesis could be tested using phenotypic engineering to manipulate parental care at the mechanistic level (Nugent et al. 2019), for example by administering an isotocin receptor antagonist, which is known to reduce direct egg care in *A. ocellaris* males and females (DeAngelis et al. 2020; see "Mechanisms Underlying Parental Care") and determining the consequences for parental interactions and fitness.

15.4 STEP-FATHERING AND ALLOPARENTAL CARE

Another evolutionary conundrum arises with the existence of alloparental care, parental care directed towards non-descendant offspring, because the benefits of care are accrued by distantly related or unrelated individuals (Emlen 1991; Wisenden 1999). Anemonefishes live in social groups composed of a breeding pair and a small number of non-breeders, setting the stage for alloparental care (see Chapter 14). However, unlike many other species that occur in such social groups, non-breeding subordinates do not participate in alloparental care when the parents are present (Buston 2004b). In anemonefishes, group members are unrelated and thus do not benefit from alloparental care in the traditional way via kin selection (Buston et al. 2007). Yet, in some anemonefishes, when one or both parents are removed while eggs are in the nest, the non-breeder will provide care (Yanagisawa and Ochi 1986; Phillips et al. 2020).

Adaptive explanations for alloparental care look for benefits, including from helping relatives, acquiring breeding experience, or improving the chance of inheriting territories (Emlen and Vehrencamp 1983; Balshine-Earn et al. 1998). One hypothesis for anemonefish alloparental care is that providing care may allow non-breeders to escape punishment by the female – indeed, in *A. clarkii*, when males were experimentally removed, females were observed head-butting and nudging the non-breeder towards the clutch, and the non-breeder began caring for the eggs (Yanagisawa and Ochi 1986). Alloparental care may be a form of reciprocal altruism (Trivers 1971), such that dominants allow non-breeders access to shelter within the anemone, and non-breeders reciprocate by caring for eggs upon the death of a breeder. In this context, punishment by females may serve as retaliation against those who fail to care for her eggs. A second hypothesis is that alloparental care strengthens the pair bond and provides parenting experience for the mate replacement (Yanagisawa and Ochi 1986; Phillips et al. 2020). Indeed, breeding experience is associated with increased male care and embryo survival (Buston and Elith 2011; Phillips et al. 2020). A third hypothesis is that the non-breeder would benefit from care even with both parents removed if the larvae produced were to return to their natal anemone and become rapid mate replacements. While returns to the natal anemone are rare, they do occur (Salles et al. 2016) and may be more common when anemones are undersaturated due to the loss of both parents (Buston 2003). Discriminating among these hypotheses will require a series of carefully designed field experiments.

Non-adaptive explanations for alloparental care propose that it represents misdirected care due to the inability to discriminate kin from non-kin (Jamieson 1989). In the lab, even when *both* biological parents are removed, *A. ocellaris* non-breeders care for eggs (Phillips et al. 2020). Thus, adaptive explanations involving strengthening the pair bond or avoiding punishment are not sufficient to explain alloparental care here (though the third adaptive hypothesis, that the non-breeders care for the eggs to rapidly produce a new mate for themselves, cannot be ruled out). Additionally, the solo non-breeder should theoretically benefit from investing in growth and development into a female, rather than gaining experience as a male (Phillips et al. 2020). The non-adaptive explanation is that caring for unrelated eggs is a side effect of selection for caring for descendant eggs and is maintained because long-term monogamy results in a low probability of a male encountering unrelated eggs.

15.5 MECHANISMS UNDERLYING PARENTAL CARE

The brain mechanisms responsible for parental care are evolutionarily ancient. All social behaviors are regulated by neuroendocrine axes which include a series of highly evolutionarily conserved interconnected sub-cortical brain areas, endocrine glands, neuro-peptides, neurotransmitters, and sex steroid hormones (Dulac et al. 2014; Rogers and Bales 2019). However, parental care is usually studied in animal models where the females are the primary caregivers of the offspring (Dulac et al. 2014). As a consequence, our understanding of paternal care at the neuroendocrine level is lacking in comparison to maternal care. While many of the substrates that regulate paternal care are likely shared with maternal care, others may differ, as males and females have sex-specific behaviors and dramatically different circulating sex steroid hormone levels that affect physiology and behavior (Ball et al. 2002; Nelson 2005). Additionally, females are primed for parenting, as changes in physiology occur during egg development, while cues for males are more subtle.

Anemonefish present an ideal model system for uncovering mechanisms of male parental (paternal) care. One of the challenges in identifying brain mechanisms of paternal care is isolating paternal care from the many other behaviors that males perform simultaneously such as territoriality and courtship (Kleszczyńska et al. 2012; O'Connell et al. 2012). In many anemonefish species, due to high dependence on host anemones for protection and consequently high social isolation from other groups, parental behavior can be examined in isolation from other confounding displays (Deangelis and Rhodes 2016). Furthermore, as described earlier, male *A. ocellaris* display alloparental care, so the entire breeding cycle can be dissociated from the display of paternal behaviors when a male that has never parented before (or is not currently parenting) is presented with eggs and they begin to display the paternal behaviors (Phillips et al. 2020). In male anemonefish, the appearance of eggs presents a rapid shift in behavior from non-parenting to parenting (Rogers and Bales 2019). This dramatic change in behavior must also be reflected within the brain.

Paternal care consists not of one single behavior but a suite of behaviors (see "Introduction"). In male false clown anemonefish (*Amphiprion ocellaris*), a father robustly cares for his offspring to promote egg development, yet when potential predators arrive, he must vigorously guard his nest against predation. Hence, fathers switch parenting tactics from egg tending to egg guarding in the presence of predators as simultaneously competitive demands necessitate parents to make prudent decisions (DeAngelis et al. 2020). These decisions to display different forms of parenting are reflected by distinct mechanisms within the brain.

Two well-known neuro-peptides involved in regulating a variety of social behaviors appear to play a critical role in dynamically regulating the switch between nurturing and aggressive defence tactics. These are isotocin (oxytocin is the mammalian homolog) and arginine vasotocin (arginine vasopressin is the mammalian homolog). These neuro-peptides are small proteins produced by neurons which act as signalling molecules within the brain (O'Connell and Hofmann 2012). Cell bodies containing these neuro-peptides reside in the preoptic area of the hypothalamus while receptors are laden throughout the brain in other evolutionarily conserved brain areas involved in regulating social behavior. Oxytocin has been well described for its role in female reproduction and maternal behavior as its release at parturition serves as a catalyst for physiological and behavioral changes (Insel 2010). While less studied in males, it likely also functions to promote paternal behavior (Figure 15.2). Arginine vasopressin has been broadly implicated in male behaviors which promote reproduction but has consistently been recognized as important for aggression in teleosts (Kleszczyńska et al. 2012; O'Connell and Hofmann 2011; Yaeger et al. 2014).

In *A. ocellaris*, pharmacological blockade of arginine vasotocin (V1a) receptors and isotocin receptors affected parental behaviors in opposite ways (DeAngelis et al. 2017). As expected, a blockade of isotocin receptors reduced the amount of nipping and fanning of the eggs without altering the time spent in the nest compared to saline controls. This is consistent with other findings in teleost fishes showing that, like mammals, isotocin signalling is critical for high levels of parental care, regardless of which sex is the primary caregiver.

More intriguing was that blockade of arginine vasotocin V1a receptors actually increased the amount of direct parental care. Given the known role of arginine vasotocin signalling in aggression in teleosts, one explanation is that vigilance was blocked, which then resulted in a greater amount of time allotted to egg care. To test this hypothesis, a follow-up experiment was conducted, in which nest predators were introduced while concurrently administering either isotocin receptor or arginine vasotocin V1a receptor antagonists to males actively fathering. Here, the isotocin receptor antagonist again reduced parental care but also increased aggression. Conversely, the V1a antagonist reduced aggression while increasing direct egg care (DeAngelis et al. 2020). These results suggest that

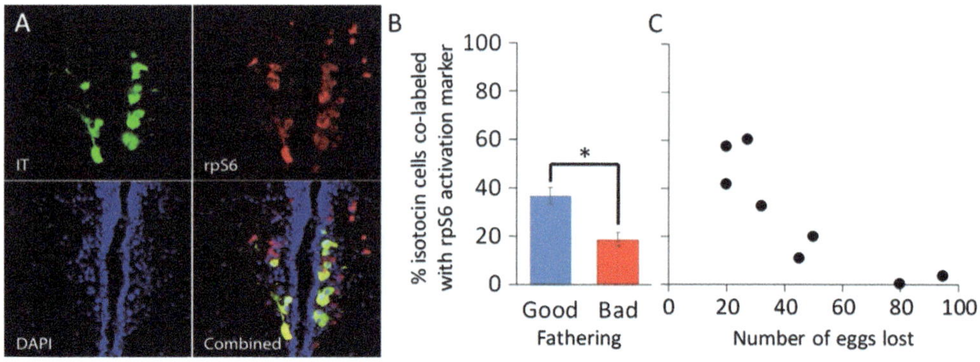

FIGURE 15.2 Activation of preoptic area isotocin neurons in good fathers. A. Representative section through the preoptic area of a fathering male *A. ocellaris* showing immunofluorescent detection of isotocin (IT, top left), the phosphorylated ribosomal protein (rpS6; top right), DAPI (bottom left), and all combined (bottom right). Performed by author. B. Percentage of IT cells co-labelled with the rpS6 activation marker in "good" and "bad" fathers (see Section 15.5) and C. as a function of the number of eggs lost. Standard errors shown. * indicates statistically significant by Fisher exact test. Methods: sexually naïve males (n=8) were given batches of eggs to step-father for 90 min. Total number of fans, nips, time in nest, and number of eggs at the beginning and end were recorded. The males were then euthanized by cervical transection and brains sectioned and stained for IT and rpS6. A total of 392 IT cells were identified in the eight fish and each IT cell was subsequently analyzed for co-expression of rpS6 by focusing through the cell on the z-axis. Good and bad fathers were identified by median split of a composite fathering score considering total number of paternal behaviors, time in nest, and number of eggs lost.

arginine vasotocin and isotocin signalling pathways act competitively in the regulation of two components of male parental care: as isotocin is blocked, more effort is allotted towards aggression versus egg nurturing, and vice versa when arginine vasotocin is blocked. These studies further suggest isotocin and arginine vasotocin act independently in the regulation of simultaneously occurring competitive demands of fathers providing offspring care, which can be both nurturing and aggressive in offspring defence, providing insight into how the paternal brain manages the tradeoffs involved in parenting.

Given the evidence cited above that isotocin has a direct role in promoting paternal care in anemonefish, males were predicted to display a greater expression of isotocin receptors in their brains than females (since males are the primary caregivers of the eggs). Furthermore, isotocin receptor expression should be upregulated during active parental care in both sexes as compared to when they are not directly caring for eggs. As predicted, males displayed greater isotocin receptor expression compared to females, while active parents, both males and females, displayed increased isotocin receptor gene expression in the brain compared to non-parental individuals (DeAngelis et al. 2018). This supports a growing body of evidence that isotocin signalling in the brain is regulated not only by the synthesis and release of isotocin from neurons but also by the density and/or distribution of isotocin receptors in the brain. These results imply that as receptor numbers increase, the signalling efficiency of isotocin also increases, promoting parental behavior. While both parents showed increased isotocin receptor gene expression in the brain, the effect was particularly pronounced in males, consistent with the observation that males are the primary caregivers of the eggs.

Taken together, these studies suggest that the ancient evolutionarily conserved signalling pathways of vasotocin and isotocin interact with steroid hormones to regulate parental behavior in anemonefish and likely across vertebrates (DeAngelis et al. 2017, 2018, 2020). However, the brain is a complex heterogeneous organ that operates on a scale from molecules, to cells, to neural circuits. Here, we have highlighted only a small portion of the mechanisms that likely orchestrate parental care and provide insight into future promising opportunities in this model system.

15.6 CONCLUSIONS AND PROSPECTS

Anemonefishes have provided novel insights into the individual, social, ecological, physiological, and evolutionary factors that influence patterns of care. Field and laboratory studies have uncovered variation within and among individuals, populations, and species, yet little is known about *why* this variation exists and what its consequences are. Future studies focusing on proximate and ultimate explanations for this variation are now possible. Studies into the mechanisms underlying care have identified the neural pathways and brain regions regulating parental care, and emerging genomic methods (see Chapter 5) have opened the way for experimental manipulations of parental care at the mechanistic level. Targeted manipulations of parental care in both males and females present a powerful tool for answering many outstanding questions, such as whether and how parents negotiate, what the function of female care is, if any, and what the consequences of variation in care are for both parents and offspring. Parental behaviors have been well-described within some species (Green and McCormick 2005; Ghosh et al. 2012; DeAngelis et al. 2017; Barbasch and Buston 2018), but interspecific variation in care has not been systematically studied. Our growing knowledge of the interspecific and intraspecific variation in ecology and social system as well as our understanding of phylogenetic relationships among species (Litsios et al. 2014; Rolland et al. 2018; see Chapter 3) allows for future comparative analyses of parental care. Finally, studies exploring parental effects on larval traits can help uncover the consequences of parental care for future generations.

REFERENCES

Allen, G. R. 1972. *The Anemonefishes: Their Classification and Biology*. Neptune City: TFH Publications.

Alonzo, S. H. 2015. Integrating the How and Why of Within-Individual and Among-Individual Variation and Plasticity in Behavior. *Current Opinion in Behavioral Sciences* 6: 69–75.

Ball, G. F., L. V. Riters, and J. Balthazart. 2002. Neuroendocrinology of Song Behavior and Avian Brain Plasticity: Multiple Sites of Action of Sex Steroid Hormones. *Frontiers in Neuroendocrinology* 23: 137–178.

Balshine-Earn, S., F. C. Neat, H. Reid, and M. Taborsky. 1998. Paying to stay or paying to breed? Field evidence for direct benefits of helping behavior in a cooperatively breeding fish. *Behavioral Ecology* 9(5): 432–438.

Barbasch, T. A., R. Branconi, R. Francis, M. Pacaro, M. Srinivasan, G. P. Jones, and P. M. Buston. 2021. Negotiations over Parental Care: A Test of Alternative Hypotheses in the Clown Anemonefish. *Behavioral Ecology* 32(6): 1256–1265.

Barbasch, T. A., and P. M. Buston. 2018. Plasticity and Personality of Parental Care in the Clown Anemonefish. *Animal Behaviour* 136: 65–73.

Barbasch, T. A., T. Rueger, M. Srinivasan, M. Y. L. Wong, G. P. Jones, and P. M. Buston. 2020. Substantial Plasticity of Reproduction and Parental Care in Response to Local Resource Availability. *Oikos* 129: 1844–1855.

Branconi, R., T. A. Barbasch, R. K. Francis, M. Srinivasan, G. P. Jones, and P. M. Buston. 2020. Ecological and Social Constraints Combine to Promote Evolution of Non-Breeding Strategies in Clownfish. *Communications Biology* 3: 1–7.

Buston, P. M. 2003. Forcible Eviction and Prevention of Recruitment in the Clown Anemonefish. *Behavioral Ecology* 14: 576–582.

Buston, P. M. 2004a. Territory Inheritance in Clownfish. *Proceedings. Biological Sciences/The Royal Society* 271 (Suppl): S252–S254.

Buston, P. M. 2004b. Does the Presence of Non-Breeders Enhance the Fitness of Breeders? An Experimental Analysis in the Clown Anemonefish *Amphiprion percula*. *Behavioral Ecology and Sociobiology* 57: 23–31.

Buston, P. M., S. M. Bogdanowicz, A. Wong, and R. G. Harrison. 2007. Are Clownfish Groups Composed of Close Relatives? An Analysis of Microsatellite DNA Variation in *Amphiprion percula*. *Molecular Ecology* 16: 3671–3678.

Buston, P. M., and J. Elith. 2011. Determinants of Reproductive Success in Dominant Pairs of Clownfish: A Boosted Regression Tree Analysis. *The Journal of Animal Ecology* 80: 528–538.

Chausson, J., M. Srinivasan, and G. P. Jones. 2018. Host Anemone Size as a Determinant of Social Group Size and Structure in the Orange Clownfish (*Amphiprion percula*). *PeerJ* 6: e5841.

Clint, E. K., E. Sober, T. Garland Jr., and Rhodes, J. S. 2012. Male Superiority in Spatial Navigation: Adaptation or Side Effect? *The Quarterly Review of Biology* 87: 289–313.

DeAngelis, R., L. Dodd, and J. Rhodes. 2020. Nonapeptides mediate trade-offs in parental care strategy. *Hormones and Behavior* 121: 104717.

DeAngelis, R., L. Dodd, A. Snyder, and J. S. Rhodes. 2018. Dynamic Regulation of Brain Aromatase and Isotocin Receptor Gene Expression Depends on Parenting Status. *Hormones and Behavior* 103: 62–70.

DeAngelis, R., J. Gogola, L. Dodd, and J. S. Rhodes. 2017. Opposite Effects of Nonapeptide Antagonists on Paternal Behavior in the Teleost Fish *Amphiprion ocellaris*. *Hormones and Behavior* 90: 113–119.

Deangelis, R. S., and J. S. Rhodes. 2016. Sex Differences in Steroid Hormones and Parental Effort Across the Breeding Cycle in *Amphiprion ocellaris*. *Copeia* 104: 586–593.

Dingemanse, N. J., and M. Wolf. 2010. Recent Models for Adaptive Personality Differences: A Review. *Philosophical Transactions of the Royal Society of London. Series B, Biological Sciences* 365: 3947–3958.

Dingemanse, N. J., and M. Wolf. 2013. Between-Individual Differences in Behavioural Plasticity within Populations: Causes and Consequences. *Animal Behaviour* 85: 1031–1039.

Dulac, C., L. A. O'Connell, and Z. Wu. 2014. Neural Control of Maternal and Paternal Behaviors. *Science* 345: 765–770.

Elliott, J. K., J. M. Elliott, and R. N. Mariscal. 1995. Host Selection, Location, and Association Behaviors of Anemonefishes in Field Settlement Experiments. *Marine Biology* 122: 377–389.

Emlen, S. T. 1991. Evolution of Cooperative Breeding in Birds and Mammals. In *Behavioral Ecology*, 2nd ed., eds J. R. Krebs and N. B. Davies, 301–337. Oxford: Blackwell.

Emlen, S. T., S. L. Vehrencamp, J. Ligon, and I. Rowley. 1983. Cooperative Breeding Strategies among Birds. In *Perspectives in Ornithology*, ed. A. Brush and G. Clark Jr, 93–120. Cambridge: Cambridge University Press.

Ghosh, S., T. T. Kumar, and T. Balasubramanian. 2012. Determining the Level of Parental Care Relating Fanning Behavior of Five Species of Clownfishes in Captivity. *Indian Journal of Geo-Marine Sciences* 41: 430–441.

Green, B. S., and M. I. McCormick. 2005. O2 Replenishment to Fish Nests: Males Adjust Brood Care to Ambient Conditions and Brood Development. *Behavioral Ecology* 16: 389–397.

Insel, T. R. 2010. The Challenge of Translation in Social Neuroscience: A Review of Oxytocin, Vasopressin, and Affiliative Behavior. *Neuron* 65: 768–779.

Jamieson, I. G. 1989. Behavioral Heterochrony and the Evolution of Birds' Helping at the Nest: An Unselected Consequences of Communal Breeding? *The American Naturalist* 133: 394–406.

Jones, G. P., S. Planes, and S. R. Thorrold. 2005. Coral Reef Fish Larvae Settle Close to Home. *Current Biology* 15: 1314–1318.

Kleszczyńska, A., E. Sokołowska, and E. Kulczykowska. 2012. Variation in Brain Arginine Vasotocin (AVT) and Isotocin (IT) Levels with Reproductive Stage and Social Status in Males of Three-Spined Stickleback (*Gasterosteus aculeatus*). *General and Comparative Endocrinology* 175: 290–296.

Leis, J. M. 2007. Behaviour as Input for Modelling Dispersal of Fish Larvae: Behaviour, Biogeography, Hydrodynamics, Ontogeny, Physiology and Phylogeny Meet Hydrography. *Marine Ecology Progress Series* 347: 185–193.

Litsios, G., P. B. Pearman, D. Lanterbecq, N. Tolou, and Salamin, N. 2014. The Radiation of the Clownfishes has Two Geographical Replicates. *Journal of Biogeography* 41: 2140–2149.

Majoris, J. E., K. A. Catalano, D. Scolaro, J. Atema, and P. M. Buston. 2019. Ontogeny of Larval Swimming Abilities in Three Species of Coral Reef Fishes and a Hypothesis for Their Impact on the Spatial Scale of Dispersal. *Marine Biology* 166: 1–14.

McNamara, J. M., C. E. Gasson, and A. I. Houston. 1999. Incorporating Rules for Responding into Evolutionary Games. *Nature* 401: 368–371.

Mousseau, T. A., and C. W. Fox. 1998. The Adaptive Significance of Maternal Effects. *Trends in Ecology and Evolution* 13: 403–407.

Moyer, J. T. 1980. Influence of Temperate Waters on the Behavior of the Tropical Anemonefish *Amphiprion clarkii* at Miyake-Jima, Japan. *Bulletin of Marine Science* 30: 261–272.

Moyer, J. T., and L. J. Bell. 1976. Reproductive Behavior of the Anemonefish *Amphiprion clarkii* at Miyake-Jima, Japan. *Japanese Journal of Ichthyology* 23: 23–32.

Moyer, J. T., and C. E. Sawyers. 1973. Territorial Behavior of the Anemonefish *Amphiprion xanthurus* with Notes on the Life History. *Japanese Journal of Ichthyology* 20: 85–93.

Nanninga, G. B., and A. Manica. 2018. Larval Swimming Capacities Affect Genetic Differentiation and Range Size in Demersal Marine Fishes. *Marine Ecology Progress Series* 589: 1–12.

Nelson, R. J. 2005. *An Introduction to Behavioral Endocrinology*, 3rd ed. Sunderland, MA: Sinauer Associates.

Nugent, B. M., K. A. Stiver, H. A. Hofmann, and S. H. Alonzo. 2019. Experimentally Induced Variation in Neuroendocrine Processes Affects Male Reproductive Behaviour, Sperm Characteristics and Social Interactions. *Molecular Ecology* 28: 3464–3481.

O'Connell, L. A., and H. A. Hofmann. 2011. The Vertebrate Mesolimbic Reward System and Social Behavior Network: A Comparative Synthesis. *Journal of Comparative Neurology* 519: 3599–3639.

O'Connell, L. A., and H. A. Hofmann. 2012. Evolution of a Vertebrate Social Decision-Making Network. *Science* 336: 1154–1157.

O'Connell, L. A., B. J. Matthews, and H. A. Hofmann. 2012. Isotocin Regulates Paternal Care in a Monogamous Cichlid Fish. *Hormones and Behavior* 61: 725–733.

Pacaro, M. 2022 Nocturnal Parental Care and the Role of Parents in Hatching Their Eggs in the Clown Anemonefish *Amphiprion percula*. [Doctoral dissertation, Boston University] https://open.bu.edu/handle/2144/44042 (Accessed June 30, 2022)

Phillips, E., R. DeAngelis, J. V. Gogola, and J. S. Rhodes. 2020. Spontaneous Alloparental Care of Unrelated Offspring by Non-Breeding Amphiprion Ocellaris in Absence of the Biological Parents. *Scientific Reports* 10: 1–11.

Rogers, F. D., and K. L. Bales. 2019. Mothers, Fathers, and Others: Neural Substrates of Parental Care. *Trends in Neuroscience* 42: 552–562.

Rolland, J., D. Silvestro, G. Litsios, L. Faye, and N. Salamin. 2018. Clownfishes Evolution Below and Above the Species Level. *Proceedings of the Royal Society B: Biological Sciences* 285: 20171796.

Ross, R. M. 1978. Reproductive Behavior of the Anemonefish *Amphiprion melanopus* on Guam. *Copeia* 1978: 103–107.

Royle, N. J., A. F. Russell, and A. J. Wilson. 2014. The Evolution of Flexible Parenting. *Science* 345: 776–781.

Rueger, T., T. A. Barbasch, M. Y. L. Wong, M. Srinivasan, G. P. Jones, and P. M. Buston. 2018. Reproductive Control via the Threat of Eviction in the Clown Anemonefish. *Proceedings of the Royal Society B* 285: 20181295.

Saenz-Agudelo, P., G. P. Jones, S. R. Thorrold, and S. Planes. 2015. Mothers Matter: Contribution to Local Replenishment is Linked to Female Size, Mate Replacement and Fecundity in a Fish Metapopulation. *Marine Biology* 162: 3–14.

Salles, O. C., G. R. Almany, M. L. Berumen, G. P. Jones, P. Saenz-Agudelo, M. Srinivasan, S. R. Thorrold, et al. 2020. Strong Habitat and Weak Genetic Effects Shape the Lifetime Reproductive Success in a Wild Clownfish Population. *Ecology Letters* 23: 265–273.

Salles, O. C., P. Saenz-Agudelo, G. R. Almany, M. L. Berumen, S. R. Thorrold, G. P. Jones, S. Planes. 2016. Genetic Tools Link Long-Term Demographic and Life-History Traits of Anemonefish to Their Anemone Hosts. *Coral Reefs* 35: 1127–1138.

Seymour, J. R., T. A. Barbasch, and P. M. Buston. 2018. Lunar Cycles of Reproduction in the Clown Anemonefish *Amphiprion percula*: Individual-Level Strategies and Population-Level Patterns. *Marine Ecology Progress Series* 594: 193–201.

Sinervo, B., and E. Svensson. 1998. Mechanistic and Selective Causes of Life History Trade-Offs and Plasticity. *Oikos* 83: 432–442.

Stearns, S. C. 1989. Trade-Offs in Life-History Evolution. *Functional Ecology* 3: 259.

Thomas, D., and C. Prakash. 2015. Spawning Behaviour and Embryonic Development in the Sebae Anemonefish *Amphiprion sebae* (Bleeker, 1853). *Indian Journal of Fisheries* 62: 58–65.

Trivers, R. L. 1971. The Evolution of Reciprocal Altruism. *The Quarterly Review of Biology* 46: 35–57.

West-Eberhard, M. J. 2003. *Developmental Plasticity and Evolution*. Oxford: Oxford University Press.

Wisenden, B. D. 1999. Alloparental Care in Fishes. *Reviews in Fish Biology and Fisheries* 9: 45–70.

Wolf, M., G. S. Van Doorn, and F. J. Weissing. 2011. On the Coevolution of Social Responsiveness and Behavioural Consistency. *Proceedings of the Royal Society B: Biological Sciences* 278: 440–448.

Yaeger, C., A. M. Ros, V. Cross, R. S. DeAngelis, D. J. Stobaugh, and J. S. Rhodes. 2014. Blockade of Arginine Vasotocin Signaling Reduces Aggressive Behavior and c-Fos Expression in the Preoptic Area and Periventricular Nucleus of the Posterior Tuberculum in Male *Amphiprion ocellaris*. *Neuroscience* 267: 205–218.

Yanagisawa, Y., and H. Ochi. 1986. Step-Fathering in the Anemonefish *Amphiprion clarkii*: A Removal Study. *Animal Behaviour* 34: 1769–1780.

Part IV

Ecology

16 Habitat Selection of Anemonefish

Kina Hayashi and James Davis Reimer

CONTENTS

16.1 INTRODUCTION

Anemonefish belong to the family Pomacentridae, of which there are 28 species in the world (Dunn 1981; Fautin and Allen 1997; Ollerton et al. 2007). The most famous feature of anemonefish is their inhabiting of sea anemones throughout their life, with the exception of the pelagic larval stage (e.g., Fautin 1991; Fautin and Allen 1992). The sea anemones used by anemonefish are called "host anemones", and there are ten species in the world (Dunn 1981; Fautin and Allen 1992). The first description of anemonefish inhabiting a host anemone was by Collingwood in 1868 (Collingwood 1868; Fautin 1991). Subsequently, Mariscal (1970) revealed that there is a mutualistic relationship between anemonefish and host anemones. Because all host anemones have nematocysts which can harm fish, anemonefish use host anemones as shelter to protect themselves from their predators (Dunn 1981; Fautin and Allen 1997). The body surface of anemonefish is covered in mucus, and this mucus prevents host anemones from stinging anemonefish (Mebs 2009). On the other hand, anemonefish prevent predators such as butterflyfish (Chaetodontidae) from feeding on the tentacles of host anemones (Moyer and Sawyers 1973) and provide ammonia for their host anemones and their endosymbiotic Symbiodiniaceae, enhancing rates of tissue growth and regeneration (Porat and Chadwick-Furman 2004).

Anemonefish are distributed from the Red Sea in the west to western Polynesia in the east and from Japan in the north to Australia in the south (Fautin and Allen 1997). The distribution range of host anemones includes, in addition to the aforementioned range, the Hawaii Islands, where anemonefish are not distributed (Fautin and Allen 1997). Each species of anemonefish is primarily found in a specific host anemone species (Fautin and Allen 1992). As well, the selection range of host anemones depends on the species of anemonefish; *Amphiprion clarkii* is a generalist that uses ten different host anemone species, whereas *A. polymnus* and *A. sandaracinos* are specialists that use only one species of host anemone (e.g., Fautin and Allen 1997; Hattori 2011; Hayashi et al. 2018, 2021). According to Litsios et al. (2014), the symbioses between anemonefish and host anemones started ten million years ago in the central Indo-Pacific region. Phylogenetic analyses suggest that such symbioses started with specialists such as *Premnas biaculeatus*/*A. ocellaris*, and then diversified with the evolution of generalists such as *A. clarkii*, and other specialists such as *A. perideraion*, *A. frenatus*, and *A. polymnus* (Elliott et al. 1999; Santini and Polacco 2006).

Anemonefish larvae initially lack resistance to host anemone venom at birth, but their skin mucus gradually acquires biochemical resistance to the host venom through acclimation behavior (Balamurugan et al. 2014). Biochemical mechanisms for protection from venom are expected to vary depending on both host species and anemonefish species. In the specialist species *A. ocellaris*, juveniles living with the natural host species grew better than individuals living with other species of host anemones (Nguyen et al. 2019). Thus, host selection is mainly determined at the larval stage, and post-maturity host recruitment is expected to be rare.

In this chapter, we reviewed field studies in the scientific literature to help answer the question "why do anemonefish choose a particular host anemone under natural (in situ) conditions?". In the first section of this work, we reviewed previous studies on factors governing host species and habitat selection by anemonefish in various regions. In the second section, we show how the species composition and density of host anemones and anemonefish vary among natural environment and human influence, based on field studies conducted in the Ryukyu Archipelago in southern Japan. The last section presents further challenges for the future in understanding the habitat selection of anemonefish and in the elucidation of appropriate conservation measures.

DOI: 10.1201/9781003125365-20

16.2 OVERVIEW OF RESEARCH ON HOST SELECTION

It has long been a matter of interest to many researchers that different species of anemonefish each use different species of host anemone. Fautin (1986) showed that five species of anemonefish exhibited host specificity for six host anemones observed around Lizard Island on the Great Barrier Reef, Australia. Based on field transplantation experiments, Fautin (1986) speculated that factors governing host specificity may include coincidence of ecological requirements, competition among anemonefish for hosts, and stochastic processes, in addition to biochemical reasons. In the following chapter, we review studies on ultimate factors (ecological meaning) rather than proximal factors (biochemical reasons or mechanisms) that play a role in host selection by anemonefish.

16.2.1 Ecological Requirements

In the natural environment, where predation pressure is high unlike in the aquarium, anemonefish should select hosts that provide high-quality hiding places, low energy consumption costs, and have suitable substrates for spawning. In other words, it is generally expected that large hosts which have ample hiding places, moderate venom strength, and are located near hard substrata would be preferred (Fautin and Allen 1997; Burke da Silva and Nedosyko 2016). It has been expected that anemonefish perform inter- and intra-specific competition to inhabit a particular preferred host anemone and acquire the right to settle on such a host through competitive exclusion (Fautin 1986).

Host anemones differ from species to species in terms of size, morphology (e.g., tentacle length and complexity of folds), venom strength, habitat substrate environment, and water environment including various factors such as water temperature and currents (e.g., Dunn 1981; Hattori 2011; Hayashi et al. 2018, 2021; Hirose 1985). Anemonefish also differ from species to species in terms of size, resistance to cnidarian venom, swimming ability, and resistance/tolerance to water environmental variables (e.g., Fautin and Allen 1992; Fautin 1991; Hayashi et al. 2018, 2019b, 2021). There are thus differences in suitable host anemones for each species of anemonefish, and in other words, resource partitioning among species of anemonefish is likely occurring (Fautin 1986).

Madang, Papua New Guinea, in the Coral Triangle, is one of the most diverse habitats for both anemone and anemonefish species, as it is the centre of distribution for both groups. Elliott and Mariscal (2001) investigated species interactions between nine species of anemonefish and ten species of host anemone, and their results supported a habitat segregation (niche differentiation) hypothesis: anemonefish each used different species of host anemones and used different zones of a reef (nearshore, mid-lagoon, outer barrier, and offshore zones) even when utilizing the same host species. Litsios et al. (2014) tested the hypothesis, explaining why generalist and specialist species coexist by analyzing a phylogeny of anemonefish and their host species. They found a negative correlation between the number of host species and environmental specificity and concluded that a tradeoff between resource use in two directions explained the coexistence of generalist and specialist. If we look at the entire distribution area, the interactions between anemonefish and their hosts are statistically significantly nested, that is, generalist fish interact with generalist and specialist anemones, and specialist fish interact with generalist anemones (Ollerton et al. 2007). However, in Manado, North Sulawesi, where seven anemonefish species interact with eight species of host anemone, Ricciardi et al. (2010) demonstrated that the interactions between anemonefish and host anemones were not significantly nested on a regional scale, and speculated that these discrepancies may be due to regional conditions in which competition forced generalist species to become more specialist in nature. In Manado, anemone species that were most abundant at each site tended to interact with more species of anemonefish, suggesting the importance of local population sizes in determining species interactions (Ricciardi et al. 2010).

16.2.2 Competition among Anemonefish

Even if there is resource partitioning, intraspecific and interspecific territorial defence behaviors are important in anemonefish successfully settling and protecting a suitable host anemone. Hattori (2002) examined competitive interactions between two species of anemonefish sharing the same host species; large *A. clarkii* and smaller *A. perideraion* using the same host anemone *H. crispa* in Okinawajima, Japan. *Amphiprion clarkii* was behaviorally dominant over *A. perideraion* in a cohabiting group, but adult *A. clarkii* emigrated from a cohabited anemone to another host anemone, probably due to the high cost of interactions with adult *A. perideraion*. Camp et al. (2016) suggested the role of cohabitation in supporting species diversity of anemonefish in the Coral Triangle, but cohabitation rarely occurs outside of the Coral Triangle, and this is likely best explained by the ratio of host anemone species to anemonefish species. The cohabiting anemonefish species finely partition their host anemone, with the subordinate species inhabiting the periphery of an anemone and rarely being attacked by the dominant species.

As the distance from the Coral Triangle increases, the numbers of species of both host anemones and anemonefish decrease, so host species selection is likely to become simpler. In the Gulf of Aqaba in the northern Red Sea, one species of anemonefish (*Amphiprion bicinctus*) is associated with two species of anemones (*Heteractis crispa* and *Entacmaea quadricolor*). Breeding adults mainly used *E. quadricolor* with long tentacles with bubbles on the tips, while juveniles occupied *H. crispa* with thinner tentacles, and anemonefish in all stages selected *E. quadricolor* over *H. crispa* under experimental conditions (Huebner et al. 2012). Huebner et al. (2012) suggested that competitive exclusion promotes habitat segregation between life stages of a single species, and that host morphology is important for host selection as a safe shelter. As the number of species living in the area decreases, interspecific competition is reduced, but there are still competitive interactions for habitat within the same species.

16.3 HOST SELECTIVITY OF ANEMONEFISH IN THE RYUKYU ARCHIPELAGO

16.3.1 Plasticity of Host Selection

The Ryukyu Archipelago, located in southern Japan, is a hotspot of host anemone and anemonefish diversity in the subtropical region (Hayashi et al. 2021). This region is near the northern limit of the distribution of anemonefish, and consists of numerous islands, with large environmental variation within the archipelago. This makes the Ryukyu Archipelago suitable for studying habitat selection in anemonefish. In addition, human impacts on coastal areas vary from island to island, making this region also important for understanding ongoing anthropogenic impacts on the anemonefish–anemone symbioses. The Ryukyu Archipelago includes the Okinawa Islands centred around Okinawa-jima Island, and the Sakishima Islands located about 300 km to the south. These two regions have very different coastal topographies. The distance from coastline to reef edge is often shorter in the Okinawa Islands than in the Sakishima Islands (Kinjo et al. 2011). Around Okinawa-jima Island, the influence of land reclamation and drainage in coastal areas have caused habitat degradation for anemonefish and host anemones, especially in places near the shore (Fujii 2001; Reimer et al. 2015; Masucci and Reimer 2019). On the other hand, the coastlines of the Sakishima Islands are comparably much more well preserved (Okinawa Prefecture 2003, 2018) (Figure 16.1). Here, we investigated how anthropogenic changes in the coastal environment have affected the species composition of host anemones, and how this in turn affects the species composition and host selection of anemonefish between these two regions.

There are seven species of host anemone and six species of anemonefish present in both the Sakishima and Okinawa Islands (Figure 16.2) (Hayashi et al. 2018, 2019a, 2021). Most observed individuals (55–66%) of host anemones and anemonefish have been shown to inhabit shallow depths (0–2 m) adjacent to the coastline in the Sakishima Islands, while they only rarely (10–14%) inhabited such shallow locations in the Okinawa Islands (Hayashi et al. 2021). Accordingly, the species composition of the host anemones differs greatly between the two regions. In the Sakishima Islands the most frequent host anemone species was *E. quadricolor* and the second-most frequent species was *S. gigantea* (Figure 16.3a), while the Okinawa Islands were dominated by *H. crispa* and *E. quadricolor* (Figure 16.3a). Differences in the abundance of each species of host anemone affected the abundance of each species of anemonefish. The most frequently observed anemonefish species was *A. frenatus* and the second-most frequent species was *A. ocellaris* in the Sakishima Islands (Figure 16.3b), and *A. clarkii* and *A. frenatus* in the Okinawa Islands (Figure 16.3b).

As *S. gigantea* prefers to inhabit shallow reef lagoons (Hattori 2011; see Chapter 17), the numbers of *S. gigantea* and those of *A. ocellaris* using it as a host were greater in the Sakishima Islands, where many natural coasts remain. *Heteractis crispa* tends to prefer rocky areas with tidal currents that face the open ocean (Hirose 1985; Hattori 2011; see Chapter 17), and thus the numbers of *H. crispa* and those of *A. clarkii* and *A. perideraion* using this species as a host (Table 16.1) were most frequent on the reef edge in the Okinawa Islands.

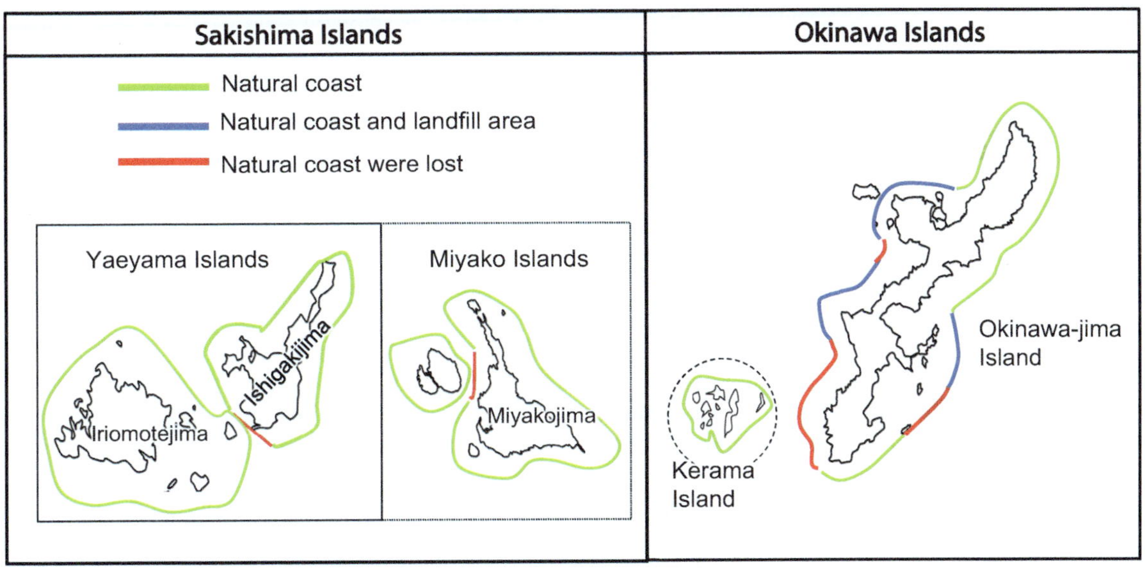

FIGURE 16.1 Natural coast distribution in Sakishima and Okinawa Islands, drawn based on data from Okinawa Prefecture (2003), in which coastlines were classified into three coastal types according to the degree of natural environment left in the coastal area.

FIGURE 16.2 Six species of anemonefish (a) and seven species of host anemone (b) observed in the Ryukyu Archipelago.

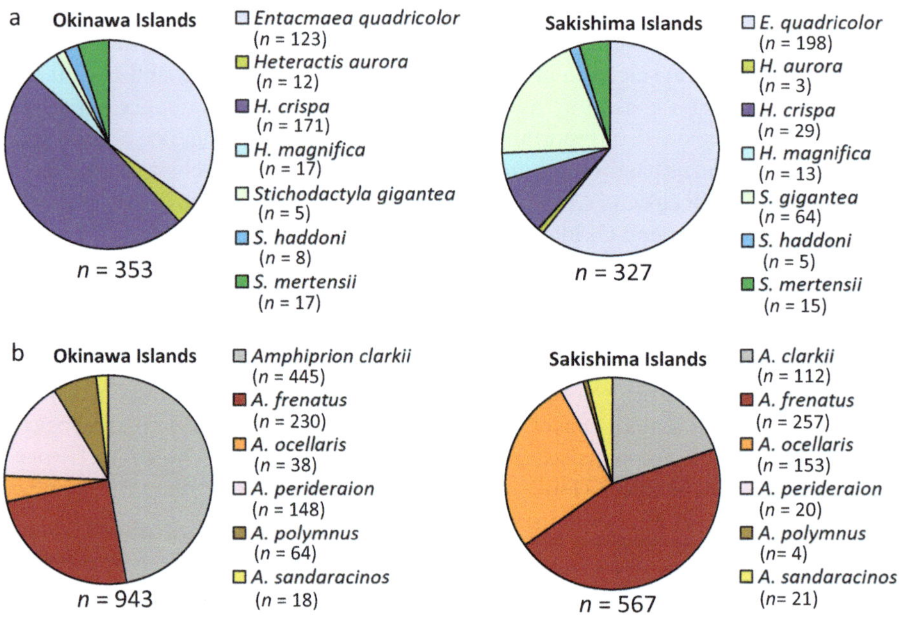

FIGURE 16.3 Differences in species composition of host anemones (a) and anemonefish (b) in the Sakishima and Okinawa Islands, based on data from Hayashi et al. (2021).

Species differences in abundances of host anemone and anemonefish affect the host selection of anemonefish. The population of *A. ocellaris* was higher on Sakishima Islands so that not only *S. gigantea* but also *H. magnifica* were all occupied by *A. ocellaris*. As a result, *A. perideraion*, which has a small population around the Sakishima Islands, was able to use only *H. crispa*. However, on the reefs of the Okinawa Islands, where *A. ocellaris* is scarce, *A. perideraion* used not only *H. crispa* but also *S. magnifica* (Table 16.1) (Hayashi et al. 2018, 2021). The combination of *A. perideraion* and *A. ocellaris* can be explained by the lottery hypothesis; a stochastic event with a chronological basis, and this may determine which species of anemonefish is present. In the Okinawa Islands, *A. ocellaris* and *A. perideraion* randomly used the same host species without any environmental segregation such as based on water depth or distance from shore (Hayashi et al. 2018) (Figure 16.4).

In addition, *A. clarkii* and *A. sandaracinos*, which use *S. mertensii*, often inhabit the same individual host anemone simultaneously, (=cohabitation) (Camp et al. 2016; Eliott and Mariscal 2001; Hayashi et al. 2018, 2021). The cohabitation ratio of *A. sandaracinos* and *A. clarkii* in the Okinawa Islands was two times higher than that of the Sakishima Islands (Hayashi et al. 2018, 2021). *Amphiprion clarkii*, which has a large population in the Okinawa Islands, tends to use not only *H. crispa*, which *A. perideraion* also uses, but also *S. mertensii* quite frequently, resulting in a high frequency of cohabitation with *A. sandaracinos* (Figure 16.3b) (Hayashi et al. 2018, 2021).

As in the Coral Triangle, in the Ryukyu Archipelago each species of anemonefish has a different host species and/or environment (niche differentiation), or two species can cohabit in a single anemone (cohabitation). However, when suitable habitat disappears due to land reclamation,

TABLE 16.1
Combinations of Host Anemone and Anemonefish Association

Anemonefish species	Anemone species									
	Cryptodendrum adhaesivum	*Entacmaea quadricolor*	*Heteractis aurora*	*H. crispa*	*H. magnifica*	*H. malu*	*Macrodactyla doreensis*	*Stichodactyla gigantea*	*S. haddoni*	*S. mertensii*
Amphiprion akallopisos										
A. akindinos										
A. allardi										
A. bicinctus										
A. barberi										
A. chagosensis										
A. chrysogaster										
A. chrysopterus										
A. clarkii		OS	OS	OS					O	OS
A. ephippium										
A. frenatus		OS								
A. fuscocaudatus										
A. latezonatus										
A. latifasciatus										
A. mccullochi										
A. melanops										
A. nigripes										
A. ocellaris					OS			OS		
A. omanensis										
A. pacificus										
A. percula										
A. perideraion				OS	O					
A. polymnus									OS	O
A. rubrocinctus										
A. sandaracinos										OS
A. sebae										
A. tricinctus										
Premnas biaculeatus										

Note: Shaded columns indicate combinations observed in previous studies conducted in all distributed areas (Burke da Silva and Nedosyko 2016; Hayashi et al. 2018, 2021). Among them, "O" means combinations observed in Okinawa Islands, and "S" in Sakishima Islands (Hayashi et al. 2018, 2021).

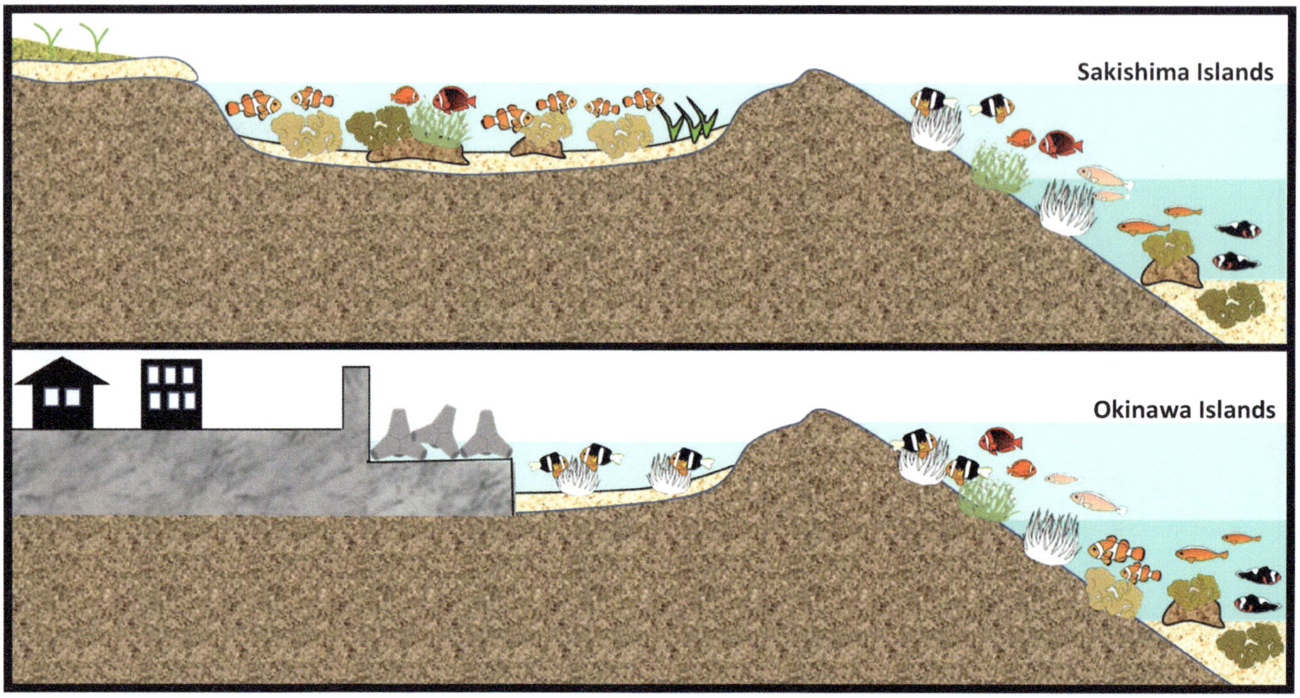

FIGURE 16.4 Schematic representation of the distribution of host anemones and anemonefish in the Sakishima and Okinawa Islands. In the Sakishima Islands, natural coastlines are preserved and *A. ocellaris* settled in *S. gigantea* along shallow coastlines. In Okinawa Islands, the coastline has been largely reclaimed and *A. ocellaris* and *S. gigantea* are rare, and most species settled on the reef edge.

populations of certain anemone and anemonefish species decreased, and niche differentiation could not be observed. In these areas, as in the lottery hypothesis, there are cases in which the anemonefish that arrives first to a host anemone becomes established there (Hayashi et al. 2018). The loss of coastal habitats and differences in distances from the coastline to reef edge may have influenced species composition of host anemones, and then the density of each species of anemonefish, and furthermore affected host selection by anemonefish.

16.3.2 Human-Induced Environmental Degradation

In recent years, several problems have arisen that threaten the survival of anemonefish and host anemones, such as land reclamation, rising seawater temperatures, and an increasing aquarium trade (e.g., Hattori 2002; Hayashi et al. 2021; Hayashi and Reimer 2020; Hobbs et al. 2013; Shuman et al. 2005). As mentioned earlier, there is a possibility that land reclamation along the coastlines of the Okinawa Islands has already affected species compositions and population densities in both host anemones and anemonefish. In addition, since anemonefish and host anemones are mainly distributed in the shallow depths of coral reefs where water temperature drastically changes, they are also seriously affected by rising temperatures. Host anemones are symbiotic with endosymbiotic Symbiodiniaeceae zooxanthellae, and they may bleach or die due to abnormally high seawater temperatures, just as zooxanthellate scleractinian corals and other host animals do (LaJeunesse et al. 2018; Muscatine and Porter 1977). It has been reported that bleaching reduces the size of host anemones (Hattori 2002; Hobbs et al. 2013; Saenz-Agudelo et al. 2011). In Sesoko-jima, the Okinawa Islands, there were many *H. crispa* reported before the large bleaching event in 1998 (Hattori 2002; Hirose, 1985, 1995). *Amphiprion perideraion* inhabited only large *H. crispa* and *A. clarkii* inhabited small and large-sized *H. crispa* before the 1998 bleaching (Hattori 2002), but after bleaching only a low number of small *H. crispa* remained and almost all *A. perideraion* had disappeared (Hattori 2002). Furthermore, an experimental study has shown that at least five species of anemonefish select unbleached hosts over bleached hosts, while the specialist species *A. latezonatus* has strong selectivity for *H. crispa*, regardless of whether it was bleached or not (Scott and Dixson 2016). The risk of local extinction of such specialist species is thus much higher than that of generalist species. In some cases, host anemones do not disappear during bleaching events, but if high water temperatures continue for longer periods of time or across consecutive years, the duration of recovery from bleaching becomes longer and the probability of disappearance increases (Hayashi and Reimer 2020). If the rise in seawater temperatures is not halted in the future, populations of both host anemones and anemonefish may decrease drastically (e.g., Hobbs et al. 2013; Hayashi and Reimer 2020).

The population study conducted at Miyagi Beach, Okinawa-jima Island, from 2014 to 2017, revealed that most immigrants were immature anemonefish, of which more than half disappeared within a month (Hayashi et al. 2019a).

Immigration of mature fish and dissolution of breeding pairs were rare. Migration (settled breeding pairs moving to different host anemones) and replacement (settled breeding pairs chased out from host anemones by intruders) were never observed during the study period (Hayashi et al. 2019a). However, in the 1990s, before bleaching occurred, the major factor implicated in dissolution at Sesoko-jima, the Okinawa Islands was typhoons, and the second factor was intraspecific interactions (Hirose 1995). Nelson et al. (1998) reported that inter-host migration of *A. ocellaris* decreased with the distance to the nearest host anemone. Mate replacement was often observed at a high-density site while it was never observed at a low-density site around Singapore's Southern Islands (Nelson et al. 1998). When the population drops below a certain level due to host bleaching and strong collecting pressure, etc., it will take time for the population to recover, even if the environment improves.

16.4 SUMMARY AND FUTURE TASKS

It has been revealed from this overview that, on an evolutionary scale, anemonefish divide resource utilization by altering the host species or environment for each species. However, many field studies conducted on a local scale have shown that behavioral competition has an important role in host selection. These results indicate that although each species of anemonefish has its own host species and environmental preferences, on a local scale there is competition for settling on more valuable host anemones. In fact, in the Ryukyu Archipelago host selection is plastic in response to the density of each species of anemonefish due to environmental changes (Hayashi et al. 2018, 2021).

However, we do not fully understand the value of host anemones for each species of anemonefish. Anemones with intermediate toxicity levels have been shown to have the highest numbers of anemonefish associates (Nedosyko et al. 2014). However, the extent to which these host characteristics alter the survival and reproductive success of anemonefish has not been fully investigated in the field. Different species of anemonefish have evolved differently from morphological, behavioral, and physiological points of view. These differences allow for the division of resource use and work to minimize costly competitive interactions. However, how species-specific traits are adapted to the host species and environment has only been studied in a piecemeal manner until now.

In addition, we do not fully understand how water temperatures, illegal collection, and landfill may affect the host selection and environmental selection of anemonefish. Human-induced changes in the environment may be beyond the limits of anemonefish and host anemone's tolerances. Determining how anemonefish species adapt to host venoms, water temperatures, and other external stresses is necessary to understand host and environmental selection, understand the distribution range of each species and possible limiting factors, as well as conserve both anemones and anemonefish from human-induced environmental degradation.

REFERENCES

Balamurugan, J., T. A. Kumar, R. Kannan, and H. D. Pradeep. 2014. Acclimation behaviour and bio-chemical changes during anemonefish (*Amphiprion sebae*) and sea anemone (*Stichodactyla haddoni*) symbiosis. *Symbiosis* 64: 127–138.

Burke da Silva, K., and A. Nedosyko. 2016. Sea anemones and anemonefish: A match made in heaven. In *The Cnidaria, past, present and future*, eds S. Goffredo and Z. Dubinsky, 425–438. Cham: Springer.

Camp, E. F., J. P. A. Hobbs, M. De Brauwer, A. J. Dumbrell, and D. J. Smith. 2016. Cohabitation promotes high diversity of clownfishes in the Coral Triangle. *Proceedings of the Royal Society B: Biological Sciences* 283: 20160277.

Collingwood, C. 1868. Note on the existence of gigantic sea anemones in the China Sea, containing within them quasiparasitic fish. *Annals and Magazine of Natural History* 4: 31–33.

Dunn, D. F. 1981. The clownfish sea anemones: Stichodactylidae (Coelenterata: Actiniaria) and other sea anemones symbiotic with pomacentrid fishes. *Transactions of the American Philosophical Society* 71: 1–115.

Elliott, J. K., S. C. Lougheed, B. Bateman, L. K. McPhee, and P. T. Boag. 1999. Molecular phylogenetic evidence for the evolution of specialization in anemonefishes. *Proceedings of the Royal Society of London. Series B: Biological Sciences* 266: 677–685.

Elliott, J. K., and R. N. Mariscal. 2001. Coexistence of nine anemonefish species: Differential host and habitat utilization, size and recruitment. *Marine Biology* 138: 23–36.

Fautin, D. G. 1986. Why do anemonefishes inhabit only some host actinians? *Environmental Biology of Fishes* 15: 171–180.

Fautin, D. G. 1991. The anemonefish symbiosis: What is known and what is not. Symbiosis 10: 23–46.

Fautin, D. G., and G. R. Allen. 1992. *Field guide to anemonefishes and their host sea anemones.* Perth, WA: Western Australian Museum.

Fautin, D. G., and G. R. Allen. 1997. *Anemonefishes and their host sea anemones.* 2nd ed. Perth, WA: Western Australian Museum.

Fujii, H. 2001. The present status of tidal flats on the Okinawa Island. *Japanese Journal of Benthology* 56: 29–31.

Hattori, A. 2002. Small and large anemonefishes can coexist using the same patchy resources on a coral reef, before habitat destruction. *Journal of Animal Ecology* 71: 824–831.

Hattori, A. 2011. Diversity of symbiotic relationship between host sea anemone and anemonefish: An ecological review of their distributions and combinations. *Japanese Coral Reef Society* 13: 1–27 (in Japanese).

Hayashi, K., and J. D. Reimer. 2020. Five-year study on the bleaching of anemonefish-hosting anemones (Cnidaria: Anthozoa: Actiniaria) in subtropical Okinawajima Island. *Regional Studies in Marine Science* 35: 101240.

Hayashi, K., K. Tachihara, and J. D. Reimer. 2018. Patterns of coexistence of six anemonefish species around subtropical Okinawa-jima Island, Japan. *Coral Reefs* 37: 1027–1038.

Hayashi, K., K. Tachihara, and J. D. Reimer. 2019a. Low density populations of anemonefish with low replenishment rates on a reef edge with anthropogenic impacts. *Environmental Biology of Fishes* 102: 41–54.

Hayashi, K., K. Tachihara, and J. D. Reimer. 2019b. Species and sexual differences in human oriented behavior of anemonefish at Okinawa Island, Japan. *Marine Ecology Progress Series* 616: 219–224.

Hayashi, K., K. Tachihara, and J. D. Reimer. 2021. Loss of natural coastline influences species diversity of anemonefish and host anemones in the Ryukyu Archipelago. *Aquatic Conservation: Marine and Freshwater Ecosystems* 31: 15–27.

Hirose, Y. 1985. Habitat, distribution and abundance of coral reef sea-anemones (Actiniidae and Stichodactylidae) in Sesoko Island, Okinawa, with notes of expansion and contraction behavior. *Galaxea* 4: 113–127.

Hirose, Y. 1995. Patterns of pair formation in protandrous anemonefishes, *Amphiprion clarkii, A. frenatus* and *A. perideraion*, on coral reefs of Okinawa, Japan. *Environmental Biology of Fishes* 43: 153–161.

Hobbs, J. P. A., A. J. Frisch, B. M. Ford, M. Thums, P. Saenz-Agudelo, K. A. Furby, M. L. Berumen. 2013. Taxonomic, spatial and temporal patterns of bleaching in anemones inhabited by anemonefishes. *PLoS One* 8: e70966.

Huebner, L. K., B. Dailey, B. M. Titus, M. Khalaf, and N. E. Chadwick. 2012. Host preference and habitat segregation among Red Sea anemonefish: Effects of sea anemone traits and fish life stages. *Marine Ecology Progress Series* 464: 1–15.

Kinjo, K., K. Arita, K. Nadaoka, and K. Nakasone. 2011. Classification of fringing coral reefs based on geomorphological characters and flow circulation patterns. *Journal of Japan Society of Civil Engineers, Series B2, Coastal Engineering* 67: 1336–1340 (in Japanese).

LaJeunesse, T. C., J. E. Parkinson, P. W. Gabrielson, H. J. Jeong, J. D. Reimer, C. R. Voolstra, and S. R. Santos. 2018. Systematic revision of Symbiodiniaceae highlights the antiquity and diversity of coral endosymbionts. *Current Biology* 28: 2570–2580.

Litsios, G., A. Kostikova, and N. Salamin. 2014. Host specialist clownfishes are environmental niche generalists. *Proceedings of the Royal Society B: Biological Sciences* 281: 20133220.

Mariscal, R. N. 1970. The nature of the symbiosis between Indo-Pacific anemone fishes and sea anemones. *Marine Biology* 6: 58–65.

Masucci, G. D., and J. D. Reimer 2019. The expanding wall and the shrinking beach: Loss of natural coastline in Okinawa Island, *Japan. Peer J.* https://doi.org/10.7717/peerj.7520.

Mebs, D. 2009. Chemical biology of the mutualistic relationships of sea anemones with fish and crustaceans. *Toxicon* 54: 1071–1074.

Moyer, J. T., and C. E. Sawyers. 1973. Territorial behavior of the anemonefish *Amphiprion xanthurus* with notes on the life history. *Japanese Journal of Ichthyology* 20: 85–93.

Muscatine, L., and J. W. Porter. 1977. Reef corals: Mutualistic symbioses adapted to nutrient-poor environments. *Bioscience* 27: 454–460.

Nedosyko, A. M., J. E. Young, J. W. Edwards, and K. Burke da Silva. 2014. Searching for a toxic key to unlock the mystery of anemonefish and anemone symbiosis. *PLoS One* 9: e98449.

Nelson, J. S., L. M. Chou, V. P. E. Phang. 1998. Inter-habitat migration of the anemonefish *Amphiprion ocellaris. Asian Journal of Tropical Biology* 3: 19–31.

Nguyen, H. T. T., A. N. T. Tran, L. T. L. Ha, D. N. Ngo, B. T. Dang, and A. J. Geffen. 2019. Host choice and fitness of anemonefish *Amphiprion ocellaris* (Perciformes: Pomacentridae) living with host anemones (Anthozoa: Actiniaria) in captive conditions. *Journal of Fish Biology* 94(6): 937–947.

Okinawa Prefecture. 2003. Ryukyu Shotou Engan Kaigann Hozen Kihon Keikaku (Ryukyu Islands coast basic plan for coastal conservation). https://www.pref.okinawa.jp/site/doboku/kaibo/kaigan/documents/honpen.pdf (in Japanese, accessed January 11, 2022).

Okinawa Prefecture. 2018. Ritou Kankei Sirhou (Materials related to remote islands). https://www.pref.okinawa.jp/site/kikaku/chiikirito/ritoshinko/documents/h31itigatuban0mokuzinado.pdf (in Japanese, accessed January 11, 2022).

Ollerton, J., D. McCollin, D. G. Fautin, and G. R. Allen. 2007. Finding NEMO: Nestedness engendered by mutualistic organization in anemonefish and their hosts. *Proceedings of the Royal Society B: Biological Sciences* 274: 591–598.

Porat, D., and N. E. Chadwick-Furman. 2004. Effects of anemonefish on giant sea anemones: Expansion behavior, growth, and survival. In *Coelenterate Biology 2003*, eds D. G. Fautin, J. A. Westfall, P. Cartwright, M. Daly and C. R. Wyttenbach, 513–520. Dordrecht: Springer.

Reimer, J. D., S. Y. Yang, K. N. White, R. Asami, K. Fujita, C. Hongo, S. Ito, et al. 2015. Effects of causeway construction on environment and biota of subtropical tidal flats in Okinawa, Japan. *Marine Pollution Bulletin* 94: 135–167.

Ricciardi, F., M. Boyer, and J. Ollerton. 2010. Assemblage and interaction structure of the anemonefish-anemone mutualism across the Manado region of Sulawesi, Indonesia. *Environmental Biology of Fishes* 87: 333–347.

Saenz-Agudelo, P., G. P. Jones, S. R. Thorrold, and S. Planes. 2011. Detrimental effects of host anemone bleaching on anemonefish populations. *Coral Reefs* 30: 497–506.

Santini, S. and G. Polacco. 2006. Finding Nemo: molecular phylogeny and evolution of the unusual life style of anemonefish. *Gene* 385: 19–27.

Scott, A., and D. L. Dixson. 2016. Reef fishes can recognize bleached habitat during settlement: Sea anemone bleaching alters anemonefish host selection. *Proceedings of the Royal Society B: Biological Sciences* 283: 20152694.

Shuman, C. S., G. Hodgson, and R. F. Ambrose. 2005. Population impacts of collecting sea anemone and anemonefish for the marine aquarium trade in the Philippines. *Coral Reefs* 24: 564–573.

17 3D Analysis of Coral Reef Informs Anemonefish Habitat

Akihisa Hattori

CONTENTS

17.1 INTRODUCTION

Coral reef fishes are among the richest animal communities in the world. Complex seascape structures, which are closely related to coral reef morphology, enhance species diversity (Pittman and Olds 2015). High species diversity and complex habitat structures complicate our understanding of ecological phenomena in situ, including habitat selection, interspecific competition, and multispecies coexistence (Chapter 16 and Chapter 18). Can we simplify this to some degree?

Anemonefishes comprise 28 species and symbiotically associate with at least ten host anemone species in coral reef regions (Dunn 1981; Fautin and Allen 1997; Ollerton et al. 2007). Among ten host species, six (*Entacmaea quadricolor, Heteractis crispa, H. magnifica, Stichodactyla gigantea, S. mertensii*, and *Cryptodendrum adhaesivum*) inhabit hard substrates, while the others (*Macrodactyla doreensis, H. malu, H. aurora, S. haddoni*) inhabit sandy bottoms. Since anemonefish spawn demersal eggs, adults inhabit anemones attached to hard substrates, except for some: *Amphiprion polymnus* inhabiting *S. haddoni* in sandy bottoms bring empty shells of bivalves as spawning sites nearby their anemones (Moyer and Steen 1979); *C. adhaesivum* inhabiting reefs are used by only juvenile *Amphiprion clarkii* because the former have very short tentacles and no space under the edge of their oral discs, which provide insufficient refuge for most all anemonefishes (Fautin and Allen 1997). Thus, *E. quadricolor, H. crispa, H. magnifica, S. gigantea*, and *S. mertensii* are major targets of interspecific competition by anemonefishes. In addition, anemonefishes include generalists that use six or more host species (e.g., *Amphiprion clarkii, A. akindynos*, and *A. chrysopterus*), specialists that use several (e.g., *A. ocellaris, A. perideraion*, and *A. polymnus*), and extreme specialists that use only one host (e.g., *A. frenatus, A. mccullochi*, and *A. biaculeatus*, see Fautin and Allen 1997; Litsios et al. 2012; Nguyen et al. 2020).

While interspecific competition effects on host selection and cohabitation by anemonefishes are often unclear (see Chapter 16 and Chapter 18), distribution patterns of anemonefishes and host anemones in relation to geomorphic zones (see below) can clarify habitat use strategies. This chapter describes distribution patterns of several species of host anemone and anemonefishes in 3D structure coral reefs, especially two fringing reefs: namely the small reef (Sesoko Reef) of Sesoko Island, among the Okinawa Islands, and the large reef (Shiraho Reef) of Ishigaki Island, among the Sakishima Islands. Both are located in Okinawa, southern Japan. Habitat use strategies of anemonefishes are discussed in relation to geomorphic zones.

17.2 GEOMORPHIC ZONE AND MAPPING ANEMONE ON AERIAL IMAGE

Reef-building corals gradually grow vertically and horizontally to form a 3D reef facilitated by endosymbiotic Symbiodiniaceae zooxanthellae. Accordingly, fringing reefs develop along coastlines, where seawater is generally clear and warm (20 to 30°C), with geomorphic zonation including an outer reef slope (facing deep open water), reef crest (transitional area between the upper reef slope and reef flat), reef flat (wave-sweeping and shallow), backreef (calm and shallow with a sandy bottom), and subtidal nearshore zones (shallow sandy bottoms, Figure 17.1). In small fringing reefs, outer sandy sea bottoms may predominate. Although the terminology is not standardized (Blanchon 2011; Bellwood et al. 2018), aerial images show geomorphic zones (Figures 17.2a–b and 17.3a–b) with unique characteristics regarding water depth (deep/shallow), wave exposure

DOI: 10.1201/9781003125365-21

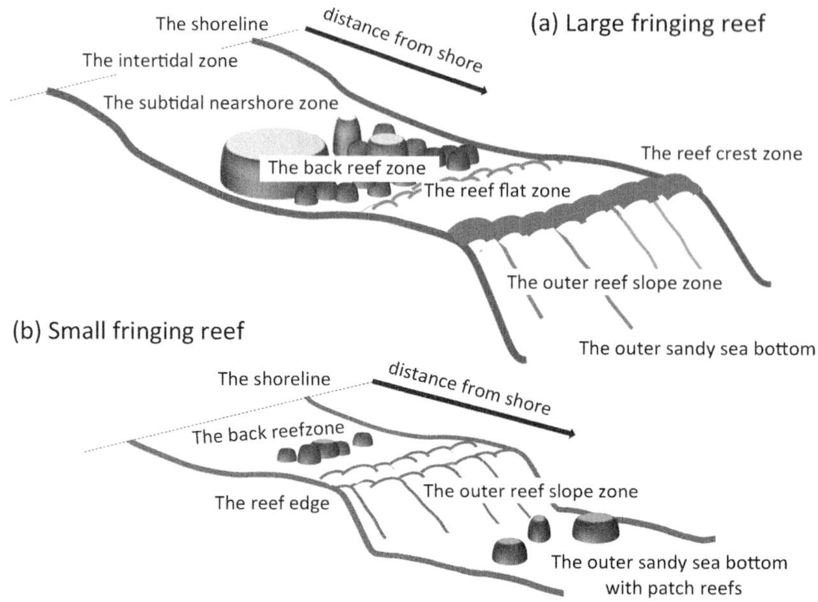

FIGURE 17.1 3D structure of fringing reefs and geomorphic zonation. a) A large fringing reef. b) A small fringing reef. See also Figure 16.4.

(high/low), water condition (fast/calm, open/sheltered), water temperature (stable/unstable), substrate (rock/sand), and crevices and holes (few/many). In fringing reefs, depth information in *centimetre* scales is critical because all but the outer reef zones are usually shallow (< 3 m deep).

Aerial images of coral reefs show seascape structures including sandy bottoms, rocky reefs, seagrass beds, and relative water depth (Figures 17.2 and 17.3). Furthermore, aerial images provide concise maps to record location information with complex coral reef structures (Hattori and Kobayashi 2009). By locating individual anemones in situ on concise maps, we can record their growth and survival, as they rarely relocate more than 0.5 m (Hattori personal observations). Maps can be subsequently updated to improve quality and to reflect changes in seascapes. Google Earth may provide high-resolution satellite images (Figure 17.2b), and more high-resolution aerial images can be obtained by low-flying drones (Figure 17.2c).

Since anemones used by anemonefish largely depend on photosynthate products produced by endosymbionts, as do reef-building corals, a larger tentacle-crown surface area facilitates capture of sunlight and prey, but the energy cost of maintaining a large body size increases in proportion to its volume (Dunn 1981; Sebens, 1982). Accordingly, anemone sizes depend on their habitat, and thus, large individuals are basically found in habitats suitable for large animals (Sebens 1982; Steen 1988). Thus, large anemones (the largest size of tentacle-crown surface area of an individual is regarded as the anemone size after two or more measurements, see Hattori 1991) are suitable hosts for anemonefishes (Fautin and Allen 1997). Information on geomorphic zones and anemone size should be incorporated when habitat use strategies of each anemonefish are discussed.

17.3 DISTRIBUTION OF ANEMONE AND ANEMONEFISHES ON A 3D REEF STRUCTURE

17.3.1 Distribution of *Heteractis Crispa* on a Small Fringing Reef

Heteractis crispa are widely distributed (Ollerton et al. 2007), and are usually abundant in the backreef, outer reef slope, and outer sandy bottom (with patch reefs) up to 12 m deep (Dunn 1981; Hayashi et al. 2021). On the small fringing reef, Sesoko Reef, Hattori (1995) plotted all anemones observed in 1988 and 1989 onto a map and measured their respective water depth while snorkelling. Their locations were replotted in 1999, 2000, and 2009 onto a high-resolution aerial image (see Figure 17.2). Of the 76 observed in 1988 (71 had been recorded in 1983 by Hirose 1985), all had perished by 2009. Nine anemones (including two of those recorded in 1983) had been observed in 2000 (Hattori 2002) of which all inhabited the outer reef slope.

In Sesoko Reef, large anemones (> 1,000 cm^2) tended to reside in the outer reef slope and in patchy reefs in the outer sandy sea bottom (beyond the reef edge zone, from 80 cm to 4 m deep at the lowest tide), while small anemones (< 500 cm^2) mainly inhabited shallow reefs (< 80 cm deep) in the reef edge, reef flat, or backreef zones. As a small fringing reef, the reef crest and subtidal nearshore zones are unclear in Sesoko Reef (Figure 17.2). Their size was positively correlated with water depth and negatively correlated with growth from 1988 to 1989 (Hattori 2006), suggesting their suitable habitats were deeper sites (> 80 cm deep). Although the area of the reef edge zone is less than those of the reef flat and backreef zones (Figure 17.2), newly appeared anemones were abundant in this reef edge zone.

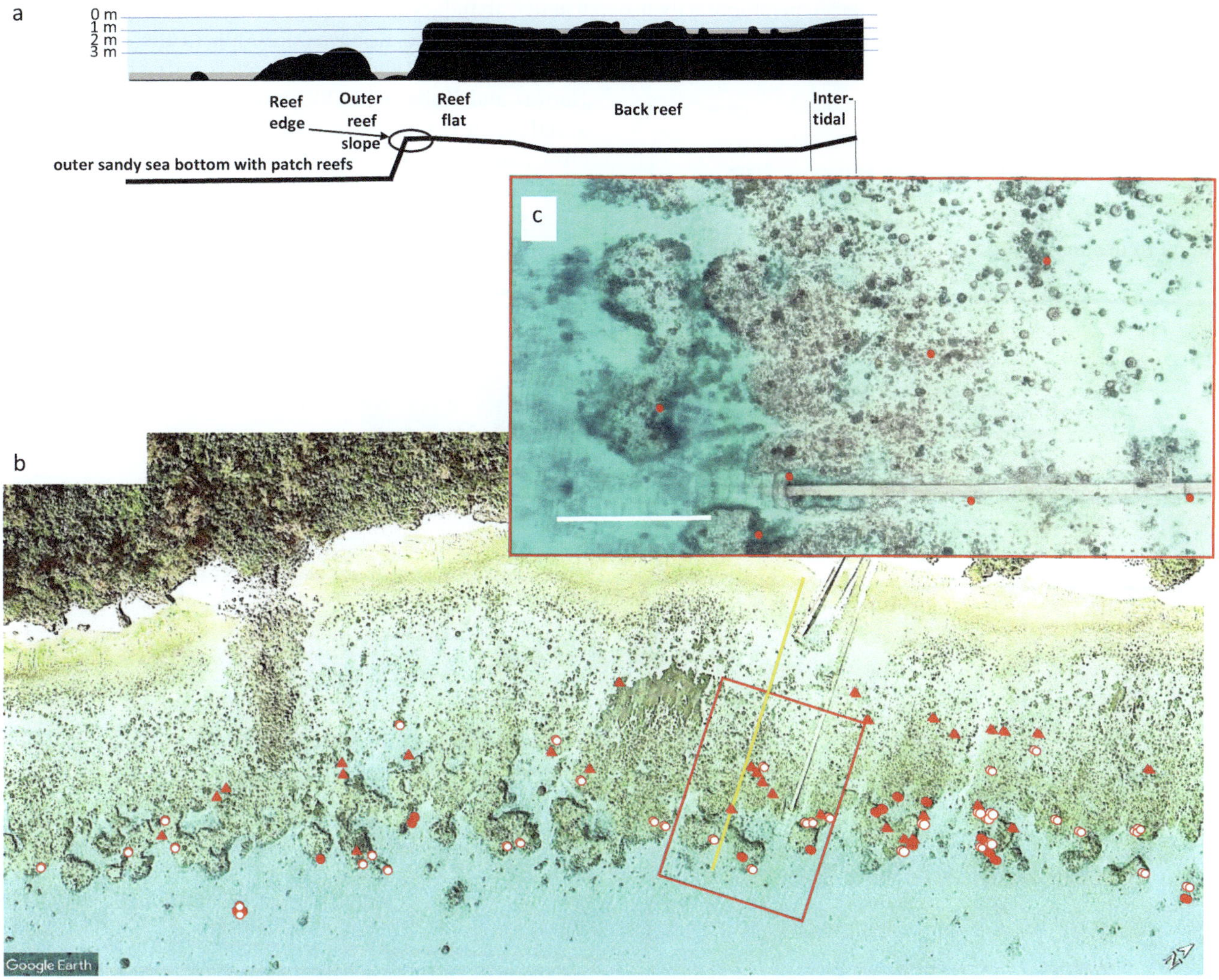

FIGURE 17.2 Geomorphic zonation of a small fringing reef (Sesoko Reef, Sesoko Island, Okinawa, Japan: 26°38'07.47" N, 127°51'56.82" E) and distributions of host anemones. a) Schematic view of the cross-section. b) Distributions of *Heteractis crispa* in 1988 shown in an aerial image from Google Earth (www.google.co.jp/intl/ja/earth/). Solid circles, open circles, and triangles indicate hosts with only *Amphiprion perideraion*, with *A. perideraion* and *A. clarkii*, and with only *A. clarkii*, respectively. Yellow line (120 m) indicates the cross-section in Figure 17.2a. Red square in Figure 17.2b indicates the range of Figure 17.2c. c) Distributions of *H. crispa* inhabited by *A. clarkii* in 2009 shown in an aerial mage taken in 2017 by a low-flying drone (DJI Phantom 4 pro plus). Red circles indicate *H. crispa*. Many massive *Porites* corals are recognizable. White bar indicates 20 m.

Their high density was thought to be attributable to the high settlement rate of larval *H. crispa* because there were almost no large anemones in the reef edge zone (Hattori 2006). Drifting larvae of anemones do not have high mobility (like zooplankton) but juvenile anemones after settlement can move to some degree (presumable in cm scales) to select better microhabitats (Scott and Harrison 2008). Abundant hard and soft corals among a topography of exposed shallow reef edges slow local currents and allow drifting larvae to settle (Loya et al. 2001). Accordingly, they might have been near the surface (Figure 17.4) and settled in the reef edge zone. *Heteractis crispa* that settled in the outer reef zones would experience high survival and growth if inhabited by anemonefish. In contrast, anemones settled in the shallow habitats with refuges are not guaranteed survival because of strong disturbances at the reef edge (waves induced by typhoons) and/or high-water temperatures in the shallow and calm backreef (causing loss of endosymbiotic zooxanthellae from anemones = bleaching, Saenz-Agudelo et al. 2011).

In the Ryukyu Islands, Okinawa, Japan, generalist *Amphiprion clarkii* and specialist *A. perideraion* use *H. crispa* (Hattori 1995, 2002; Hayashi et al. 2021). Since *A. clarkii* has high mobility, adult pairs often use two or more hosts in close proximity, whereas small juveniles use only one host until they move as adults to take breeding posts in large hosts (Hattori 1994; see Chapter 16). However, plots of *H. crispa* with inhabiting anemonefish clearly show that *A. perideraion* in the backreef zone are all located not near the shore but near the reef edge (Figure 17.2b). As generalists are better migrants (Hattori 2002), they can temporarily use unsuitable habitats, because they can change hosts during growing up (Hattori 1994).

17.3.2 Distribution of *Stichodactyla gigantea* on a Large Fringing Reef

Stichodactyla gigantea are often found on the outer reef slope to around 5 m deep (Dunn 1981), although they are usually abundant in the subtidal nearshore zone (Mitchell 2005; Hattori and Kobayashi 2009). In the subtidal nearshore zone of Shiraho Reef (a large fringing reef), in 2003, Hattori and Kobayashi (2009) plotted *S. gigantea* onto an aerial image (Figure 17.3c–d). They measured the tentacle-crown surface area, water depth, and distance from dense seagrass beds, and examined their disappearance over three years. In this zone, *S. gigantea* was abundant at the sandy bottom, residing at 20 to 60 cm depths (at the lowest

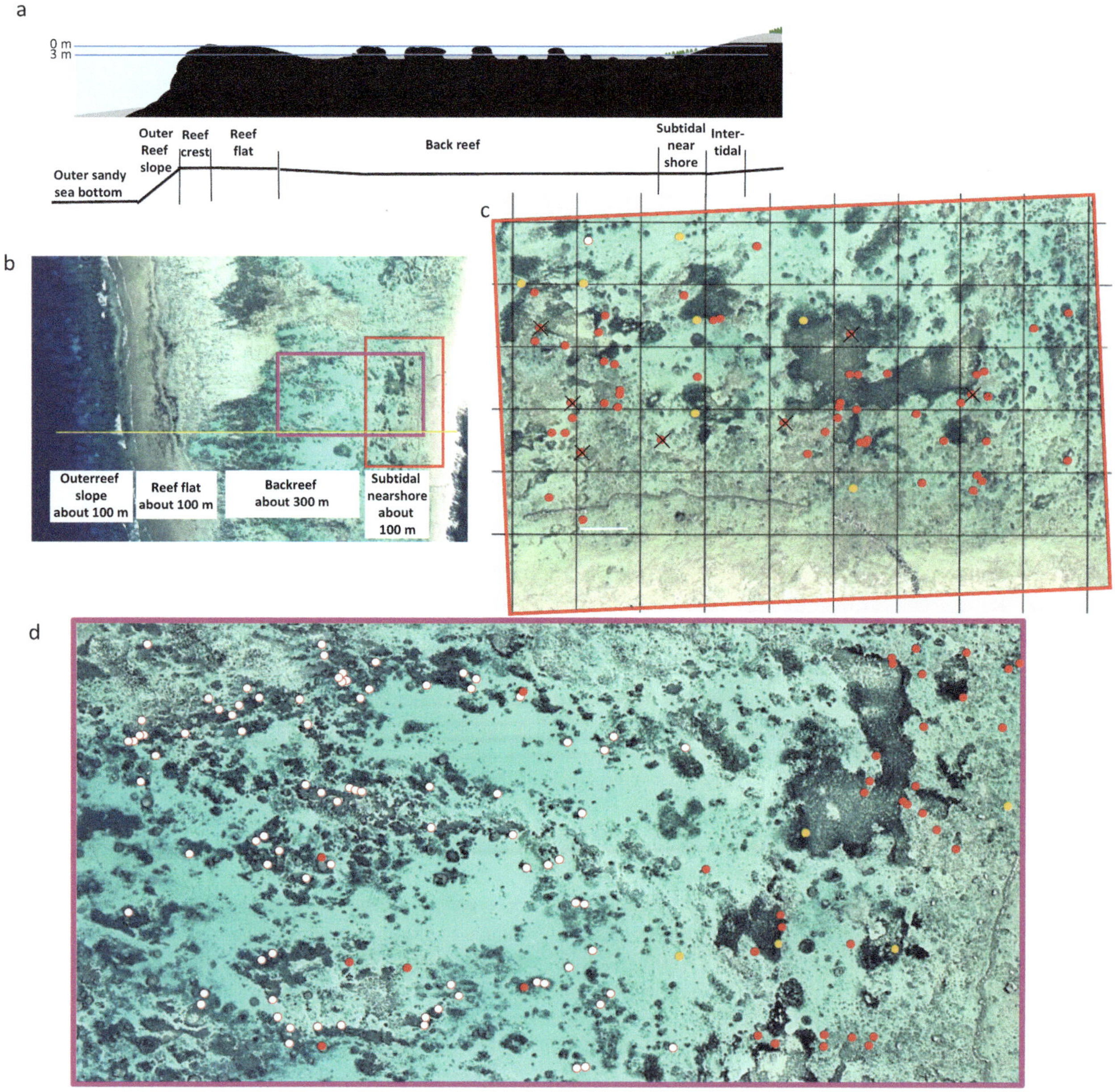

FIGURE 17.3 Geomorphic zonation of a large fringing reef (Shiraho Reef, Ishigaki Island, Okinawa, Japan: 24°22'19.22" N, 124°15'13.40" E) and distributions of host sea anemones. a) Schematic view of the cross-section and geomorphic zones. b) High-resolution aerial image of the reef taken on 20 September 2006 by PASCO Co. Japan (Ishigaki C-19-1608, Geospatial Information Authority of Japan, 2006, 2,540 dpi, altitude 1,500 m). Yellow line (890 m) indicates the cross-section in Figure 17.3a. Red square and pink square in Figure 17.3b indicate the range of Figure 17.3c and Figure 17.3d, respectively. c) Distributions of *Stichodactyla gigantea* and *S. haddoni* in 2003 shown in the aerial image. X indicates that anemone disappeared by 2006. Red circles and orange circles indicate *S. gigantea* with *A. ocellaris* and *S. haddoni* with *A. clarkii*, respectively. White bar indicates 20 m. White circle indicates *Entacmaea quadricolor*. d) Distributions of *E. quadricolor* (clonal assemblage), *S. gigantea*, and *S. haddoni* in 2002 shown in a high-resolution aerial image taken in 2007.

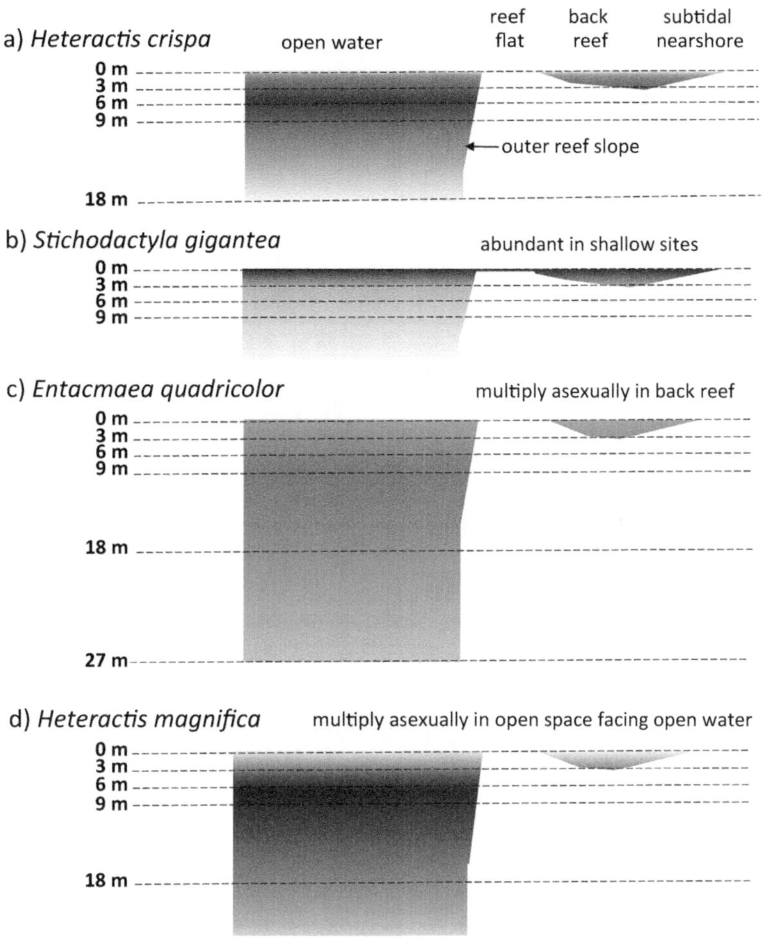

FIGURE 17.4 Hypothetical distributions of drifting larvae of four host anemone species outside and inside of a large fringing reef. Gradation indicates the relative density of larvae in the water column. Higher settlements of anemones are assumed to be closely related to higher density of larvae.

tide), about 0 to 6 m from dense seagrass beds, but were less abundant in small reefs (i.e., emergent rocks on sandy bottoms). Furthermore, there were almost no individuals on sandy bottoms deeper than 60 cm or within dense seagrass beds (< 20 cm deep). Although individuals inhabiting small reefs were few, they were larger, and their disappearance rate over three years was lower than those inhabiting sandy habitats (Hattori and Kobayashi 2009). Thus, small reefs are more suitable for *S. gigantea*. This distribution pattern of *S. gigantea* can be explained by two things: (1) higher settlement rates in the edge zone (20 to 60 cm deep) from dense seagrass beds (total area of this zone is larger than that of small reefs), and (2) low disappearance rate in small reefs (sandy bottoms are unstable). Drifting larvae of this anemone probably settle on hard substrates when the tidal current slows or stops at the lowest tide. The drifting larvae near the water surface likely explain the distribution pattern of *S. gigantea*, which are found in shallow sites in the nearshore, backreef, and outer reef slope zones (Figure 17.4).

This anemone is mainly used by *Amphiprion ocellaris* (Hayashi et al. 2021). However, this specialist mainly uses *H. magnifica* in relatively deep places (i.e., in the outer reef slope and outer sandy sea bottoms with patch reefs) without competitors (Ricciardi et al. 2010). Reproductive ecology of *A. ocellaris* has been studied among *S. gigantea* in the subtidal nearshore zone, where sandy bottoms and seagrass beds predominate (Mitchell 2005; Hattori 2012). *Stichodactyla gigantea* require rocky substrates beneath sandy bottoms for adhesion, and *A. ocellaris* rely on emergent hard substrates as spawning sites, which are scarce in this zone (Hattori and Kobayashi 2009).

In the subtidal nearshore zone of Shiraho Reef, a data logger (Onset Hobo StowAway Tidbit Temperature Data Logger) recorded temperatures of approximately 13°C on 18 and 22 December and 38°C on 4 July and 1 August (Figure 17.5a; Hattori 2011), and indicated a fluctuation of water temperature within 24 h approximately from 0°C on 10, 16, 17 July, and 5 August to 11°C on 4 July and 13 May (Figure 17.5b). As *A. ocellaris* and *A. clarkii* inhabit the subtidal nearshore zone, they tolerate a wide range of temperatures. Low water temperatures are critical in limiting the distribution ranges of tropical anemonefish, leading to local adaptation to temperate waters (Clark et al. 2021), where host anemones for specialists *A. ocellaris* are not distributed (Fautin and Allen 1997).

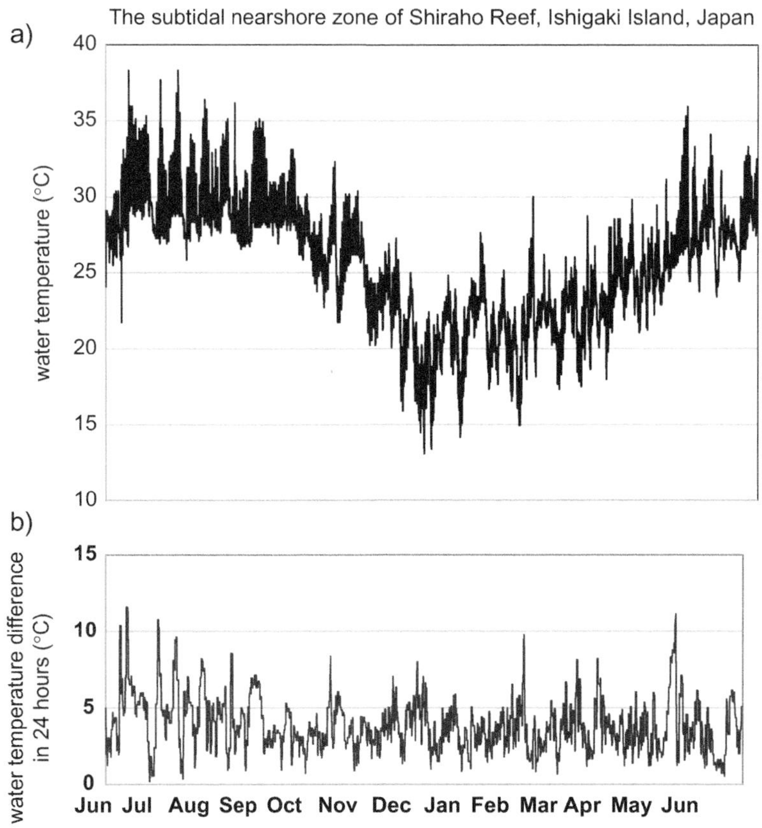

FIGURE 17.5 Change of water temperatures in the subtidal nearshore zone (not intertidal pool; at a depth of about 20 cm approximately 30 m from the shoreline at the lowest tide) of Shiraho Reef in the 12-month period 21 June 2005 to 20 June 2006. a) Water temperatures in every 30 min. b) Ranges of water temperatures (max–min in 24 h) for every 30 min. Water from intertidal pools may influence the water temperature at the subtidal nearshore zone.

17.3.3 Distribution of *Entacmaea Quadricolor* on a Large Fringing Reef

Entacmaea quadricolor are widely distributed and can reproduce both sexually and asexually (Dunn 1981). They tend to form colonies, or clonal assemblages, comprising small individuals in shallow sites or large individuals in deep sites (Dunn 1981; Scott and Harrison 2007). They are not rare in the outer reef slope down to around 20 m deep and can even be distributed to 60 m deep (Bridge et al. 2012). Since larger individuals tend to be found in habitats suitable for large animals (Sebens 1982; Steen 1988), the shallow sites may be unsuitable for *E. quadricolor*.

At Shiraho Reef, Hattori and Kobayashi (2007) and Hattori (2017) plotted *E. quadricolor* onto aerial images in 2001 and 2014 (Figure 17.3d). After the infamous bleaching occurred in 1998, bleaching of *E. quadricolor* was found again in 2007 (Harii et al. 2014). This anemone was most abundant at sites from 10 to 60 cm deep and less abundant at sites deeper than 120 cm (Hattori and Kobayashi 2007). No individuals were found in the intertidal zone or the subtidal nearshore zone. The number of anemones increased from 821 in 2001 to 864 in 2014 but the number of clonal assemblages decreased from 93 to 67 (Hattori 2017) in the same period. Surviving *E. quadricolor* in 2007 increased in number by 2014. Statistical analysis of the original data obtained by Hattori and Kobayashi (2007) in 2001 revealed that neither total tentacle-crown surface area ($r=0.05$, $t_s=0.5$, $P=0.58$, $n=93$) nor size of an assemblage was related to water depth ($r=0.11$, $t_s=1.1$, $P=0.25$), but the two features were significantly correlated ($r=0.795$, $t_s=12.5$, $P<0.000001$). *Entacmaea quadricolor* is thought to multiply in empty spaces, where bleaching caused coral mortality, and then surviving *E. quadricolor* increased in number. The number of *E. quadricolor* in nine quadrats (area of a quadrat was 50 m×63 m) was correlated to the area of 3D small patchy reefs detected in aerial images, excluding bare flat substrates (Hattori and Kobayashi 2007; Hattori 2017). While anemonefish depletion reduces survival of *E. quadricolor* (Frisch et al. 2016), bleached *E. quadricolor* may be recovered by resident anemonefish, as evidenced by bleached *H. magnifica* recovered by resident *A. chrysopterus* (Cortese et al. 2021).

Entacmaea quadricolor are inhabited by 16 species of anemonefish (Fautin and Allen 1997). However, in the coral reefs of the Ryukyu Islands, Japan, only members of extreme specialist *A. frenatus* occupy this anemone, although it is inhabited by only the generalist *A. clarkii* in the temperate rocky reefs of southern Japan (Hattori 2011; Clark et al. 2021). In the Manado region, Sulawesi,

Indonesia, individuals in deep sites (> 9 m deep) are all inhabited by generalist *A. clarkii*, and those in shallow sites are used by the specialist *A. melanopus* or the extreme specialist *A. biaculeatus* (Ricciardi et al. 2010).

These distribution patterns of *E. quadricolor* can be explained by two things: (1) drifting larva use a large range of water depth, unlike *S. gigantea* and *H. crispa* (Figure 17.4); (2) if they settle in reefs with many crevices and deep holes, they reproduce asexually to occupy many narrow vacant sites, but if they settle in reefs with large holes or large vacant sites, they do not reproduce asexually because larger body sizes produce more eggs. As *H. magnifica* also reproduce both sexually and asexually (Dunn 1981) but do not adhere to crevices or deep holes, they multiply asexually in open spaces facing open water at shallow sites around 50 cm deep (Fricke 1979), or at deep sites without high coral coverage up to 40 m deep (Brolund et al. 2004), where spaces are sufficient for multiple large animals, unlike *E. quadricolor.*

Larvae of *H. magnifica* and *E. quadricolor* might use deeper water on average than those of *S. gigantea* and *H. crispa*; and *E. quadricolor* has been shown to inhabit a larger range of water depth (0.5 to 60 m deep, Bridge et al. 2012). Speculated depth distributions of anemone larvae (Figure 17.4) remain to be investigated.

17.4 GEOMORPHIC ZONE AND STRATEGY OF HABITAT USE

Water depths of host anemones and their distances from shore are often used by researchers in the analysis of habitat selection by anemonefishes (Chapter 16; Hayashi et al. 2021). However, these data have different meanings in the context of the 3D structure of coral reefs because each geomorphic zone has a unique depth profile and environmental characteristics (see Figure 17.1).

The outer reef slope is characterized by a wide range of water depths and wave exposure and hard substrates. Suitable hosts for anemonefish, *E. quadricolor, H. crispa, H. magnifica, S. gigantea*, and *S. mertensii*, are all found in this zone, because of the availability of hard substrates and stable water temperatures. Here, both specialist and generalist species may be found. In this zone, interspecific competition should be potentially intense and cohabitation by two anemonefishes may be temporarily found (see Chapter 18 and Chapter 16). The two largest host species, *S. mertensii* and *H. magnifica*, are mostly distributed in this zone. Ricciardi et al. (2010) reported that (probably beyond the reef edge) in the Manado region, Sulawesi, Indonesia, generalist *A. clarkii* and specialist *A. sandaracions* use *S. mertensii*, and specialist *A. perideraion* use *H. magnifica* and *H. crispa*, while *A. clarkii* use these all-host species (though 3D information is unavailable). *Amphiprion sandaracions* and *A. perideraion* may be superior competitors because each species often cohabits with *A. clarkii* but occupy different host species (see Hayashi et al. 2021). *Amphiprion clarkii* may avoid interspecific competition after temporary cohabitation (see Hattori 2002). The width of this zone is measurable. In the large fringing reef (Figure 17.3), for instance, the width of the outer reef *slope* zone was calculated as 116.6 m (assuming 60 m deep and 100 m wide measured in aerial image), which is larger than the subtidal nearshore zone (100 m) or the reef flat zone (100 m) but may be smaller than the backreef zone (300 m, although about half of the backreef zone is a sandy sea bottom).

The subtidal nearshore zone must be large in large fringing reefs and characterized by shallowness, a very large fluctuation of water temperatures and unstable sandy bottoms (shapes of seagrass beds change over several years). This zone may not be distinct in small fringing reefs. It provides a unique habitat for specialist *A. ocellaris* with *S. gigantea*, which may have high thermal change tolerance. Although *A. clarkii* was sometimes found with *S. haddoni* in this zone, this generalist does not reproduce there because it lives without hard substrates and cannot move small shells, unlike specialist *A. polymnus* (Moyer and Steen 1979). In contrast, *A. ocellaris* reproduce at some *S. gigantea* in this zone, as mentioned earlier. Reproducing in this zone needs unique abilities for anemonefishes. The area of potential habitats for *S. gigantea* in this zone is measurable with image analysis software (Hattori and Kobayashi 2009).

The backreef may be the largest zone in large fringing reefs (see Figure 17.3). This zone consists of sandy bottoms, patchy reefs, and continuous reefs connected to the reef flat. Clonal assemblages of *E. quadricolor* are most abundant in this zone. Hattori (2017) compared two indicators of the backreef habitats for *E. quadricolor* and *A. frenatus:* total area of dark-colored patch reefs detectable in an aerial image with image analysis software, and total area of tall patch reef (> 1.5 m in height) detectable in stereoscopic aerial images with a stereoscope (SOKKIA, Mirror Stereoscope, Model MS27). Unexpectedly, the total area of dark-colored patch reef was the better indicator for habitation. As many crevices and holes in reefs are recognizable as blackish colors in aerial images, we can estimate the total area of 3D rugose reefs, excluding barren flat substrates. Rugose reefs in this zone provide major habitats for *E. quadricolor* and inhabitants *P. biaculeatus, A. melanopus, A. ephippium*, and *A. rubrocinctus*, as well as *A. frenatus*, which are specialists or extreme specialists. These species should have high thermal change tolerance.

Reef flats provide scarce habitat for host anemones because these seem to be intertidal flat pavement reefs in aerial images. However, deep channels cutting through the reef flat provide suitable habitats for host anemones, like the outer reef slope zone. In small fringing reefs, the reef crest and upper reef slope connect and are difficult to distinguish, and the edge zone provides habitats for host anemone and anemonefish (see Figure 17.2).

In conclusion, information regarding geomorphic zones is crucial to a valid discussion of interspecific competition and host selection strategies by anemonefishes. In overlapping geographic ranges, where several competing anemonefish species prefer one host species, specialists and extreme specialists should be abundant in a particular zone, often adapting to a particular habitat. Since generalists should have high mobility with broad environmental tolerance, they must be found ubiquitously, avoiding interspecific competition, which may complicate the determination of host selection and cohabitation patterns by anemonefishes.

ACKNOWLEDGEMENTS

I am grateful to Vincent Laudet, James Davis Reimer, David Price, Kina Hayashi, and anonymous reviewers, for their valuable advice regarding the article. Fumitoshi Iwasaki and Hitoshi Hasegawa provided valuable information.

REFERENCES

Bellwood, D. R., S. B. Tebbett, O. Bellwood, M. Mihalitsis, R. A. Morais, R. P. Streit, and C. J. Fulton. 2018. The role of the reef flat in coral reef trophodynamics: Past, present, and future. *Ecology and Evolution* 8: 4108–4119.

Blanchon, P. 2011. Geomorphic zonation. In *Encyclopedia of Modern Coral Reefs: Structure, Form and Process*, ed. D. Hopley, 469–486. Dordrecht: Springer.

Bridge, T., A. Scott, and D. Steinberg. 2012. Abundance and diversity of anemonefishes and their host sea anemones at two mesophotic sites on the Great Barrier Reef, Australia. *Coral Reefs* 31: 1057–1062.

Brolund, T. M., A. Tychsen, L. E. Nielsen, and M. Arvedlund. 2004. An assemblage of the host anemone *Heteractis magnifica* in the northern Red Sea, and distribution of the resident anemonefish. *Journal of the Marine Biological Association of the United Kingdom* 84: 671–674.

Clark, R. D., M. L. Aardema, P. Andolfatto, P. H. Barber, A. Hattori, J. A. Hoey, H. R. Montes Jr, and M. L. Pinsky. 2021. Genomic signatures of spatially divergent selection at clownfish range margins. *Proceedings of the Royal Society B: Biological Sciences* 247: 20210407.

Cortese, D., T. Norin, R. Beldade, A. Crespel, S. S. Killen, and S. C. Mills. 2021. Physiological and behavioural effects of anemone bleaching on symbiont anemonefish in the wild. *Functional Ecology* 35: 663–674.

Dunn, D. F. 1981. The clownfish sea anemones: Stichodactylidae (Coelenterata: Actiniaria) and other sea anemones symbiotic with pomacentrid fishes. *Transactions of the American Philosophical Society* 71: 1–115.

Fautin, D. G., and G. R. Allen. 1997. *Anemonefishes and Their Host Sea Anemones*. 2nd ed. Perth, WA: Western Australian Museum.

Fricke, H. W. 1979. Mating system, resource defense and sex change in the anemonefish *Amphiprion akallopisos*. *Zeitschrift für Tierpsychologie* 50: 313–326.

Frisch, A., J. R. Rizzari1, K. P. Munkres, and J.-P. A. Hobbs. 2016. Anemonefish depletion reduces survival, growth, reproduction and fishery productivity of mutualistic anemone–anemonefish colonies. *Coral Reefs* 35: 375–386.

Harii, S., C. Hongo, M. Ishihara, Y. Ide, and H. Kayanne. 2014. Impacts of multiple disturbances on coral communities at Ishigaki Island, Okinawa, Japan, during a 15 year survey. *Marine Ecology Progress Series* 509: 171–180.

Hattori, A. 1991. Socially controlled growth and size-dependent sex change in the anemonefish *Amphiprion frenatus* in Okinawa, Japan. *Japanese Journal of Ichthyology* 38: 165–177.

Hattori, A. 1994. Inter-group movement and mate acquisition tactics of the protandrous Anemonefish, *Amphiprion clarkii*, on a coral reef Okinawa, Japan. *Japanese Journal of Ichthyology* 41: 159–165.

Hattori, A. 1995. Coexistence of two anemonefish, *Amphiprion clarkii* and *A. perideraion*, which utilize the same host sea anemone. *Environmental Biology of Fishes* 42: 345–353.

Hattori, A. 2002. Small and large anemonefishes can coexist using the same patchy resources on a coral reef, before habitat destruction. *Journal of Animal Ecology* 71: 824–831.

Hattori, A. 2006. Vertical and horizontal distribution patterns of the giant sea anemone *Heteractis crispa* with symbiotic anemonefish on a fringing coral reef. *Journal of Ethology* 24: 51–58.

Hattori, A. 2011. Diversity of symbiotic relationship between host sea anemone and anemonefish: An ecological review of their distributions and combinations. *Japanese Coral Reef Society* 13: 1–27 (in Japanese with English summary).

Hattori, A. 2012. Determinants of body size composition in limited shelter space: Why are anemonefishes protandrous? *Behavioural Ecology* 23: 512–520.

Hattori, A. 2017. Aerial images can detect 3D small patch reefs that are potential habitats for anemonefish *Amphiprion frenatus*. *Ecological Research* 32: 943–949.

Hattori, A., and M. Kobayashi. 2007. Configuration of small patch reefs and population abundance of a resident reef fish in a complex coral reef landscape. *Ecological Research* 22: 575–581.

Hattori, A., and M. Kobayashi. 2009. Incorporating fine-scale seascape composition in an assessment of habitat quality for the giant sea anemone *Stichodactyla gigantea* in coral reef shore zone. *Ecological Research* 24: 415–422.

Hayashi, K., K. Tachihara, and J. D. Reimer. 2021. Loss of natural coastline influences species diversity of anemonefish and host anemones in the Ryukyu Archipelago. *Aquatic Conservation: Marine and Freshwater Ecosystems* 31: 1–13.

Hirose, Y. 1985. Habitat, distribution and abundance of coral reef sea-anemones (Actiniidae and Stichodactylidae) in Sesoko Island, Okinawa, with notes of expansion and contraction behavior. *Galaxea* 4: 113–127.

Litsios, G., C. A. Sims, R. O. Wüest, P. B. Pearman, N. E. Zimmermann, and N. Salamin. 2012. Mutualism with sea anemones triggered the adaptive radiation of clownfishes. *BMC Evolutional Biology* 12: 212.

Loya, Y., K. Sakai, K. Yamazato, Y. Nakano, H. Sambali, and R. van Woesik, 2001. Coral bleaching: The winners and the losers. *Ecology Letters* 4: 122–131.

Mitchell, J. S. 2005. Queue selection and switching by false clown anemonefish, *Amphiprion ocellaris*. *Animal Behaviour* 69: 643–652.

Moyer, J. T., and R. C. Steen. 1979. Nesting-behavior of the anemonefish *Amphiprion polymnus*. *Japanese Journal of Ichthyology* 26: 209–214.

Nguyen, H. T. T., A. N. T. Tran, L. T. L. Ha, D. N. Ngo, B. T. Dang, and A. J. Geffen. 2020. Host choice and fitness of anemonefish *Amphiprion ocellaris* (Perciformes: Pomacentridae) living with host anemones (Anthozoa: Actiniaria) in captive conditions. *Journal of Fish Biology* 94: 937–947.

Ollerton, J., D. McCollin, D. G. Fautin, and G. R. Allen. 2007. Finding NEMO: Nestedness engendered by mutualistic organization in anemonefish and their hosts. *Proceedings of the Royal Society B: Biological Sciences* 247: 591–598.

Pittman, S. J., and A. D. Olds. 2015. Seascape ecology of fishes on coral reefs. In *Ecology of Fishes on Coral Reefs*, ed. C. Mora, 274–282. Cambridge: Cambridge University Press.

Ricciardi, F., M. Boyer, and J. Ollerton. 2010. Assemblage and interaction structure of the anemonefish-anemone mutualism across the Manado region of Sulawesi, Indonesia. *Environmental Biology of Fishes* 87: 333–347.

Saenz-Agudelo, P., G. P. Jones, S. R. Thorrold, and S. Planes. 2011. Detrimental effects of host anemone bleaching on anemonefish populations. *Coral Reefs* 30: 497–506.

Scott, A., and P. L. Harrison. 2007. Embryonic and larval development of the host sea anemones *Entacmaea quadricolor* and *Heteractis crispa*. *The Biological Bulletin* 213: 110–121.

Scott, A., and P. L. Harrison. 2008. Larval settlement and juvenile development of sea anemones that provide habitat for anemonefish. *Marine Biology* 154: 833–839.

Sebens, K. P. 1982. The limits to indeterminate growth: An optimal size model applied to passive suspension feeders. *Ecology* 63: 209–222.

Steen, G. 1988. The bioenergetics of symbiotic sea anemones (Anthozoa: Actiniaria). *Symbiosis* 5: 103–142.

18 Cohabitation and Competition in Anemonefishes

Patterns and Consequences

Maya Srinivasan and Geoffrey P. Jones

CONTENTS

18.1 INTRODUCTION

Anemonefishes most commonly live in small conspecific groups that cohabit within and are confined to a single anemone host (Verwey 1930; Mariscal 1970; Allen 1972; Fautin and Allen 1992; Burke da Silva and Nedoskyo 2016). Early researchers discovered their complex group structure and mating system, which consists of a pair of breeding adults, the largest individual the female and the second largest the male, and a variable number of progressively smaller juveniles (Fricke and Fricke 1977; Moyer and Nakazono 1978; Fricke 1979). The group is characterized by a strong social hierarchy that controls juvenile growth, maturation, and sex change from male to female in the adults (Allen 1972; Fricke and Fricke 1977; Ross 1978; Fricke 1979). Social group sizes vary depending upon the anemonefish species and their host, and larger social groups and individuals are associated with larger anemones or clusters (Ross 1978; Fricke 1979; Fautin 1992). Following on from this pioneering work, we now have a much more nuanced understanding of the rules under which individuals of a single species can cohabit and avoid eviction from the confined space of their anemone (Buston 2003a, b, c, 1984a, 1984b; Buston and Cant 2006; Rueger et al. 2018). In contrast to intraspecific cohabitation, the presence of more than one anemonefish species has historically been viewed as a rare event (Fautin and Allen 1992). However, there has been recent interest in high levels of interspecific cohabitation between pairs of species with overlapping geographic ranges and patterns of host use in the coral triangle (Hattori 2000; Camp et al. 2016; De Brauwer et al. 2016). This has led to the hypothesis that cohabitation is a significant process contributing to local diversity and coexistence in anemonefish species.

Ever since the discovery of the anemonefish–anemone association, it has always been assumed that there is intense competition within and among anemonefishes for anemone hosts and the resources they provide (Allen 1972; Ross 1978; Fricke 1979; Fautin 1992). Competition clearly occurs among individuals within anemones, determining patterns in group size and structure, but at a larger scale may limit population sizes and anemonefish species distributions among anemone species and habitats. Intraspecific competition is often inferred because almost all anemones are usually occupied (Allen 1972; Ross 1978; Fricke 1979; Fautin 1986, 1992), because of the positive relationships between group size, fish size, and anemone size (Fautin 1992), and because of a commonly positive relationship between anemone and anemonefish abundances at a larger scale (Ross 1978; Fricke 1979). More recently, it has also been inferred from declines in group size or body size in stressed anemones (Saenz Agudelo et al. 2011; Burke da Silva and Nedoskyo 2016; Beldade et al. 2017) and declines in abundance in response to sharp drops in anemone numbers (Hattori 2002; Howell et al. 2016). Historically, interspecific competition has been inferred from observations of interspecific aggressive behavior (Fautin 1986a, 1986b), high levels of habitat and host partitioning among anemonefish species, and consequently, the low levels of interspecific cohabitation (Dunn 1981; Fautin 1985, 1986; Elliott and Mariscal 2001). An early conclusion was that highly specialized anemonefish associated with a single host have evolved mechanisms to outcompete the more generalist species (Fautin 1985, 1986, 1991; Burke da Silva and Nedoskyo 2016). However, without critical experiments being undertaken, it is not clear how many assumptions about the importance of competition still hold.

DOI: 10.1201/9781003125365-22

After almost 50 years since Allen's seminal book on anemonefishes (Allen 1972), which contained the first experimental test of the effects of adults on juvenile growth, a re-evaluation of the importance of cohabitation and competition within and among anemonefish species is warranted. Numerous ecological and behavioral studies have now been published, including work on many of the 28 recognized species (Chapter 3) and study sites that span the entire distribution of anemonefishes, from the Red Sea to the eastern Pacific. The increasing information continues to indicate strong intra- and interspecific competition for anemone habitat and reproductive status within anemones, which in some cases, appears to explain patterns of distribution, abundance, and habitat use on larger scales. There is no evidence that cohabitation is promoted by any direct mutualistic interactions within or among species, and we argue interspecific cohabitation should be viewed as a consequence of high biodiversity, rather than a mechanism promoting coexistence.

18.2 INTRASPECIFIC COHABITATION

Group sizes vary among anemonefishes, with some species only found in pairs (e.g., *Premnas biaculeatus*; Fautin 1986; Srinivasan et al. 1999), others in larger groups on single anemones (e.g., *Amphiprion percula* and *A. perideraion*; Fautin 1992), and others in much larger groups on colonial anemones (e.g., *A. melanopus*; Ross 1978; Fautin 1986; Srinivasan et al. 1999). Group sizes can also vary geographically (Fautin 1992), and for host generalists, among anemone species (Allen 1972). However, an anemone only ever supports one breeding pair. In larger colonies, individuals are territorial and breed in monogamous pairs, with non-reproductive individuals often restricted to the edge of the colony (Ross 1978; Fautin 1986). Solitary individuals can be found in small or less preferred anemones.

What determines the number of cohabiting individuals in an anemone? Group size and the body sizes of individuals in groups are tightly controlled by the size of the host and behavioral interactions within the group (Buston 2003a, b, c; Chausson et al. 2018). Group members control when larval settlers can join a group. In *Premnas*, new settlers are excluded by the reproductive pair and larval settlement is only observed when one of the adults disappears or is experimentally removed (Dixson and Jones 2018). In small groups of *A. percula*, larvae cannot successfully settle until the smallest resident juvenile is above a certain size threshold (Buston 2003b). In this species, the number of fish in the group is strongly correlated with anemone size (Fautin 1992; Buston 2003b; Chausson et al. 2018), and the social hierarchy enforces a fixed size difference between group members of adjacent rank (Buston 2003a; Buston and Cant 2006). While the largest breeding individual ultimately determines the growth of subordinates, it appears that lower-ranked individuals control their growth rates, so as not to be evicted by individuals above them in the social hierarchy (Buston 2003b; Rueger et al. 2018). There is increasing evidence that anemone size may not affect group size directly, but rather indirectly, by allowing dominant fish to reach larger sizes on larger anemones (Fautin 1992; Mitchell and Dill 2005; Kobayashi and Hattori 2006; Chausson et al. 2018). For example, work by Chausson et al. (2018) examining the relationships among anemone size, depth, distance from shore, social group size, and body size in *A. percula* showed that anemone surface area had a strong, direct effect on the size of the dominant female, which had strong, direct effects on male size and the size of subadults (Figure 18.1).

18.3 INTRASPECIFIC COMPETITION

The importance of anemone size in determining group size, the strong social hierarchy, and regulation of size structure and growth, are all indicative of intense intraspecific competition at the scale of the individual anemone. However, evaluating what individuals are actually competing for, whether it be social status or access to resources such as food, is complex. Within an anemone, it is clear that individuals compete

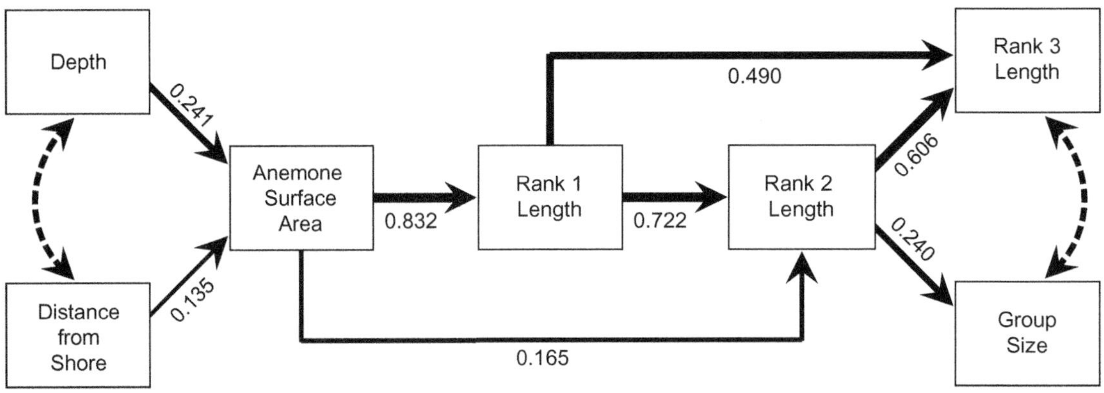

FIGURE 18.1 Path analysis showing the relative importance of environmental and social variables on social group size and structure. Direct effects are indicated by solid lines, with line thicknesses indicating the strength of the effect. Path coefficients, the standardized versions of linear regression weights, are indicated below each solid line. Covariances are indicated by dotted lines. Figure from Chausson et al. (2018).

for reproductive status as the breeding population is limited to two individuals. Allen (1972) undertook the first experiments demonstrating intraspecific competition between adults and juveniles within anemones. Experimental removal of the dominant fish in *A. perideraion* social groups resulted in a rapid increase in the growth of juveniles. He inferred juvenile "stunting" was due to competition for food, with aggression by the dominant fish reducing the foraging area and foraging time of juveniles.

Reproductive pairs limit breeding population size by preventing other adults from coexisting in the same anemone through aggression and territorial defence. It seems obvious that they are competing for the anemone habitat as an ideal safe haven from predators. The size and isolation of the anemone relative to the area requirements of the fish mean that each anemone is a defendable resource. Adult density manipulations in *P. biaculeatus* and *A. melanopus* by Srinivasan et al. (1999) showed that larger individuals would always exclude conspecific adults from both solitary and colonial *E. quadricolor* anemones. For each species, two conspecific pairs were placed next to either a solitary anemone or several colonial anemones in large fibreglass tanks (approximately 1,150 mm diameter, 600 mm water depth) and the fish were observed for one hour to determine which individuals occupied the anemone(s). In *Premnas*, the largest female and male would always displace the smaller female and male (Figure 18.2A), with partner swapping occurring in nine out of 20 trials. In *A. melanopus*, the outcome ranged from just one individual (the largest) occupying the anemone in all trials on solitary anemones, and in colonial anemones, between one and all four individuals occupied anemones with the largest individual occupying the largest anemone (Figure 18.2B).

More recent work suggests the juveniles and subadults within social groups are competing for social rank, so as to eventually inherit reproductive status (Buston 2003c; Reed et al. 2019). Aggression is most intense between individuals of adjacent rank and similar size (Fricke and Fricke 1977; Wong et al. 2016). Recent experiments showing juvenile fish do not grow faster when food is added suggest they control their own growth when adults are present, to avoid competition and eviction (Reed et al. 2019). Rueger et al. (2018) showed for *A. percula* that small, mature individuals are much more likely to be evicted than similar-sized immature individuals. Although the group size and number of breeding individuals may not directly be food limited, the reproductive success of adult pairs may well be. The fact that the size of dominant fish increases with anemone size, which allows more fish to join the social group, suggests that resources like food are ultimately important in determining female fecundity. Experiments have shown that food supplementation can dramatically increase egg production in female *A. percula* (Rueger et al. 2018; Barbasch et al. 2020).

On a larger scale, it is clear that the anemone habitat itself is a resource that limits breeding populations as a whole, especially for species that have near 100% occupancy by reproductive pairs (Fautin 1992). This is supported by many studies showing positive relationships between anemonefish population size and anemone abundance or cover (Richardson 1999; Ricciardi et al. 2010; Sato et al. 2014; Steinberg et al. 2020), although this is not universal (Kobaysahi and Hattori 2006). Habitat limitation has been tested experimentally, with Schmitt and Holbrook (2000) manipulating anemone densities and finding that adult abundance and recruitment increase in proportion to anemone area. Dramatic declines in anemone numbers always result in massive declines in anemonefish densities (Hattori 2002; Howell et al. 2016). Additional evidence comes from the commonly observed intraspecific habitat segregation.

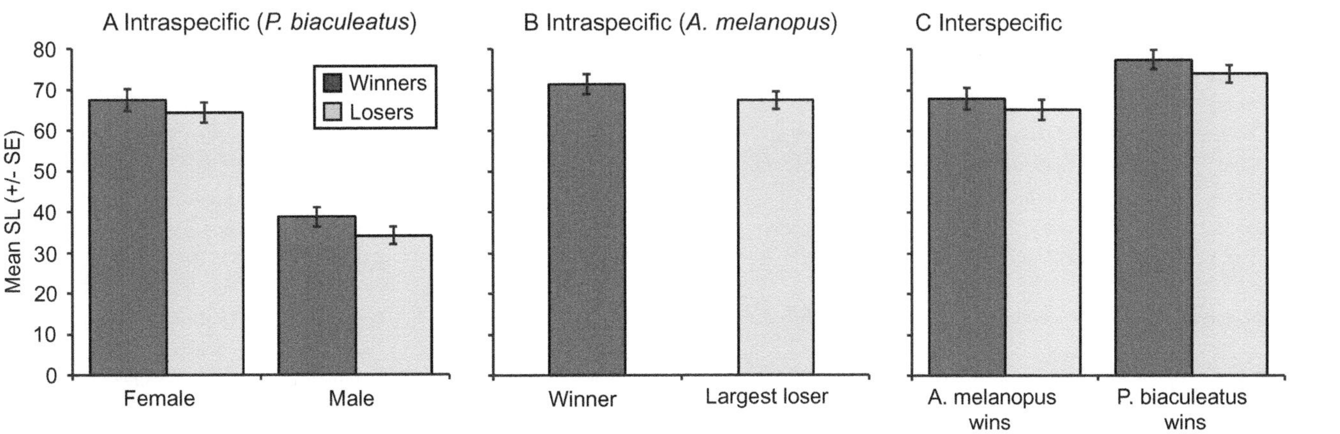

FIGURE 18.2 Mean sizes of winners and losers for (A) intraspecific competition trials between two pairs of *P. biaculeatus* (no significant difference between size of largest winner and size of largest loser); (B) intraspecific competition trials between two pairs of *A. melanopus* (significant difference in size between the winner and the largest of the losers, paired t-test); and (C) interspecific competition trials between a pair of *A. melanopus* and a pair of *P. biaculeatus* (significant differences in size between females and between males, paired t-tests). In each case, trials using solitary and colonial *E. quadricolor* (ten trials per anemone morph) were combined. Adapted from Srinivasan et al. (1999).

In species occupying more than one anemone species that vary in quality, adults often dominate the preferred species and restrict juveniles to marginal habitat. For example, adult *A. bicinctus* in the Red Sea prefer *Entacmaea quadricolor* over *Heteractis crispa* (Chadwick and Arvedlund 2005; Heubner et al. 2012). Experiments show juveniles are confined to *H. crispa* by competitive exclusion, but prefer *E. quadricolor* when it is available (Heubner et al. 2012). Both adults and juveniles survive better on *E. quadricolor*, presumably because it offers better shelter (Howell et al. 2016).

18.4 INTERSPECIFIC COHABITATION

Early papers only mention cohabitation between two anemonefish species as rare and transitory events of little significance (Elliott and Mariscal 2001). However, there is an increasing list of species pairs that have been documented living in the same anemone (Table 18.1). It appears to be most common in the coral triangle where there is a high diversity of species with highly overlapping geographic ranges and host preferences (Hattori 2000; Camp et al. 2016; De Bauwer et al. 2016; Hayashi et al. 2018) (Table 18.1). Camp et al. (2016) show frequency of cohabitation declines in proportion to the decline in anemonefish diversity as a function of distance from the centre of biodiversity. Cohabitation is most common for pairs of species with large geographic ranges and a high degree of overlap in the Coral Triangle, including *A. clarkii* and *A. perideraion*, *A. melanopus* and *Premnas biaculeatus*, and *A. chrysopterus* and *A. sandaracinos* (Figure 18.3, Table 18.1). Interestingly, high levels of cohabitation appear to be restricted to core areas within the region of overlap.

In the most comprehensive study to date, Camp et al. (2016) found up to 57% of all anemonefish individuals cohabiting with other species at Hoga Island in Indonesia. Cohabitation was so common that the authors advanced the view that cohabitation is a mechanism promoting species co-existence and local diversity. Close inspection shows most cases of cohabitation at Hoga Island (51/55) were between two anemonefish species, *A. clarkii* and *A perideraion* on *Heteractis crispa* (Figure 18.4A), the most commonly inhabited anemone at the site (Camp et al. 2016, Supplementary Material). Furthermore, in 47 of the 51 cases, there were adult *A. clarkii* present, but only juvenile *A. perideraion*, with the juvenile often restricted to the margins. Earlier, Hattori (2000) showed that these two species are never observed to interbreed on the same anemone. *A. perideraion* juveniles only mature and breed after *A. clarkii* individuals are removed or disappear. This does not suggest that the two species amicably cohabit the same anemone in a manner that mutually benefits their breeding populations. *A. sandaracinos* is another species that occasionally cohabits with *A. clarkii*, but only as juveniles (Bos et al. 2011; Hayashi et al. 2018). Most of the other reported cases of cohabitation are similar in that just one of the species breeds and the other is only present as a juvenile, often restricted to the periphery of the anemone (Bos et al. 2011; Camp et al. 2016; De Brauwer et al. 2016). Given that cohabitation does not promote the breeding population size of either species,

TABLE 18.1
Published Examples of Two Anemonefish Species Occupying the Same Individual Anemone

Species 1	Species 2	Anemone	Location	% overlap	Occurrence	Reference
A. clarkii	*A. percula*	*Heteractus magnifica*	Madang	100	Rare	Elliott and Mariscal 2001
A. chrysopterus	*A. percula*	*Heteractus magnifica*	Madang	75.7	Rare	Elliott and Mariscal 2001
A. clarkii	*A. perideraion*	*Heteractus magnifica*	Madang	83.5	Rare	Elliott and Mariscal 2001
A. chrysopterus	*A. perideraion*	*Heteractus magnifica*	Madang	90.1	Rare	Elliott and Mariscal 2001
A. clarkii	*A. sandaracinos*	*Stichodactyla mertensii*	Madang	100	Rare	Elliott and Mariscal 2001
A. chrysopterus	*A. sandaracinos*	*Stichodactyla mertensii*	Madang	48.8	Common	Elliott and Mariscal 2001
A. clarkii	*A. melanopus*	*Entacmaea quadricolor*	Madang	80.2	Rare	Elliott and Mariscal 2001
A. clarkii	*A. perideraion*	*Heteractis crispa*	Okinawa	83.5	Common	Hattori 2002
A. clarkii	*A. sandaracinos*	*Stichodactyla mertensii*	Philippines	100	Rare	Bos 2011
A. clarkii	*P. biaculeatus*	*Entacmaea quadricolor*	Indonesia	100	Rare	De Brauwer et al 2016, Cant et al. 2016
A. melanopus	*P. biaculeatus*	*Entacmaea quadricolor*	Indonesia	61.6	Rare	De Brauwer et al 2016, Cant et al. 2016
A. clarkii	*A. perideraion*	*Heteractis crispa*	Indonesia	83.5	Common	De Brauwer et al 2016, Cant et al. 2016
A. clarkii	*A. sandaracinos*	*Stichodactyla mertensi*	Indonesia	100	Rare	De Brauwer et al 2016, Cant et al. 2016
A. chrysopterus	*A. sandaracinos*	*Stichodactyla mertensii*	PNG, Solomon Is.	48.8	Common	Gainsford et al. 2014, 2020

Note: The table includes anemone species cohabited, the % overlap in the two anemonefish species geographic range (mean of % of species with smallest range overlapping with species with the largest range – calculated from fishbase), how common co-occurrence is, and the source. The average pairwise overlap among all anemonefish species is only ~19% and the majority of species pairs have little or no opportunity to cohabit.

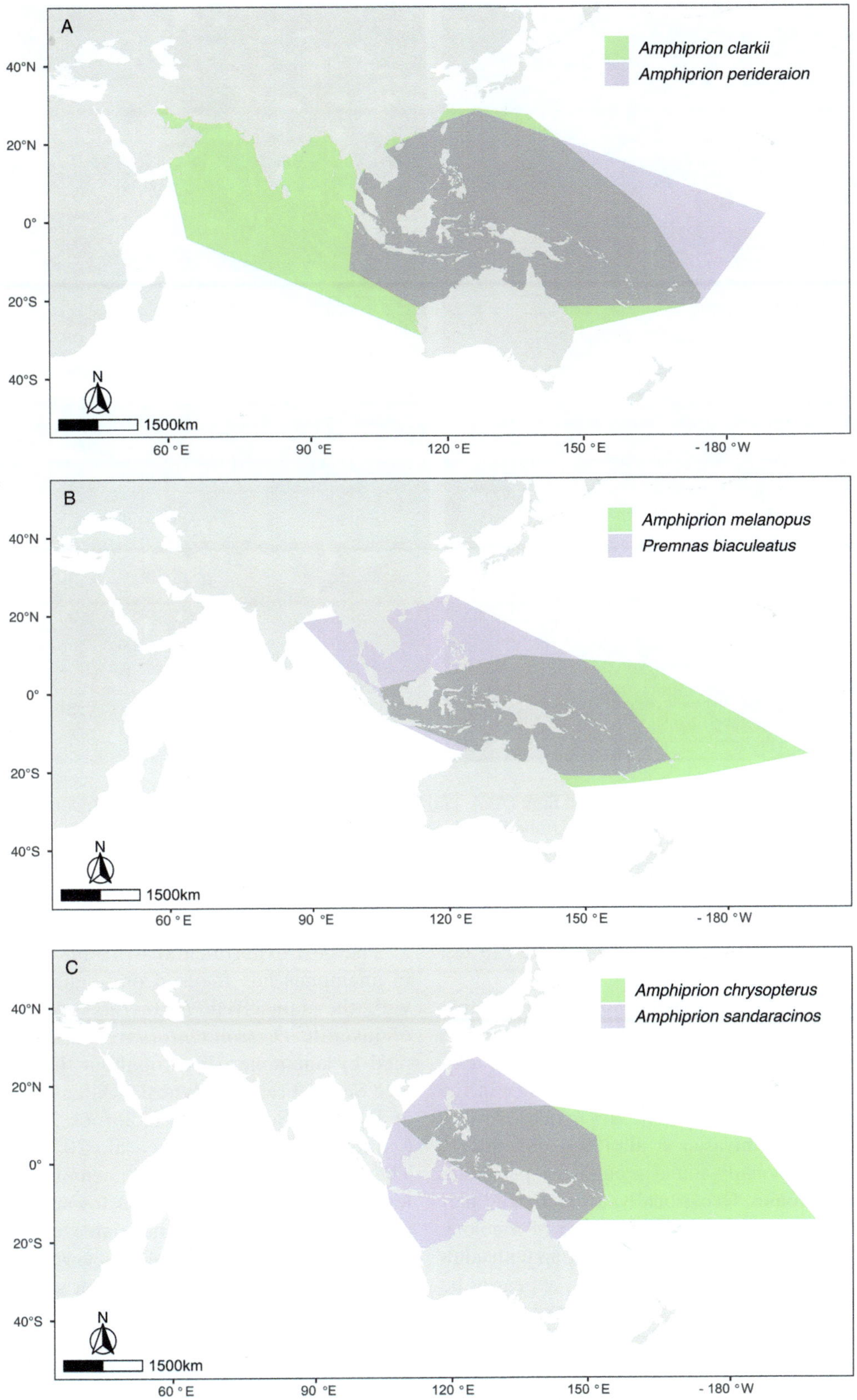

FIGURE 18.3 Distribution maps of three pairs of species observed to cohabit the same anemones, showing high geographic overlap in the Coral Triangle: (A) *A. clarkii* and *A. perideraion*; (B) *A. melanopus* and *P. biaculeatus*; and (C) *A. chrysopterus* and *A. sandaracinos*.

FIGURE 18.4 Examples of interspecific cohabitation: (A) *A. clarkii* and *A. perideraion* at Hoga Island, Indonesia; (B) *A. melanopus* and *P. biaculeatus* on colonial *E. quadricolor* at Kimbe Bay, Papua New Guinea; (C) *A. chrysopterus* and *A. sandaracinos* at Kimbe Bay, Papua New Guinea; and (D) *A. polymnus* and *D. trimaculatus* at Bootless Bay, Papua New Guinea.

we argue it may simply be a consequence of the high diversity in the coral triangle, rather than a mechanism promoting it. However, further research will be required to test these alternative hypotheses.

A. melanopus and *Premnas biaculeatus* are two other species often recorded as cohabiting. Both are specialists on the anemone *Entacmaea quadricolor*, but *A. melanopus* is almost always restricted to shallow water in the colonial form, while *P. biaculeatus* occurs in deeper water in the solitary form (Srinivasan et al. 1999). Juvenile *P. biaculeatus* are occasionally found around the perimeter of *A. melanopus* colonies. Occasionally, a super-sized pair of *P. biaculeatus* can be found in the centre of a colony (Figure 18.4B), but their territories do not overlap with adult *A. melanopus*. Hence, the two species do not appear to be cohabiting the same space within larger anemone colonies.

The only example of two anemonefish species breeding on the same anemone is *A. chryspoterus* and *A. sandaracinos* (Figure 18.4C). In this case, the two species breed together producing a hybrid form, formerly recognized as a separate species, *A. leucokranos* (Gainsford et al. 2014, 2020). This hybridization only occurs in a narrow band of geographic overlap, only on the anemone species on which they co-occur, and usually when one of the parent species is rare. Because back-crossing between hybrids and parent species only appears to occur with *A. sandaracinos*, the two species have remained distinct.

The best experimental work on cohabitation of fish in anemones has focused on anemonefish interactions with the damselfish *Dascyllus trimaculatus*. Schools of juvenile *D. trimaculatus* colonize anemones occupied by anemonefish throughout their range, from the Red Sea to French Polynesia (Allen et al. 1972; Brolund et al. 2004; Schmitt and Holbrook 2003; Holbrook and Schmitt 2004; Hayashi et al. 2020) (Figure 18.4D). There is a difference among anemonefish species in the level of aggression displayed towards *D. trimaculatus* and the degree to which they can exclude them (Hayashi et al. 2020). In Moorea, *A. chrysopterus* can competitively exclude *D. trimaculatus* on small anemones, but because the presence of anemonefish actually increases the size of the anemones they occupy, this can indirectly promote juvenile *D. trimaculatus* numbers. While *A. chrysopterus* is the dominant competitor, it appears the gregarious settlement of *D. trimaculatus* makes competitive exclusion too costly. It is worth noting that *D. trimaculatus* migrate away from anemones as adults and breed on large patch reefs, so as with most cases of cohabitation, this is not a case of two species with coexisting breeding populations.

18.5 INTERSPECIFIC COMPETITION

Most of the aforementioned examples of interspecific cohabitation suggest some level of interspecific competition rather than any mutualistic relationships between fish species. In fact, cohabitation at some point is a necessary prerequisite for interspecific competition, unless there is no overlap between two species' geographic ranges, their depth distributions, or their anemone hosts. So, what is the evidence for intense interspecific competition among anemonefishes? Most studies directly addressing this topic have focused on locations where species have been observed occupying the same anemone species, but not necessarily at the same time. The first work was Fautin's (1985, 1986) "challenge" experiments designed to examine competitive hierarchies at Lizard Island on the Great Barrier Reef. Fautin transferred individuals of three species, *Premnas biaculeatus*, *A. melanopus*, and *A. akyndynos* to anemones of *Entacmaea quadricolor* occupied by one of the other species. There appeared to be a competitive hierarchy, with *Premnas* dominating *A. melanopus*, and *A. melanopus* dominating *A. akyndynos* (although *E. quadricolor* is marginal habitat for the latter). Similarly, it was concluded that *A. perideraion* dominates *A. percula*, which in turn dominates *A. akindynos* on *Heteractis magnifica*. Competitive outcomes were largely predicted on the basis of body size and it was hypothesized that the more specialized the anemonefish species, the more likely it would be to dominate when on its preferred host.

Few studies have examined the potential effect of one species on another via more rigorous competition experiments (see Bonin et al. 2015). Fautin (1992) used removal experiments to examine potential interactions between *A. percula* and *A. perideraion*, which both occupy *H. magnifica* at Madang, PNG. Resident adults of both species restricted the settlement of their own species and the other species. When whole groups were removed, either species could recolonize the anemone, suggesting chance and competitive equality may be important in explaining coexistence. There has been only one properly controlled competition experiment in which potentially competing individuals were transferred to the same anemone, either from the same species or from two species, to tease apart the effects of intraspecific and interspecific competition. Srinivasan et al. (1999) did this for *Premnas* and *A. biaculeatus* on *Entacmaea* at Lizard Island and found strict competitive exclusion within anemones. The experiment showed that these two species never cohabit. There was no strict competitive hierarchy, so either species could be the winning adult pair (Figure 18.2C). This strongly suggests that interspecific competition limits the number of species within anemones. However, experiments on habitat preferences suggest the different depth and habitat distributions of the two species are largely explained by innate habitat preferences, rather than by competitively induced habitat partitioning.

It is common in multispecies anemonefish studies at any location to interpret differences or similarities in host use, or depth distributions, as evidence either for or against interspecific competition (Richardson 1999; Elliott and Mariscal 2001; Ricciardi et al. 2010; Sato et al. 2014; Steinberg et al. 2020). Invariably, where differences are found on a local scale, an interspecific competition hypothesis is invoked. Nestedness in community structure is interpreted as greater competition and greater habitat partitioning among specialized species, with generalists more likely to come into contact and compete with one another (Ollerton et al. 2007). In regions of diverse anemonefish assemblages, researchers have tried to classify pairwise patterns in resource use in terms of different non-competition-based hypotheses (cohabitation model) and competition-based explanations (niche differentiation, patch heterogeneity, lottery hypothesis). For example, Hattori (2002) argued *A. perideraion* is a competitive dominant over *A. clarkii*, but the two species can coexist because the latter can occupy more anemone species, smaller anemones, and can disperse more widely in a patchy environment. Hayashi et al. (2018, 2019) suggested competitively induced niche differentiation at Okinawa applied to most species pair combinations, except for *A. clarkii* and *A. sandaracinos*, which they argued supports the cohabitation hypothesis, and *A. ocellaris* and *A.perideraion*, which appears to conform to the lottery hypothesis. All of these interpretations, while they cannot be dismissed, seem rather weak inferences in the absence of well-designed experiments to test the defining hypotheses from each of the alternative models (Bonin et al. 2015).

18.6 CONCLUSIONS

Anemonefishes have clearly been model species for our understanding of the intense intraspecific competitive processes operating in small, single-species social groups in confined habitats. Because of their restricted movements and ease of capture, sophisticated field experiments have been undertaken to test not only how intraspecific competition is occurring but the resources fishes are competing for. Within anemones, larger individuals are clearly competing for breeding status, while cohabiting juveniles compete for rank, in order to assume breeding status when the opportunity arises. Anemonefishes occupy a unique predator-free and patchily distributed habitat which is readily defended by an adult pair. However, being restricted to anemones comes at the cost of it being a small home where food resources are likely to be restricted. Occupying larger anemones enables the adult female to access more food, reach larger sizes, and achieve greater reproductive success. This enables the second-ranked individual, the male, to grow larger as well. Individuals of lower rank adjust their growth accordingly, but only enough to avoid conflict and potential eviction. Competition within groups has large-scale consequences for population size and intraspecific habitat segregation, with adult numbers limited by anemone availability and juveniles often restricted to less preferred anemone species.

Anemonefishes have the potential to be model species for our understanding of interspecific cohabitation and competition as they occur in discrete social groups which can be manipulated, but as yet, experimentation has been limited. More density manipulations of pairs of cohabiting species are needed to assess long-term detrimental effects of interspecific competition on growth and reproduction. Interspecific cohabitation and competition have recently been viewed as mutually exclusive, opposite mechanisms that allow anemonefish species to coexist (Hayashi 2018, 2019), but this may not be the case. While cohabitation could occur without competition, this would not constitute a mechanism of coexistence. On the other hand, you cannot have interspecific competition without some level of cohabitation, because individuals of different species have to come into contact to compete. All the evidence suggests that there is strong competition between cohabiting species for breeding status, the outcome of which is that only the larger species gets to breed. At this stage, there is no evidence of mutualistic interactions between two anemonefish species that may constitute an alternative mechanism to explain cohabitation. Interspecific competition is also likely to have ramifications in diverse areas for species distributions, local abundance, and habitat use. However, after 50 years of ecological research, we only have a preliminary understanding of these larger-scale and longer-term phenomena. More sophisticated experiments are needed to test alternative mechanisms of coexistence such as resource partitioning, competitive exclusion, or lottery competition.

In the future, we clearly need a better understanding of the costs and benefits for dominant individuals to tolerate or exclude other individuals. We currently have a poor understanding of the critical resources of anemonefishes such as food and shelter, how these affect anemonefishes, and how anemonefishes affect them. Few animal species have the capacity that anemonefishes have to live in a predator-free refuge, to protect that refuge from predators and even promote the growth and size of their habitat patch. There is clearly much more to be done, but if there is to be one unifying theme in studies of anemonefish cohabitation and competition it will be the need to understand who gets to breed and who does not.

ACKNOWLEDGEMENTS

We would like to acknowledge the College of Science and Engineering and the ARC Centre of Excellence for Coral Reef Studies at James Cook University for logistic support. We also thank the Mahonia Na Dari Research and Conservation Centre and Walindi Plantation Resort for supporting our research in Papua New Guinea. We thank Peter Buston and an anonymous reviewer for constructive comments on the manuscript. Special thanks to Gemma Galbraith for calculating species range overlaps, and to Emma Camp, Tane Sinclair-Taylor, and Serge Planes for permission to use their photographs. Finally, we would like to thank each other as, for us, this chapter marks a return to where it all began.

REFERENCES

Allen, G. R. 1972. *The Anemonefishes: Their Classification and Biology*. Neptune City, NJ: TFH Publications.

Barbasch, T. A., T. Rueger, M. Srinivasan, M. Y. L. Wong, G. P. Jones, and P. M. Buston. 2020. Substantial plasticity of reproduction and parental care in response to local resource availability in a wild clownfish population. *Oikos* 129: 1844–1855.

Beldade, R., A. Blandin, R. O'Donnell, and S. C. Mills. 2017. Cascading effects of thermally-induced anemone bleaching on associated anemonefish hormonal stress response and reproduction. *Nature Communications* 8: 716.

Bonin, M. C., L. Boström-Einarsson, P. L. Munday, and G. P. Jones. 2015. The prevalence and importance of competition among coral reef fishes. *Annual Reviews in Ecology, Evolution and Systematics* 46: 169–190.

Bos, A. R. 2011. Clownfishes *Amphiprion clarkii* and *A. sandaracinos* (Pomacentridae) coexist in the sea anemone *Stichodactyla mertensii*. *Coral Reefs* 30: 369.

Brolund, T. M., A. Tychsen, L. E. Nielsen, and M. Arvedlund. 2004. An assemblage of the host anemone *Heteractus magnifica* in the northern Red Sea, and their distribution of resident anemonefish. *Journal of the Marine Biological Association of the United Kingdom* 84: 671–674.

Burke da Silva, K., and A. Nedoskyo. 2016. Sea anemones and anemonefishes: A match made in heaven. In *The Cnidaria, Past, Present and Future*, eds S. Goffredo, and Z. Dubinsky, 425–438. New York: Springer.

Buston, P. 2003a. Size and growth modification in clownfish. *Nature* 424: 145–146.

Buston, P. 2003b. Forcible eviction and prevention of recruitment in the clown anemonefish. *Behavioural Ecology* 14: 576–582.

Buston, P. M. 2003c. Mortality is associated with social rank in clown anemonefish (*Amphiprion percula*). *Marine Biology* 143: 811–815.

Buston, P. M., and M. A. Cant. 2006. A new perspective on size hierarchies in nature: Patterns, causes and consequences. *Oecologia* 149: 362–372.

Camp, E. F., J. P. A. Hobbs, M. De Brauwer, A. J. Dumbrell, and D. J. Smith. 2016. Cohabitation promotes high diversity of clownfishes in the coral triangle. *Proceedings of the Royal Society B* 283: 20160277.

Chadwick, N. E., and M. Arvedlund. 2005. Abundance of giant sea anemones and patterns of association with anemonefish in the northern Red Sea. *Journal of the Marine Biological Association of the United Kingdom* 85: 1287–1292.

Chausson, J., M. Srinivasan, and G. P. Jones. 2018. Host anemone size as a determinant of social group size and structure in the orange clownfish (*Amphiprion percula*). *PeerJ* 6: e5841.

Dunn, D. F. 1981. The clownfish sea anemones: Stichodactylidae (Coelenterata: Actiniaria) and other sea anemones symbiotic with pomacentrid fishes. *Transactions of the American Philosophical Society* 71: 1–115.

De Brauwer, M., E. Camp, J. Jompa, and D. J. Smith. 2016. High levels of heterospecific cohabitation among anemonefishes in Hoga Island, Indonesia. *Marine Biodiversity* 46: 19–20.

Dixson, D. L., and G. P. Jones. 2018. Influence of prior residents on settlement preferences in the anemonefish, *Premnas biaculeatus*. *Coral Reefs* 37: 519–526.

Elliott, J. K., and R. N. Mariscal. 2001. Coexistence of nine anemonefish species: Differential host and habitat utilization, size and recruitment. *Marine Biology* 138: 23–36.

Fautin, D. G. 1985. Competition by anemone fishes for host actinians. *Proceedings of the 5th International Coral Reef Congress* 5: 373–377.

Fautin, D. G. 1986. Why do anemonefishes inhabit only some host actinians. *Environmental Biology of Fishes* 15: 171–180.

Fautin, D. G. 1991. The anemonefish symbiosis: What is known and what is not? *Symbiosis* 10: 23–46.

Fautin, D. G. 1992. Anemonefish recruitment: The roles of order and chance. *Symbiosis* 14: 143–160.

Fautin, D. G., and G. R. Allen. 1992. *Anemone Fishes and Their Host Sea Anemones*. Perth, WA: Western Australian Museum.

Fricke, H., and S. Fricke. 1977. Monogamy and sex change by aggressive dominance in coral reef fish. *Nature* 266: 830–832.

Fricke, H. W. 1979. Mating system, resource defence and sex change in the anemonefish *Amphiprion akallopisos*. *Zeitschrift für Tierpsychologie* 6: 71–77.

Gainsford, A., G. P. Jones, J. P. A. Hobbs, F. M. Heindler, and L. van Herwerden. 2020. Species integrity, introgression, and genetic variation across a coral reef fish hybrid zone. *Ecology and Evolution* 10: 11998–12014.

Gainsford, A., L. van Herwerden, and G. P. Jones. 2014. Hierarchical behaviour, habitat use and species size differences shape evolutionary outcomes of hybridization in a coral reef fish. *Journal of Evolutionary Biology* 28: 205–222.

Hattori, A. 2000. Social and mating systems of the protandrous anemonefish *Amphiprion perideraion* under the influence of a larger congener. *Austral Ecology* 25: 187–192.

Hattori, A. 2002. Small and large anemonefishes can coexist using same patchy resources on a coral reef, before habitat destruction. *Journal of Animal Ecology* 71: 824–831.

Hattori, A. 2006. Vertical and horizontal distribution patterns of the giant sea anemone *Heteractis crispa* with symbiotic anemonefish on a fringing reef. *Journal of Ethology* 24: 51–57.

Hayashi, K., K. Tachihara, and J. D. Reimer. 2018. Patterns of coexistence of six anemonefish species around subtropical Okinawa-Jima Island, Japan. *Coral Reefs* 37: 1027–1038.

Hayashi, K., K. Tachihara, and J. D. Reimer. 2019. Loss of natural coastline influences species diversity of anemonefish and host anemones in the Ryuku Archipelago. *Aquatic Conservation: Marine and Freshwater Ecosystems* 31: 15–27.

Hayashi, K., K. Tachihara, and J. D. Reimer. 2020. Anemonefish aggressiveness affects the presence of *Dascyllus trimaculatus* co-existing with host anemones. *Marine Biology* 167: 84.

Holbrook, S. J., and R. J. Schmitt. 2004. Population dynamics of a damselfish: Effects of a competitor that is also an indirect mutualist. *Ecology* 85: 979–985.

Howell, J., T. L. Goulet, and D. Goulet. 2016. Anemonefish musical chairs and the plight of the two-band anemonefish, *Amphiprion bicinctus*. *Environmental Biology of Fishes* 99: 873–886.

Huebner, L. K., B. Dailey, B. M. Titus, M. Khalaf, and N. E. Chadwick. 2012. Host preference and habitat segregation among Red Sea anemonefish: Effects of sea anemone traits and fish life stages. *Marine Ecology Progress Series* 464: 1–15.

Kobayashi, M., and A. Hattori. 2006. Spacing pattern and body size composition of the protandrous anemonefish *Amphiprion frenatus* inhabiting colonial host anemones. *Ichthyological Research* 53: 1–6.

Mariscal, R. N. 1970. A field and laboratory study of the symbiotic behavior of fishes and sea anemones from the tropical Indo-Pacific. *University of California Publications in Zoology* 91: 1–43.

Mitchell, J. S., and L. M. Dill. 2005. Why is group size correlated with the size of the sea anemone in the false clown anemonefish? *Canadian Journal of Zoology* 83: 372–376.

Moyer, J. T., and A. Nakazono. 1978. Protandrous hermaphroditism in six species of the anemonefish genus *Amphiprion* in Japan. *Japanese Journal of Ichthyology* 25: 101–106.

Ollerton, J., D. McCollin, D. G. Fautin, and G. R. Allen. 2007. Finding NEMO: Nestedness engendered by mutualistic organization in anemonefish and their hosts. *Proceedings of the Royal Society B* 274: 591–598.

Reed, C., R. Branconi, C. Johnson, and P. Buston. 2019. Competitive growth in a social fish. *Biology Letters* 15: 20180737.

Ricciardi, F., Boyer, M., and Ollerton, J. 2010. Assemblage and interaction structure of the anemonefish-anemone mutualism across the Manado region of Sulawesi, Indonesia. *Environmental Biology of Fishes* 87: 333–347.

Richardson, D. L. 1999. Correlates of environmental variables with patterns in the distribution and abundance of two anemonefishes (Pomacentridae: *Amphiprion*) on an eastern Australian sub-tropical reef system. *Environmental Biology of Fishes* 55: 255–263.

Ross, R. M. 1978. Territorial behavior and ecology of the anemonefish *Amphiprion melanopus* on Guam. *Zeitschrift für Tierpsychologie* 46: 71–83.

Rueger, T., T. A. Barbasch, M. Y. L. Wong, M. Srinivasan, G. P. Jones, and P. M. Buston. 2018. Reproductive control via the threat of eviction in the clown anemonefish. *Proceedings of the Royal Society B* 285: 20181295.

Saenz-Agudelo, P., G. P. Jones, S. R. Thorrold, and S. Planes. 2011. Detrimental effects of host anemone bleaching on anemonefish populations. *Coral Reefs* 30: 497–506.

Sato, M., K. Honda, K. O. Bolisay, Y. Nakamura, M. D. Fortes, and M. Nakaoka. 2014. Factors affecting the local abundance of two anemonefishes (*Amphiprion frenatus* and *A. perideraion*) around a semi-closed bay in Puerto Galera, the Philippines. *Hydrobiologia* 733: 63–69.

Schmitt, R. J., and S. J. Holbrook. 2000. Habitat-limited recruitment of coral reef damselfish. *Ecology* 81: 3479–3494.

Schmitt, R. J., and S. J. Holbrook. 2003. Mutualism can mediate competition and promote coexistence. *Ecology Letters* 6: 898–902.

Srinivasan, M., G. P. Jones, and M. J. Caley. 1999. Experimental evaluation of the roles of habitat selection and interspecific competition in determining patterns of host use by two anemonefishes. *Marine Ecology Progress Series* 186: 283–292.

Steinberg, R. K., M. H. van der Meer, M. S. Pratchett, L. van Herwerden, and J. P. A. Hobbs. 2020. Keep your friends close and your anemones closer – ecology of endemic wideband anemonefish, *Amphiprion latezonatus*. *Environmental Biology of Fishes* 103: 1513–1526.

Verwey, J. 1930. Coral reef studies. I. The symbiosis between damselfishes and sea anemones in Batavia Bay. *Treubia* 12: 305–366.

Wong, M. Y. L., C. Uppaluri, A. Medina, J. Seymour, and P. M. Buston. 2016. The four elements of within-group conflict in animal societies: An experimental test using the clown anemonefish, *Amphiprion percula*. *Behavioural Ecology and Sociobiology* 70: 1467–1475.

19 No Place Like Home
Can Omics Uncover the Secret behind the Sea Anemone and Anemonefish Symbiotic Relationship?

Cassie M. Hoepner, Emily K. Fobert, Catherine A. Abbott, and Karen Burke da Silva

CONTENTS

19.1 INTRODUCTION

First recorded in 1868 (Collingwood 1868), anemonefish and anemones have one of the most well-known and iconic symbiotic relationships (Hobbs et al. 2012; Mebs 2009; Nedosyko et al. 2014). There are 28 different species of anemonefish that form associations with only ten species of host anemones (Fautin and Allen 1992). Although the association between anemonefish and sea anemones has existed for at least 12 million years (Marcionetti et al. 2019), this symbiotic relationship is quite rare, occurring in only ten out of over 1,200 species of anemones. Anemones also form a tripartite symbiosis with zooxanthellae that provide up to 85% of their daily nutrient budget (Lonnstedt and Frisch 2014). The symbiotic relationship with anemonefish has likely evolved three times amongst three unrelated anemone families (Thalassianthiade, Actinidae, Stichodactylidae) (Titus et al. 2019), with two genera contributing seven species (*Heteractis* – four species; *Stichodactyla* – three species) (Fautin 1991). In comparison, the anemonefish mutualism with anemones is thought to be present in the common ancestor of all anemonefish (Litsios et al. 2012). The evolution and diversification of anemonefish have benefited from their associations with host anemones, through increased rates of species diversification and morphological evolution in comparison to other coral reef fish without anemone associations (Litsios et al. 2012). The majority of anemonefish diversity is thought to have occurred in the last five million years, with 25 of the 28 species evolving during that time.

The mutualistic nature of the anemone and anemonefish symbiosis indicates that both organisms provide and receive a variety of benefits. For anemonefish, the toxic anemone provides a safe site for reproduction and protection from predation (Holbrook and Schmitt 2004). In return, anemonefish aid the growth, reproduction, and survival of anemones by providing nutrients (such as nitrogen and carbon) via faeces, increasing oxygenation by swimming amongst the tentacles, and actively defending their host anemone from various predators such as chaetodontid

DOI: 10.1201/9781003125365-23

fishes and sea turtles (Godwin and Fautin 1992; Nedosyko et al. 2014; Frisch et al. 2016; Mariscal 1970a). This unique symbiosis has allowed anemonefish to develop a range of life-history traits that can be attributed to their close association with venomous host anemones. For example, evidence suggests that anemonefish have exceptionally long lifespans for a reef fish of their size, living up to 30-plus years (Buston and Garcia 2007), compared to five to ten years for other similarly sized reef fish (Sale 1980). Anemonefish are also unusually bold and aggressive for their size. For example, when approached, anemonefish will swim out of their anemone towards the threat, rather than retreating to safety (Godwin and Fautin 1992). These traits are not seen in closely related damselfish or other similar-sized reef fishes, thus providing anemonefish with a unique ecological advantage (Marcionetti et al. 2018).

Although the ecological success of both anemone and anemonefish is clearly enhanced due to the evolution of their symbiotic relationship, the mechanism enabling anemonefish resistance to anemone venom remains unclear. Exactly how anemonefish glean such significant fitness advantages that improve their lifespan and potential reproductive success is not yet fully understood; however, it is widely believed that anemonefish have a unique mucus layer covering their scales that is somehow involved in enabling the formation and existence of their symbiotic relationship with sea anemones. Despite decades of study, there are still many more questions that remain unanswered such as: how do anemonefish live unharmed amongst the anemone's tentacles? How did this symbiotic relationship first evolve? And how do anemonefish pick the best anemone host?

In this chapter we (1) present an overview of the symbiotic relationship between anemones and anemonefish, including the factors that influence host selection; (2) present current hypotheses and discuss the existing evidence within the literature with a particular focus on the advances omics techniques have provided; (3) explore anemone venom research and discuss how toxin resistance in other model systems can be applied to further our understanding of the anemonefish and anemone symbiosis; and (4) discuss how omics can be applied in the future to help answer the remaining questions surrounding this symbiotic relationship.

19.2 INFLUENCES ON ANEMONEFISH HOST SELECTION

The relationship between different anemonefish species and anemone host species follows a unique and organized pattern that is not yet fully understood, with new associations being discovered even now (Bennett-Smith et al. 2021) (Table 19.1). Anemonefish can be classified as host generalists; for example, Clark's anemonefish (*Amphiprion clarkii*), is the only anemonefish species to form associations with all ten species of host anemones. In contrast, there are nine species of anemonefish that are host specialists (*A. frenatus*, *A. chagosensis*, *A. pacificus*, *A. fuscocaudatus*, *A. latifasciatus*, *A. mccullochi*, *A. nigripes*, *A. sebae*, and *A. biaculeatus*), forming associations with only a single anemone species (Burke da Silva and Nedosyko 2016). Despite co-existing within the same geographic region, there are a large number of anemone hosts with which anemonefish species do not associate (Table 19.1). This pattern demonstrates that geographic range is not the factor that determines which anemonefish and anemone species associate (Fautin 1986). Other factors that may contribute to which species form associations could include: (1) anemone morphology, (2) anemone toxicity, and/or (3) intraspecific competition amongst anemonefish species.

19.2.1 Anemone Morphology

Anemone tentacle length may provide a selective advantage to anemonefish by concealing them from predators (Huebner et al. 2012). Anemone species with longer tentacles can provide a larger surface area for anemonefish to hide in and thus reduce the visibility of the anemonefish to predators (Huebner et al. 2012). Stevens and Merilaita (2009) hypothesized that anemonefish stripes act to break up the body shape, making it more difficult for predators to detect the anemonefish amongst the tentacles and thus enhancing the anemone's protective features at varying distances. The number of stripes on anemonefish was found to be correlated to the length of their host anemone's tentacle; anemonefish species with two to three stripes form relationships with anemone species that had longer tentacle morphology, compared to anemonefish species with one or no stripes (Merilaita and Kelley 2018) (Table 19.1). Merilaita and Kelley (2018) also found that anemonefish with fewer stripes formed associations with a smaller number of host anemone species compared to anemonefish species with more stripes.

Furthermore, the morphology of anemone tentacles may make a species attractive as hosts for anemonefish. For example, the beaded anemone (*Heteractis aurora*) and bubble-tip anemone (*Entacmaea quadricolor*) have unique tentacle shapes that increase the surface area in which the anemonefish can hide, with dense beaded or bulb-like tentacles (Figures 19.1a,b). The magnificent anemone (*Heteractis magnifica*) has the unique ability to enclose all its tentacles within its soft body by contracting inwards when disturbed (Figure 19.1c), providing increased protection to the anemonefish who can hide inside the anemone body during this dangerous time. As anemonefish rely on their anemone host for protection, anemonefish may favour hosts whose morphological traits offer them better shelter or protection from predators.

19.2.2 Anemone Toxicity

Host anemone species range in the potency of their venom, from low to high haemolytic and neurotoxic toxicities (Nedosyko et al. 2014). Interestingly, host anemones with higher haemolytic and neurotoxic toxicities have shorter tentacles (< 20 mm) compared with anemones with

TABLE 19.1
Matrix of Anemonefish and Host Sea Anemone Associations According to Toxicity

***Amphiprion* spp:**	***Heteractis malu***	***Macrodactyla doreensis***	***Heteractis crispa***	***Heteractis aurora***	***Entacmaea quadricolor***	***Heteractis magnifica***	***Stichodactyla mertensii***	***Stichodactyla haddoni***	***Stichodactyla gigantea***	***Cryptodendrum adhaesivum***	**Number of hosts**	**Number of stripes**
clarkii	X	X	X	X	X	X	X	X	X	X	10	3
akindynos	Ø	Ø	X	X	X	X	X	X	X	Ø	7	2
chrysopterus	Ø	X	X	X	X	X	X	X	Ø	Ø	7	2
bicinctus			X	X	X	X	X	X	X	Ø	7	2
tricinctus	Ø		X	X	X	Ø	X	X	Ø	Ø	5	3
chrysogaster				X		X	X	X			4	3
perideraion	Ø	X	X	Ø	X	X	Ø	Ø	X	Ø	4	2
melanopus	Ø	Ø	X	Ø	X	X	Ø	Ø	X	Ø	4	1
allardi			Ø	X	X	Ø	X	Ø	Ø	Ø	3	2
ocellaris	Ø	Ø	Ø	Ø	Ø	X	X	Ø	X	Ø	3	3
percula	Ø	Ø	X	Ø	Ø	X		Ø	X	Ø	3	3
polymnus	Ø	X	X	Ø	Ø	Ø	Ø	X	Ø	Ø	3	3
omanensis			X	Ø	X	Ø	Ø	X	Ø	Ø	3	2
rubrocinctus	Ø		Ø	Ø	X	Ø	Ø	Ø	X	Ø	2	1
sandaracinos	Ø		X	Ø	Ø	Ø	X	Ø	Ø	Ø	2	1
akallopisos	Ø	Ø	Ø	Ø	Ø	X	X	Ø	Ø	Ø	2	1
barberi			X		X						2	1
ephippium	Ø		X	Ø	X	Ø	Ø	Ø	Ø	Ø	2	0
latezonatus			X		X			Ø	Ø		2	4
frenatus	Ø		Ø	Ø	X	Ø	Ø	Ø	Ø	Ø	1	1
chagosensis			Ø	Ø	X	Ø	Ø	Ø	Ø	Ø	1	2
pacificus	Ø	Ø	Ø	Ø	Ø	X	Ø	Ø	Ø	Ø	1	1
fuscocaudatus		Ø		Ø	Ø	Ø	X	Ø			1	3
latifasciatus					Ø	Ø	X	Ø			1	2
mccullochi			Ø		X			Ø	Ø		1	1
nigripes			Ø	Ø	Ø	X	Ø	Ø	Ø	Ø	1	1
sebae	Ø		Ø	Ø	Ø	Ø	Ø		Ø	Ø	1	2
biaculeatus	Ø	Ø	Ø	Ø	X (solitary)	Ø	Ø	Ø	Ø	Ø	1	3
Number of associates	1	4	14	7	17	12	12	9	8	1		
Tentacle length (mm)	40	175	75	50	100	100	20	10	10	5		
Toxicity rank	1	2	3	4	5	6	6	7	7	8		

Note: X indicates species that associate, Ø indicates species that are in the same marine province (Litsios et al. 2012), but do not associate. Association matrix updated from Burke da Silva and Nedosyko (2016) via Bennett-Smith et al. (2021), Allen et al. (2010), Allen et al. (2008), Fautin and Allen (1992), Hobbs et al. (2014), Scott et al. (2015); Pryor et al. (2022). Tentacle length and toxicity data from Merilaita and Kelley (2018), Nedosyko et al. (2014).

FIGURE 19.1 Various morphology of anemone hosts that aid in camouflaging anemonefish. A) Beaded tentacles of *Heteractis aurora*, B) bulb-like tentacles of *Entacmaea quadricolor*, C) retraction of tentacles by *Heteractis magnifica*. Images: Emily Fobert.

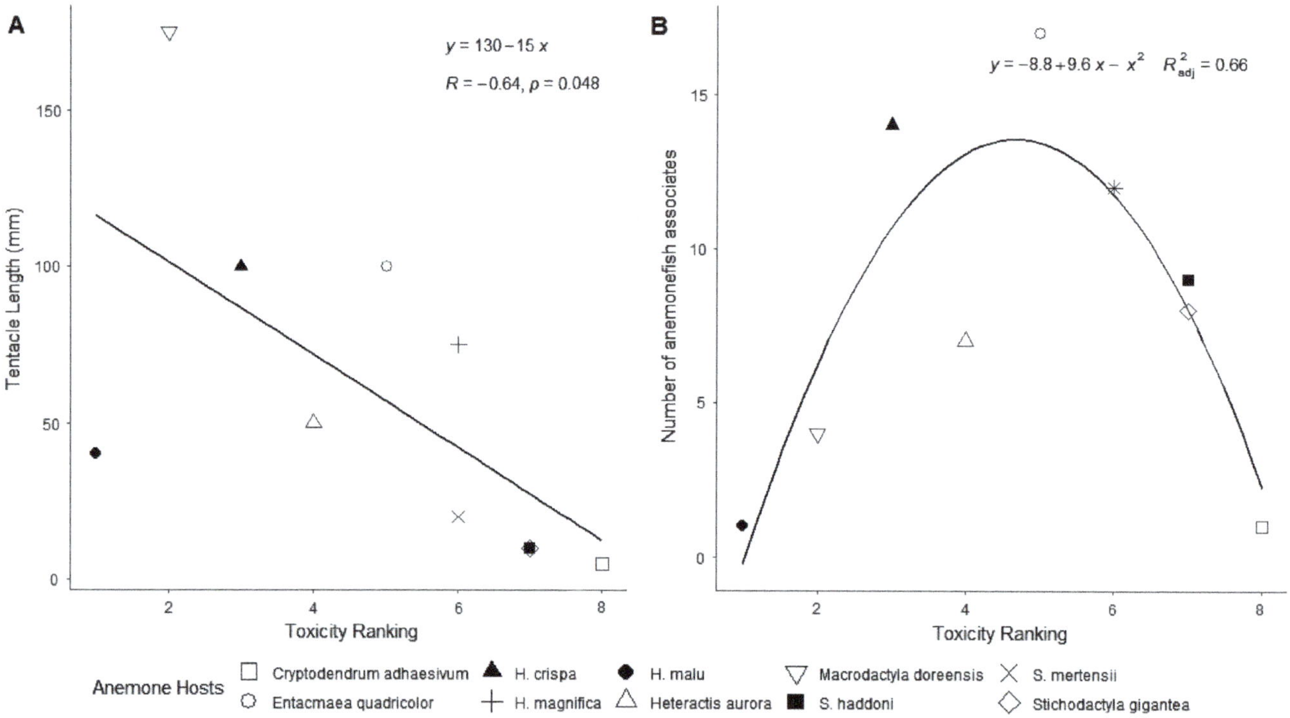

FIGURE 19.2 A) Negative relationship between anemone tentacle length (mm) and overall host anemone toxicity ranking (Merilaita and Kelley 2018). B) Relationship between number of anemonefish associates and overall host anemone toxicity ranking. Updated from Merilaita and Kelley (2018), Nedosyko et al. (2014).

mid-range or low toxicities (Figure 19.2a) (Merilaita and Kelley 2018). This creates a protective trade-off, where anemones with higher toxicity levels are potentially better able to protect their anemonefish through their venom and thus do not need to invest in increased tentacle length to provide shelter for the anemonefish. Less toxic anemone hosts may use a combination of a low toxicity venom and a longer tentacle length to provide better shelter for anemonefish, than low toxicity alone would. The corkscrew anemone (*Macrodactyla doreensis*) is a key example of this trade-off, having the second-lowest toxicity level but the longest tentacles of any host anemone (175 mm) (Fautin and Allen 1992).

A study by Nedosyko et al. (2014) found a relationship between host anemone haemolytic and neurotoxic toxicity and anemonefish preference (Figure 19.2b). Host anemones that fell into the mid-range toxicity had the highest number of anemonefish species as symbiotic partners. These results suggest toxicity may be an important factor in anemonefish host preference and that anemone toxicity and the fitness costs associated with withstanding toxin is an important aspect of anemonefish and anemone symbiosis. Forming an association with an anemone species that has low toxicity may provide a small fitness advantage to anemonefish by helping them to gain protection from predators; however, evolving resistance to an anemone species that has high toxicity may require large energetic costs, which could also have negative impacts on anemonefish fitness. Thus, anemone species with mid-range toxicity may provide the best protection per energetic cost, and ultimately be preferred by more anemonefish species (Nedosyko et al. 2014). The anemone species *E. quadricolor*, which has a mid-range haemolytic and neurotoxic toxicity, forms associations with 16 of the 28 species of anemonefish, whereas the delicate anemone (*Heteractis malu*) with lowest toxicity and the pizza anemone (*Cryptodendrum adhaesivum*) with highest haemolytic and neurotoxic toxicity form associations with only a single anemonefish species (Fautin and Allen 1992). These association patterns provide support for the suggestion that toxicity plays a key role in the establishment and maintenance of symbiotic relationships between different anemone and anemonefish species (Nedosyko et al. 2014; Burke da Silva and Nedosyko 2016).

19.2.3 Interspecific Competition amongst Anemonefish Species

Interspecific competition for anemone host species can be an indicator of host quality or host preference by anemonefish. Fautin (1986) defined preferred hosts as those harbouring many anemonefish associates. Anemonefish are known to be aggressive, which is needed to maintain ownership of their anemones, as well as the social hierarchies within the anemone, to exclude or eliminate individuals from the anemone, or for larger more dominant species to obtain a preferred or occupied anemone (Burke da Silva and Nedosyko 2016; Buston 2003). Competitive exclusion between anemonefish species for preferred hosts is thought to be a key factor influencing which associations are found between anemone hosts and the different anemonefish species (Srinivasan 1999; Burke da Silva and Nedosyko 2016). The maroon clownfish (*Amphiprion biaculeatus*) is thought to be competitively dominant over all other anemonefish species (Srinivasan 1999) and is an anemone specialist only found in the anemone species *E. quadricolor*, the anemone in the mid-toxicity range. Similarly, other large anemonefish species such as *A. melanopus* are also generally found specializing in preferred mid-toxicity range host anemones, particularly when there is competition with other smaller anemonefish species on the same reef (Fautin 1986). As climate change continues to impact host quality and availability, it is likely that an increase in competitive exclusion by larger dominant anemonefish species may occur, leaving smaller anemonefish species vulnerable to predation (Saenz-Agudelo et al. 2011; Scott and Hoey 2017; Hoepner and Fobert 2022).

19.3 CURRENT HYPOTHESES AND OMICS APPLICATIONS TO UNCOVER THE MECHANISM BEHIND THE ANEMONE AND ANEMONEFISH SYMBIOSIS

Despite decades of research, the exact mechanism that enables anemonefish to live within the toxic environment of their host anemone has yet to be resolved. Several studies have found the mucus layer of anemonefish to be chemically different to that of other coral reef fish (Abdullah and Saad 2015; Balamurugan et al. 2015; Lubbock 1980), concluding that the anemonefish mucus layer may be the key to their protection. However, there are now new technologies available to help us investigate the mechanism(s) behind anemonefish resistance to anemone venom. Advancements in omics techniques such as genomics, transcriptomics, and proteomics will enable the exploration of this symbiotic relationship at a molecular level and may provide insights not previously attainable. In recent years, omics techniques have started to be used to tackle questions related to the symbiotic relationship between sea anemones and anemonefish, with a focus on metagenomics and genomics. Four main hypotheses have been proposed to explain how the anemonefish mucus layer can provide anemonefish with unique protection from the anemone venom. These hypotheses are summarized in Table 19.2, and each is discussed in the following with a focus on areas where omics technologies have currently been applied.

19.3.1 Hypothesis 1: Anemonefish Are Innately Protected from Anemone Venom

An early hypothesis was that anemonefish are born protected and therefore are innately immune to anemone venom (Elliot and Mariscal 1996; Miyagawa and Hidaka 1980). This research focused on the anemonefish species *A. clarkii*, which is able to form associations with all ten species of host anemones and can enter anemones with little or no acclimation time (Miyagawa and Hidaka 1980). Through multiple laboratory experiments, focusing on a number of different anemonefish species, it was noted that fish require an acclimation period in order to fully enter and remain within a host anemone (Balamurugan et al. 2015; Brooks and Mariscal 1984; Mebs 1994; Davenport and Norton 1958; Mariscal 1970a). This acclimation period can vary between anemonefish species, ranging from minutes to days before the fish can comfortably exist within the anemone (Balamurugan et al. 2015; pers obv; Miyagawa and Hidaka 1980). Anemonefish perform a range of specific behaviors – including touching anemone tentacles with their tail, biting the tentacle tips, and continuous fanning of tentacles with their pectoral fins – to acclimate and then enter the anemone (Balamurugan et al. 2015). Furthermore, anemonefish also lose their protection when isolated from their anemone host for more than 21 hours and are required to reacclimate (Mariscal 1970b). Overall, the experimental evidence clearly indicates that anemonefish require an acclimation period to form symbiosis with a host anemone. The acclimation period may activate the expression of novel genes that have been inherited from the one common anemonefish ancestor as the anemonefish species diversified (Litsios et al. 2012), allowing for the anemonefish to switch on their resistance to anemone venom.

TABLE 19.2
Previous Research into the Mechanism behind the Anemonefish Symbiosis with Anemones Fits into Four Main Hypotheses

	Hypothesis	Status	Reference	Methodology
1a	**Anemonefish are innately protected from anemone venom**	Rejected	Miyagawa and Hidaka 1980	Forced contact
			Elliot and Mariscal 1996	Forced contact
			Davenport and Norris 1958	Observation of behaviors
b	**Anemonefish gain protection through an acclimation period**	Supported	Mariscal 1970a, b	Observation of behaviors
			Brooks and Mariscal 1984	Acclimation time to surrogate anemones
			Mebs 1994	Ichthyotoxic activity
			Balamurugan et al. 2015	Observation of behaviors
2	**Anemonefish have a thicker mucus layer than other fish**	Insufficient evidence	Lubbock 1980	Nomarski optics (*A. clarkii*)
3	**Anemonefish mucus molecularly mimics the composition of anemone mucus**	Insufficient evidence	Schlichter 1976	Electrophoresis/radiolabelled mucus
			Elliot et al. 1994	Antibody assays
4	**Anemonefish mucus does not trigger firing of the anemone's nematocysts**	Insufficient evidence	Lubbok 1980	Nematocysts per cm^2
			Abdullah and Saad 2015	N-acetylneuraminic Acid Detection

19.3.2 Hypothesis 2: Anemonefish Have a Thicker Mucus Layer Than Other Fish

Another key hypothesis is that anemonefish have a thicker mucus layer than other coral reef fish species that cannot enter an anemone (Lubbock 1980). By having a thicker mucus layer, it is thought that the anemonefish are better able to withstand the sting of the anemone, or that the nematocysts – the firing cells that deliver the anemones' sting – are unable to penetrate the skin due to the mucus barrier. Lubbock (1980) showed that *A. clarkii* mucus was three to four times thicker than that of other coral reef fish species, but that there was no significant difference in mucus thickness when the anemonefish were associated with an anemone host (*S. haddoni* or *E. quadricolor*), compared to anemonefish separated from the anemone host for five months. As anemonefish are initially stung upon entering the anemone (Balamurugan et al. 2015; Brooks and Mariscal 1984; Mebs 1994; Davenport and Norton 1958; Mariscal 1970a) and the mucus thickness does not change with acclimation, it is unlikely that mucus thickness is the sole mechanism for anemonefish toxin resistance. Furthermore, only one of 28 species of anemonefish have been examined for mucus thickness; therefore it is currently unclear if all anemonefish species have thicker mucus layers than other coral reef fish.

19.3.3 Hypothesis 3: Anemonefish Mucus Molecularly Mimics the Composition of Anemone Mucus

One of the most popular hypotheses is that the anemonefish cover themselves in anemone mucus to molecularly disguise themselves and live undetected amongst the anemone's tentacles, referred to as molecular mimicry (Schlichter 1976; Elliot et al. 1994). It is proposed that the anemonefish cover their body in anemone mucus, thus inhibiting the firing of anemone nematocysts, via the same mechanism anemones use to recognize their own tentacles and prevent firing nematocysts at themselves. This is referred to as self-/non-self-recognition and anemone antigens (proteins or peptides) are thought to be involved in this self-recognition process (Elliot et al. 1994).

A study by (Elliot et al. 1994) found that anemonefish (*A. clarkii*) living within an anemone host (*H. crispa* and *S. haddoni*) had anemone antigens in their mucus, whereas these anemone antigens were not found in the mucus of *A. clarkii* that were separated from the anemone and only sharing an aquarium separated by a partition. Previously, Pantin (1942) found that anemones did not fire nematocysts at food sources covered in their own mucus, whereas it has been shown that anemones will fire nematocysts when presented with the mucus of another anemone species (Ertman and Davenport 1981). This evidence suggests that molecular mimicry likely plays a role in anemonefish protection from their host anemone.

There are three ways in which anemonefish may acquire anemone peptides or proteins in their mucus: (1) anemonefish may cover themselves with a coat of the anemone's mucus during brief contact with the anemone tentacles during the acclimation period; (2) some anemone surface antigens may be incorporated into the mucus coating of the anemonefish (Elliot et al. 1994); or (3) anemonefish produce their own proteins, molecularly similar to anemone proteins that they embed in their mucus layer when in contact with a host anemone. To date, there is no experimental evidence that discerns between these three possible mechanisms behind the molecular mimicry that allows the anemone to recognize the anemonefish as self, facilitating their symbiosis. However, metagenomics studies have found that the microbiome of anemone and anemonefish mucus can converge during association (Pratte et al. 2018; Roux et al. 2019; Titus et al. 2020), demonstrating the potential for microbial proteins to be involved in molecular mimicry or the facilitation of the symbiotic relationship.

19.3.3.1 Omics Application: Metagenomics

Three recent studies have investigated the diversity of the mucus microbiome from anemonefish and their symbiotic sea anemone hosts (Pratte et al. 2018; Roux et al. 2019; Titus et al. 2020). All three studies found similar results, that the microbiomes of anemones and anemonefish were different when not in direct contact, and that direct microbial transfer or a shift in diversity occurs, making the microbiomes more similar when anemonefish and anemones are in contact. Specifically, Titus et al. (2020) found that the microbiomes of anemones (*C. adhaesivum, E. quadricolor, H. aurora, H. magnifica, and S. mertensii*) that were hosts to the same species of anemonefish (*A. nigripes* or *A. clarkii*) were more similar to each other than to that of anemones that were hosts to different species of anemonefish, or no fish at all. Pratte et al. (2018) also found that the microbiome of *A. clarkii* reverted back to a pre-association state after removal from the anemone *E. quadricolor*. The study by Roux et al. (2019) suggested that the convergence of microbiomes that occurred during anemone *H. magnifica* and the false clownfish (*A. ocellaris*) association could play a role in the establishment of their symbiosis. Bacteria in the mucus could allow for the transfer or processing of proteins and metabolites between the species, for example, to allow for the anemonefish to withstand the anemone's venom (Roux et al. 2019). This gives support to the hypothesis that anemonefish molecularly (or at least bacterially) mimic the anemone to disguise themselves amongst the anemone tentacles.

19.3.4 Hypothesis 4: Anemonefish Mucus Lacks the Trigger for Firing the Anemone's Nematocysts

The final key hypothesis is that the anemonefish mucus layer lacks the trigger for the anemone to fire nematocysts. Lubbock (1980) qualitatively observed the behavioral response of Haddon's anemone (*Stichodactyla haddoni*) to different mucus types on a glass rod (response categories: no response, poor response, strong response). *Amphiprion clarkii* mucus in contact with *S. haddoni* did not elicit a behavioral response (10/10) and *A. clarkii* mucus isolated from a host anemone also did not elicit a behavioral response (37/45), whereas mucus from closely related damselfishes elicited strong responses in all instances from *S. haddoni* – humbug damselfish (*Dascyllus aruanus*) (25/25), black-and-gold chromis (*Paraglyphidodon nigroris*) (5/5), and blue-green chromis (*Chromis caerulea*) (5/5). Lubbock (1980) also found that there was no difference between the number of nematocysts fired by the anemone at gelatine-covered coverslips in the presence or absence of anemonefish (10^4 capsules/mm^2) (Lubbock 1980), demonstrating that anemonefish presence does not impact the ability of the anemone to fire nematocysts at external stimuli. There is no study to date that has quantified the nematocyst firing response of a host anemone when presented with anemonefish mucus. However, the use of genomics has increased our understanding of the potential proteins utilized in the prevention of nematocyst discharge.

19.3.4.1 Omics Application: Genomics

A study by Marcionetti et al. (2019) identified the first candidate genes that may have evolved to grant anemonefish protection from anemone venom. This study utilized whole-genome assemblies from ten anemonefish species (*A. biaculeatus, A. ocellaris, A. perideraion, A. akallopisos, A. polymnus, A. sebae, A. melanopus, A. bicinctus, A. nigripes*, and *A. frenatus*) and applied molecular evolutionary analysis to uncover specific genes that were positively selected for during the evolution of symbiosis. Seventeen genes were identified as being under positive selection at the origin of anemonefish, which later switched to purifying selection. When advantageous traits evolve, they are usually positively selected for and then there is a switch to purifying selection to maintain these traits in descendants (Marcionetti et al. 2019).

Versican Core Protein was one of the genes identified and is particularly interesting due to its link to the anemone nematocyst firing mechanism. Nematocysts are highly specialized cells that distribute the anemone's venom by piercing the skin of predators or prey. The discharge of the nematocyst is controlled by chemosensory, mechanosensory, and endogenous pathways that respond to sensory stimulation (Anderson and Bouchard 2009). Anemones possess chemoreceptors for N-acetylneuraminic acid (Neu5Ac), a type of salic acid and a common carbohydrate side chain of glycoproteins found in fish mucus. Binding of the chemoreceptor to sugars in the mucus, specifically the acidic side chain of glycoproteins, triggers a multi-signal pathway that causes the nematocyst to fire (Anderson and Bouchard 2009; Ozacmak et al. 2001). Mucus from many coral reefs species has been shown to contain Neu5Ac; however, Neu5Ac has been found to be significantly lower in the mucus of *A. ocellaris* (Abdullah and Saad 2015). Abdullah and Saad (2015) found that *A. ocellaris* lacked Neu5Ac (1.6 mg/mL), in comparison to other non-symbiotic fishes such as the scissor-tailed sergeant (*Abudefduf sexfasciatus*) (50.4mg/mL) and moon wrasse (*Thalassoma lunare*) (71.9 mg/mL). Lubbock (1980) also showed that the mucus of *A. clarkii* was chemically different to other coral reef fish that are unable to enter host anemone species. The mucus of Clark's anemonefish (*A. clarkii*) mainly consisted of neutral glycoproteins, which could be produced by a lack of an acidic side chain on the N-acetylated sugars that is normally present in fish mucus glycoproteins (Abdullah and Saad 2015). Versican core protein found to be expressed in the epidermis of *A. ocellaris* is thought to potentially bind to N-acetylated sugars, masking their detection by anemone chemoreceptors and thus failing to trigger nematocyst firing. Protein O-GlcNAse was also found to be positively selected for, and this protein has the potential to cleave the acidic side chain creating a neutral glycoprotein that does not stimulate the anemone chemoreceptors (i.e., does not trigger) (Marcionetti et al. 2019) providing support for the hypothesis that anemonefish mucus lacks the trigger for anemone nematocyst firing.

19.3.5 Hypotheses for Future Research

Of the hypotheses presented, only two hypotheses stand out as possible mechanistic explanations of anemonefish toxin resistance: firstly hypothesis (3), the anemonefish mucus molecularly mimics the composition of the anemone's mucus to inhibit nematocyst firing, and secondly hypothesis (4), the anemonefish mucus does not trigger the firing of anemone nematocysts. While these two concepts may seem similar, we are defining (1) inhibits firing: as mucus properties that bind to receptors preventing the nematocysts' firing (Elliot, Mariscal, and Roux 1994' Lubbock 1980) and (2) does not trigger: as mucus composition that lacks the trigger to stimulate the anemone's nematocyst firing. Ultimately, as the anemonefish need to perform acclimation behaviors in order to enter the anemone, there must be a change in the anemonefish's mucus at the molecular level. Moving forward in this chapter we will explore how lessons from venom research and toxin resistance in other species can be used as a model to better understand the anemone and anemonefish symbiosis and how omics have been used in these models and can inform future investigation into the potential mechanisms behind these hypotheses at the molecular level.

19.4 LESSONS FROM OTHER MODEL SYSTEMS

To uncover the mechanism(s) that anemonefish use to withstand the anemone's venom, we need to better understand the evolution of the anemone's venom itself. Anemone venom is a complex and diverse mixture of a variety of toxic components, including cytolysins (toxins that cause cell lysis), neurotoxins (toxins that damage or impair the nervous system), and phospholipases (enzymes which cause inflammation and pain) amongst many others (Anderluh and Macek 2002; Frazao et al. 2012; Madio et al. 2019). Furthermore, cnidarians (corals, anemones, and jellyfish) are the only organisms that do not have a centralized venom gland like other venomous organisms (e.g., snakes); instead, the venom is produced in tissues throughout their body via nematocytes and ectodermal gland cells (Madio et al. 2019). Nematocysts, which are found in the anemone tentacles, are highly specialized cells that venom is packaged into. Nematocysts consist of a capsule with an inverted tubule, which when triggered expels the tubule that disperses the venom by piercing the skin of predators or prey. The discharge of nematocysts is controlled by chemosensory, mechanosensory, and endogenous pathways that respond to external sensory stimulation (Anderson and Bouchard 2009). When predators or prey come into contact with the anemone, the anemone is able to chemically detect the response required and act accordingly. Ectodermal gland cells allow for the secretion of a larger volume of venomous mucus over the anemone, however, it is unclear if the venom composition of the mucus is the same or different to the venom packaged into the nematocysts (Madio et al. 2019), or if ectodermal gland cells are present in host anemones. While each component of the venom has a specific role, there are generally a few that contribute to the major lethality effect (Arbuckle et al. 2017). Potential symbiotic partners can benefit from this by attempting to evolve toxin resistance to the venom as a whole, rather than evolving resistance to each single component in the venom. This would enable partner species to selectively evolve resistance to the most lethal components or the most functionally similar elements, enabling multiple venom proteins to be treated as one for resistance purposes (Arbuckle et al. 2017).

While it is yet not clear how anemonefish are able to live within the toxic environment of anemones, we can look to other species and the mechanisms of toxin resistance utilized for new research avenues to explore in the anemone and anemonefish system. Resistance to toxins has evolved on multiple occasions across a wide variety of phyla, from mammals to fish and insects (Arbuckle et al. 2017). There are three main mechanisms that have been put forward to broadly explain the evolution of toxin resistance (Holding et al. 2016; Arbuckle, Rodriguez de la Vega, and Casewell 2017):

(1) Venom inhibitors: inhibitor proteins can inhibit the function of major toxic proteins found in venom through direct interaction, and are often members of large/old gene families. Venom inhibitors have been identified in at least 30 mammal species from six orders. Toxin-neutralizing serum factors, such as α1B-glycoprotein found in opossums and mongoose can neutralize snake venom metalloendopeptidases (SVMPs) and phospholipases (Holding et al. 2016; Voss and Jansa 2012). Venom inhibitors can also allow species such as snakes, for example, to be resistant to their own venom (Bastos et al. 2016). We know that anemones have self-recognition abilities which prevent the firing of nematocysts when their tentacles touch (Elliot et al. 1994). Proteins may have potentially evolved in anemonefish that can be used to disrupt or prevent the firing of nematocysts thus working as venom inhibitors. Versican core protein (Marcionetti et al. 2019), may be an example of this as it is thought to bind to N-acetylated sugars, masking their detection by anemone chemoreceptors.

(2) Target alteration: toxic proteins found in venom bind to a receptor protein in a prey species to elicit a toxic action. Thus, a small number of amino acid mutations in the receptor protein found in the prey can change it such that the toxin can no longer bind, while the receptor protein still maintains its

original physiological function. Target alterations are often members of small gene families, or even encoded by single genes. In cobras, binding resistance to alpha neurotoxins from snakes is caused by single amino acid substitutions that lead to glycosylation of the target protein that then prevents the binding of the toxin (Takacs et al. 2001). The evolution of the protein O-GlcNAse gene and the expression of this protein in the anemonefish epidermis potentially allow for the cleavage of the acidic side chain on glycoproteins in the anemonefish mucus (Marcionetti et al. 2019). This may be an example of target alteration as the nematocyst firing would no longer be triggered by the now neutral glycoproteins.

(3) Repurposed toxins: is the binding of venom proteins to an untargeted receptor, blocking the effects of the venom components that cause pain or other lethal actions. These can also occur with just a single amino acid replacement (Arbuckle et al. 2017). Grasshopper mice, who eat and are often stung by scorpions, are the only known example of a species that has evolved the use of repurposed toxins. This response results in the binding of the toxin to a downstream sodium channel rather than the targeted sodium channel, resulting in numbness in the mice rather than pain (Rowe et al. 2013). Anemonefish go through an acclimation process to associate with the anemone, however, what exactly happens at the molecular level during this acclimation is currently unknown. Repurposing of toxins to untargeted receptors could be activated during this process resulting in anemonefish no longer feeling the sting of their anemone host.

In general, in predator/prey relationships, prey species often evolve a biochemical defence or resistance to a predator's venom, triggering an increase in venom toxicity by the predator. Prey resistance will then also increase, resulting in a coevolutionary chemical arms race (Brodie III and Brodie Jr. 1999). In contrast, in a symbiotic relationship, where the aim is to maintain mutualistic benefit, a balance between maintaining venom toxicity level but still enabling a symbiotic partner to interact is important. However, the toxicity must remain at a level that can continue to benefit the toxic species. The anemone and anemonefish mutualistic relationship requires a balance of toxin resistance and venom strength, rather than an arms race of increasing toxin and resistance levels. Research by Nedosyko et al. (2014) supports this concept as they showed that host anemones with mid-range toxicity had the highest number of anemonefish associates, demonstrating that there is a trade-off between producing a venom that is too venomous or not venomous enough and being able to host anemonefish.

In recent years progress in deciphering the mechanisms behind the anemonefish and anemone symbiosis has stalled, despite technological development. Just as we use anemonefish as a model species for other research applications, study into this symbiotic relationship may benefit from the application of concepts and knowledge from venom transcriptomic and proteomic studies (Sunagar et al. 2016; Madio, Undheim, and King 2017) and the study of evolution of toxin resistance in other species, particularly of prey to snake venoms (Gibbs et al. 2020).

19.5 FUTURE USE OF OMICS

While researchers have begun to use omics to investigate a mechanistic explanation for anemone and anemonefish symbiosis, there is a wide array of omics techniques that could still be applied, particularly focusing on the fish mucus layer and how it acts to protect the anemonefish from the anemone venom. Fish mucus is comprised of a combination of proteins, lipids, and glycoproteins, all of which can be analyzed via omics to test the two leading hypotheses for the mechanism(s) behind this symbiosis: (1) hypothesis 3: anemonefish mucus molecularly mimics the anemone's mucus and (2) hypothesis 4: anemonefish mucus prevents the nematocysts firing. For example, given the importance of glycoproteins for triggering nematocysts response in anemones (hypothesis 3), analyzing the mucus layer of anemonefish using glycomics could provide insight into the side chain structure of the glycoproteins present in the anemonefish mucus and would provide support for the genomic research by Marcionetti et al. (2019). Additionally, utilizing proteomics, proteins from the anemone mucus can be identified and matched to proteins in the anemonefish mucus after association, which could determine if the anemonefish molecularly mimic anemone mucus (hypothesis 3). The merging of mucus microbiomes between anemones and anemonefish during association suggests that it is possible that mucus molecular composition will also show similarities during symbiosis. Further proteomics and transcriptomics studies of both fish and anemones under controlled experimental conditions or in the wild could be used to look to see if proteins targeted by venom components are altered and/or whether toxins are able to bind to decoy receptors as discussed earlier as mechanisms of resistance to snake venoms, to explore the possibility of anemonefish deploying these strategies.

Previously, the research into anemone and anemonefish symbiosis has focused solely on the anemonefish and how they adapt to live in the toxic environment of their anemone host. As this is a mutualistic relationship where both anemonefish and anemone gain fitness benefits from their association, the anemones' role in the formation of this symbiosis should also be explored. A combined transcriptomic and proteomic approach is becoming more popular when studying venom as it allows for a holistic view of venom composition (Madio et al. 2017). Using this approach, Madio et al. (2017) discovered 12 new families of venom proteins and peptides in Haddon's anemone (*S. haddoni*). Currently, research into anemone venoms focuses on novel toxin identification for drug discovery and medical applications, rather than an ecological focus (Hoepner et al. 2019); however, applying widely used techniques that explore drug discovery to an ecological-based venom question could allow for the exploration of the mechanism behind the anemone and anemonefish symbiosis from a perspective not yet explored. For example, the bubble-tip anemone (*E. quadricolor*) is the most popular host of anemonefish (Nedosyko et al. 2014), yet research into its venom composition is very limited. A combined transcriptomic and proteomic approach to investigate the venom composition of *E. quadricolor* will allow for the comparison of the venom to other anemone hosts as well as non-host anemones and could identify potential unique features of the venom that lend itself to symbiosis with anemonefish. Analysis of venom before and after forming associations with anemonefish could also uncover any changes in the anemone venom or production that could enable or enhance the association with anemonefish. Omics is a promising field for investigating how anemonefish mucus layer interacts with anemone venom at the molecular level and closely interrogating hypotheses posed for future research.

19.6 CONCLUSION

Despite decades of research, we are still exploring and discovering exactly how the anemonefish can withstand the venomous sting of their anemone hosts and live harmoniously for mutual benefit. Of the numerous hypotheses explored, there are two main frontrunners that could explain the mechanisms of anemonefish resistance to anemone venom: (1) hypothesis 3: the anemonefish mucus molecularly mimics the composition of the anemone's mucus to inhibit nematocyst firing and (2) hypothesis 4: the anemonefish mucus does not trigger the firing of anemone nematocysts. These hypotheses do have areas of overlap and it may be a combination of both mechanisms that results in overall protection. The application of omics techniques, such as transcriptomics, proteomics, and metabolomics, as well as learnings from other model systems to this ecological question, may provide the molecular insight needed to finally uncover the secrets behind the anemone and anemonefish symbiosis.

REFERENCES

Abdullah, N. S., and S. Saad. 2015. Rapid detection of N-acetylneuraminic acid from false clownfish using HPLC-FLD for symbiosis to host sea anemone. *Asian Journal of Applied Sciences* 3 (5): 858–864.

Allen, G. R., J. Drew, and D. Fenner. 2010. *Amphiprion pacificus*, a new species of anemonefish (Pomacentridae) from Fiji, Tonga, Samoa, and Wallis Island. *aqua, International Journal of Ichthyology* 16 (3): 129–138.

Allen, G. R., J. Drew, and L. Kaufman. 2008. *Amphiprion barberi*, a new species of anemonefish (Pomacentridae) from Fiji, Tonga, and Samoa. *aqua, International Journal of Ichthyology* 14 (3): 105–114.

Anderluh, G., and P. Macek. 2002. Cytolytic peptide and protein toxins from sea anemones (Anthozoa: Actiniaria). *Toxicon* 40 (2): 111–124.

Anderson, P. A., and C. Bouchard. 2009. The regulation of cnidocyte discharge. *Toxicon* 54 (8): 1046–1053.

Arbuckle, K., R. C. Rodriguez de la Vega, and N. R. Casewell. 2017. Coevolution takes the sting out of it: Evolutionary biology and mechanisms of toxin resistance in animals. *Toxicon* 140: 118–131.

Balamurugan, J., T. T. Ajith Kumar, R. Kannan, and H. D. Pradeep. 2015. Acclimation behaviour and bio-chemical changes during anemonefish (*Amphiprion sebae*) and sea anemone (*Stichodactyla haddoni*) symbiosis. *Symbiosis* 64 (3): 127–138.

Bastos, V. A., F. Gomes-Neto, J. Perales, A. G. Neves-Ferreira, and R. H. Valente. 2016. Natural inhibitors of snake venom metalloendopeptidases: History and current challenges. *Toxins (Basel)* 8 (9): 250.

Bennett-Smith, M. F., J. E. Majoris, B. M. Titus, and M. L. Berumen. 2021. Clownfish hosting anemones (Anthozoa, Actiniaria) of the Red Sea: New associations and distributions, historical misidentifications, and morphological variability. *Marine Biodiversity Records* 14 (1): 22.

Brodie III, E. D., and E. D. Brodie Jr. 1999. Costs of exploiting poisonous prey: Evolutionary trade-offs in a predator-prey arms race. *Evolution* 53: 626–631.

Brooks, R., and R. N. Mariscal. 1984. Acclimation of anemonefishes to sea anemones protection by changes in the fish's mucus coat. *Journal of Experimental Marine Biology and Ecology* 81: 277–285.

Burke da Silva, K., and A. Nedosyko. 2016. Sea anemones and anemonefish: A match made in heaven. In *The Cnidaria, Past, Present and Future*, eds S. Goffredo and Z. Dubinsky, 425–438. New York: Springer.

Buston, P. M. 2003. Forcible eviction and prevention of recruitment in the clown anemonefish. *Behavioral Ecology* 14 (4): 576–582.

Buston, P. M., and M. B. Garcia. 2007. An extraordinary life span estimate for the clown anemonefish *Amphiprion percula*. *Journal of Fish Biology* 70 (6): 1710–1719.

Collingwood, C. 1868. IV.—Note on the existence of gigantic sea-anemones in the China Sea, containing within them quasi-parasitic fish. *Journal of Natural History Series* 4 (1): 31–33.

Davenport, D., and K. S. Norton. 1958. Observations on the symbiosis of the sea anemone stoichactis and the pomacentrid fish *Amphiprion percula*. *The Biological Bulletin* 115 (3): 397–410.

Elliot, J. K., and R. N. Mariscal. 1996. Ontogenetic and interspecific variation in the protection of anemonefishes from sea anemones. *Journal of Experimental Marine Biology and Ecology* 208: 57–72.

Elliot, J. K., R. N. Mariscal, and K. H. Roux. 1994. Do anemonefishes use molecular mimicry to avoid being stung by host anemones? *Journal of Experimental Marine Biology and Ecology* 79 (1): 99–113.

Ertman, S. C., and D. Davenport. 1981. Tentacular nematocyte discharge and "self-recognition" in anthopleura elegantissima brandt. *Biological Bulletin* 161 (3): 366–370.

Fautin, D. G. 1986. Why do anemonefishes inhabit only some host actinians? *Environmental Biology of Fishes* 15 (3): 171–180.

Fautin, D. G. 1991. The anemonefish symbiosis: What is known and what is not. *Symbiosis* 10: 23–46.

Fautin, D. G., and G. R. Allen. 1992. *Field Guide to Anemonefishes and Their Host Sea Anemones*. Perth, WA: Western Australian Museum.

Frazao, B., V. Vasconcelos, and A. Antunes. 2012. Sea anemone (Cnidaria, Anthozoa, Actiniaria) toxins: An overview. *Marine Drugs* 10 (8): 1812–1851.

Frisch, A. J., J. R. Rizzari, K. P. Munkres, and J. P. A. Hobbs. 2016. Anemonefish depletion reduces survival, growth, reproduction and fishery productivity of mutualistic anemone–anemonefish colonies. *Coral Reefs* 35 (2): 375–386.

Gibbs, L. H., L. Sanz, A. Pérez, A. Ochoa, A. T. B. Hassinger, M. L. Holding, and J. J. Calvete. 2020. The molecular basis of venom resistance in a rattlesnake-squirrel predator-prey system. *Molecular Ecology* 29 (15): 2871–2888.

Godwin, J., and D. G. Fautin. 1992. Defense of host actinians by anemonefish. *Copeia* 3: 902–908.

Hobbs, J. P. A., M. Beger, M. De Brauwer, and M. J. Emslie. 2014. North-eastern range extension of the anemone *Stichodactyla haddoni* to the Marshall Islands represents a new record of host use by the endemic anemonefish *Amphiprion tricinctus*. *Marine Biodiversity Records* 7: e106.

Hobbs, J. P., A. J. Frisch, B. M. Ford, M. Thums, P. Saenz-Agudelo, K. A. Furby, and M. L. Berumen. 2012. Taxonomic, spatial and temporal patterns of bleaching in anemones inhabited by anemonefishes. *PLoS One* 8 (8): e70966.

Hoepner, C. M., C. A. Abbott, and K. Burke da Silva. 2019. The ecological importance of toxicity: Sea anemones maintain toxic defence when bleached. *Toxins (Basel)* 11 (5): 266.

Hoepner, C. M., and E. K. Fobert. 2022. Just keep swimming: Long-distance mobility of tomato clownfish following anemone bleaching. *Ecology* 103(3): e3619.

Holbrook, S. J., and R. J. Schmitt. 2004. Growth, reproduction and survival of a tropical sea anemone (Actiniaria): Benefits of hosting anemonefish. *Coral Reefs* 24 (1): 67–73.

Holding, M. L., D. H. Drabeck, S. A. Jansa, and H. L. Gibbs. 2016. Venom resistance as a model for understanding the molecular basis of complex coevolutionary adaptations. *Integrative and Comparative Biology* 56 (5): 1032–1043.

Huebner, L. K., B. Dailey, B. M. Titus, M. Khalaf, and N. E. Chadwick. 2012. Host preference and habitat segregation among Red Sea anemonefish: Effects of sea anemone traits and fish life stages. *Marine Ecology Progress Series* 464: 1–15.

Litsios, G., C. A. Sims, R. O. Wüest, P. B. Pearman, N. E. Zimmermann, and N. Salamin. 2012. Mutualism with sea anemones triggered the adaptive radiation of clownfishes. *BMC Evolutionary Biology* 12: 212.

Lonnstedt, O. M., and A. J. Frisch. 2014. Habitat bleaching disrupts threat responses and persistence in anemonefish. *Marine Ecology Progress Series* 517: 265–270.

Lubbock, R. 1980. Why are clownfishes not stung by sea anemones? *Proceedings of the Royal Society B: Biological Sciences* 207: 35–61.

Madio, B., G. F. King, and E. A. B. Undheim. 2019. Sea anemone toxins: A structural overview. *Mar Drugs* 17 (6): 325.

Madio, B., E. A. B. Undheim, and G. F. King. 2017. Revisiting venom of the sea anemone Stichodactyla haddoni: Omics techniques reveal the complete toxin arsenal of a well-studied sea anemone genus. *Journal of Proteomics* 166: 83–92.

Marcionetti, A., V. Rossier, J. A. M. Bertrand, G. Litsios, and N. Salamin. 2018. First draft genome of an iconic clownfish species (*Amphiprion frenatus*). *Molecular Ecology Resources* 18 (5): 1092–1101.

Marcionetti, A., V. Rossier, N. Roux, P. Salis, V. Laudet, and N. Salamin. 2019. Insights into the genomics of clownfish adaptive radiation: Genetic basis of the mutualism with sea anemones. *Genome Biology and Evolution* 11 (3): 869–882.

Mariscal, R. N. 1970a. The nature of the symbiosis between Indo-Pacific anemonefishes and sea anemones. *Marine Biology* 6: 58–65.

Mariscal, R. N. 1970b. An experimental analysis of the protection of amphiprzon xanthurus cuvier & valenciennes and some other anemone fishes from sea anemones. *Journal of Experimental Marine Biology and Ecology* 4: 134–149.

Mebs, D. 1994. Anemonefish symbiosis: Vulnerability and resistance of fish to the toxin of the sea anemone. *Toxicon* 32 (9): 1059–1068.

Mebs, D. 2009. Chemical biology of the mutualistic relationships of sea anemones with fish and crustaceans. *Toxicon* 54 (8): 1071–1074.

Merilaita, S., and J. L. Kelley. 2018. Scary clowns: Adaptive function of anemonefish coloration. *Journal of Evolutionary Biology* 31 (10): 1558–1571.

Miyagawa, K., and T. Hidaka. 1980. *Amphiprion clarkii* juvenile innate protection against and chemical attraction by symbiotic sea anemones. *Proceedings of the Japan Academy Series B-Physical and Biological Sciences* 56 (6): 356–361.

Nedosyko, A. M., J. E. Young, J. W. Edwards, and K. Burke da Silva. 2014. Searching for a toxic key to unlock the mystery of anemonefish and anemone symbiosis. *PLoS One* 9 (5): e98449.

Ozacmak, V. H., G. U. Thorington, W. H. Fletcher, and D. A. Hessinger. 2001. N-acetylneuraminic acid (nana) stimulates in situ cyclic amp production in tentacles of sea anemone (*Aiptasia pallida*): Possible role in chemosensitization of nematocyst discharge. *The Journal of Experimental Biology* 204: 2011–2020.

Pantin, C. F. A. 1942. The excitation of nematocysts. *Journal of Experimental Biology* 19: 294–310.

Pratte, Z. A., N. V. Patin, M. E. McWhirt, A. M. Caughman, D. J. Parris, and F. J. Stewart. 2018. Association with a sea anemone alters the skin microbiome of clownfish. *Coral Reefs* 37 (4): 1119–1125.

Pryor, S. H., H. A. Malcolm, B. P. Kelaher, K. L. Davis, and A. Scott. 2022. Habitat limits the poleward establishment of anemonefishes in a climate change hotspot. *Estuarine, Coastal and Shelf Science* 264: 107662.

Roux, N., R. Lami, P. Salis, K. Magre, P. Romans, P. Masanet, D. Lecchini, et al. 2019. Sea anemone and clownfish microbiota diversity and variation during the initial steps of symbiosis. *Scientific Reports* 9 (1): 19491.

Rowe, A. H., Y. Xiao, M. P. Rowe, T. R. Cummins, and H. H. Zakon. 2013. Voltage-gated sodium channel in grasshopper mice defends against bark scorpion toxin. *Science* 342 (6157): 441–446.

Saenz-Agudelo, P., G. P. Jones, S. R. Thorrold, and S. Planes. 2011. Detrimental effects of host anemone bleaching on anemonefish populations. *Coral Reefs* 30 (2): 497–506.

Sale, P. F. 1980. *The Ecology of Fishes on Coral Reefs*. San Diego, CA: Academic Press.

Schlichter, D. 1976. *Macromolecular Mimicry: Substances Released by Sea Anemones and Their Role in the Protection of Anemonefishes*. New York: Plenum Press.

Scott, A., and A. S. Hoey. 2017. Severe consequences for anemonefishes and their host sea anemones during the 2016 bleaching event at Lizard Island, Great Barrier Reef. *Coral Reefs* 36 (3): 873–873.

Scott, A., K. J. W. Rushworth, S. J. Dalton, and S. D. A. Smith. 2015. Subtropical anemonefish *Amphiprion latezonatus* recorded in two additional host sea anemone species. *Marine Biodiversity* 46 (2): 327–328.

Srinivasan, M., G. P. Jones and M. J. Caley. 1999. Experimental evaluation of the roles of habitat selection and interspecific competition in determining patterns of host use by two anemonefishes. *Marine Ecology Progress Series* 186: 283–292.

Stevens, M., and S. Meritaila. 2009. Defining disruptive coloration anddistinguishing its functions. *Philosophical Transactions of The Royal Society B* 364: 423–427

Sunagar, K., D. Morgenstern, A. M. Reitzel, and Y. Moran. 2016. Ecological venomics: How genomics, transcriptomics and proteomics can shed new light on the ecology and evolution of venom. *Journal of Proteomics* 135: 62–72.

Takacs, Z., K. C. Wilhelmsen, and S. Sorota. 2001. Snake α-Neurotoxin binding site on the Egyptian Cobra (*Naja haje*) nicotinic acetylcholine receptor Is conserved. *Molecular Biology and Evolution* 18 (9): 1800–1809.

Titus, B. M., C. Benedict, R. Laroche, L. C. Gusmao, V. Van Deusen, T. Chiodo, C. P. Meyer, et al. 2019. Phylogenetic relationships among the clownfish-hosting sea anemones. *Molecular Phylogenetics and Evolution* 139: 106526.

Titus, B. M., R. Laroche, E. Rodriguez, H. Wirshing, and C. P. Meyer. 2020. Host identity and symbiotic association affects the taxonomic and functional diversity of the clownfish-hosting sea anemone microbiome. *Biology Letters* 16 (2): 20190738.

Voss, R. S., and S. A. Jansa. 2012. Snake-venom resistance as a mammalian trophic adaptation: Lessons from didelphid marsupials. *Biological Reviews* 87 (4): 822–837.

20 Larval Dispersal in Anemonefish Populations

Self-Recruitment, Connectivity, and Metapopulation Dynamics

Geoffrey P. Jones, Hugo B. Harrison, Michael L. Berumen, Serge Planes, and Simon R. Thorrold

CONTENTS

20.1 INTRODUCTION

Determining how far marine larvae disperse is a difficult but essential task for understanding how marine populations persist (Hastings and Botsford 2006; Botsford et al. 2009a), how marine species adapt to their environment (Sultan and Spencer 2002; Conover et al. 2006; Sotka 2012), and how they would be best conserved or sustainably exploited (Palumbi 2003; Sale et al. 2005; Botsford et al 2009b). For coral reef fishes, early dogma held that populations of reef fish relied on dispersal from distant populations (Sale 1991; Roberts 1997) with largely "open" dynamics. In these circumstances, self-recruitment, or *the proportion of settlers in a local population that are the progeny of adults in the same population*, was not considered a major factor for most populations (Jones et al. 2009). However, these views were based on entrenched assumptions rather than hard facts about where larvae come from or where they go. Since 2000, studies on anemonefish have been at the forefront of a host of new discoveries about the sources and destinations of larval fishes (see reviews by Jones et al. 2009; Jones 2015). They have been embraced as model species to test and apply numerous established and novel techniques to estimate larval fish dispersal on coral reefs. Anemonefishes were an ideal choice because they exist in small predictable habitat patches, they are easy to sample, survey, and estimate population parameters, and they have relatively short larval durations for reef fish (10–12 d). The logic was simple – if you cannot determine larval dispersal patterns for anemonefishes, then in all likelihood, it cannot be done.

The last three decades of research have not only shown that it can be done, but new techniques have delivered information once never thought possible. There is a long list of firsts. Studies on anemonefishes were among the first to show relatively local levels of population differentiation using traditional tools of population genetics (Bell et al. 1982; Doherty et al. 1995; Nelson et al. 2000). The dual application of larval tagging and genetic parentage analysis for *Amphiprion polymnus* in Papua New Guinea was the first to show self-recruitment in local populations and the smallest scale of dispersal known for a marine fish (Jones et al. 2005). Almany et al. (2007) were the first to apply maternally transmitted stable isotope markers to demonstrate self-recruitment in an isolated *A. percula* population. Soon after, genetic parentage analysis was used for the first time to track larvae dispersing from one population to another (Planes et al. 2009; Saenz Agudelo et al. 2009). Pinsky et al. (2010) were the first to use isolation-by-distance (IBD) approaches to estimate average dispersal distances in a marine fish, using *A. clarkii* as their model species. Beldade et al. (2012) linked the quality of parents to the success of local contribution to recruitment in *A. chrysopterus* in French Polynesia. Simpson et al. (2014) used an assignment test to measure long-dispersal between distant populations of *A. omanensis*. Lett et al. (2015) were the first to use the results of parentage analysis to calculate

DOI: 10.1201/9781003125365-24

the parameters necessary for applying marine metapopulation models for understanding population persistence. In 2016, an anemonefish study produced the first pedigree for a marine fish population, with family trees spanning five generations (Salles et al. 2016a). One year later, *A. percula* provided the first full empirical description of a dispersal kernel for a marine fish (Almany et al. 2017), which estimates the likelihood of dispersal over a given distance. Salles et al. (2020) went on to provide the first measure of Darwinian fitness in a marine population, by tracing reproductive success among generations. Most recently, we have seen the first full integration of parentage data and a metapopulation model for the anemonefish *A. clarkii* (Catalano et al. 2020; Dedrick et al. 2021).

We have now reached the stage where we have good estimates of self-recruitment, average and maximum dispersal distances, and a range of dispersal metrics for half of the 28 recognized anemonefish species (Figure 20.1). These studies, mostly carried out over the last 20 years, span the entire distribution of anemonefishes, from the Red Sea to French Polynesia (Figure 20.1), including both widely distributed Indo-Pacific species and small-range endemics at the subtropical margins of the anemonefish range. The aim of this chapter is to review the methodologies and their contributions, summarize the emerging patterns in self-recruitment and larval connectivity, and discuss the ecological, evolutionary, and management implications of these findings. We will finish by speculating on new research directions that will provide the list of "firsts" we are likely to see in the next 20 years.

20.2 MEASURING DISPERSAL

There have been a number of approaches to measuring larval dispersal in anemonefishes on ecological time scales, including chemical tagging (Jones et al. 2005; Thorrold et al. 2006; Almany et al. 2007), individual-based genetic parentage analysis (e.g., Jones et al. 2005; Planes et al. 2009; Saenz Agudelo et al. 2009) and population-based genetic approaches based on genetic differentiation between distant populations (van der Meer et al. 2012a; Gainsford et al. 2015; Huyghe and Kochzius 2017; O'Donnell et al. 2017;

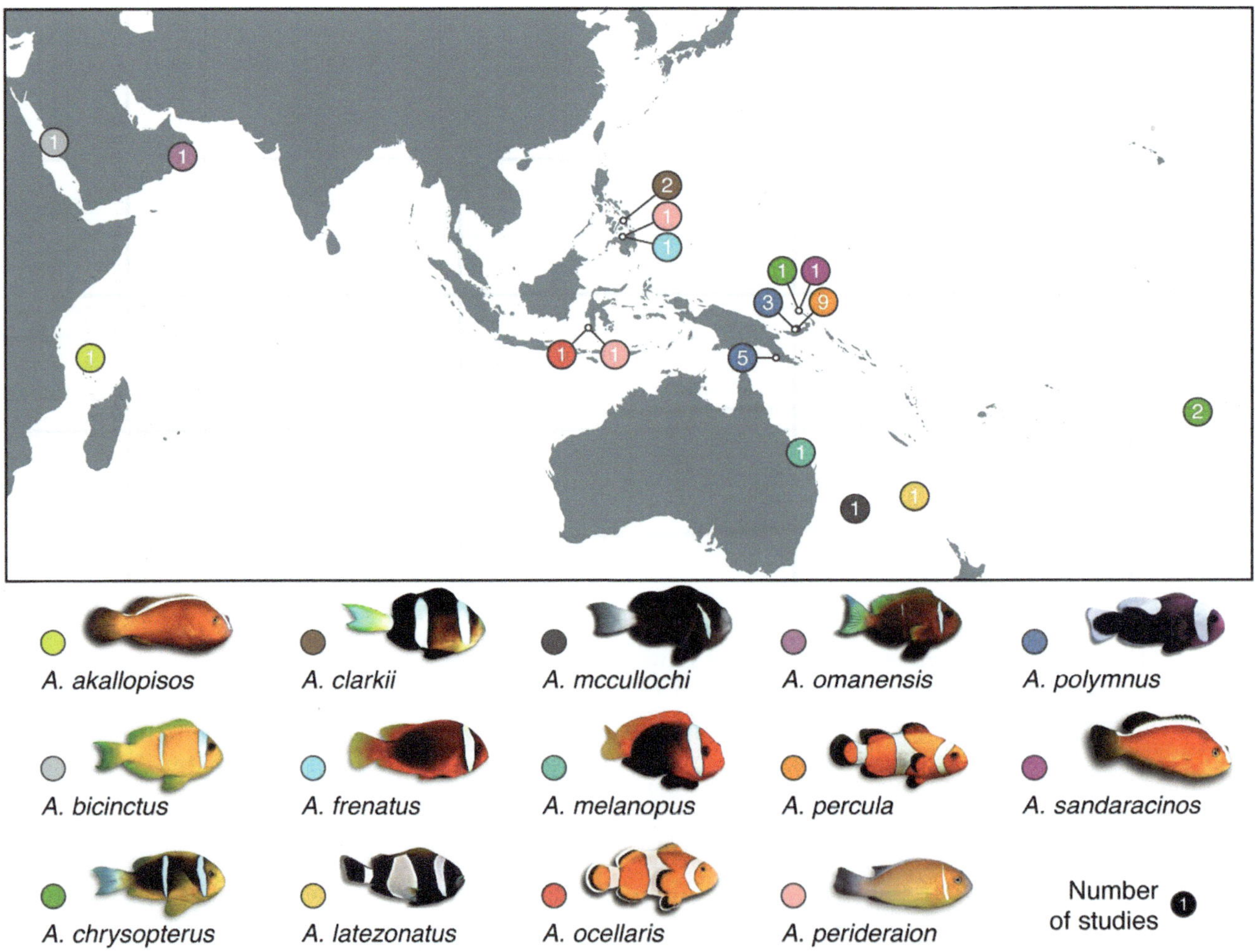

FIGURE 20.1 Geographic distribution of 33 studies that report estimates of larval dispersal and self-recruitment in 14 *Amphiprion* spp. (f. Pomacentridae) spanning the Indo-Pacific region. Circles indicate the number of studies in each location. Images are not proportional to size.

Timm et al. 2017), sometimes combined with assignment tests (e.g., Saenz Agudelo et al. 2009; Steinberg et al. 2016) and isolation by distance (IBD) estimates of larval dispersal kernels (Pinsky et al. 2010). Further application of these approaches has now provided information on dispersal for 14 species, including multiple studies on the same species (Figure 20.2a). Since the very first study of self-recruitment in *A. polymnus*, which used tetracycline tagging of embryos to validate parentage analysis (Jones et al. 2005), matching juvenile recruits to putative parents has dominated studies of larval dispersal. At first, the cost of developing new microsatellite libraries limited its use to only a few species. Early studies relied on successful cross-amplification from a single microsatellite library for *A. polymnus* (Quenouille et al. 2004). However, the rapid development of next-generation sequencing was a gateway for parentage analyses to become the dominant method to measure self-recruitment and connectivity in new species of anemonefish (Figure 20.2a). From 2012 to 2017, parentage studies were done on eight species of anemonefish (Figure 20.2b), mainly using microsatellite markers (Figure 20.2c).

The great advantages of genetic parentage analysis are that it provides an individual-based dispersal vector linking the location of offspring to the location of one or both parents, and it is easy to scale up projects in time and space. Parentage assignments are extremely reliable for anemonefishes because a large proportion, if not all, of the parents in a source population can be sampled, with studies repeatedly assigning juveniles to both parents in the population. This provides a high level of confidence when distinguishing true-parent offspring pairs from unrelated individuals in the population, provided sufficient and diverse markers are used (Harrison et al. 2013). The fact that anemonefish rarely venture far from their host anemones contributes to the relative ease of sampling potential parents and quantifying the proportion of a population sampled. Gradually, one-off studies on levels of self-recruitment are being enhanced by longer-term studies that have enabled us to quantify temporal variation in dispersal patterns and distances (Figure 20.2d) or track individuals through multiple generations from individual genotypes. While the earliest, and to date, most, of the studies have focussed on self-recruitment or dispersal on scales of less than a kilometre, over time the spatial scale of studies has expanded, and so too has our appreciation of the maximum dispersal range of anemonefish species (Figure 20.2e).

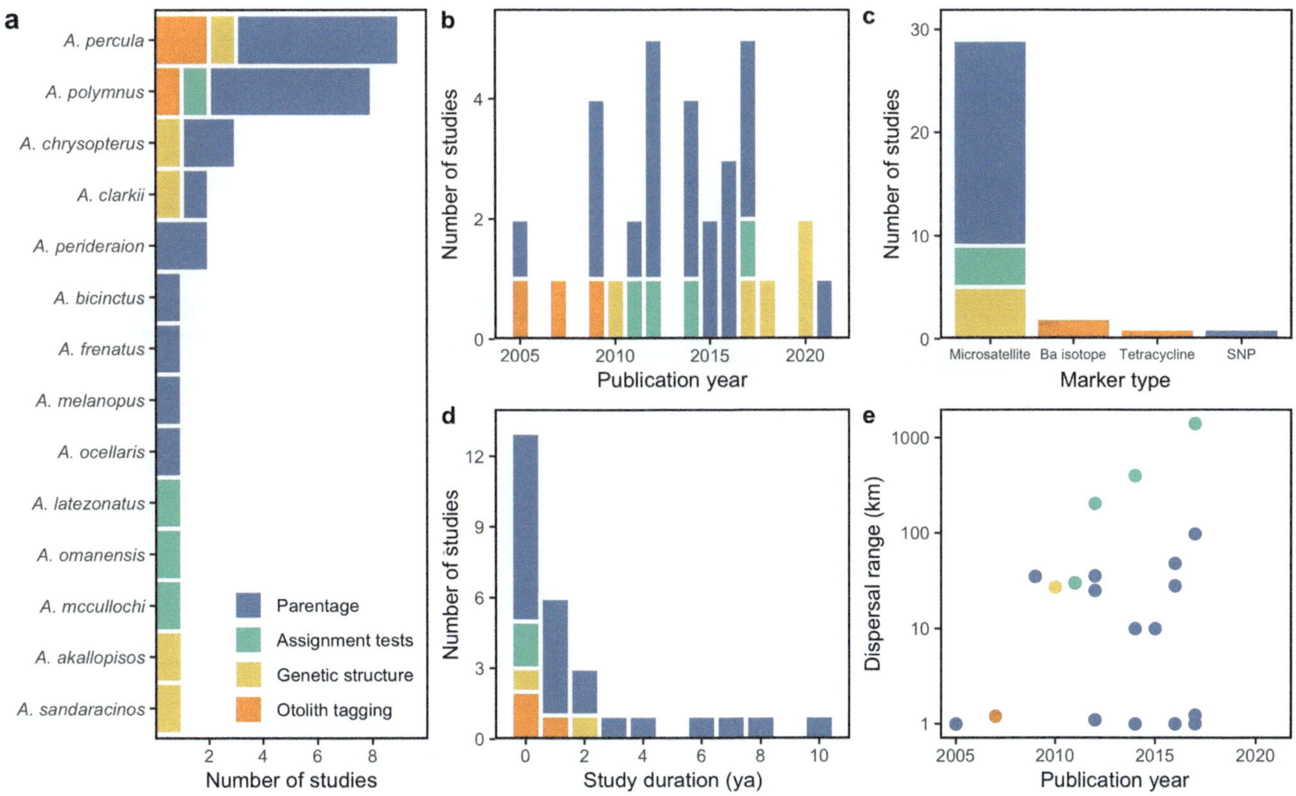

FIGURE 20.2 Four approaches have dominated dispersal and recruitment studies in *Amphiprion* spp. (a) Parentage analysis, assignment tests, and population genetic structure are DNA-based methods, while otolith tagging methods are based on the analysis of biomarkers. The first of these studies was published in 2005 (b) and the genus has remained a popular study organism since then. Dispersal and recruitment studies in *Amphiprion* have employed a diversity of methodologies (c). Microsatellite and SNP studies are molecular techniques while Ba (barium) isotopes and tetracycline are chemical markers used in otolith studies. The length of the individual studies range from several weeks to several years (d). The results of these studies show a wide range of maximum dispersal distances increasing as studies grew larger in scale (e).

The application of multiple methodologies to quantify larval retention and connectivity in anemonefish has been the foundation of a wider application to other marine fishes (Jones 2015). While methods such as chemical tags have largely been superseded, one of the strengths of this field has been the cross-validation of dispersal estimates using different approaches (Jones et al. 2005; Pinsky et al. 2017). At some point, the sampling effort required for large-scale parentage studies becomes logistically prohibitive, costly, and the results potentially inaccurate due to increasing levels of genetic differentiation among populations (Saenz-Agudelo et al. 2009). However, alternative methods such as assignment tests and IBD become much more reliable at these scales, providing estimates of average dispersal distances (Pinsky et al. 2010, 2017) and allowing rare individuals that have dispersed long distances to be recognized (Simpson et al. 2014; Steinberg et al. 2016). Fortunately, there appears to be an overlapping spatial scale where both individual and population-based genetics can work together to cross-validate estimates of dispersal metrics. For example, in Kimbe Bay, an enormous sampling effort (tens of thousands of individuals) to capture parentage assignments across the bay at a scale of ~100 km was successful in validating average dispersal distances using IBD, most likely at the minimum spatial scale it can be applied (Pinsky et al. 2017). At these scales, dispersal is inherently rare, making parentage studies logistically challenging, but sufficiently common to preclude assignment tests from accurately distinguishing between populations. At even larger scales (>100 km), interspecific comparisons of genetic differentiation between populations have shown marked differences among anemonefish species, which testifies to differences among species in long-distance dispersal abilities (van der Meer et al. 2012a, 2012b; Huyghe and Kochzius 2018; Gainsford et al. 2020) and their capacity for hybridization (Gainsford et al. 2015, 2020; Saenz-Agudelo et al. 2015a)

20.3 EMERGING PATTERNS

20.3.1 Increasing Spatial and Temporal Scales

Geographically, studies of larval dispersal and recruitment have grown from a single reef over a few 100 metres (Jones et al. 2005) to studies that span complex reef networks and up to 10,000 km^2 (Almany et al. 2017) (Figure 20.3a). Maximum dispersal ranges reflect the different scales of the studies and the techniques applied, rather than intrinsic differences among the species examined. The most

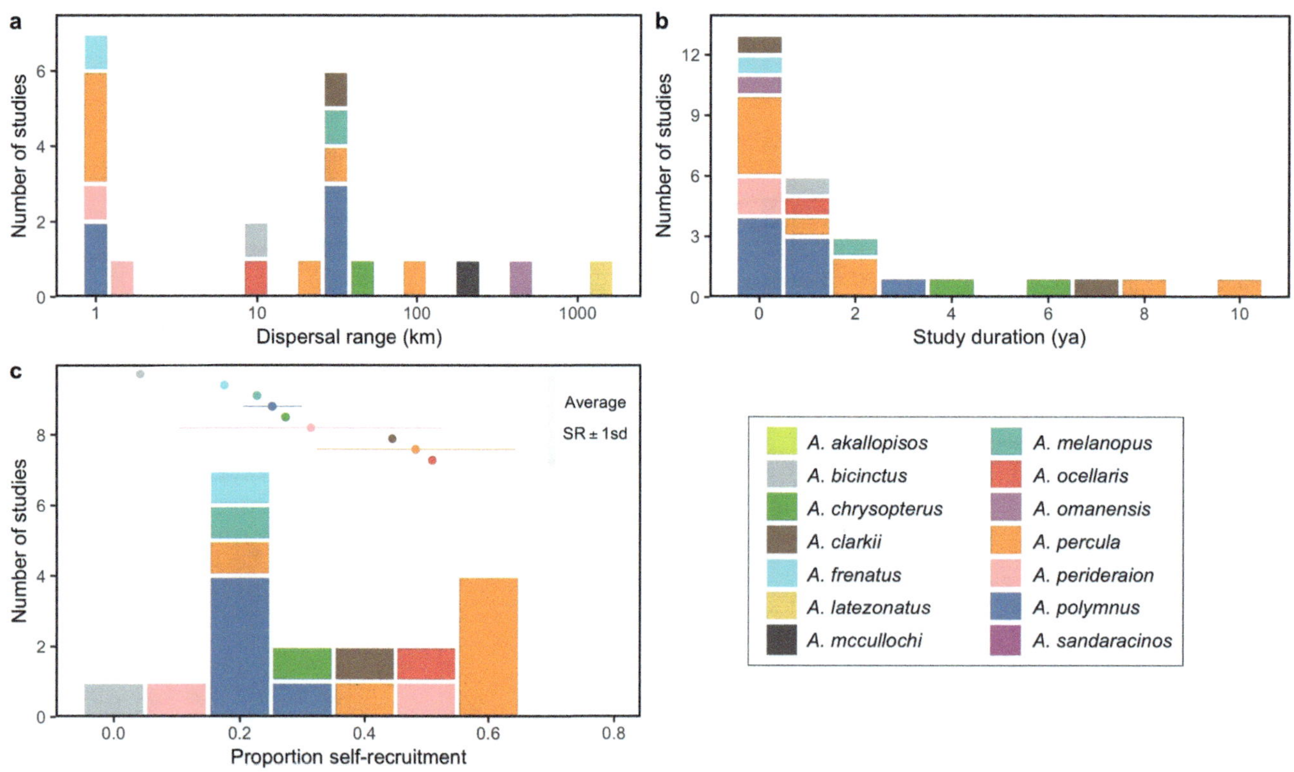

FIGURE 20.3 Species and locations lend themselves to different scales, with most studies focused on scales relevant to self-recruitment (< 1 km) and larval dispersal (< 50 km). Few studies have captured temporal patterns, with the large majority of studies capturing dispersal and self-recruitment over periods of one year or less (b). Estimates of self-recruitment vary between species, which may indicate different life-history strategies that favour self-recruitment, or the spatial and temporal scale over which they are measured. Habitat availability, population sizes, oceanographic patterns, and the presence/absence of nearby populations may also be key factors in determining the likelihood of self-recruitment and contribute to the variability observed within and between species (c). Standard deviation (SD) in self-recruitment was measured only where multiple studies were available for the same species.

common scales are < 1 km and < 50 km reflecting the shift in focus from self-recruitment to connectivity among reefs, respectively. These studies have captured larval fish recruiting within metres of their natal anemone (Bonin et al. 2016) and some dispersing up to 130 km (Almany et al. 2017). Recruitment to the natal anemone appears to be extremely rare, but pedigree analyses for *A. percula* at Kimbe Island show that this can occur on rare occasions (Salles et al. 2016). To date, empirical measures of larval dispersal at scales over 100 km remain rare. They require either very unique systems (Simpson et al. 2014; Steinberg et al. 2016), where assignment tests can capture migrant individuals, or large sampling efforts to improve the likelihood of parentage assignment at these scales (Almany et al. 2017). Beyond, at scales of 1,000 km or more, measures of connectivity may no longer be demographically relevant but can inform broad biogeographic processes and detect barriers to dispersal (Doherty et al. 1995; Timm et al. 2012; Dohna et al. 2015; Saenz-Agudelo et al. 2015a).

Temporally, studies of larval dispersal and recruitment have grown from capturing a single recruitment event over a few months (Jones et al. 2005) to spanning ten years and five generations within a single population (Salles et al. 2016a; Figure 21.3b). Long-term studies of larval dispersal in *A. percula* (Salles et al. 2016a) and *A. clarkii* (Catalano et al. 2020) demonstrate that dispersal is extremely variable. These studies are consistent in that discrete populations almost always have some level of self-recruitment and some level of connectivity to nearby populations, and show connectivity is often multi-directional in nature.

20.3.2 Levels of Self-Recruitment, Dispersal Distances, and the Shape of the Dispersal Kernel

Combining all studies that have used parentage analysis and otolith tagging to capture empirical patterns of larval dispersal, estimates of self-recruitment (%SR = the percentage of juveniles that recruit to a population that come from parents in the same population [Jones et al. 1999]) in anemonefish species are on average ~33% (Table 20.1). This ranges from 13% to 68% of the population in studies that specifically measure variance in %SR, either spatially or temporally. Parentage studies show self-recruitment varies among species, from < 1% for *A. bicinctus* in the Red Sea (Nanninga et al. 2015) to > 50% for the two sister species, *A. percula* and *A. ocellaris* in the coral triangle (Almany et al. 2007; Planes et al. 2009; Madduppa et al. 2014; Salles et al. 2015) (Figure 20.3c). Population assignments at larger spatial scales can show much larger estimates of %SR, for example, 89% in remote island populations of the small-range species *A. latezonatus* on the eastern Australian coast, Lord Howe Island, and Norfolk Island. Temporal studies show a remarkable consistency in estimates of %SR. For Kimbe Island, self-recruitment of *A. percula* ranges from 39% to 68% (average 56%) (Salles et al. 2015). In Ormoc

TABLE 20.1
Selection of Larval Dispersal Studies That Report Empirical Estimates of Larval Dispersal Distance and % Self-Recruitment (SR) in Anemonefish Species

Species	PLD[a]	Study duration (ya)	Dispersal range (km)	% SR[b] (min–max)	Reference
A. polymnus	9–12	1.5	0–1	(16–32)	Jones et al. 2005
A. percula	10–12	0.25	0–1.2	60	Almany et al. 2007
A. polymnus	9–12	0.58	0–9	25	Saenz-Agudelo et al. 2009
A. polymnus	9–12	0.25	0–1	32	Saenz-Agudelo et al. 2009
A. percula	11	1.25	15–35	42	Planes et al. 2009
A. polymnus	9–12	0.33	Jan–30	18	Saenz-Agudelo et al. 2011
A. polymnus	9–12	3	1.2–35.5	(18–23)	Saenz-Agudelo et al. 2012
A. percula	10–13	0.25	0–25	64	Berumen et al. 2012
A. ocellaris	8–12	0.75	0–10	52 (44–65)	Madduppa et al. 2014
A. perideraion	18	0.1	0–1	47	Madduppa et al. 2014
A. bicinctus	11	1	0–10	0.6	Nanninga et al. 2015
A. percula	10	8	0–1	56 (39–68)	Salles et al. 2015
A. melanopus	8–14	1.8	0–28	21	Bonin et al. 2016
A. percula	10	10	0–1	56	Salles et al. 2016a
A. chrysopterus	14–17	4	0–48	26 (25–27)	Beldade et al. 2016
A. frenatus	7–9	0.25	0–1	15	Sato et al. 2017
A. perideraion	10–12	0.25	0–1.2	14	Sato et al. 2017
A. percula	10–12	2	0–98	(13–20)	Almany et al. 2017
A. clarkii	15–16	7	0–28	45 (33–50)	Catalano et al. 2020

[a] PLD: Pelagic Larval Duration
[b] %SR: percent self recruitment (the percentage of juveniles that recruit to a population that come from parents in the same population)

Bay, the Philippines, self-recruitment of *A. clarkii* ranges from 33% to 50% (average 44%) (Catalano et al. 2020). This is quite remarkable given the potential for variation in hydrodynamic processes to cause fluctuations in larval trajectories.

While there are no doubt species-specific differences in typical %SR estimates (Table 20.1), it is difficult to assess these against a backdrop of studies conducted at different spatial and temporal scales and in reef ecosystems that vary greatly in terms of reef size, shape, and geographic arrangement. There is clearly a range in the levels of reliance on self-recruitment at the population level. Habitat availability, population sizes, oceanographic patterns, and the presence/absence of nearby populations may be key factors in determining the likelihood of self-recruitment. It seems likely that isolated populations of small-range endemics are going to be very good at returning to natal populations. Low levels of %SR in *A. bicinctus* (Nanniga et al. 2015) appear to be linked to strong hydrodynamic regimes and biophysical gradients in the Red Sea (Nanninga et al. 2014; Saenz Agudelo et al. 2015a; Raitsos et al. 2017). Levels of self-recruitment are often linked to hydrodynamic retention mechanisms. For example, Beldade et al. (2016) show more self-recruitment in *A. chrysopterus* at Moorea from spawners in lagoons than in passes, where larvae may be advected from the reef.

Parentage studies also show a large difference among species in maximum dispersal distances and these estimates also vary greatly among studies of the same species (Figure 20.3a). As for self-recruitment estimates, it is difficult to separate species-specific differences in the propensity to disperse from other factors, such as the spatial scale of the individual studies. There is no obvious relationship between %SR and maximum dispersal distance (Figure 20.3c), which makes looking for generalities in the propensity to settle locally or disperse long distances even more of a mystery.

Assignment tests and studies of population differentiation provide some insight into long-distance dispersal, congruent with the broad geographic distribution of some species of anemonefishes. The longest reported distances travelled are from *A. omanensis*, an endemic of the southern coast of Oman, restricted to two populations separated by > 400 km (Simpson et al. 2014). The direction of dispersal was consistent with the main direction of the water currents along the coast. The broad range of some species across vast oceanic systems suggests that larvae may be capable to disperse even further. In the western Pacific, the widely distributed *A. chrysopterus* has high genetic similarity throughout its range, while in the same area, *A. sandaracinos* exhibits high levels of population subdivision, suggesting the latter is much less able to disperse long distances (Gainsford et al. 2020). It could be hypothesized that *A. clarkii*, the anemonefish species with the broadest geographic range, should be a good disperser. The study by Catalano et al. (2020) that captures dispersal events at a scale of up to 30 km shows that a high proportion of juveniles do not disperse far, but populations do appear to be reliant on immigration from distant sources (Dedrick et al. 2021).

Clearly, a comprehensive picture of localized and long-distance dispersal requires a full empirical description of the dispersal kernel and an estimate of the likelihood of dispersal over a given distance from the natal origin. The first description of a dispersal kernel in marine fishes was for *A. percula* in Kimbe Bay, Papua New Guinea, where seven populations were sampled from one side of the bay to the other, over a distance of 120 km (Almany et al. 2017). This also enables average dispersal distances to be calculated, which are likely to be of greater demographic significance than the two extremes. Dispersal distance is initially a steeply declining function with distance, but with a long tail extending just over 100 km. Over two years of sampling in 2009 and 2011, estimated mean dispersal distances were 13–19 km, with 90% of settlement occurring within 31–43 km. These dispersal distances are much shorter than expected based on the distribution of distances between all areas of suitable habitat (Almany et al. 2017). Across most studies that report average dispersal distances, estimates are concentrated around ~10 km, though depend largely on the scale of the study, heterogeneity of available habitat and the exhaustiveness of sampling. The few other dispersal kernels available provide a more comprehensive view of the likelihood of dispersal, with 50% of larval recruiting within 8–15 km of their natal origin, and 90% within 15–157 km (Catalano et al. 2020). Collectively, they suggest that long-distance dispersal occurs, but is rare and most likely of little short-term demographic importance.

20.3.3 Population Persistence and Metapopulation Dynamics

Our knowledge of larval dispersal patterns remains in the early stages of being applied to understand the primary demographic drivers of population size and persistence, both at the local level and across whole metapopulations. Persistence of connected populations depends on whether the sum of larval recruitment by means of local retention and connectivity from distant patches exceeds that which is necessary to sustain population dynamics over multiple generations (Hastings and Botsford 2006). High levels of %SR alone do not assure population persistence, which is related more to % local retention (%LR = the ratio of locally produced settlement to local egg production) (Burgess et al. 2014; Lett et al. 2015). This is much harder to calculate as it requires long-term estimates of egg production; however, complete dispersal kernels allow this to be estimated (Almany et al. 2017). Studies to date suggest a strong positive relationship between %SR and %LR (Lett et al. 2015; Almany et al. 2017). Hence, once this relationship is known, one metric can be calculated from the other, which will greatly facilitate calculating levels of %LR needed to sustain local populations and levels of dispersal needed to sustain a metapopulation.

Salles et al. (2015) showed that for a discrete population of ~200 reproductive pairs of *A. percula* at Kimbe Island, self-recruitment in the order of 40–60% was sufficient to maintain the population at the carrying capacity of the island (Salles et al. 2015). This seems likely the case for species in which there is a high level of local retention, where all suitable anemones are occupied by a reproductive pair and where there is a large pool of non-reproductive juveniles queuing for reproductive status within social groups (Buston 2003). Dedrick et al. (2021) came to the opposite conclusion for *A. clarkii* in the Philippines. They showed that despite stable population numbers and enough surviving offspring to ensure replacement, the local population was unlikely to persist in isolation without immigrants from other locations. It is worth noting that unlike *A. percula*, which is clearly habitat-limited, in the *A. clarkii* situation, anemones are undersaturated and often vacant, suggesting that the subpopulation is recruitment-limited.

Understanding how anemonefish metapopulations work requires a good empirical description of connectivity matrices, which is a function of both dispersal distances and directions. In the case of *A. percula*, levels of %SR are highly variable and connectivity is generally multi-directional, with links greatest between adjacent subpopulations and from larger populations (Buston et al. 2012; Berumen et al. 2012; Almany et al. 2017). Ensuring local and metapopulation persistence in the face of increasing human pressures may rely on the ability to identify and protect key larval sources, which may be larger, centrally located, or self-sustaining subpopulations. Metapopulation models also need to take into account variation in habitat quality, as the contribution to the next generation may come from relatively few individuals at prime locations (Salles et al. 2016b, 2020).

20.3.4 Reproductive Success and Local Adaptation

Parentage analysis, and the ability to link offspring to their parents, have allowed studies of anemonefishes to venture into the area of estimating the relative contributions of particular individuals to the next and future generations. In the past, marine ecologists have been limited to making assumptions about Darwinian fitness, based simply on fecundity, spawning success, or other proxy metrics. Parentage studies have confirmed that larger, more fecund females do contribute more offspring to the next generation than smaller females, at least in terms of the contribution to the local population (Beldade et al. 2012; Saenz Agudelo et al. 2015b). Salles et al. (2016b) showed that for *A. percula*, which occupies two anemone species at Kimbe Island, individuals occupying one species of host anemone made a greater contribution to the next generation than those associated with the other species. Salles and co-workers have been able to take this further to look at reproductive success over multiple generations and assess what ecological and genetic factors account for large family trees (Salles et al. 2020). The long-term study of Kimbe Island shows that there is an extremely low level of heritability of fitness, with reproductive success linked to where fish are in the local population, rather than the genetic qualities of parents. While this may sound an alarm in the sense that anemonefishes may have a limited ability to genetically adapt to local environmental change, it does suggest a high level of phenotypic plasticity in fitness traits, which may enable individuals to rapidly respond to change on ecological time frames.

20.4 CONCLUSIONS AND FUTURE DIRECTIONS

Anemonefishes have proven to be a valuable model system for testing new methodologies that have provided the first direct estimates of marine larval fish dispersal. Parentage studies have dominated the study of larval dispersal, providing unprecedented details into the most elusive life stage of a marine fish. This approach has stimulated numerous studies where researchers have taken up the challenge of investigating larval dispersal in more difficult, but often important fishery species (e.g., Harrison et al. 2012; Almany et al. 2013; Williamson et al. 2016; Le Port et al. 2017; Harrison et al. 2020; Hamilton et al. 2021). Next-generation sequencing technologies have already revolutionized the field, simplifying the process of genotyping thousands of individuals and increasing the accuracy of parentage analysis and dispersal studies (Anderson 2012; Harrison et al. 2013; Baetscher et al. 2019). These advances ensure that we can expand the scale of sampling in space and time to provide the full dispersal kernels we need to understand anemonefish metapopulation dynamics and the best approaches to manage them in the context of increasing local and global threats.

Although we only have information on half of the known species of anemonefishes, there have been many significant findings that have changed our way of thinking about how marine populations work. The discoveries that self-recruitment is common (on average 33%) and average dispersal distances are short (on average ~10 km) are very different to what might have been predicted 20 years ago. While it may have been predicted that dispersal distances would decline as a function of distance from source, few would have anticipated the relatively short spatial scale over which this occurs. The frequency that larvae settle locally in combination with an ability to disperse great distances has huge implications, not only for the short-term persistence and resilience of local populations but also for the long-term roles of local adaptation and phenotypic plasticity in responses to environmental change (Jones 2015). Addressing such questions is likely to dominate research on anemonefish over the next 20 years, as we grapple with how they will respond to climate change, fishing pressure, pollution, and degradation of the habitat and decline in the health of anemones in which they reside.

A number of future directions must be highlighted. Firstly, while some interesting generalizations have emerged with the study of dispersal in 14 anemonefish species, it may take studies on the other half of the known anemonefish species to fully appreciate their diversity of larval dispersal patterns. Given the confounding effects of geographic location and time, however, concurrent studies on different species at the same location in the centre of anemonefish biodiversity are needed to tease apart the relative roles of species-specific and location-specific dispersal patterns. More work is also needed at the fringes of the anemonefish distribution, where the population retention mechanisms of numerous small-range and geographically isolated species have yet to be studied. These species are particularly vulnerable, given their extremely small area of occurrence (see Chapter 25).

Secondly, despite the enormous difficulty in expanding sampling effort even more, we do need studies at larger spatial scales to fully describe self-recruitment and patterns of larval connectivity for more species and at the scale of whole metapopulations. Expanding the temporal scales will also be needed to fully understand the temporal stability of connectivity patterns and extend ecological time frames to measure short-term evolutionary responses to rapid environmental change. This scaling up in space and time will better inform how marine protected areas work, not only for protecting individual populations over multiple generations but enhancing the long-term persistence and growth of populations over networks of reserves, via portfolio effects (Harrison et al. 2020) and the protection of critical sources (Jones et al. 2007). The development of metapopulation models to simulate population dynamics may also be greatly improved by the rapid development of network analyses that capture the multi-directional nature of connectivity that is not currently considered in dispersal kernels (Rollins et al. 2012). These may help identify regions for network connectivity (Watson et al. 2011), identify seascapes features that regulate connectivity patterns (Alberto et al. 2011), and inform conservation strategies (Andrello et al. 2015). However, these will require careful consideration of how the degree of connectivity between populations is estimated and the uncertainty associated with these measures.

A third line of research will be a great focus on biophysical modelling to provide a more general picture of local and large-scale dispersal patterns and an ability to predict these in the future. This is one area in which work on anemonefish dispersal has lagged behind that for other marine fishes, but it will be important, especially for species or at scales where parentage studies are impractical. The development and validation of biophysical models will also provide a means to explore the ecological and evolutionary consequences of dispersal and simulate the impacts of changing oceanic conditions on larval survivorship, dispersal patterns and population persistence. Predicting the variability and strength of connectivity patterns in complex seascapes may be integral to optimizing the design of a global network of marine protected areas and measuring the risk of recruitment failure in fisheries. While further validation of biophysical models is needed, this will not necessarily have to rely on in-depth empirical studies (e.g., Bode et al. 2019). If biophysical models can accurately represent an isotropic dispersal kernel, this may represent connectivity at spatial and temporal scales that is sufficient to inform conservation and fisheries objectives.

Fourthly, the further application of validated demographic models to evaluate population persistence and response to environmental changes, including declining habitat availability and changing oceanography, is absolutely essential. This is another area in which only a handful of studies are available. How these changes affect anemonefish dispersal remains largely unanswered and will require careful studies on reproductive outputs, dispersal kernels, and processes affecting the magnitude of recruitment to anemones. Even the most comprehensive studies that systematically sample every individual within a population and explain up to 70% of local recruitment have yet to fully determine what this means for the persistence of populations. In addition, extending our understanding of responses into evolutionary time frames will require further studies on intergenerational pedigrees and the factors that account for successful lineages. Understanding the role of natural selection in rapid adaptation in anemonefishes will be enhanced by the recent development of genomics, with nine *Amphiprion* genomes now available (Lehmann 2019; Marcionetti et al. 2019). Among other things, these genomes will provide new opportunities to identify selection for larval traits that may favour different dispersal strategies.

In conclusion, solutions to the increasing scale and severity of human impacts on anemonefishes and their habitat will hinge on a continuation of an era of new discoveries. There is every reason to be pessimistic, given the life history and ecological features of anemonefishes that expose them to these threats, including extreme habitat specialization, a centre of biodiversity in a region of uncontrolled habitat degradation, extreme sensitivity to temperature and acidification, and a thriving aquarium fish collecting industry (see Chapter 25). However, optimism must surely come from our increasing understanding of anemonefish dispersal patterns, which indicates a much-needed level of resilience to local and large-scale disturbances. The secrets to their survival will come from an understanding of how they navigate their way back to natal reefs or switch to finding suitable habitats hundreds of kilometres from home.

ACKNOWLEDGEMENTS

We would like to acknowledge the College of Science and Engineering and the ARC Centre of Excellence for Coral Reef Studies at James Cook University for logistic support. We also thank the Mahonia Na Dari Research and Conservation Centre and the Walindi Plantation Resort for supporting our research in Papua New Guinea. We dedicate this chapter to our dear friend Glenn Almany who is still with us in spirit.

REFERENCES

Alberto, F., P. T. Raimondi, D. C. Reed, J. R. Watson, D. A. Siegel, S. Mitarai, N. Coelho, and E. A. Serrao. 2011. Isolation by oceanographic distance explains genetic structure for *Macrocystis pyrifera* in the Santa Barbara Channel. *Molecular Ecology* 20: 2543–2554.

Almany, G. R., M. L. Berumen, S. R. Thorrold, S. Planes, and G. P. Jones. 2007. Local replenishment of coral reef fish populations in a marine reserve. *Science* 316: 742–744.

Almany, G. R., R. J. Hamilton, M. Matawai, M. Bode, T. Potuku, P. Saenz Agudelo, S. Planes, M. L. Berumen, K. L. Rhodes, S. R. Thorrold, G. P. Jones, and G. R. Russ. 2013. Dispersal of grouper larvae drives local resource sharing in a coral reef fishery. *Current Biology* 23: 626–630.

Almany, G. R., S. Planes, S. R. Thorrold, M. L. Berumen, M. Bode, P. Saenz Agudelo, V. Messmer, et al. 2017. Larval fish dispersal in a coral reef seascape. *Nature Ecology and Evolution* 1: 0148.

Anderson, E. C. 2012. Large-scale parentage inference with SNPs: An efficient algorithm for statistical confidence of parent pair allocations. *Statistical Applications in Genetics and Molecular Biology* 11: 5.

Andrello, M., M. N. Jacobi, S. Manel, W. Thuiller, and D. Mouillot. 2015. Extending networks of protected areas to optimize connectivity and population growth rate. *Ecography* 38: 273–282.

Baetscher, D. S., E. C. Anderson, E. A. Gilbert-Horvath, D. P. Malone, E. T. Saarman, M. H. Carr, and J. C. Garza. 2019. Dispersal of a nearshore marine fish connects marine reserves and adjacent fished areas along an open coast. *Molecular Ecology* 28: 1611–1623.

Beldade, R., S. J. Holbrook, R. J. Schmitt, S. Planes, and G. Bernardi. 2016. Spatial patterns of self-recruitment of a coral reef fish in relation to island-scale retention mechanisms. *Molecular Ecology* 25: 5203–5211.

Beldade, R., S. J. Holbrook, R. J. Schmitt, S. Planes, D. Malone, and G. Bernardi. 2012. Larger female fish contribute disproportionately more to self-replenishment. *Proceedings of the Royal Society B* 279: 2116–2121.

Bell, L. J., J. T. Moyer, and K. Numachi. 1982. Morphological and genetic variation in Japanese populations of the anemonefish *Amphiprion clarkii*. *Marine Biology* 72: 99–108.

Berumen, M. L., G. R. Almany, S. Planes, G. P. Jones, P. Saenz-Agudelo, and S. R. Thorrold. 2012. Persistence of self-recruitment and patterns of larval connectivity in a marine protected area network. *Ecology and Evolution* 2: 444–452.

Bode, M., J. M. Leis, L. B. Mason, D. H. Williamson, H. B. Harrison, S. Choukroun, and G. P. Jones. 2019. Successful validation of a larval dispersal model using genetic parentage data. *PLoS Biology* 17(7): e3000380.

Bonin, M. C., H. B. Harrison, D. H. Williamson, A. J. Frisch, P. Saenz Agudelo, M. L. Berumen, and G. P. Jones. 2016. The role of marine reserves in the replenishment of a locally-impacted population of anemonefish on the Great Barrier Reef. *Molecular Ecology* 25: 487–499.

Botsford, L. W., J. W. White, M.-A. Coffroth, C. B. Paris, S. Planes, T. L. Shearer, S. R. Thorrold, and G. P. Jones 2009a. Measuring connectivity and estimating resilience of coral reef metapopulations in MPAs: Matching empirical efforts to modelling needs. *Coral Reefs* 28: 327–337.

Botsford, L. W., D. R. Brumbaugh, C. Grimes, J. B. Kellner, J. Largier, M. R. O'Farrell, S. Ralston, E. Soulanille, and V. Wespestad. 2009b. Connectivity, sustainability, and yield: Bridging the gap between conventional fisheries management and marine protected areas. *Reviews in Fish Biology and Fisheries* 19: 69–95.

Burgess, S. C., K. J. Nickols, C. D. Griesemer, L. A. K. Barnett, A. G. Dedrick, E. V. Satterthwaite, L. Yamane, et al. 2014. Beyond connectivity: How empirical methods can quantify population persistence to improve marine protected-area design. *Ecological Applications* 24: 257–270.

Buston, P. 2003. Size and growth modification in clownfish. *Nature* 424: 145–146.

Buston, P. M., G. P. Jones, S. Planes, and S. R. Thorrold. 2012. Probability of successful dispersal declines fivefold over one kilometer in a coral reef fish. *Proceedings of the Royal Society, Series B* 279: 1883–1888.

Catalano, K. A., A. G. Dedrick, M. R. Stuart, J. B. Puritz, H. R. Jr. Montes, and M. L. Pinsky. 2020. Quantifying dispersal variability among nearshore marine populations. *Molecular Ecology* 30: 2366–2377.

Conover, D. O., L. M. Clarke, S. B. Munch, and G. N. Wagner. 2006. Spatial and temporal scales of adaptive divergence in marine fishes and the implications for conservation. *Journal of Fish Biology* 69: 21–47.

Dedrick, A. G., K. A. Catalano, M. R. Stuart, J. W. White, H. R. Jr. Montes, and M. L. Pinsky. 2021. Persistence of a reef fish metapopulation via network connectivity: Theory and data. *Ecology Letters* 24: 1121–1132.

Doherty, P. J., S. Planes, and P. Mather. 1995. Gene flow and larval duration in seven species of fish from the Great Barrier Reef. *Ecology* 76: 2373–2391.

Dohna, T. A., J. Timm, L. Hamid, and M. Kochzius. 2015. Limited connectivity and a phylogenetic break characterize populations of the pink anemonefish, *Amphiprion perideraion*, in the Indo-Malay Archipelago: Inferences from mitochondrial and microsatellite loci. *Ecology and Evolution* 5: 1717–1733.

Gainsford, A., L. van Herwerden, and G. P. Jones. 2015. Hierarchical behaviour, habitat use and species size differences shape evolutionary outcomes of hybridization in a coral reef fish. *Journal of Evolutionary Biology* 28: 205–222.

Gainsford, A., G. P. Jones, J.-P. A. Hobbs, F. M. Heindler, and L. van Herwerden. 2020. Species integrity, introgression and genetic variation across a coral reef anemonefish hybrid zone. *Ecology and Evolution* 10: 11998–12014.

Hamilton, R.J., D. Lozano-Cortés, M. Bode, G.R. Almany, H.B. Harrison, J. Pita, P. Saenz-Agudelo, et al. 2021. Larval dispersal and fishing pressure influence recruitment in a coral reef fishery. *Journal of Applied Ecology* 58: 2924–2935.

Harrison, H. B., M. Bode, D. H. Williamson, M. L. Berumen, and G. P. Jones. 2020. A connectivity portfolio effect stabilizes marine reserve performance. *Proceedings of the National Academy of Science* 117: 25595–25600.

Harrison, H. B., P. Saenz-Agudelo, S. Planes, G. P. Jones, and M. L. Berumen. 2013. Relative accuracy of three common methods of parentage analysis in natural populations. *Molecular Ecology* 22: 1158–1170.

Harrison, H. B., D. H. Williamson, R. D. Evans, G. R. Almany, S. R. Thorrold, G. R. Russ, K. A. Feldheim, et al. 2012. Larval export from marine reserves and the recruitment benefit for fish and fisheries. *Current Biology* 22: 1023–1028.

Hastings, A., and L. W. Botsford. 2006. Persistence of spatial populations depends on returning home. *Proceedings of the National Academy of Science* 103: 6067–6072.

Huyghe, F., and M. Kochzius. 2017. Highly restricted gene flow between disjunct populations of the skunk clownfish (*Amphiprion akallopisos*) in the Indian Ocean. *Marine Ecology* 38 (2017): e12357.

Huyghe, F., and M. Kochzius. 2018. Sea surface currents and geographic isolation shape the genetic population structure of a coral reef fish in the Indian Ocean. *PLOS One* 13(3): e193825.

Jones, G. P. 2015. Mission impossible: Unlocking the secrets of dispersal in coral reef fishes. In *Ecology of Fishes on Coral Reefs: The Functioning of an Ecosystem in a Changing World*, ed. C. Mora, 16–27. Cambridge University Press.

Jones, G. P., G. R. Almany, G. R. Russ, P. F. Sale, R. S. Steneck, M. J. H. van Oppen, and B. L. Willis. 2009. Larval retention and connectivity among populations of corals and reef fishes: History, advances and challenges. *Coral Reefs* 28: 307–325.

Jones, G. P., M. J. Milicich, M. J. Emslie, and C. Lunow. 1999. Self-recruitment in a coral reef fish population. *Nature* 402: 802–804.

Jones, G. P., S. Planes, and S. R. Thorrold. 2005. Coral reef fish larvae settle close to home. *Current Biology* 15: 1314–1318.

Jones, G. P., M. Srinivasan, and G. R. Almany. 2007. Population connectivity and conservation of marine biodiversity. *Oceanography* 20: 100–111.

Le Port, A., J. C. Montgomery, A. N. H. Smith, A. E. Croucher, I. M. McLeod, and S. D. Lavery. 2017. Temperate marine protected area provides recruitment subsidies to local fisheries. *Proceedings of the Royal Society, Series B* 284: 20171300.

Lehmann, R., D. J. Lightfoot, C. Schunter, C. T. Michell, H. Ohyanagi, K. Mineta, S. Foret, et al. 2019. Finding Nemo's genes: A chromosome-scale reference assembly of the genome of the orange clownfish *Amphiprion percula*. *Molecular Ecology Resources* 19: 570–585.

Lett, C., T. Nguyen-Huu, M. Cuif, P. Saenz-Agudelo, and D. M. Kaplan. 2015. Linking local retention, self-recruitment, and persistence in marine metapopulations. *Ecology* 96: 2236–2244.

Madduppa, H. H., J. Timm, and M. Kochzius. 2014. Interspecific, spatial and temporal variability of self-recruitment in anemonefishes. *Plos One* 9(2): e90648.

Marcionetti, A., V. Rossier, J. A. M. Bertrand, G. Litsios, and N. Salamin. 2019. First draft genome of an iconic clownfish species (*Amphiprion frenatus*). *Molecular Ecology Resources* 18: 1092–1101.

Nanninga, G. B., P. Saenz-Agudelo, A. Manica, and M. L. Berumen. 2014. Environmental gradients predict the genetic population structure of a coral reef fish in the Red Sea. *Molecular Ecology* 23: 591–602.

Nanninga, G. B., P. Saenz-Agudelo, P. Zhan, L. Hoteit, and M. L. Berumen. 2015. Not finding Nemo: Limited reef-scale retention in a coral reef fish. *Coral Reefs* 34: 383–392.

Nelson, J. S., R. J. Hoddell, L. M. Chou, W. K. Chan, and V. P. E. Phang. 2000. Phylogeographic structure of false clownfish, *Amphiprion ocellaris*, explained by sea level changes on the Sunda shelf. *Marine Biology* 137: 727–736.

O'Donnell, J. L., R. Beldade, S. C. Mills, H. E. Williams, and G. Bernardi. 2017. Life history, larval dispersal, and connectivity in coral reef fish among the Scattered Islands of the Mozambique Channel. *Coral Reefs* 36: 223–232.

Quenouille, B., Y. Bouchenak-Kelladi, C. Hervet, and S. Planes. 2004. Eleven microsatellite loci for the saddleback clownfish *Amphiprion polymnus* (Telestei: Pomacentridae). *Molecular Ecology Notes* 4: 291–293.

Palumbi, S. 2003. Population genetics, demographic connectivity, and the design of marine reserves. *Ecological Applications* 13: S146–S158.

Pinsky, M. L., H. R. Montes, and S. R. Palumbi. 2010. Using isolation by distance and effective density to estimate dispersal scales in anemonefish. *Evolution* 64: 2688–2700.

Pinsky, M. L., P. Saenz-Agudelo, O. Salles, G. R. Almany, M. Bode, M. L. Berumen, S. Andréfouët, et al. 2017. Congruence of marine dispersal scales over evolutionary and ecological time. *Current Biology* 27: 149–154.

Planes, S., S. R. Thorrold, and G. P. Jones. 2009. Larval dispersal connects fish populations in a network of marine protected areas. *Proceedings of the National Academy of Science* 106: 5693–5697.

Raitsos, D. E., R. J. W. Brewin, P. Zhan, D. Dreano, Y. Pradhan, G. B. Nanninga, and I. Hoteit. 2017. Sensing coral reef connectivity pathways from space. *Scientific Reports* 7: 9338.

Roberts, C. M. 1997. Connectivity and management of Caribbean coral reefs. *Science* 278: 1454–1457.

Rollins, L. A., L. E. Browning, C. E. Holleley, J. L. Savage, A. F. Russell, and S. C. Griffith. 2012. Building genetic networks using relatedness information: A novel approach for the estimation of dispersal and characterization of group structure in social animals. *Molecular Ecology* 21: 1727–1740.

Saenz-Agudelo, P., J. D. DiBattista, M. Piatek, M. R. Gaither, H. B. Harrison, G. B. Nanninga, and M. L. Berumen 2015a. Seascape genetics along environmental gradients in the Arabian Peninsula: Insights from ddRAD sequencing of anemonefishes. *Molecular Ecology* 24: 6241–6255.

Saenz Agudelo, P., G. P. Jones, S. R. Thorrold, and S. Planes. 2009. Estimating connectivity in marine populations: An empirical evaluation of assignment tests and parentage analysis under different gene flow scenarios. *Molecular Ecology* 18: 1765–1776.

Saenz-Agudelo, P., G. P. Jones, S. R. Thorrold, and S. Planes. 2011. Connectivity dominates larval replenishment in a coastal coral reef fish metapopulation. *Proceedings of the Royal Society, Series B* 278: 2954–2961.

Saenz-Agudelo, P., G. P. Jones, S. R. Thorrold, and S. Planes. 2012. Patterns and persistence of larval retention and connectivity in a marine fish metapopulation. *Molecular Ecology* 21: 4695–4705.

Saenz-Agudelo, P., G. P. Jones, S. R. Thorrold, and S. Planes. 2015b. Mothers matter: Contribution to local replenishment is linked to female size, mate replacement and fecundity in a fish metapopulation. *Marine Biology* 162: 3–14.

Sale, P. F. 1991. Reef fish communities: Open, nonequilibrial systems. In *The Ecology of Fishes on Coral Reefs*, ed. P. F. Sale, 564–598. Elsevier.

Sale, P. F., R. K. Cowen, B. S. Danilowicz, G. P. Jones, J. P. Kritzer, K. C. Lindeman, S. Planes, et al. 2005. Critical science gaps impede use of no-take fishery reserves. *Trends in Ecology and Evolution* 20: 74–80.

Salles, O. C., G. R. Almany, M. L. Berumen, G. P. Jones, P. Saenz-Agudelo, M. Srinivasan, S. R. Thorrold, et al. 2020. Strong habitat and weak genetic effects shape the lifetime reproductive success in a wild clownfish population. *Ecology Letters* 23: 265–273.

Salles, O. C., J. A. Maynard, M. Joannides, C. M. Barbu, P. Saenz-Agudelo, G. R. Almany, M. L. Berumen, et al. 2015. Coral reef fish populations can persist without immigration. *Proceedings of the Royal Society, Series B* 282: 20151311.

Salles, O. C., B. Pujol, J. A. Maynard, G. R. Almany, M. L. Berumen, G. P. Jones, P. Saenz-Agudelo, et al. 2016a. First genealogy for a wild marine fish population reveals multi-generational philopatry. *Proceedings of the National Academy of Science* 113: 13245–13250.

Salles, O. C., P. Saenz-Agudelo, G. R. Almany, M. L. Berumen, S. R. Thorrold, G. P. Jones, and S. Planes. 2016b. Genetic tools link long-term demographic and life-history traits of clownfish to their anemone hosts. *Coral Reefs* 35: 1127–1138.

Sato, M., K. Honda, W. H. Uy, D. I. Baslot, T. G. Genovia, Y. Nakamura, L. P. C. Bernardo, et al. 2017. Marine protected area restricts demographic connectivity: Dissimilarity in a marine environment can function as an ecological barrier. *Ecology and Evolution* 7: 7859–7817.

Simpson, S. D., H. B. Harrison, M. R. Claereboudt, and S. Planes. 2014. Long-distance dispersal via ocean currents connects Omani clownfish populations throughout entire species range. *PLOS One* 9(9): e107610.

Sotka, E.E. 2012. Natural selection, larval dispersal, and the geography of phenotype in the sea. *Integrative and Comparative Biology* 52: 538–545.

Steinberg, R., M. van der Meer, E. Walker, M. L. Berumen, J.-P. A. Hobbs, and L. van Herwerden. 2016. Genetic connectivity and self-replenishment of inshore and offshore populations of the endemic anemonefish, *Amphiprion latezonatus*. *Coral Reefs* 35: 959–970.

Sultan, S. E., and H. G. Spencer. 2002. Metapopulation structure favors plasticity over local adaptation. *American Naturalist* 160: 271–283.

Thorrold, S. R., G. P. Jones, S. Planes, and J. A. Hare. 2006. Transgenerational marking of embryonic otoliths in marine fishes using barium stable isotopes. *Canadian Journal of Fisheries and Aquatic Science* 63: 1193–1197.

Timm, J., M. Kochzius, H. H. Madduppa, A. I. Neuhaus, and T. Dohna. 2017. Small scale genetic population structure of coral reef organisms in Spermonde Archipelago, Indonesia. *Frontiers in Marine Science* 4: 294.

Timm, J., S. Planes, and M. Kochzius. 2012. High similarity of genetic population structure in the false clown anemonefish (*Amphiprion ocellaris*) found in microsatellite and mitochondrial control region analysis. *Conservation Genetics* 13: 693–706.

van der Meer, M. H., J.-P. A. Hobbs, G. P. Jones, and L. van Herwerden. 2012a. Genetic connectivity among and self-replenishment within island populations of a restricted range subtropical reef fish. *PLOS One* 7: e49660.

van der Meer, M. H., L. van Herwerden, G. P. Jones, J.-P. A. Hobbs. 2012b. Historic hybridisation and introgression between two iconic Australian anemonefishes and contemporary patterns of population connectivity. *Ecology and Evolution* 2: 1592–1604.

Williamson, D. H., H. B. Harrison, G. R. Almany, M. L. Berumen, M. C. Bonin, S. Choukroun, P. J. Doherty, et al. 2016. Large-scale, multi-directional larval connectivity among grouper populations in the Great Barrier Reef Marine Park. *Molecular Ecology* 25: 6039–6054.

Watson, J. R., D. A. Siegel, B. E. Kendall, S. Mitarai, A. Rassweiller, and S. D. Gaines. 2011. Identifying critical regions in small-world marine metapopulations. *Proceedings of the National Academy of Sciences of the United States of America* 108: E907–E913.

Part V

Human Impact and Conservation

21 The Impact of Popular Film on the Conservation of Iconic Species

Anemonefishes in the Aquarium Trade

Carmen R. B. da Silva, Cassie M. Hoepner, Manon Mercader, Vincent Laudet, and Karen Burke da Silva

CONTENTS

21.1 INTRODUCTION

Even before the popular movies *Finding Nemo* (2003) and *Finding Dory* (2016), clownfish were undoubtedly the most charismatic and sought-after marine aquarium species. Their bold colors, long life span, and gregarious personality make them ideal pets. These fishes were therefore much discussed in terms of their acquisition, trade, and production in aquaculture. Very clearly, their popularity has raised concerns about the status of their natural populations in the Pacific and Indian Oceans and their influence in the ornamental fish trade. Recent research has pointed out a striking lack of knowledge for many essential questions regarding the trade of these iconic species, thus highlighting a need for improved monitoring in the environment, fisheries, importation, and trade.

In this chapter, we first describe what is known about the trade of not just clownfish but all anemonefish species: where are the commercialized fish coming from? In what numbers are they traded? What is the importance and impact of aquaculture? What was the impact of the popular films on anemonefish trade? Second, we discuss the impacts and types of wild fish harvesting (including impacts of targeting rare color variants) and their associated mortality occurrences. Lastly, we present some initiatives used to promote conservation and public awareness regarding the conservation of these iconic fishes, coral reefs, and marine environments more generally.

DOI: 10.1201/9781003125365-26

21.2 THE AQUARIUM TRADE

21.2.1 Global Trade

Global ornamental fish trade started in the 1930s, but really developed in the 1950s, and is now a multimillion-dollar industry becoming increasingly difficult to map and quantify (Rhyne et al. 2012a, 2012b, 2017b; Biondo and Burki 2020). Before specifically focusing on anemonefish, it is important to provide some background information on the global ornamental fish trade and to provide answers to four central questions: what are the main countries involved in exporting and importing marine aquarium species? How many fish species are involved in the trade? How many individual fish are traded annually? What is the economic value of the marine ornamental fish trade? As there is no unified tariff system and no systematic global monitoring system, it remains very difficult to answer these four questions with precision, even though such answers are needed to ascertain the sustainability of the marine ornamental trade (Biondo and Burki 2020).

21.2.1.1 What Are the Main Countries Involved in Exporting and Importing Marine Aquarium Species?

In a recent systematic review analysing 546 publications from 1975 to 2019, Biondo and Burki (2020) found 48 exporting and 38 importing countries involved in the freshwater and marine ornamental fish trade. These numbers exclude the countries trading exclusively freshwater fish and are therefore likely to be a good approximation of the number of countries involved in the marine fish trade. Furthermore, these findings are consistent with an analysis focusing only on fish imported into the US, suggesting that 40 countries are involved with the marine ornamental fish trade (Rhyne et al. 2012b). Most collection of marine aquarium fish occurs in Southeast Asia, primarily the Philippines and Indonesia (Tissot et al. 2010; Rhyne et al. 2017b), but South America, Oceania, and Africa also are involved. Within these statistics, we find that some countries export a very large number of species (up to 1,000), and some species are exported by many countries (e.g., the common damselfish *Chromis viridis* is exported by 29 countries) (Rhyne et al. 2012b). The US and Europe are the greatest importers of ornamental fish, but the number of importing countries involved is increasing, especially across Asia (Rhyne et al. 2017b). In addition, many countries (including Australia, Singapore, India, Costa Rica, and France) are both exporters and importers of ornamental fish. However, as ornamental fish trade reporting is not mandatory in many countries, it is exceedingly difficult to estimate the true volume and diversity of fish traded globally (Rhyne et al. 2012b; Rhyne et al. 2017b).

21.2.1.2 How Many Fish Species Are Involved in the Aquarium Trade?

As there is limited global fish trade reporting, the best analyses to date are those focusing on the import of fish into the US market. This is because they receive the largest volume of ornamental fish annually and maintain records of all wildlife imports and exports (Romagosa 2014; Stringham et al. 2021). Importation into Europe remains poorly reported (Rhyne et al. 2012b, 2017b; Biondo and Calado 2021), and little information is available for Switzerland (Biondo 2018), Australia (Trujillo Gonzalez and Militz 2019), Kenya (Okemwa et al. 2016), Papua New Guinea (Militz et al. 2018b), or India (Prakash et al. 2017), specifically. By developing an online publicly available Marine Aquarium Biodiversity and Trade flow database (www.aquariumtradedata.org), Rhyne et al. (2012b, 2017b) came to the conclusion that *ca.* 1,800 marine fish species (and 550 invertebrate species) are traded annually, with 2,300 fish species cumulatively traded over a three-year period. They found that fish within the Pomacentridae family, including anemonefish, are the most frequently imported fish into the US (more than 50% in volume for 170 species) with Labridae (228 species), Pomacanthidae (66 species), and Gobiidae (138 species) also traded in high volumes. Furthermore, only 20 species make up 52% of the total volume of fish imported to the US, indicating that a large proportion of the total fish trade is driven by a small number of species (Rhyne et al. 2017b).

21.2.1.3 How Many Individual Fish Are Traded Annually?

In 2003, it was estimated that 46 million individual marine ornamental organisms were sold annually to approximately two million hobbyists worldwide (Wabnitz et al. 2003). In a more recent systematic review, Biondo and Burki (2020) provide figures ranging from 15 to 26 million individual coral reef fish being traded annually but suggest that the value could be as high as 150 million. Studies by Rhyne et al. (2012b, 2017b), focusing on trade in the US between 2000 and 2011, provide numbers ranging from 6.9 to 11.2 million, with a decreasing trend over time suggesting that marine ornamental trade peaked in 2005, a year and a half after the release of *Finding Nemo* in August 2003. Despite recommendations for more accurate monitoring through the use of trade invoices, the importation of marine aquarium organisms continues with little or no accountability (Rhyne et al. 2012a; Biondo and Calado 2021), thus the impact on marine ecosystems is difficult to ascertain.

21.2.1.4 What Is the Economic Value of the Aquarium Trade?

The marine ornamental aquarium trade is a multimillion-dollar economy, but precise and recent values are almost impossible to estimate. The most recent figures range from 800 million to 30 billion dollars in 2004 which includes all supporting elements of the trade, such as food, tanks, drugs, etc. for the global market (Biondo and Burki 2020). In 2014, the export value for both freshwater and marine aquarium fish was estimated at 347.5 million dollars and the import value was estimated at 290 million dollars (Palmtag 2017; Biondo and Burki 2020). While the economic value of the marine aquarium trade is small compared to food fisheries, estimated at 100 billion dollars annually (Lam et al. 2016), many exporting countries rely on this economy as a significant source of income and revenue. Thus, policy changes to regulate trade should be carefully designed with local governments to reduce negative economic impacts on local collectors.

21.2.1.5 What Is the Value of Anemonefishes in the Marine Aquarium Trade?

An analysis of the 20 most traded aquarium species in the USA (which makes up over 50% of the total fish traded) during 2008, 2009, and 2011 shows that species importation rankings remain fairly consistent across the years (Rhyne et al. 2017b). *Chromis viridis* is by far the most imported species making up 10% of the total fish traded, while the clownfish *A. ocellaris* was present each year in this list at the sixth or seventh position representing 3% of the total fish traded (Rhyne et al. 2017b). A list of the number of individuals imported into the US in 2011 for each of the 26 species of anemonefish according to the Marine Aquarium Biodiversity and Trade flow database is shown in Table 21.1 (Rhyne et al. 2012b, 2017b). This table shows that the trade of all anemonefish species combined represented 6.1% of all ornamental marine fishes traded in the US in 2011. Furthermore, the clownfish *A. ocellaris*, *P. biaculeatus*, and *A. percula* are represented in the top three positions, and *A. ocellaris* and *A. percula* are the two species that most resemble the cartoon character Nemo and are almost impossible to differentiate. The two species are so closely related that hybrids of the two species are a regular occurrence (Balamurugan et al. 2017). The Philippines and Indonesia are the two major countries where collection of *A. ocellaris* and *A. percula* occurs, with the Philippines representing 51.1% of US imports and Indonesia 35.6% (Figure 21.1).

The large number of anemonefish trades as shown in Table 21.1 (more than 200,000 *A. ocellaris* individuals) highlights the popularity of clownfish and the economic value related to the global trade market. However, many of these fish are no longer being taken from the wild but are

TABLE 21.1
Number and Percentage of Each Anemonefish Imported into the US in 2011

Genus	Species	Imports (# of fishes)	% of anemonefishes	% of all marine fishes
Amphiprion	*ocellaris*	208,920	49.9	3.03
Premnas	*biaculeatus*	44,714	10.7	0.65
Amphiprion	*percula*	44,123	10.5	0.64
Amphiprion	*frenatus*	33,799	8.1	0.49
Amphiprion	*sebae*	24,066	5.8	0.35
Amphiprion	*clarkii*	12,918	3.1	0.19
Amphiprion	*polymnus*	11,363	2.7	0.16
Amphiprion	*perideraion*	6,742	1.6	0.10
Amphiprion	*bicinctus*	5,510	1.3	0.08
Amphiprion	*allardi*	4,477	1.1	0.06
Amphiprion	*latezonatus*	4,341	1.0	0.06
Amphiprion	*melanopus*	4,060	1.0	0.06
Amphiprion	*akallopisos*	3,035	0.7	0.04
Amphiprion	*akindynos*	2,453	0.6	0.04
Amphiprion	*sandaracinos*	2,035	0.5	0.03
Amphiprion	*chrysopterus*	1,924	0.5	0.03
Amphiprion	*ephippium*	1,670	0.4	0.02
Amphiprion	*nigripes*	662	0.2	0.01
Amphiprion	*chrysogaster*	611	0.1	0.01
Amphiprion	*fuscocaudatus*	457	0.1	0.01
Amphiprion	*tricinctus*	438	0.1	0.01
Amphiprion	*leucokranos*	13	0.0	0.00
Amphiprion	*rubrocinctus*	1	0.0	0.00
Amphiprion	*chagosensis*	0	0.0	0.00
Amphiprion	*mccullochi*	0	0.0	0.00
Amphiprion	*omanensis*	0	0.0	0.00
All anemonefishes		**418,332**	100.0	**6.07**
All marine fishes		**6,892,960**		

Source: retrieved from the database aquariumtradedata.org (Rhyne et al. 2017b) on 2 February 2022.

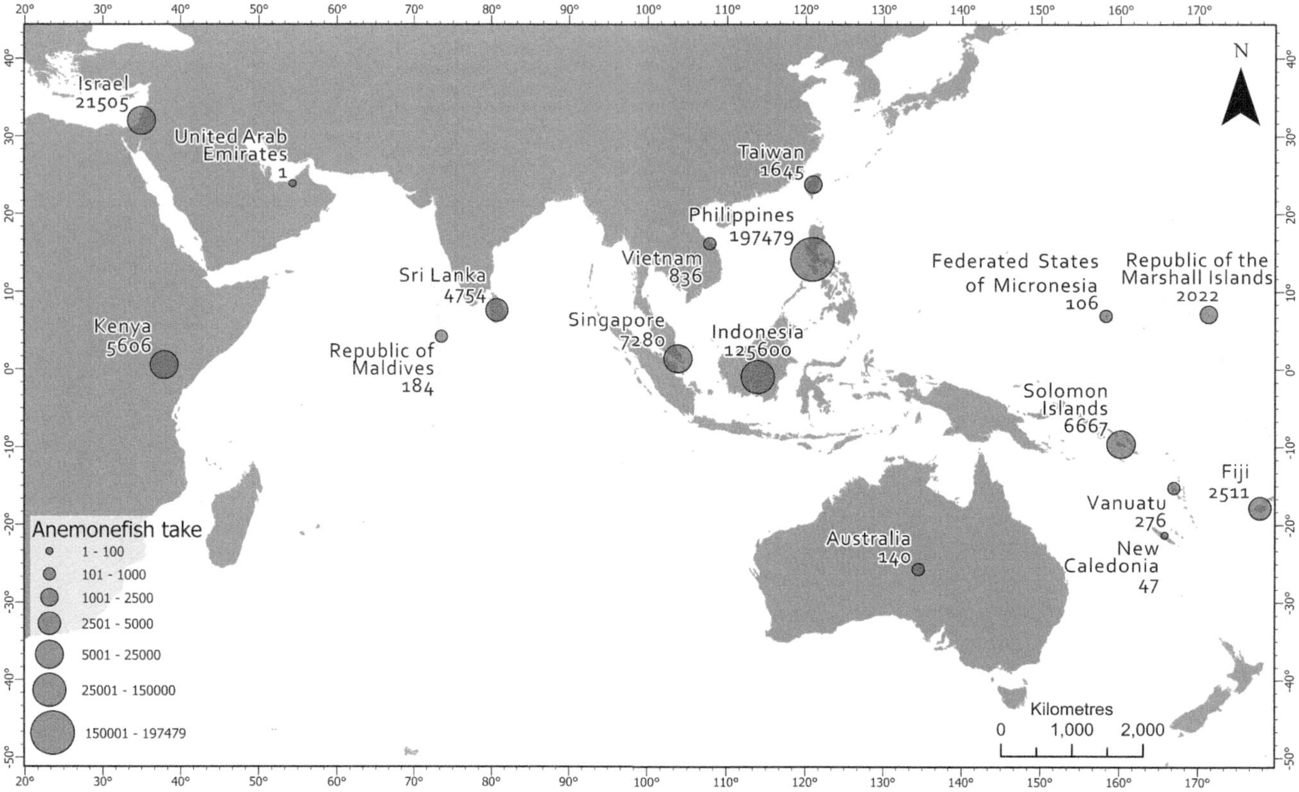

FIGURE 21.1 Locations of anemonefish collected for export to the US in 2011 as retrieved from the database aquariumtradedata.org (Rhyne et al. 2017b) on 2 February 2022.

now being produced in aquaculture facilities. For example, the majority of the *A. latezonatus* reported in Table 21.1 (4,341 individuals) that are imported to the US come from the Philippines and must be captive-bred as they are endemic to Australia.

21.2.2 Fisheries and Impact on Natural Populations

21.2.2.1 Anemonefish Fisheries

Due to a deficit of collection reporting data, it is difficult to estimate the total number of anemonefish traded annually across the globe. However, we can examine case studies to examine trends in anemonefish collection in different countries. For example, in 2012 Papua New Guinea reported that the most fished and exported marine aquarium species was *Amphiprion percula,* making up 32.8% of the total fish collected (15,615 individuals) (Militz et al. 2018b). Furthermore, *A. biaculeatus* and *A. clarkii* were also popular fish to collect and export and each made up 1.8% of the total fish collected (260 individuals per species) (Militz et al. 2018b). *Amphiprion percula* represented 84.8% of the Total Allowable Catch (TAC) for the species in Papua New Guinea, suggesting that the level of collection pressure is likely to indicate overfishing of this species, as actual catch numbers are usually higher than the reported figures. Overfishing has also been documented to occur in the Philippines (Shumanet al. 2005) and Indonesia (Madduppa et al. 2014, 2018). In India, within the Gulf of Mannar region, an analysis reveals that *Amphiprion sebae* was the most collected marine ornamental species, making up 19.95% (33,197 individuals) of the total fish collected (Prakash et al. 2017). In addition, *A. clarkii* made up 5.01% (8,391 individuals) of the total fish collected and exported for the aquarium trade (Prakash et al. 2017).

While many of the anemonefish exported globally may be produced through captive breeding, the data above shows that localized pressure on some species could lead to overexploitation. Given that anemonefishes are also affected by other stressors (climate change impacts on anemone bleaching and fish physiology, habitat destruction, pollution; see Chapter 25 "Saving Nemo"; Jones et al. 2008; Norin et al. 2018; Hoepner et al. 2019) more studies are urgently needed to precisely define the status of these species in the wild. To our knowledge, only one global assessment of wild anemonefish population numbers has been conducted, where the National Oceanic and Atmospheric Administration – US Department of Commerce (NOAA) found that 13 to 18 million individuals of *A. percula* are likely to exist globally. Thus, they were given a good risk score of "the overall extinction risk to *A. percula* is low both now and in the foreseeable future" (Maison and Graham 2016). However, we currently have a very limited understanding of the population sizes of the other anemonefish species.

Several authors have used Productivity Susceptibility Analysis (PCA), a semi-quantitative risk assessment tool that incorporates variables such as life history characteristics to evaluate whether species are able to rebound from overcollection and susceptibility to adverse environmental effects such as poor fishing practices. Indeed, anemonefish have a risky life history strategy (planktonic larval phase, size-based reproduction, recruitment, and long lifespans) that contribute to their vulnerability to wild collection (Shuman et al. 2005). The strict size-based hierarchy in which only the largest anemonefish is a breeding female, the second largest individual is the breeding male, and all other individuals are competitively maintained as smaller subordinate immature males means that only one female breeds in each anemone. Therefore, overcollection of anemonefish, particularly of the breeding pair, can result in the loss of a generation of anemonefish. If one or both breeding adults are removed from the anemone it takes at least 50 days for the next largest juvenile to physiologically adjust their gonads to become a reproductive individual (Molloy et al. 2007; Casas et al. 2016). Thus, future research that examines the impact of anemonefish collection on reproductive rates will be important for identifying what level of collection for the aquarium trade is sustainable.

The vulnerability and risk of overcollection of coral reef fish species imported into the USA have been assessed using PSA (Dee et al. 2014, 2019b). In this work, *A. biaculeatus* appeared fourth in the top ten most vulnerable species and five other species of *Amphiprion* (*A. clarkii*, *A. ephippium*, *A. frenatus*, *A. melanopus*, *A. ocellaris*, and *A. polymnus*) came in 12th, 13th, 14th, and 17th position (*A. ocellaris*, *A. frenatus*, *A. polymnus*, and *A. ephippium* respectively) (Dee et al. 2014, 2019b). Other analyses based on PSA have reached similar conclusions. For example, Kenyan fisheries ranked *A. allardi* and *A. akallopisos* as high risk (Okemwa et al. 2016), and Fujita et al. (2013) ranked *A. ocellaris*, *A. percula*, and *A. clarkii* as highly vulnerable in their analysis of Indonesian ornamental reef fish susceptibility. However, a more recent analysis by Baillargeon et al. (2020) classified *A. biaculeatus*, *A. ocellaris*, and *A. percula* as "sustainable" and *A. clarkii* as "highly sustainable" using their new algorithm. The major difference between this study and the others was based not on the productivity scores, but on how susceptibility is estimated and geographic location. Thus, there remains the need to standardize vulnerability assessments and for additional studies to quantify the vulnerability of these species. However, it is interesting to note that despite differences in interpretation across studies, there is a convergence on the finding that the group of clownfish, *A. biaculeatus, A. ocellaris, and A. percula*, are the most vulnerable species when compared to other marine ornamental species as they always have the highest collection rates, indicating that particular group of anemonefish species should be targeted for further vulnerability assessments.

Some initiatives have been developed to try to reduce overexploitation of wild fish populations. In French Polynesia, trials were undertaken to capture coral reef fish at the larval stage using nets on reef crests. After capture, larvae were reared in aquaria until they reached a suitable size before entering the world aquarium market (Fujita et al. 2013; Lecaillon 2017; Lecchini et al. 2006). Indeed, capturing larvae that are 90 times more abundant than adult stock puts far less pressure on wild populations. Unfortunately, these post-larval capture and culture methods have not proven to be highly economically effective, and the practice has not been fully established. The majority of larval fishes collected were of low commercial value and the cost of growing them to market size was considered too high for the practice to be deemed economically viable (Ellis 2010; Lecaillon 2017).

While collection pressure is impacting anemonefish populations, other anthropogenic factors are also impacting their reproduction and likely population sizes. For example, a recent study found that anemonefish are vulnerable to the negative effects of artificial light at night (ALAN) (Forbert et al. 2019). All anemonefish species are limited to relatively shallow coral reef distributions due to the light requirements of their sea anemone hosts and thus are exposed to ALAN across much of their distribution. Fobert et al. (2019) found that clownfish exposed to ALAN in the laboratory resulted in complete reproductive failure with no eggs hatching. In addition, clownfish exposed to different light spectra of ALAN also had significantly reduced reproductive success (Fobert et al. 2021) meaning that this important factor should be taken into consideration as ALAN continues to increase across the globe with growing human population sizes infringing upon anemonefish habitat.

21.2.2.2 Giant Sea Anemone Fisheries

Anemonefish can survive and reproduce well in captivity without their sea anemone host, but they are at high risk of predation if they are without an anemone in the wild (Feeney et al. 2019; Roux et al. 2020). Therefore, the sea anemone host must be taken into consideration when examining the impact of the aquarium trade. Giant sea anemones are in themselves a popular organism sought after by marine aquarium enthusiasts but remain much less sought after compared to anemonefish. In the analysis conducted by Rhyne et al. (2017b), only a single giant sea anemone species, *Heteractis malu*, appeared in the 20 most US-imported invertebrate species. It is likely that the lower level of interest by hobbyists for giant sea anemones is due to the difficult requirements needed to maintain them in captivity as they require higher levels of aquarist skills and specialized equipment (e.g., lights emitting adequate photosynthetically active radiation) to keep them healthy. Anemones in captivity can undergo asexual propagation which has also been shown experimentally with good success rates (Scott et al. 2014; Moore et al. 2020). However, sexual reproduction in anemones is less well understood and it appears that only two of the ten host anemone species, *Entacmaea quadricolor* and *Heteractis crispa*, have been bred in a controlled environment in captivity (Scott and Harrison 2005, 2007, 2009). Despite patenting the method for breeding *E. quadricolor* in 2006, production has yet to be reported and no captive-bred anemones are available on a commercial scale.

Harvesting of anemones for the aquarium trade, in combination with climate change and increased bleaching events, is likely to lead to a significant reduction of habitat availability for anemonefish in the wild. Anemone bleaching reduces the size of available shelter for anemonefish and, although anemonefish numbers may not decline when in a bleached anemone, anemonefish egg production decrease by 38% compared to egg production in unbleached anemones (Saenz-Agudelo et al. 2011). In addition, if populations of anemones reduce to a point where the distance between available anemones becomes too great for the movement of fish between anemones or ultimately recruitment of new juveniles (ca. 45 km; Almany et al. 2017), the likelihood of local extinction will increase.

In conclusion, while more and more anemonefish (and in particular the iconic clownfishes *A. ocellaris*, *A. percula*, and *A. biaculeatus*) are being bred in captivity, the risk of overexploitation levels in certain geographic areas remains a concern. Combined with other stresses, decreased population sizes could lead to reduced genetic diversity (Madduppa et al. 2018) or even local extinction. It is critically important to monitor and quantify active collecting pressure and environmental threats in order to predict future population declines of anemonefish and host anemone populations in the wild so that conservation measures can be put in place.

21.2.3 Conservation Aquaculture

More than 90% of freshwater aquarium species are now bred in captivity (Tlusty 2002; Moorhead and Zeng 2010). Conversely, marine species remain mostly wild-caught, with 338 captive-bred fish species reported in 2017 and, among them, only 18% were regularly available in the market (Wabnitz et al. 2003; Moorhead and Zeng 2010; Murray and Watson 2014; Pouil et al. 2020). A recent book by Calado et al. (2017) provides a detailed and specific account of aquaculture practises of marine ornamental species so we will only discuss aspects of anemonefish aquaculture here. One important general point is that aquaculture of marine ornamental fishes is on the rise where new techniques are being employed, including improved larval rearing so that a number of species that previously were found to be too difficult to breed are now being bred in captivity (Pouil et al. 2020). This is excellent news and should make a positive impact on overcollection in the wild, however, captive breeding does have some negative environmental aspects that must be considered and managed accordingly. For example, pollution, overfishing for feed, and an increase in invasive species are all current environmental threats caused by the aquaculture industry (Calado et al. 2017; Lockwood et al. 2019; Pouil et al. 2020).

Pomacentrids are excellent species for conservation aquaculture as they breed well and live long lives in captivity. The three iconic clownfish species (*A. ocellaris*, *A. percula*, and *A. biaculeatus*) in particular, make up the bulk of aquaculture species, yet the proportion of fish from aquaculture compared to the number of fish being taken from the wild remains unknown.

A sector of aquaculture on the rise and of importance for anemonefish is the breeding of new pigmentation variants (see Klann et al. [2021] as well as Chapter 7). New variants started primarily with the remarkable color morphotypes developed in *A. ocellaris* and *A. percula*, which have launched a "designer clownfish" era referred to as "guppification" of the anemonefish (Calado et al. 2017). An example of this is the lightning maroon variant of *A. biaculeatus* developed from variant specimens captured in the wild and now is a major source of revenue for the breeders. It is important to mention that as discussed in Chapter 7 (Salis et al.), pigmentation strains are in fact genetic mutants and could consequently be a beneficial resource for future research in identifying genes that underlie different traits relative to pigmentation (Klann et al. 2021).

21.3 THE CONDITION OF THE AQUARIUM TRADE

The collection methodology used for obtaining marine aquarium fish varies from highly controlled practices to extremely detrimental methods that harm individual fish as well as their environments (Palmtag 2017; Rhyne et al. 2017a). For example, the Great Barrier Reef (GBR), Australia, is one of the best-managed coral reefs in the world – where quota restrictions, voluntary stewardship agreements, and no-take zones are closely monitored (Evans et al. 2006; Pendleton et al. 2019). Most commonly, fish are collected via SCUBA using hand nets and barrier nets that reduce physical impact and stress to the fish. Yet, even with these controls in place, many populations are still declining (Jones et al. 2008; Frisch et al. 2009; Scott and Baird 2015). As an economically wealthy country, Australia should be leading the conservation management strategies for both conservation aquaculture and collection and import of aquarium species from the wild. Although importation records of aquarium species entering Australia are very difficult to ascertain (as the species of fish is not required in documentation), large numbers of aquarium fish continue to be imported into Australia.

21.3.1 Cyanide Fishing

In contrast to highly managed fisheries, there remains significant concern in the Indo-Pacific where cyanide fishing, an illegal and highly destructive practice, has continued since the 1970s (Burke et al. 2011). Local fishers spray concentrated cyanide solutions to stun fish to make them easier to capture (Pomeroy and Balboa 2004; Millar 2013; reviewed by Madeira and Calado, 2019). This results in high levels of coral and fish mortality (Cervino et al. 2003). Recent studies pointed out that nearly 15% of fish in the aquarium trade screened with a non-invasive method (namely testing positive for the presence of the thiocyanate anion in their urine) displayed physiological evidence of being illegally collected using cyanide poisoning (Vaz et al. 2017). This occurrence is similar to data reported nearly two decades ago and indicates that the efforts to eradicate cyanide use from the marine aquarium trade have not been effective. Of note, even though this study doesn't detail the data for anemonefish specifically, the percentage of positive cyanide fish was approximately 23% for Pomacentrids, suggesting that anemonefishes are still captured using this method.

Higher profit margins seem to be driving the practice of cyanide use, as it is a significant time-saving method, making it attractive to collectors who, very often, are locals working at subsistence levels. Although the Marine-Life Alliance (IMA) has trained collectors in the Philippines and Indonesia to use barrier nets effectively and efficiently, it has not been widely adopted due to the economic pressure on the fishers (Rubec 1986; Rubec et al. 2001). Using cyanide as a collecting method also results in high mortality rates for fish (up to 50% – Hess et al. 2015) and produces significant degradation of the coral reef habitat (Bruckner and Roberts 2008). These practices coupled with other anthropogenic disturbances such as climate change and pollution may lead to population effects and even local extinctions, as shown for other organisms (Madeira et al. 2020 and references therein), ultimately reducing the financial viability of the initial short-term economic gains from cyanide use (Madeira et al. 2020). Other anaesthetics, either synthetic (AQUI-S®, 2-phenoxyethanol, tricaine methane-sulfonate (MS-222), benzocaine, metomidate, etomidate) or natural (i.e., plant-based; clove, mint, thyme) are extensively used in aquaculture and scientific studies to collect fish (either in the lab or in the field) and although their efficiency and side effects are still being studied (Readman et al. 2017; Aydin and Barbas 2020), they are considered safer than cyanide. More studies are needed to fully assess the impact of these substances on reef ecosystems, but AQUI-S® and clove oil are both easy to acquire, of low cost, and relatively environmentally friendly and thus are a good potential alternative to cyanide to capture fish for the aquarium trade (Javahery et al. 2012). Unfortunately, thus far all efforts to ban cyanide use in the Indo-Pacific have remained unsuccessful (National Intelligence Council 2016).

21.3.2 Mortality Data and the Impact of Mortality

In addition to the vast number of fishes exported for sale, fishes collected for the aquarium trade are subject to a process of quality control and may be discarded along the supply chain if aesthetically impaired, of inappropriate size, or unhealthy (Militz et al. 2016). For example, up to 25% of the total fish collected for the aquarium trade in Papua New Guinea were rejected over a six-month period and were never reported in buyer invoices (Militz et al. 2016). The proportion of discarded catch can vary greatly and examples with less mortality exist (e.g., Queensland fishery; Roelofs and Silcock 2008; Dee et al. 2014). However, a study based on a single export enterprise containing several branches in Bali and throughout Indonesia found very high mortality rates, between 10% to 40% before exportation (Schmidt and Kunzmann 2005). Similar numbers have been reported for *A. biaculeatus* (29 fish of 96 fish caught, 30%; MAMTI 2006). It is likely that the total number of fish deaths is often much greater than recorded or acknowledged, thus more studies are needed to understand the full cost of mortality associated with the aquarium trade.

Even the wild-caught fish that do survive and make it to the consumer end up having a much low survival rate than fish from the wild, in many cases living only one to two years (Millar 2013; Biondo and Burki 2020). Reduced life spans of wild-caught fish sold in the aquarium trade lead to more frequent replacement, thus precipitating the need for higher collection from the wild. Decreased abundance of aquarium fish in marine environments was noted as early as 1978 in Hawaii (Walsh 1978) and follow-up studies confirmed that collection of aquarium fish does impact numbers in the wild (Tissot and Hallacher 2003; see also the recent analysis by Schaar and Cox 2021). For example, in the Philippines, the abundance of both anemones and anemone fish was found to be much reduced in exploited areas compared to unexploited areas (Shuman et al. 2005). Furthermore, visual surveys on the Great Barrier Reef highlighted that regions with the highest densities of anemonefish were found in areas that were protected and closed to fishing for the aquarium trade (Jones et al. 2008). This finding is not surprising but does indicate that collection pressure clearly impacts population sizes. The question of concern, and for which we still lack clear data, is whether a stable fishery over time could maintain effective population sizes, or if population declines will continue, leading to an eventual population collapse.

While collection of marine aquarium fish affects population sizes and abundance, fishing in general has been shown to impact average fish body size, with the number of large fish declining in marine environments overall, especially in low latitude regions (Bianchi and Morri 2000). This finding is relevant because fish size is related to reproductive output, where larger females produce significantly more offspring than smaller females (Hixon et al. 2014; Barneche et al. 2018). Furthermore, collection of fish from marine environments (including fishing for food consumption) is known to have far-reaching negative cascading effects throughout ecosystems (Salomon et al. 2008), and the extraction of fish that contribute most towards population reproduction should be ceased to avoid collapse of food webs and loss of ecosystem function. Coral reef fish are amongst the most severely overexploited and depleted from oceanic environments (for food and the aquarium trade) (Musick et al. 2000; Dankel et al.

2008), and organisms that inhabit tropical environments are predicted to be among the most vulnerable to warming climates due to living in warm environmental conditions that are already close to their upper thermal limits (Somero 2010). Thus, it is important to conserve tropical marine ornamental fish as they are exposed to multiple negative stressors.

21.3.3 New and Rare Variants: A Saviour or a Risk Factor?

Naturally rare species, or morphs of species, may suffer more from exploitation, as consumer demand makes them more attractive than standard varieties (Courchamp et al. 2006; Hall et al. 2008; Dee et al. 2019a). Price premiums also reflect perceived rarity and thus can drive markets for collection in the wild (Rhyne et al. 2012b). A study by Militz et al. (2018a) in Papua New Guinea found that collectors and exporters received export prices that were exponentially related to the rarity of two different color morphs of two anemonefish species, *A. perculua* (grey circles) and *A. biaculeatus* (black circles) color morphs. Even with the increase in economic value of individual morphs, they did not find that overexploitation of the fishery as fishers collected more opportunistically rather than focusing the time needed to find the rare morphs. This makes economic sense as the authors also found that the rare morphs were mostly randomly distributed, however, when a morph was non-randomly distributed the regional morph was at very high risk of overexploitation. Militz et al. (2018a) concluded that when targeted and opportunistic collection strategies are employed, the resulting practice can be unsustainable, at least within their study, not only affecting biodiversity but also economic loss to fishers. A similar study with two anemonefish species, *A. leucokranos* and *A. theilei*, found similar results (Gainsford et al. 2020; Chapter 25). To pre-empt overexploitation of rare species, policy management requirements need to be put in place. Exportation bans on rare morphs or species or collection limits together with the implementation of new protection zones such as marine parks can all reduce the extent of overcollection in the wild (Branch et al. 2013), however, the effectiveness of these practices around the world has been limited (Militz et al. 2018a).

A mechanism to maintain rarity and unusual color morphs may be better attained through conservation aquaculture and crossing pre-existing variants or rare mutants in captivity (Klann et al. 2021). Following the spread of rare morphs within the freshwater aquarium industry (betta fighting fish for example), a focus on understanding the genetics of anemonefish could not only produce economically viable rare morphs but also reduce pressure on rare wild variants. This, of course, may not occur for many years as aquaculture for marine ornamental species is only in its infancy and consumer demand remains high for exotic or rare morphotypes. Shifting consumer demand may be the only current option available.

21.4 THE NEED TO IMPROVE CONSUMER AWARENESS

21.4.1 The Effect of a Movie

Motion pictures are increasingly important in shaping our perceptions and understanding of the natural world as well as creating strong emotional connections with individual species (Visch et al. 2010; Sandbrook et al. 2015; Militz and Foale 2017). The film *Bambi* (1942) is an early example of how emotive films can be in shaping an audience's perceptions and changing views, with strong attitudinal changes towards deer hunting witnessed after its release (Hastings 1996). Another example is the film *Happy Feet* (2006) which provided a strong conservation message around overfishing and plastic pollution (Silk et al. 2018).

Films highlighting natural environments and charismatic species can also influence consumers' behavior and demand for pets (reviewed in Yong et al. 2011). Following the success of the film series *Free Willy* (1993, 1995, 1997), the whale watching industry grew to more than one billion dollars (Hoyt 2001; Lawrence and Phillips 2004). The rise in popularity of species or breeds driven by media or the film industry can therefore have both positive and negative effects. Indeed, *Finding Nemo* has educated audiences worldwide about Australian coral reef species. From a conservation perspective, public interest and enthusiasm towards clownfish are exactly what is needed to engage and educate people about the detrimental circumstances surrounding the marine aquarium trade. However, while the film promotes a strong conservation message, "Don't take Nemo from the sea and put him in an aquarium", Nemo's charm could have had the opposite effect by increasing the demand for anemonefish as pets.

Following the release of the Disney film *Finding Nemo* in 2003, several members of the aquarium trade suggested that the consumer demand for clownfish increased (McClenachan et al. 2012), yet, other evidence indicates no clear global increase in fishery catches of *A. ocellaris* and *A. percula* after the movie (Militz and Foale 2017). The same finding occurred with the Pixar character, Dory, the blue tang (*Paracanthurus hepatus*) featured in *Finding Dory*. Although there was an increase in global online search for this species two to three weeks after the release of the film, no substantial evidence exists to support an increase in importation of this species into the US (Veríssimo et al. 2020).

21.4.2 Public Perceptions and Awareness

Public awareness regarding conservation and environmental issues has grown over the past few decades (Du et al. 2018; Rousseau and Deschacht 2020). Knowing the public's perceptions and their understanding of current threats will help conservationists develop education and sustainability programs that target lack of knowledge or misconceptions. For example, in 2017, the conservation and education program Saving Nemo was launched to provide education and information to the public to ultimately reduce pressure on harvesting anemonefish and conserve coral reefs more broadly. The website SavingNemo.org informs readers of the problem and sets out solutions, many of which involve public engagement.

Despite the potential issues it might cause (discussed in Section 21.2.1), captive breeding of marine organisms for the aquarium trade is the most beneficial solution to reduce the risk of overcollecting from the wild, especially at a time when coral reefs are facing additional anthropogenic threats. Unfortunately, many people do not realise that marine aquarium species are often collected from the wild and they assume that like most freshwater fish species, they are bred in captivity. It is therefore imperative that consumers and distributors understand where their aquarium fish are coming from. Even if decision-making is a multifaceted process, consumers with appropriate education and understanding will make informed decisions about their product and even pay more for a sustainably produced fish. Additionally, it is likely that anemonefish bred in captivity will make better pets as aquarists report they live longer lives, they have less stress and disease, and they tolerate captive conditions more favourably than wild-caught fish. Stayman (1999), for example, reports that fish bred in captivity exhibit fewer stereotypes, such as pacing, hiding, or refusing to eat when they are introduced to a new home environment.

21.4.3 Citizen Science Initiatives and Relationships with Local Actors to Protect Anemonefish

Citizen science projects can increase public awareness and action for conservation. The citizen science project IC-ANEMONE is a branch of SavingNemo.org that provides citizens with an opportunity not only to "Find Nemo", but to learn about anemonefish biology and the threats to their persistence. Citizen scientists collect data at monitoring stations (affiliated with resorts and dive shops) in the form of snorkel/dive trails, each targeting different species of anemonefish and anemones (see Orca Nation at Rawa Island for example). As well as contributing high-quality consistent data to researchers, this program also provides financial incentive to the resorts, as they can charge for this unique activity.

A partnership with i-naturalist.org (a community for citizen scientists to report data) enables sharing of data between both publicly available websites. Quantifying broad-scale patterns in nature can require a vast amount of data collected across a large network of locations spanning many years. Involving citizen scientists provides the person-power needed for global monitoring which in turn provides data to inform policy for the management of these and other threatened species. The presence of iconic species on coral reefs ensures that tourists are not disappointed and continue to spend their travel money at coral reef destinations. Therefore, once aware of the ecology and threats that marine organisms face, hotels, dive shops, and other marine-related service providers should be more willing to assist in the conservation effort of marine species and habitats.

Another example is Seragaki Island Clownfish Restoration Project in Okinawa, Japan (Figure 21.2). This project is led by the Hyatt Regency (a luxurious hotel complex by the ocean) in collaboration with the Okinawa Institute of Science and Technology (OIST) and aims to

FIGURE 21.2 A. The Hyatt Hotel; clownfish are released in the shallow water surrounding the hotel. B. Scientific divers releasing a new pair of clownfish in *Stichodactyla gigantea* host anemone. C. A newly released clownfish pair with its tag for monitoring and tourists' information. Pictures courtesy of Yoko Shintani (A) and Erina Kawai (B, C).

conserve and restore *A. ocellaris* populations and develop eco-tourism programs. The scientific team survey local populations in their natural habitat to assess the best release sites for juvenile fish. They are also conducting a long-term genetic survey to ensure that the introduction of new juveniles will not cause unnatural genetic variations within the local population. Lastly, released and naturally present fish are monitored to assess if a self-sustaining population can result from this program (Figure 21.2). The hotel provides financial support as well as information panels and snorkelling activities for their guests, bringing awareness of fish ecology and the threats to their persistence.

21.5 CONCLUSION: THE FLAG FISH SPECIES CONUNDRUM

A recent study shows that our subconscious biases the choice of species that researchers choose to study (Bellwood et al. 2020). By surveying the literature on coral reef fishes, they conclude that the most studied species are predominantly yellow in color, behaviorally bold, and live in warm, calm, attractive locations, all of which correspond to anemonefish and their habitat (albeit a bit restrictive on color pigmentation). Other authors have reached similar conclusions in terms of conservation effort and suggest that species' "aesthetic value" is disconnected from their ecological value and could be misleading for conservation purposes (Tribot et al. 2018). We agree that both research and conservation efforts are likely affected by this bias. However, we also believe (at least in both Australia and Japan) that the popularity of species such as the clownfish can have a large impact on increasing public awareness and conservation efforts that support broad communities of organisms. As discussed recently, flagship species do not necessarily impede cost-effective conservation and can raise funds for conservation and help target resources most appropriately to conserve biodiversity (McGowan et al. 2020). The iconic relationship between anemonefish and their sea anemone host provides an excellent platform to build a discourse that illustrates the complexity of life in coral reef ecosystems and promotes public understanding of the importance of preserving biodiversity.

REFERENCES

Almany, G. R., S. Planes, S. R. Thorrold, M. L. Berumen, M. Bode, P. Saenz-Agudelo, M. C. Bonin, et al. 2017. Larval Fish Dispersal in a Coral-Reef Seascape. *Nature Ecology and Evolution* 1 (6): 0148.

Aydin, B., and L. A. L. Barbas. 2020. Sedative and Anesthetic Properties of Essential Oils and Their Active Compounds in Fish: A Review. *Aquaculture* 520: 734999.

Baillargeon, G., M. Tlusty, E. Dougherty, and A. Rhyne. 2020. Improving the Productivity-Susceptibility Analysis to Assess Data-Limited Fisheries. *Marine Ecology Progress Series* 644 (June): 143–56.

Balamurugan, J., T. T. Ajith Kumar, K. Kathiresan, and B. Meenakumari. 2017. Determination of Growth, Colour and Other Traits in F1 Hybrid of *Amphiprion Percula* (Male) × *A. Ocellaris* (Female). *Aquaculture Research* 48 (6): 2989–3003.

Barneche, D. R., D. R. Robertson, C. R. White, and D. J. Marshall. 2018. Fish Reproductive-Energy Output Increases Disproportionately with Body Size. *Science* 360 (6389): 642–45.

Bellwood, D. R., C. R. Hemingson, and S. B. Tebbett. 2020. Subconscious Biases in Coral Reef Fish Studies. *BioSciences* 70: 621–27.

Bianchi, C N., and C. Morri. 2000. Marine Biodiversity of the Mediterranean Sea: Situation, Problems and Prospects for Future Research. *Marine Pollution Bulletin* 40 (5): 10.

Biondo, M. V. 2018. Importation of Marine Ornamental Fishes to Switzerland. *Global Ecology and Conservation* 15: e00418.

Biondo, M. V., and R. P. Burki. 2020. A Systematic Review of the Ornamental Fish Trade with Emphasis on Coral Reef Fishes—An Impossible Task. *Animals* 10: 2014.

Biondo, M. V., and R. Calado. 2021. The European Union Is Still Unable to Find Nemo and Dory-Time for a Reliable Traceability System for the Marine Aquarium Trade. *Animals* 11: 1668.

Branch, T. A., A. S. Lobo, and S. W. Purcell. 2013. Opportunistic Exploitation: An Overlooked Pathway to Extinction. *Trends in Ecology and Evolution* 28: 409–13.

Bruckner, A. W., and G. Roberts (editors). 2008. Proceedings of the International Cyanide Detection Testing Workshop. NOAA Technical Memorandum NMFS-OPR-40, Silver Spring, MD, 164 pp. https://www.coris.noaa.gov/activities/cyanide/cyanide_rpt.pdf (accessed March 29, 2022).

Burke, L., K. Reytar, M. Spalding, and A. Perry. 2011. *Reefs at Risk Revisited*. Washington, DC: World Resources Institute.

Calado, R., I. Olivotto, M. P. Oliver, and J. Holt, eds. 2017. *Marine Ornamental Species Aquaculture*. Chichester, UK: Wiley-Blackwell.

Casas, L., F. Saborido-Rey, T. Ryu, C. Michell, T. Ravasi, and X. Irigoien. 2016. Sex Change in Clownfish: Molecular Insights from Transcriptome Analysis. *Scientific Reports* 6 (1): 35461.

Cervino, J. M., R. L. Hayes, M. Honovich, T. J. Goreau, S. Jones, and P. Rubec. 2003. Changes in Zooxanthellae Density, Morphology, and Mitotic Index in Hermatypic Corals and Anemones Exposed to Cyanide. *Marine Pollution Bulletin* 46: 573–86.

Courchamp, F., E. Angulo, P. Rivalan, R. J. Hall, L. Signoret, L. Bull, and Y. Meinard. 2006. Rarity Value and Species Extinction: The Anthropogenic Allee Effect. *PLoS Biology* 4: e415.

Dankel, D. J., D. W. Skagen, and Ø. Ulltang. 2008. Fisheries Management in Practice: Review of 13 Commercially Important Fish Stocks. *Reviews in Fish Biology and Fisheries* 18 (2): 201–33.

Dee, L. E., J. Cowles, F. Isbell, S. Pau, S. D. Gaines, and P. B. Reich. 2019a. When Do Ecosystem Services Depend on Rare Species? *Trends in Ecology and Evolution* 34: 746–58.

Dee, L. E., S. S. Horii, and D. J. Thornhill. 2014. Conservation and Management of Ornamental Coral Reef Wildlife: Successes, Shortcomings, and Future Directions. *Biological Conservation* 169: 225–37.

Dee, L. E., K. A. Karr, C. J. Landesberg, and D. J. Thornhill. 2019b. Assessing Vulnerability of Fish in the U.S. Marine Aquarium Trade. *Frontiers Marine Science* 5: 527.

Du, Y., X. Wang, D. Brombal, A. Moriggi, A. Sharpley, and S. Pang. 2018. Changes in Environmental Awareness and Its Connection to Local Environmental Management in Water Conservation Zones: The Case of Beijing, China. *Sustainability* 10 (6): 2087.

Ellis, S. 2010. *Development of Post-Larval Capture and Culture (PCC) in Pohnpei for the Ornamental Fish Market.* Coral Reef Initiatives for the Pacific, Technical Report. www.meripmicronesia.org/wp-content/uploads/2012/01/Merip-PCC-Report-2010.pdf (Accesse June 28., 2022)

Evans, L. S., P. Fidelman, C. Hicks, C. Morgan, A. L. Perry, and R. Tobin. 2006. *Limits to Climate Change Adaptation in the Great Barrier Reef: Scoping Ecological and Social Limits.* Gold Coast, QLD: National Climate Change Adaptation Research Facility.

Feeney, W. E., R. M. Brooker, L. N. Johnston, J. D. J. Gilbert, M. Besson, D. Lecchini, D. L. Dixson, P. F. Cowman, and A. Manica. 2019. Predation Drives Recurrent Convergence of an Interspecies Mutualism. *Ecology Letters* 22 (2): 256–64.

Fobert, E. K., K. Burke da Silva, and S. E. Swearer. 2019. Artificial Light at Night Causes Reproductive Failure in Clownfish. *Biology Letters* 15: 20190272.

Fobert, E. K., K. Schubert, and K. Burke da Silva. 2021. The Influence of Spectral Composition of Artificial Light at Night on Clownfish Reproductive Success. *Journal of Experimental Marine Biology and Ecology* 540: 151559.

Frisch, A. J., J.-P. A. Hobbs, and Great Barrier Reef Marine Park Authority. 2009. *Rapid Assessment of Anemone and Anemonefish Populations at the Keppel Islands: A Report to the Great Barrier Reef Marine Park Authority.* Townsville, QLD: Great Barrier Reef Marine Park Authority.

Fujita, R., D. J. Thornhill, K. Karr, C. H. Cooper, and L. E. Dee. 2013. Assessing and Managing Data-Limited Ornamental Fisheries in Coral Reefs. *Fish and Fisheries* 15 (4): 661–75.

Gainsford, A., G. P. Jones, J-P. A. Hobbs, F. M. Heindler, and L. Herwerden. 2020. Species Integrity, Introgression, and Genetic Variation across a Coral Reef Fish Hybrid Zone. *Ecology and Evolution* 10 (21): 11998–14.

Hall, R. J., E. J. Milner-Gulland, and F. Courchamp. 2008. Endangering the Endangered: The Effects of Perceived Rarity on Species Exploitation. *Conservation Letters* 1: 75–81.

Hastings, A. W. 1996. Bambi and the Hunting Ethos. *Journal of Popular Film and Television* 24 (2): 53–59.

Hess, S., A. S. Wenger, T. D. Ainsworth, and J. L. Rummer. 2015. Exposure of Clownfish Larvae to Suspended Sediment Levels Found on the Great Barrier Reef: Impacts on Gill Structure and Microbiome. *Scientific Reports* 5 (1): 10561.

Hixon, M. A., D. W. Johnson, and S. M. Sogard. 2014. BOFFFFs: On the Importance of Conserving Old-Growth Age Structure in Fishery Populations. *ICES Journal of Marine Science* 71 (8): 2171–85.

Hoepner, C. M., C. A. Abbott, and K. Burke da Silva. 2019. The Ecological Importance of Toxicity: Sea Anemones Maintain Toxic Defence When Bleached. *Toxins (Basel)* 11: 266.

Hoyt, E. 2001. *Whale Watching: Worldwide Tourism Numbers, Expenditures, and Expanding Socioeconomic Benefits.* International Fund for Animal Welfare. Yarmouth Port, MA: International Fund for Animal Welfare.

Javahery, S., H. Nekoubin, and A. H. Moradlu. 2012. Effect of Anaesthesia with Clove Oil in Fish. *Fish Physiology and Biochemistry* 38: 1545–52.

Jones, A. M., S. Gardner, and W. Sinclair. 2008. Losing 'Nemo': Bleaching and Collection Appear to Reduce Inshore Populations of Anemonefishes. *Journal of Fish Biology* 73 (3). 753–61.

Klann, M., M. Mercader, L. Carlu, K. Hayashi, J. D. Reimer, and V. Laudet. 2021. Variation on a Theme: Pigmentation Variants and Mutants of Anemonefish. *EvoDevo* 12 (1): 8.

Lam, V. W. Y., W. W. L. Cheung, G. Reygondeau, and U. Sumaila. 2016. Projected Change in Global Fisheries Revenues under Climate Change. *Scientific Reports* 6 (1): 32607.

Lawrence, T. B., and N. Phillips. 2004. From Moby Dick to Free Willy: Macro-Cultural Discourse and Institutional Entrepreneurship in Emerging Institutional Fields. *Organization* 11 (5): 689–711.

Lecaillon, G. 2017. Post-Larval Capture and Culture of Ornamental Fishes. In *Marine Ornamental Species Aquaculture*, eds R. Calado et al. 327–46. Hoboken: Wiley-Blackwell.

Lecchini, D., S. Polti, Y. Nakamura, P. Mosconi, M. Tsuchiya, G. Remoissenet, and S. Planes. 2006. New Perspectives on Aquarium Fish Trade. *Fisheries Science* 72 (1): 40–47.

Lockwood, J. L, D. J. Welbourne, C. M. Romagosa, P. Cassey, N. E. Mandrak, A. Strecker, B. Leung, et al. 2019. When Pets Become Pests: The Role of the Exotic Pet Trade in Producing Invasive Vertebrate Animals. *Frontiers in Ecology and the Environment* 17 (6): 323–30.

Madduppa, H. H., J. Timm, and M. Kochzius. 2018. Reduced Genetic Diversity in the Clown Anemonefish *Amphiprion ocellaris* in Exploited Reefs of Spermonde Archipelago, Indonesia. *Frontiers in Marine Science* 5 (March): 80.

Madduppa, H. H., K. von Juterzenka, M. Syakir, and M. Kochzius. 2014. Socio-Economy of Marine Ornamental Fishery and Its Impact on the Population Structure of the Clown Anemonefish Amphiprion Ocellaris and Its Host Anemones in Spermonde Archipelago, Indonesia. *Ocean and Coastal Management* 100 (November): 41–50.

Madeira, D., J. Andrade, M. C. Leal, V. Ferreira, R. J. M. Rocha, R. Rosa, and R. Calado. 2020. Synergistic Effects of Ocean Warming and Cyanide Poisoning in an Ornamental Tropical Reef Fish. *Frontiers in Marine Science* 7 (April): 246.

Madeira, D., and R. Calado. 2019. Defining Research Priorities to Detect Live Fish Illegally Collected Using Cyanide Fishing in Indo-Pacific Coral Reefs. *Ecological Indicators* 103 (August): 659–64.

Maison, K. A., and K. S. Graham. 2016. *Status Review Report: Orange Clownfish (Amphiprion Percula).* NOAA technical memorandum NMFS-PIFSC; 52. https://repository.library.noaa.gov/view/noaa/10369 (accessed March 29, 2022).

MAMTI. 2006. *Report on Roving Collectors: Case Studies from Indonesia and the Philippines.* Marine Aquarium Council. Makati: Marine Aquarium Management Transformation Initiative.

McClenachan, L., A. B. Cooper, K. E. Carpenter, and N. K. Dulvy. 2012. Extinction Risk and Bottlenecks in the Conservation of Charismatic Marine Species. *Conservation Letters* 5 (1): 73–80.

McGowan, J., L. J. Beaumont, R. J. Smith, A. L. M. Chauvenet, R. Harcourt, S. C. Atkinson, J. C. Mittermeier, et al. 2020. Conservation Prioritization can Resolve the Flagship Species Conundrum. *Nature Communications* 11: 994.

Militz, T. A., and S. Foale. 2017. The 'Nemo Effect': Perception and Reality of Finding Nemo's Impact on Marine Aquarium Fisheries. *Fish and Fisheries* 18 (3): 596–606.

Militz, T. A., S. Foale, J. Kinch, and P. C. Southgate. 2018b. Natural Rarity Places Clownfish Colour Morphs at Risk of Targeted and Opportunistic Exploitation in a Marine Aquarium Fishery. *Aquatic Living Resources* 31: 18.

Militz, T. A., J. Kinch, S. Foale, and P. C. Southgate. 2016. Fish Rejections in the Marine Aquarium Trade: An Initial Case Study Raises Concern for Village-Based Fisheries. *PLOS ONE* 11 (3): e0151624.

Militz, T. A., J. Kinch, D. S. Schoeman, and P. C. Southgate. 2018a. Use of Total Allowable Catch to Regulate a Selective Marine Aquarium Fishery. *Marine Policy* 90 (April): 160–67.

Millar, N. 2013. 98% of Marine Fish Headed for the Aquarium Trade Die within a Year in the Philippines. https://news.mongabay.com/2013/10/98-of-marine-fish-headed-for-the-aquarium-trade-die-within-a-year-in-the-philippines/ (accessed March 29, 2022).

Molloy, P. P., N. B. Goodwin, I. M. Côté, M. J. G. Gage, and J. D. Reynolds. 2007. Predicting the Effects of Exploitation on Male-First Sex-Changing Fish. *Animal Conservation* 10 (1): 30–38.

Moore, A. M., I. Yasir, R. A-Rappe, S. Ndobe, and J. Jompa. 2020. Asexual Propagation of Two Sea Anemone Taxa for Banggai Cardinalfish Microhabitat Enhancement. *IOP Conference Series: Earth and Environmental Science* 473 (1): 012011.

Moorhead, J. A., and C. Zeng. 2010. Development of Captive Breeding Techniques for Marine Ornamental Fish: A Review. *Reviews in Fisheries Science* 18 (4): 315–43.

Murray, J. M., and G. J. Watson. 2014. A Critical Assessment of Marine Aquarist Biodiversity Data and Commercial Aquaculture: Identifying Gaps in Culture Initiatives to Inform Local Fisheries Managers. *PLoS ONE* 9 (9): e105982.

Musick, J. A., G. Burgess, G. Cailliet, M. Camhi, and S. Fordham. 2000. Management of Sharks and Their Relatives (Elasmobranchii). *Fisheries* 25 (3): 9–13.

National Intelligence Council. 2016. Global Implications of Illegal, Unreported, and Unregulated (IUU) Fishing. USA. https://www.dni.gov/files/documents/Newsroom/Reports%20and%20Pubs/NIC_White_Paper_on_IUU_Fishing.pdf (accessed March 15, 2022).

Norin, T., S. C. Mills, A. Crespel, D. Cortese, S. S. Killen, and R. Beldade. 2018. Anemone Bleaching Increases the Metabolic Demands of Symbiont Anemonefish. *Proceedings of the Royal Society B*, 285: 20180282.

Okemwa, G. M., B. Kaunda-Arara, E. N. Kimani, and B. Ogutu. 2016. Catch Composition and Sustainability of the Marine Aquarium Fishery in Kenya. *Fisheries Research* 183 (November): 19–31.

Palmtag, M. 2017. The Marine Ornamental Species Trade. In *Marine Ornamental Species Aquaculture*, ed. R. Calado et al., 3–14. Hoboken: Wiley-Blackwell.

Pendleton, L., O. Hoegh-Guldberg, R. Albright, A. Kaup, P. Marshall, N. Marshall, S. Fletcher, G. Haraldsson, and L. Hansson. 2019. The Great Barrier Reef: Vulnerabilities and Solutions in the Face of Ocean Acidification. *Regional Studies in Marine Science* 31 (September): 100729.

Pomeroy, R. S., and C. Balboa. 2004. The Financial Feasibility of Small-Scale Marine Ornamental Aquaculture in the Philippines. *Asian Fisheries Science* 17 (4): 365–376.

Pouil, S., M. F. Tlusty, A. L. Rhyne, and M. Metian. 2020. Aquaculture of Marine Ornamental Fish: Overview of the Production Trends and the Role of Academia in Research Progress. *Reviews in Aquaculture* 12 (2): 1217–30.

Prakash, S., T. Thangappan, A. Kumar, R. Raghavan, A. Rhyne, M. F. Tlusty, and T. Subramoniam. 2017. Marine Aquarium Trade in India: Challenges and Opportunities for Conservation and Policy. *Marine Policy* 77 (March): 120–29.

Readman, G. D., S. F. Owen, T. G. Knowles, and J. C. Murrell. 2017. Species Specific Anaesthetics for Fish Anaesthesia and Euthanasia. *Scientific Reports* 7 (1): 7102.

Rhyne, A. L., M. F. Tlusty, and L. Kaufman. 2012a. Long-Term Trends of Coral Imports into the United States Indicate Future Opportunities for Ecosystem and Societal Benefits: Trade of Aquarium and Curio Corals. *Conservation Letters* 5 (6): 478–85.

Rhyne, A. L., M. F. Tlusty, P. J. Schofield, L. Kaufman, J. A. Morris, and A. W. Bruckner. 2012b. Revealing the Appetite of the Marine Aquarium Fish Trade: The Volume and Biodiversity of Fish Imported into the United States. Edited by Julian Clifton. *PLoS ONE* 7 (5): e35808.

Rhyne, A. L., M. F. Tlusty, and J. T. Szczebak. 2017a. Early Culture Trials and Overview on U.S. Marine Ornamental Species Trade. In *Marine Ornamental Species Aquaculture*, ed. R. Calado et al. 51–69. Hoboken, New Jersey: Wiley-Blackwell.

Rhyne, A. L., M. F. Tlusty, J. T. Szczebak, and R. J. Holmberg. 2017b. Expanding Our Understanding of the Trade in Marine Aquarium Animals. *PeerJ* 5: e2949.

Roelofs, A., and R. Silcock. 2008. *A Sustainability Assessment of Marine Fish Species Collected in the Queensland Marine Aquarium Trade.* Brisbane, QLD: Department of Primary Industries and Fisheries.

Romagosa, C. 2014. Patterns of Live Vertebrate Importation into the United States: Analysis of an Invasion Pathway. In *Invasive Species in a Globalized World: Ecological, Social, and Legal Perspectives on Policy*, eds R. P. Keller et al, pp. 115–146. Chicago, IL: University of Chicago Press.

Rousseau, S., and N. Deschacht. 2020. Public Awareness of Nature and the Environment During the COVID-19 Crisis. *Environmental and Resource Economics* 76 (4): 1149–59.

Roux, N., P. Salis, S-H. Lee, L. Besseau, and V. Laudet. 2020. Anemonefish, a Model for Eco-Evo-Devo. *EvoDevo* 11 (1): 20.

Rubec, P. J. 1986. The Effects of Sodium Cyanide on Coral Reefs and Marine Fish in the Philippines. In *The First Asian Fisheries Forum*, eds J. L. Maclean, et al. 297–302. Manila. Philippine: Asian Fisheries Society.

Rubec, P. J., F. Cruz, V. Pratt, R. Oellers, B. McCullough, and F. Lallo. 2001. Cyanide-Free Net-Caught Fish for the Marine Aquarium Trade. *Aquarium Sciences and Conservation* 3 (1/3): 37–51.

Saenz-Agudelo, P., G. P. Jones, S. R. Thorrold, and S. Planes. 2011. Detrimental Effects of Host Anemone Bleaching on Anemonefish Populations. *Coral Reefs* 30 (2): 497–506.

Salomon, A. K., N. T. Shears, T. J. Langlois, and R. C. Babcock. 2008. Cascading Effects of Fishing Can Alter Carbon Flow through a Temperate Coastal Ecosystem. *Ecological Applications* 18 (8): 1874–87.

Sandbrook, C., W. M. Adams, and B. Monteferri. 2015. Digital Games and Biodiversity Conservation: Digital Games and Biodiversity Conservation. *Conservation Letters* 8 (2): 118–24.

Schaar, S. I., and L. J. Cox. 2021. The Future for Hawai'he Future for Hawaiampleery: A Cost Benefit Analysis Compared to an Environmental Impact Assessment. *Marine Policy* 127: 104429. https://doi.org/10.1016/j.marpol.2021.104429.

Schmidt, C., and A. Kunzmann. 2005. Post-Harvest Mortality in the Marine Aquarium Trade: A Case Study of an Indonesian Export Facility. *ASPC Live Reef Fish Information Bulletin* 13: 10.

Scott, A., and A. H. Baird. 2015. Trying to Find Nemo: Low Abundance of Sea Anemones and Anemonefishes on Central and Southern Mid-Shelf Reefs in the Great Barrier Reef. *Marine Biodiversity* 45 (2): 327–31.

Scott, A., J. M. Hardefeldt, and K. C. Hall. 2014. Asexual Propagation of Sea Anemones that Host Anemonefishes: Implications for the Marine Ornamental Aquarium Trade and Restocking Programs. *PLoS ONE* 9 (10): e109566.

Scott, A., and P. L. Harrison. 2005. Synchronous Spawning of Host Sea Anemones. *Coral Reefs* 24 (2): 208–208.

Scott, A., and P. L. Harrison. 2007. Embryonic and Larval Development of the Host Sea Anemones *Entacmaea quadricolor* and *Heteractis crispa. The Biological Bulletin* 213 (2): 110–21.

Scott, A., and P. L. Harrison. 2009. Gametogenic and Reproductive Cycles of the Sea Anemone, Entacmaea Quadricolor. *Marine Biology* 156 (8): 1659–71.

Shuman, C. S., G. Hodgson, and R. F. Ambrose. 2005. Population Impacts of Collecting Sea Anemones and Anemonefish for the Marine Aquarium Trade in the Philippines. *Coral Reefs* 24 (4): 564–73.

Silk, M. J., S. L. Crowley, A. J. Woodhead, and A. Nuno. 2018. Considering Connections between Hollywood and Biodiversity Conservation. *Conservation Biology* 32 (3): 597–606.

Somero, G. N. 2010. The Physiology of Climate Change: How Potentials for Acclimatization and Genetic Adaptation Will Determine 'Winners' and 'Losers.' *Journal of Experimental Biology* 213 (6): 912–20.

Stayman, A. P. 1999. A Policy Maker's Viewpoint: The Marine Species in Your Tank—Where Do They Come From? *Tropical Fish Hobbyist* 9: 30–47.

Stringham, O. C., P. Garcia-Diaz, A. Toomes, L. Mitchell, J. V. Ross, and P. Cassey. 2021. Reptile Smuggling is Predicted by Trends in the Legal Exotic Pet Trade. *Conservation Letters* 45: 1130–39.

Tissot, B. N., B. A. Best, E. H. Borneman, A. W. Bruckner, C. H. Cooper, H. D'Agnes, T. P. Fitzgerald, et al. 2010. How U.S. Ocean Policy and Market Power Can Reform the Coral Reef Wildlife Trade. *Marine Policy* 34 (6): 1385–88.

Tissot, B. N., and L. E. Hallacher. 2003. Effects of Aquarium Collectors on Coral Reef Fishes in Kona, Hawaii. *Conservation Biology* 17 (6): 1759–68.

Tlusty, M. 2002. The Benefits and Risks of Aquacultural Production for the Aquarium Trade. *Aquaculture* 205 (3–4): 203–19.

Tribot, A. S., Q. Carabeux, J. Deter, T. Claverie, S. Villéger, and N. Mouquet. 2018. Confronting Species Aesthetics with Ecological Functions in Coral Reef Fish. *Scientific Reports* 8: 11733.

Trujillo-González, A., and T. A. Militz. 2019. Taxonomically Constrained Reporting Framework Limits Biodiversity Data for Aquarium Fish Imports to Australia. *Wildlife Research* 46: 355.

Vaz, M. C. M., V. I. Esteves, and R. Calado. 2017. Live Reef Fish Displaying Physiological Evidence of Cyanide Poisoning Are Still Traded in the EU Marine Aquarium Industry. *Scientific Reports* 7 (1): 6566.

Veríssimo, D., S. Anderson, and M. Tlusty. 2020. Did the Movie Finding Dory Increase Demand for Blue Tang Fish? *Ambio* 49 (4): 903–11.

Visch, V. T., E. S. Tan, and D. Molenaar. 2010. The Emotional and Cognitive Effect of Immersion in Film Viewing. *Cognition and Emotion* 24 (8): 1439–45.

Wabnitz, C., M. Taylor, E. Green, and T. Razak, eds. 2003. *From Ocean to Aquarium: The Global Trade in Marine Ornamental Species.* World Conservation Monitoring Centre. UNEP-WCMC Biodiversity Series 17. Cambridge, UK: UNEP World Conservation Monitoring Centre.

Walsh, W. J. 1978. Aquarium Reef Fish Collecting: Promise of Problem? In Papers and Comments on Tropical Reef Fish. *Working Paper* 34: 8–12. Sea Grant College Program, University of Hawaii.

Yong, D. L., S. D. Fam, and S. Lum. 2011. Reel Conservation: Can Big Screen Animations Save Tropical Biodiversity? *Tropical Conservation Science* 4 (3): 244–53.

22 Anemonefish Husbandry

Jennifer M. Donelson, Pascal Romans, Sakuto Yamanaka, Masato Kinoshita, and Natacha Roux

CONTENTS

22.1 INTRODUCTION

Unlike most reef fish families, which are broadcast spawners, damselfishes and anemonefishes are benthic spawners. Indeed, as mentioned in Chapter 6, anemonefishes live in the vicinity of their sea anemone host and lay eggs on a substrate nearby. Due to many of their characteristics including small body size (> 20 cm), naturally small territories, sexual dimorphism, sequential hermaphroditism, visually attractive, laying easily accessible eggs, and habituating well to captivity (Thresher 1984), anemonefish have rapidly become one of the most popular marine ornamental fish traded worldwide since the 1950s (Wabnitz 2003; Wittenrich 2007). Such popularity is also due to the fact that once a stable breeding pair of anemonefishes is held in captivity, they reproduce all year long with a frequency of two clutches (200–800 eggs) per month. This is extremely valuable for commercial purposes as it allows a constant stock without having to induce reproduction hormonally or by manual gamete extraction. Such characteristics are also likely to be one of the main reasons why anemonefish have also become popular for scientists from various fields of research (Roux et al. 2020). In the last 20 years, we have seen numerous papers in which researchers describe their own rearing methods to avoid field collection or no longer depend on aquaculture companies (e.g., Gopakumar et al. 1999; Divya et al. 2011; Ghosh et al. 2011; Anil et al. 2012; Ghosh et al. 2012; Kumar et al. 2012; Mitchell et al. 2021; Roux et al. 2021; Yamanaka et al. 2021). Development of rearing methods suitable at a laboratory research scale is extremely valuable for scientists interested in using anemonefish as a model organism for ecology, evolution, and development (eco-evo-devo). In this chapter, we will first give an overview of the rearing systems enabling the reproduction of anemonefish at laboratory scale and we will emphasize the methods recently developed specifically for experimental purposes (low volume rearing method and egg development without parental care).

22.2 ANEMONEFISH REARING SYSTEM AT LABORATORY SCALE

22.2.1 Broodstock System

The first consideration for scientists in installing systems for anemonefish husbandry is access to seawater. Since most research facilities and laboratories will not be near a constant source of seawater, which is required for an open system, most of the published methods are developed using a closed system (Table 22.1). When laboratories are located near, but not adjacent to the sea, they can have the opportunity to use natural seawater, but on a closed system. As the distance to a seawater source increases, researchers will commonly choose to work with artificial seawater in a closed recirculating system as in the one described in Roux et al. (2021) (Figure 22.1). Fortunately, anemonefish species are robust to seawater type (natural or artificial) and either an open or closed circulatory system if water quality is maintained. All system designs inherently have their pros and cons in terms of costs and concerns that will be situation dependent. If using natural seawater, the most important parameters to consider are the location of the pumping system in the sea, the quality of the filtration system, and the ambient water temperature (relative to the desire for research). Indeed, as this water will be used for maintaining breeding pairs and for larval rearing, it is best to avoid any disruption in water quality due to mechanical or environmental influence (e.g., failure in the pumping system,

DOI: 10.1201/9781003125365-27

TABLE 22.1
Summary of Anemonefish Broodstock Husbandry Systems, Including Tank Size, Seawater Source, and Water Parameters

Species	Origin	Sea anemone	Breeding pair tank volume	Recirculating system type	Seawater type	Water exchange	Salinity (ppt)	Temperature (°C)	Photoperiod (L/D)	pH	Reference
A. akallopisos	OFT	Yes	400 L	Closed	Estuarine filtered seawater	50% once a week	25	27	12/12	8	Dhaneesh et al. 2012a
A. chrysogaster	W	Yes	–	–	–	–	–	–	–	–	Gopakumar et al. 1999
A. clarkii	OFT	Yes	750 L	Closed	Brackish water	–	26	–	–	–	Ghosh et al. 2011
A. clarkii	OFT	Yes	750 L	Closed	Brackish water	–	26	–	–	–	Ghosh et al. 2012
A. clarkii	OFT	No	200 L	Closed	–	–	30	28	14:10	8–8.2	Olivotto et al. 2008
A. melanopus	–	No	–	–	–	–	–	28	–	–	Green and McCormick 2001
A. nigripes	W	No	500 L	Closed	Filtered seawater	10% everyday	30–32	27	–	8–8.2	Anil et al. 2012
A. nigripes	W	Yes	1000 L	Closed	Filtered seawater	50% once a week	34–35	26	13/11	88.3	Kumar et al. 2012
A. ocellaris	W	Yes	500 L	Closed	Filtered seawater	25% once a week	32 to 36	26– 9	–	88.9	Madhu et al. 2006a
A. ocellaris	OFT	Yes	750 L	Closed	Filtered estuarine seawater	10–20% water renewing every 2 days	–	25–32	12/12	–	Kumar and Balasubramanian 2009
A. ocellaris	OFT-W	No	60 L	Closed	Artificial seawater	30% once a month	34	26	14/10	8–8.5	Roux et al. 2021
A. ocellaris	–	No	200 L	–	–	–	30	28	13/11	8–8.5	Avella et al. 2007
A. percula	OFT	Yes	1000 L	Closed	Filtered brackish water	20% once a week	24	24 to 28	12/12	–	Dhaneesh et al. 2012b
A. sebae	OFT	Yes	1000 L	Closed	Filtered estuarine seawater	20–40% regularly	22–24	28	12/12	7.–8.2	Kumar et al. 2010
A. sebae	W	Yes	1000 L	–	–	–	33–35	28–32	–	–	Igniatus et al. 2001
P. biaculeatus	–	Yes	500 L	Closed	–	–	–	27–29	–	–	Madhu et al. 2006b

Notes: OFT: ornamental fish trade, W: wild, L: light, D: dark, ppt: part per thousands (1 ppt = 1 g/L).

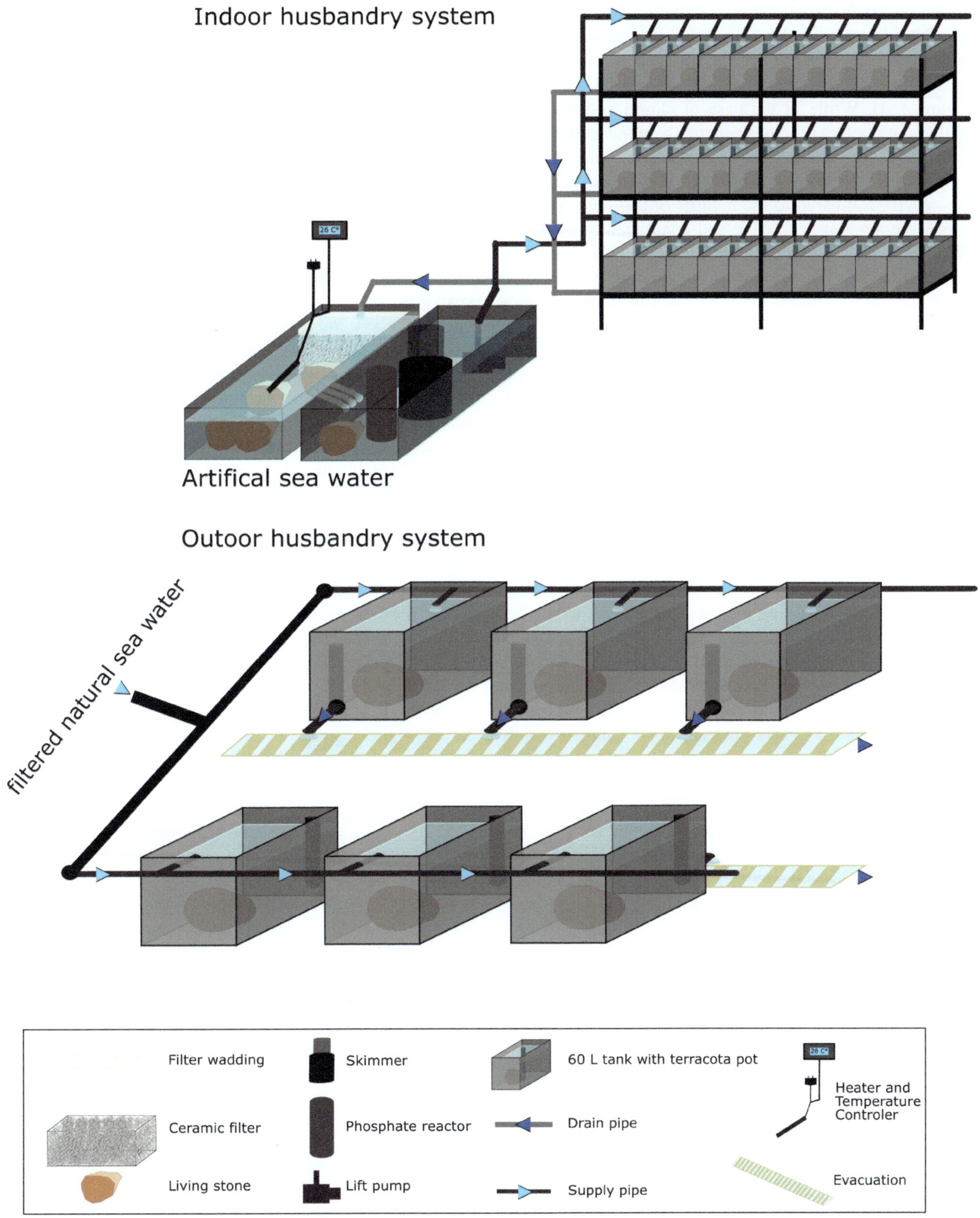

FIGURE 22.1 Schematic diagram illustrating two examples of husbandry systems at the laboratory scale. Indoor closed husbandry system with artificial seawater and a filtration tank on the bottom left, which can include both mechanical and biological filtration as well as temperature control. Outdoor open husbandry system with constant supply of filtered natural seawater (adapted from Roux et al. 2021). If temperature control was required, it could be achieved with inline heating/cooling or in-tank heaters.

presence of parasites, turbidity, rainfall events). If the quality of the source of seawater can be maintained, then the main advantage of working with an open system is that it does not require manual water exchanges (see Table 22.1, Figure 22.1). While in the case of closed systems water quality parameters need to be regularly monitored to determine the frequency of the water changes. This exchange frequency will be highly dependent on the total volume of the system, compared to the number of fish held (i.e., biomass), the amount of feeding and consequently excretion, and the filtration power of the system (mechanical and biological filtration). These water changes aim to mainly avoid accumulation of nitrates or phosphates in the broodstock system.

The next main consideration when designing a husbandry system is the maintenance of seawater temperature, as it constrains reproductive activity. Generally, anemonefish naturally spawn during spring to summer on a lunar cycle (Richardson et al. 1997; Madhu and Madhu 2007; Dhaneesh et al. 2012a), but in some locations, breeding can occur all year round (Thresher 1984). In some circumstances, laboratories may decide to be directly dependent on the natural seawater temperature seasonally, and as such will have fluctuations in the spawning rate as described in Dhaneesh et al. (2012b) and Madhu et al. (2006a) (Table 22.2). Such a setup may also encounter difficulties in tracking the developmental speed of embryos, as water temperature will likely vary diurnally, between days and weeks within a season. Potential variation in embryonic development will require higher surveillance from the technical personnel to estimate precisely when a clutch will be ready to hatch by observing the color of the eyes of embryos (becoming silvery and shiny the day of hatching) (Madhu et al. 2006b; Kumar et al. 2012). Additionally, if ambient local temperatures are not suitable for anemonefish reproduction, inline heating, heating in header tanks, or heating within broodstock tanks will be required to warm water to a suitable temperature for breeding.

22.2.2 Broodstock Maintenance

Breeding pairs can be directly purchased from the wild (with legal authorization), from ornamental fish traders, or from captive stock by selecting a larger and a smaller sexually immature juvenile. The first element to take into consideration when bringing new fish into an aquarium system is a minimum quarantine period of 15–30 days (or more depending on the fish health state) in tanks located on a separate closed system or isolated from the main system to avoid any risk of contamination. Many research facilities suggest treatment of fish prior to entering the quarantine tanks to remove any potential pathogens (e.g., freshwater and/or formalin baths). During the quarantine period, fish should be monitored for evidence of disease, treated as appropriate, and the quarantine period extended as required. When fish are collected from the wild, first spawning will generally occur between three and six months after pair formation. First spawning can be as early as two months for *Amphiprion nigripes* (Anil et al. 2012; Table 22.2) or can take up to two years before first spawning as observed in *A. ocellaris* (Roux et al. 2021; Table 22.2). This variability is likely linked to the age, size, and maturity of the selected individuals, the time needed to acclimatize to the new environment and the compatibility of paired individuals. Indeed, if no spawning has occurred after two years it is advised to change partners or remove males and add smaller sexually immature juveniles to try to form new pairs.

Multiple conditions must be met to ensure optimal spawning in terms of frequency and quality. First, the husbandry system ideally should be in a quiet space as advised in Roux et al. (2021). Second, it is important to separate breeding pairs from each other by placing them in separate tanks or using mesh to avoid contact and aggression between them. If the husbandry system is composed of glass tanks, it is advisable to prevent pairs from seeing their neighbours or their reflection with opaque white or light-colored panels between each tank. Alternatively, algae could be allowed to colonize tank walls to avoid mirror effects. An additional benefit of algal growth is that it has a positive effect on water quality due to the consumption of nitrates and phosphates for their growth (Roux et al. 2021). If other materials than glass are used for the tanks, the color of the walls does not make a difference in the maintenance success of breeding pairs. When fish are collected from the wild and have already reached their maximum size it is better to consider volume bigger than 60 L for the biggest species (as mentioned in Table 22.1) but if they measure less than 10 cm they will perfectly adapt to 60 L tank and able to reproduce in such volume as it has been observed for *A. clarkii*, *P. biaculatus*, and *A. frenatus* in Roux et al. (2021) husbandry (Romans and Roux, personal observations).

As mentioned earlier, anemonefish breeding pairs are relatively robust and can reproduce under a range of seawater conditions. The breeding temperature in the literature ranges between 24 and 29°C (summarized in Table 22.1). The anemonefish species are collected from tropical regions, and as such, the temperature range represents spring to summer temperatures (when anemonefish tend to reproduce in the wild). When applying the knowledge summarized in this book chapter to different anemonefish species, including those with subtropical distributions, information on the natural water temperature for the species and collection location should be used. In systems where temperature is maintained for spawning all year long, breeding pairs can become exhausted, leading to a reduction in reproductive success (Roux et al. 2021). If this occurs, it is advisable to decrease the system temperature by one or two degrees for several weeks to give fish a break before returning to warmer breeding temperatures. The literature suggests that water salinities between 24 and 36 ppt can be used for breeding anemonefish, however, caution should be applied when using lower salinities as it can negatively impact

TABLE 22.2
Summary of Anemonefish Broodstock Feed Types and the Resulting Reproductive Parameters

Species	Feed (mixtures of listed)	Feeding frequency (times/day)	First spawning after pair formation	Spawning frequency	Clutch size (egg number)	Nest site	Hatching rate	Reference
A. akallopisos	Shrimp, boiled oyster, and clam meat	3	3 months	–	300–400	Ceramic tiles or live rocks	–	Dhaneesh et al. 2012a
A. chrysogaster	Minced beef and boiled mussel	2	–	Between 10 to 45 days	300–800	Earthen pots or granite stone	90%	Gopakumar et al. 1999
A. clarkii	Boiled oyster meat and prawn	3	–	Twice a month	513 average	Ceramic tiles	93%	Ghosh et al. 2011
A. clarkii	Boiled oyster meat and prawn	3	3 months	Twice a month, 1month break	513 average	–	93%	Ghosh et al. 2012
A. clarkii	Frozen *Artemia*, frozen plankton, chopped fish, and shrimp	2	12 months	Every 12 days	350–400	Terracotta pot	96%	Olivotto et al. 2008
A. melanopus	–	–	–	–	500–1,000	Terracotta pot	–	Green et McCormick 2001
A. nigripes	Pellets, boiled mussel, *Artemia nauplii*, mysids, copepods	3	74 days	Every 12 to 16 days	350–450	Clay pots	94%	Anil et al. 2012
A. nigripes	Tuna egg mass, boiled clam, squid, and octopus	3	–	–	300–700	Ceramic tiles or live rocks	–	Kumar et al. 2012
A. ocellaris	Shrimp, mussel, and clam, formulated feed every 3 hours during the day, *B. plicatilis* and *Artemia nauplii* at night (enriched with *N. oculata*, *P. lutheri*, *I. galbana*, *D. inornata*, *C. pleoides*, *C. marina*, and *C. gracilis*)	–	4 to 6 months	Every 12 to 15 days	300–1,000	Tiles or earthen pot	–	Madhu et al. 2006a
A. ocellaris	Prawn, mussel, squid, and live *Acetes* sp.	3	4 months	Average of 2.4 clutch per month (increase in summer)	400–800	Broken coral pieces, dead shells, and live rocks	90–95%	Kumar et Balasubramanian 2009
A. ocellaris	Squid, mussels, shrimps, pellets, and nori algae	–	6 to 24 months	Every 12 to 15 days with one/two months break	100–500	Terracotta pot	–	Roux et al. 2021
A. ocellaris	Frozen mysids and krill	2	–	Every 15 days	–	Terracotta pot	–	Avella et al. 2007

(*Continued*)

TABLE 22.2 (CONTINUED)
Summary of Anemonefish Broodstock Feed Types and the Resulting Reproductive Parameters

Species	Feed (mixtures of listed)	Feeding frequency (times/day)	First spawning after pair formation	Spawning frequency	Clutch size (egg number)	Nest site	Hatching rate	Reference
A. percula	Live *Acetes* sp., boiled clam meat, trash fish, and fry feed	3	–	Every 10 to 21 days	110–276	Ceramic tiles, dead coral pieces, and live rocks	–	Dhaneesh et al. 2012b
A. sebae	Live *Acetes* sp., trash fish, clam, mussels, squids	3	4 months	Every 7 to 15 days with one month break	400–2,000	Tile, earthen pot, PVC pipe	–	Kumar et al. 2010
A. sebae	Fish, bivalve and polychaete worms	2	3 months	Every 10 days	300–600	Piece of asbestos sheet	70%	Igniatus et al. 2001
P. biaculeatus	Mussel meat, shrimp and clam meat, live feed (adult *Artemia* and rotifers enriched with vitamins, minerals, and fatty acids)	4	–	Every 13 to 15 days	115–1,000	Earthen pots	–	Madhu et al. 2006b

Note: *Brachionus* sp., *Nannochloropsis* sp., *Isochrysis* sp., *Dicrateria* sp., *Chromulina* sp., *Chlorella* sp., *Chaetoceros* sp.

larval survival (see Chapter 23 for further discussion). Both photoperiod (12/12 to 14/10, L/D) and pH (7.5 to 8.9) are quite flexible (Table 22.1). However, nitrogen compounds (NH_4, NO_2, NO_{3-} should remain at low levels (0–0.5 mg/L, 0–0.2 mg/L, 0–5 mg/L respectively) to preserve breeding pair health and the quality of clutches (Edwards et al. 2005; Callan 2007).

Several aspects, while not essential, are worth considering to ensure optimal spawning. One of which is the quality and quantity of the food provided, as it is known that highly nutritious food increases fecundity and embryo viability (Callan 2007). Consistently, the literature suggests not predominantly using aquaculture feed like pellets, but instead a variety of whole foods including mussels, oyster, clams, squid, and octopus, as well as live *Acetes* and *Artemia* (see Table 22.2; Wittenrich 2007). Food is generally given *ad libitum* two to three times per day. The importance of nutritious food was highlighted by Dhaneesh et al. (2012a) who demonstrated that the reproductive success of *A. percula* (number of eggs produced) can be doubled with the use of more nutritive food such as live prey *Acetes* sp. (276 ± 22.3 eggs) and clam meat (204 ± 16.4 eggs) compared to dry food (110 ± 10 eggs). In addition, food quality may also directly influence larval survival as has been observed in *A. sebae* (Varghese 2009). Indeed, breeding pairs fed with deep-sea prawns led to higher survival rates (62.7% ± 6.7%) compared to mature mussel meat (60.3% ± 2.1%), squid meat (59.0% ± 5.3%), cuttlefish meat (54.7% ± 11.2%), and immature mussel meat (44.3% ± 5.7%) (Varghese, 2009). While the presence of a sea anemone is not a requirement for anemonefish reproduction, they have almost always been included in the published methods (see Table 22.1). The absence of an anemone can facilitate the design and maintenance of the husbandry system, especially in closed systems (Roux et al. 2021) or when sufficient natural or artificial lighting is not possible. Indeed, anemones are very delicate invertebrates which need specific light features and strong water circulation. Another critical element to take into consideration when breeding anemonefish is the provision of a suitable substrate for the clutch. Diverse material types can be used as a nest site, including ceramic or clay tiles, live rocks, terracotta pots, or dead coral pieces (see Table 22.2). In the absence of a nest site, breeding pairs will lay eggs on the wall of the tank which forces technical staff to come on the night of hatching to capture larvae and transfer them to a rearing tank. This process is not ideal as it can cause stress to the larvae and result in reduced survival also this has never been properly demonstrated. If larvae are not collected, they will be trapped in the filtration system.

22.2.3 Classic Larval Rearing Systems

Anemonefish larval rearing always occurs in a different tank than the parent, but the seawater parameters should be the same. It is common to fill the larval rearing tank with the seawater directly from the tank of the parents or the same system, as a change in water quality at the sensitive hatching stage could cause mortality due to stress. The volume of the rearing tanks will depend on the space available for the rearing system and can range from 30 to 100 L (see Table 22.3). Tank shape can vary from round, round with a conical base, square or rectangular. Square or rectangular need special attention as corners may prevent homogeneous mixing of water but this can be easily solved by adding two air diffusers on opposite angles of the tanks (Wittenrich 2007; Roux et al. 2021). Larval rearing tanks should ideally be dark in color and non-transparent (e.g., black plastic wrapping glass tanks). Dark tank colors allow a background of contrast compared to prey items aiding in feeding efficiency (Wittenrich 2007), and if possible, a white tank bottom eases the visual monitoring of larvae. A semi-opaque black lid or shade cloth can cover the tank until completion of metamorphosis to reduce illumination and distracting light reflections at the surface, which can attract larvae and cause stress, oxygen depletion, reduce feeding, starvation, and thus death (Tucker 1958; Calado et al. 2017), and is especially important if green water is not used (see the following for further discussion).

Hatching always occurs at night, generally shortly after dusk. It is common practice to carefully remove the substrate with eggs attached from the parent tank, transport it fully covered in water, and gently place it in the larval rearing tank. As breeding pairs are not transported with the clutch, oxygenation must be provided on the eggs to replace parental care (e.g., fanning) and ensure proper hatching. This is easily performed by using air diffusers below the eggs, with finer air bubbles preferred and airflow should be adjusted to ~15 L/h (Roux et al. 2021). Too low or high air flow rate, as well as air getting trapped around the eggs, will prevent proper hatching. Most of the methods run each larval rearing tank closed, without any filtration system and with aeration but no flow from a main water system (Table 22.3). This system setup means that tanks require manual daily water changes to remove dead larvae, surplus feed, faeces, and waste compounds. The amount of manual daily water exchange can be as low as 25% every three days but is commonly ten to 60% of the total volume per day (Table 22.3).

However, it is also possible to create a semi-closed system, where tanks are closed during daylight hours when green water and food are provided and then automatically flushed with filtered seawater at night (Munday et al. 2009; Dixson et al. 2010; Dixson et al. 2011). This automated flushing is common in the experimental literature when water conditions are manipulated to simulate climate change. When completing any water exchanges it is advised that new seawater should be supplied with a gentle flow (e.g., through narrow 4–5 mm tubing) so that there are no large amounts of water movement, which can lead to mortality, and larvae can slowly acclimate to the new water conditions (Roux et al. 2021).

Newly hatched individuals are generally kept in the larval rearing tank until 20–30 dph, as this coincides with the

TABLE 22.3
Summary of Anemonefish Larval Rearing Systems

Species	Larval tank volume (L)	Water exchange	Photoperiod (L/D)	Reference
A. akallopisos	100	10%/day	12/12	Dhaneesh et al. 2012b
A. chrysogaster	100–200	Constant (recirculating system)	–	Gopakumar et al. 1999
A. clarkii	45	10%/day	12/12	Ghosh et al. 2011
A. clarkii	45	–	12/12	Ghosh et al. 2012
A. clarkii	16	100%/night	14/10	Anto et al. 2009
A. clarkii	20	10 times/day by dripping system	14/10	Olivotto et al. 2008
A. frenatus	16	100%/night	14/10	Anto et al. 2009
A. melanopus	70	100%/night	13/11	Dixson et al. 2011
A. melanopus	40	100%/night	–	Hess et al. 2019
A. melanopus	100	100%/night	13/11	Jarrold and Munday 2019
A. melanopus	200	–	14/10	Job and Bellwood 2000
A. nigripes	50	–	–	Anil et al. 2012
A. nigripes	25	10%/day	24/00	Kumar et al. 2012
A. ocellaris	100	–	–	Madhu et al. 2006a
A. ocellaris	250	10–15 /day	12/12	Kumar et Balasubramanian 2009
A. ocellaris	30	30%/day	14/10	Roux et al. 2021
A. ocellaris	20	2 times/hour by dripping system	24/00	Avella et al. 2007
A. ocellaris	38	30%/day	–	Frakes et Hoff 1982
A. percula	30–50	10% /day	16/8	Dhaneesh et al. 2012a
A. percula	70	100%/night	13/11	Munday et al. 2009; Dixson et al. 2010, 2011
A. percula	100	100%/night	12/12	McMahon et al. 2019
A. perideraion	38	–	12/12	Coughlin et al. 1992
A. sebae	250	–	12/12	Kumar et al. 2010
A. sebae	250	50%/day	12/12	Divya et al. 2011
A. sebae	100–250	25% every 3 days	–	Igniatus et al. 2001
P. biaculeateus	200	–	14/10	Job et Bellwood 2000
P. biaculeateus	100	25%/day	–	Madhu et al. 2006b
P. biaculeateus	40	60–70%/day	–	Donelson 2015

Note: L: litre, L/D: light/dark.

end of metamorphosis. When they reach this age, young juveniles are transferred (after an acclimation period of 20 minutes minimum) into juvenile rearing tanks to continue their growth. The tank shape, color, and dimensions are more flexible than during the larval phase, and requirements are like those described for adult broodstock. It is also important to note that juveniles can become very aggressive and thus constant fighting may induce high levels of mortality (up to 90%, personal observations). To avoid this, it is recommended to raise juveniles at densities between 37 and 300 fish per 100 L to have survival rates exceeding 80% (Chambel et al. 2015; Pietoyo et al. 2020). From our own experience, it is also advised to avoid shelter in the tank, to prevent the young fish from settling and establishing a distinct territory (Roux et al. 2021).

22.2.4 Larval Food

To ensure a high larval survival rate for anemonefish, it is essential to feed them with live prey throughout their development as predatory behavior needs to be stimulated (Calado et al. 2017). Proper management of live prey is thus an essential prerequisite to rear anemonefish larvae but can be a time-consuming component of larval husbandry. Rotifers are required to be maintained alive in bulk stocks in preparation for larval feeding, with production most reliable in stable thermal conditions. Maintaining rotifer cultures requires abundant feeding of microalgae and regular water exchanges to maintain water quality (Roux et al. 2021). *Artemia* is more easily prepared as required since cysts can be stored long-term in the fridge or freezer and hatched to be ready for feeding within 24 hours. Batch cultures of *Artemia* sp. can also be maintained, however, care should be taken not to position rotifer tanks near *Artemia*, or else risk *Artemia* contaminating and consuming the rotifer culture.

It is important that correct prey size is provided to match the stage of larval development (i.e., visual acuity, mouth gape, swimming ability) and relative to the attractiveness for larvae at different developmental stages, which both will be species specific (Figure 22.2). Generally, from hatching larval feeding begins with *Brachionus* sp. (i.e.,

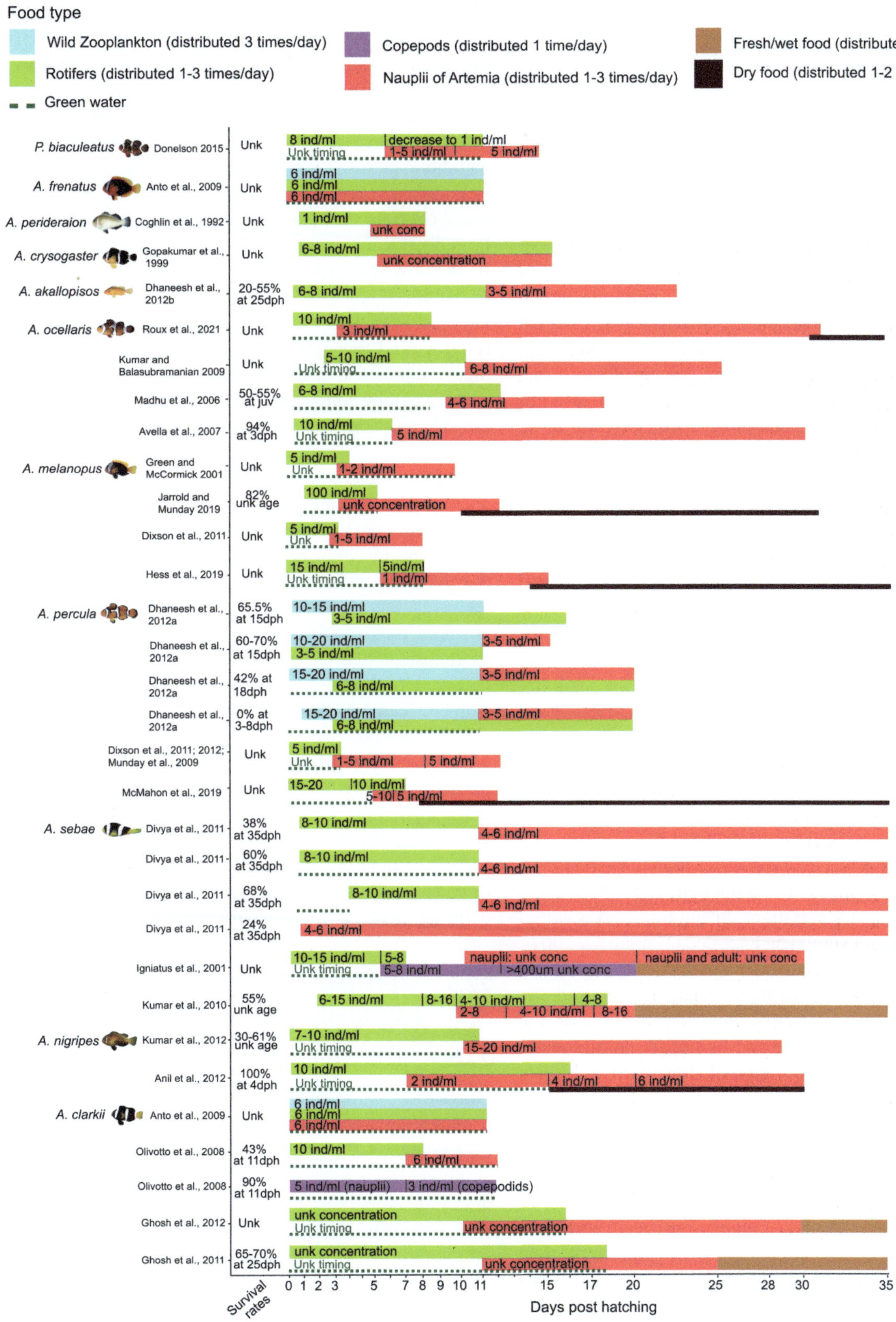

FIGURE 22.2 Graphic summarizing the sequence of food distribution during anemonefish larval development for different species. (dph: days post hatching, ind/mL: individual per mL, unk: unknown).

Rotifers; ~50–200 μm), followed by *Artemia* sp. *nauplii* (~400 μm) and finally to a fine particle dry food that is of similar size (200–400 μm). The transitions between food sizes are generally overlapped for a few days to a week allowing larvae ample time to transition (Figure 22.2) and allowing flexibility depending on the larval size diversity within a clutch. In some cases, wild zooplankton can be collected and distributed to the larvae (Dhaneesh et al. 2012b), though this would only be possible in cases where laboratories are located near the ocean. As the nutritive value of live prey is critical to ensuring the survival and growth of larvae, they are sometimes enriched with polyunsaturated or highly unsaturated fatty acids (PUFA, HUFA) (Sargent et al. 1997; Sargent et al. 1999). This involves preparing rotifers and *Artemia* in advance and soaking feeds in an enrichment solution, allowing for the absorption of additional nutritional value (Avella et al. 2007; Olivotto et al. 2008; Ghosh et al. 2011; Kumar et al. 2012). Indeed, Kumar et al. (2012) showed that rotifers fed with microalgae or PUFA significantly increase survival rates of *A. nigripes* larvae compared to unfed rotifers (50–55% versus 20% of survival rate).

Microalgae (*Nanochloropsis* sp., *Chlorella* sp., *Isochrysis* sp.), often called green water, is also used in rearing tanks during the early period of larval development to reduce the light penetration in the tank (due to the high light sensitivity of early larval development) and to enhance the visibility of prey by providing a background (Naas et al. 1992; Naas et al. 1996). Microalgae are also known to enhance survival rates through the preservation of rotifers and *Artemia nauplii* nutritional quality and maintain water quality, even though they are not directly eaten by anemonefish larvae (Figure 22.2). Microalgae can be cultured live or purchased commercially frozen. However, close attention should be paid to the quality of purchased microalgae as this can be altered by temperature variations during transport. Correct levels of green water should be provided (Wittenrich 2007), as an overdose can reduce water quality (e.g., ammonia), especially in cases where the frozen dead algae is used.

An important thing to take into consideration is the nutritive value of larval food and its impact on pigmentation pattern. Since the colouration of clownfish has a great influence on their commercial value, how to grow visually appealing individuals (with striking colors and well-formed white bars) has been well studied, mainly focusing on their feeding habitat. In aquaculture, live prey plays a critical role in fish colouration and metamorphosis (Vissio et al. 2021). Among these studies, colouration (hue, saturation, brightness) and irregular white bars (miss-bar) are of great interest to researchers. On the contrary, there are exceptions where miss-bars are preferred, like the Snowflake mutant of *A. ocellaris*. Many researchers have thus been interested in investigating the role of carotenoids (especially astaxanthin) in anemonefish colouration, and its effect on raising their body pigmentation (Tanaka et al. 1992; Yasir and Qin 2007; Yasir and Qin 2010; Ho et al. 2013a; Ho et al. 2013b; Díaz-Jiménez et al. 2021) as observed in other aquatic animals (Lim et al. 2018). When anemonefishes (*A. ocellaris*, *P. biacleatus*, *A. frenatus*, and *A. clarkii*) are fed diets containing astaxanthin or esterified astaxanthin, the amount of carotenoids in the epidermis increases and the colouration becomes more reddish accordingly. It has also been suggested that anemonefish may have a reductive metabolic pathway for astaxanthin (Tanaka et al. 1992; Ho et al. 2013a; Yasir and Qin 2010). Previously, only oxidative metabolism of carotenoids was known in fish, but a similar reductive metabolism pathway of carotenoids was found in rainbow trout (Schiedt et al. 1985) and chum salmon (Kitahara 1983). The second effect is the link between fatty acids and especially highly unsaturated fatty acids (HUFA) and their impact on irregular white bar patterning. Anemonefish have a variety of skin patterns, and some species exhibit many color pattern polymorphisms within the species. Some of these polymorphisms are genetically determined, while others are determined by environmental factors (Klann et al. 2021). A certain percentage of anemonefish juveniles in a typical rearing tank is miss-bar. Therefore, the nutritional condition and stress of the larvae and juveniles during the first month after hatching may have a significant impact on the bar formation. In studies using flatfish, the level of HUFAs in the diet affects not only the growth but also the pigmentation and metamorphosis of the larvae (Copeman et al. 2002; Bell et al. 2003). Although few studies have examined the relationship between miss-bar and diet in anemonefish, Avella et al. (2007) showed that HUFA-enriched rotifer and *Artemia* were effective in growth and inhibiting miss-bar in *A. ocellaris* larvae. Furthermore, Chambel et al. (2015) showed that the percentage of miss-bar *A. percula* larvae increased with increasing stocking density but was not affected by the protein content of the diet. Further studies are now needed to understand how HUFA may interact with the signalling pathway involved in white bar formation in anemonefishes.

22.3 METHODS DEDICATED FOR EXPERIMENTAL PURPOSES

22.3.1 Low Volume Rearing Method

One of the main interests of model organisms in certain research domains, such as ecotoxicology and developmental biology, is the possibility to expose larvae to various compounds such as pesticides, microplastics, endocrine disruptors, and other pharmacological treatments to assess their toxic effects or to help gain insight on the role of specific genes or pathways (e.g., effects of thyroid hormones on anemonefish white bar formation; Salis et al. 2021). One of the main constraints of such experiments is that they often require expensive synthetic compounds that are often directly injected. However, given the small size of anemonefish larvae, injections are not an option, and chemical treatments by balneation in classic larval rearing tanks (30 l and greater) cannot be considered either. For these reasons, Roux et al. (2021) developed a low-volume rearing method using small-volume glass beakers (< 1 l) to enable pharmacological treatments on anemonefish larvae.

This method consists of using the same principles from a classical rearing system outlined earlier (i.e., same temperature, photoperiod and light system, salinity) but the tank size and number of larvae within each tank are scaled down. Specifically, no more than ten larvae are placed in transparent glass beakers (500 mL or 800 mL) in a dark-colored temperature-controlled water bath (equipped with a heater and water pump). Since the beakers are transparent, the water bath is covered by a semi-opaque lid to decrease the light intensity (as in classic rearing conditions). Green water is still supplied to also assist with reducing light and providing enhanced prey capture. Air is supplied constantly by glass pipettes connected to an air pump allowing gentle aeration of the seawater at a rate of ~150 bubbles/min (Figure 22.3). Larval hatching occurs in a classic rearing tank (30 L) and then larvae are partitioned into the smaller rearing vessels. Larvae collection must be gentle and conducted with a glass container to limit their stress. Rearing methods associated with feeding and water quality are similar to classic rearing methods including daily water exchange (100 mL per day ~13–20%) and feeding with living prey and green algae with dosage rates adapted to the rearing volume. This method has been successful by allowing the exposure of *A. ocellaris* larvae to pharmacological compounds such as T3 (a thyroid hormone) and MPI (a mix of methimazole, potassium perchlorate, and iopanoic acid) to understand the role of TH on anemonefish metamorphosis (Salis et al. 2021). Thus, it is very likely that the low-volume rearing method will be used to conduct further functional experiments on anemonefishes.

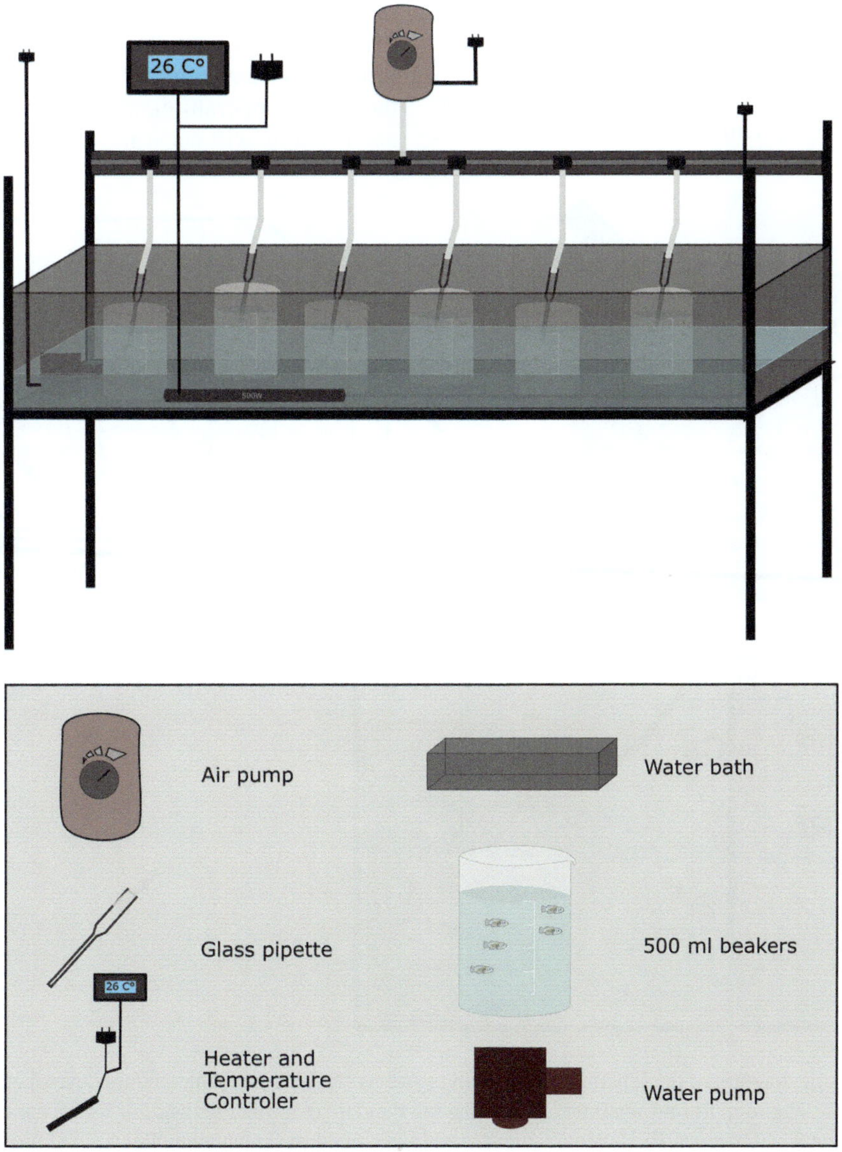

FIGURE 22.3 Schematic diagram of the low volume rearing setup. It illustrates the water baths and the experimental beakers, as well as the set-up of the air system (from Roux et al. 2021).

22.3.2 Rearing and Hatching Method without Parental Care

There is one aspect of anemonefish biology that remains challenging for developmental biologists wanting to use methods such as CRISPR (see Chapter 5). While benthic spawning of anemonefish has been an asset allowing suitability for aquarium rearing, it is also an issue for the use of techniques that require easy access and manipulation of the eggs for microinjection. Separation of the eggs is not a problem, and this can be done easily by breaking the egg nest support (e.g., tiles and clay pots) into small pieces (2 × 4 cm) using a hammer and a chisel (Mitchell et al. 2021). A second option consists of separating the eggs from the nest support by using forceps and gently detaching the eggs without squeezing them or damaging the envelope (Yamanaka et al. 2021). Instead of forceps, a sharp scalpel blade can also be used to cut the adhesive disc that attaches the egg, allowing separation from the support (personal observations).

Once the eggs are separated from the parents or detached from the support, the main issue is how to ensure their correct development. Parental care is essential to ensure the successful development as fanning and mouthing the eggs provides oxygenation and cleaning (Green 2004). If eggs are still attached to their support, Mitchell et al. (2021) advise placing the small support pieces in an upright position in a small plastic container with holes (like the zebrafish removable insert – Figure 22.4A) in a tank of similar volume to what is usually used for larval rearing (36 L). This incubation tank should be filled with seawater, maintained at the same temperature as the natal tank, and an anti-fungal/antibacterial treatment (e.g., methylene blue) should be added to prevent egg mortality. Parental care is replaced by the addition of a fine bubble air diffuser (e.g., wooden) placed 10 cm away from the eggs and below the support pieces, and airflow is easily adjusted by connecting an inline attenuator (Figure 22.4A). If placed too close or bubbles are too vigorous it can cause mechanical stress to the eggs. Regular care must be provided to remove any dead embryos and avoid rot propagation among others by adding disinfectants such as methylene blue (Mitchell et al. 2021). The day of hatching (when embryos have shiny eyes), support pieces are transferred into a new tank filled with clean seawater (always at the same temperature), airflow over the eggs is increased, and hatching will occur normally at night.

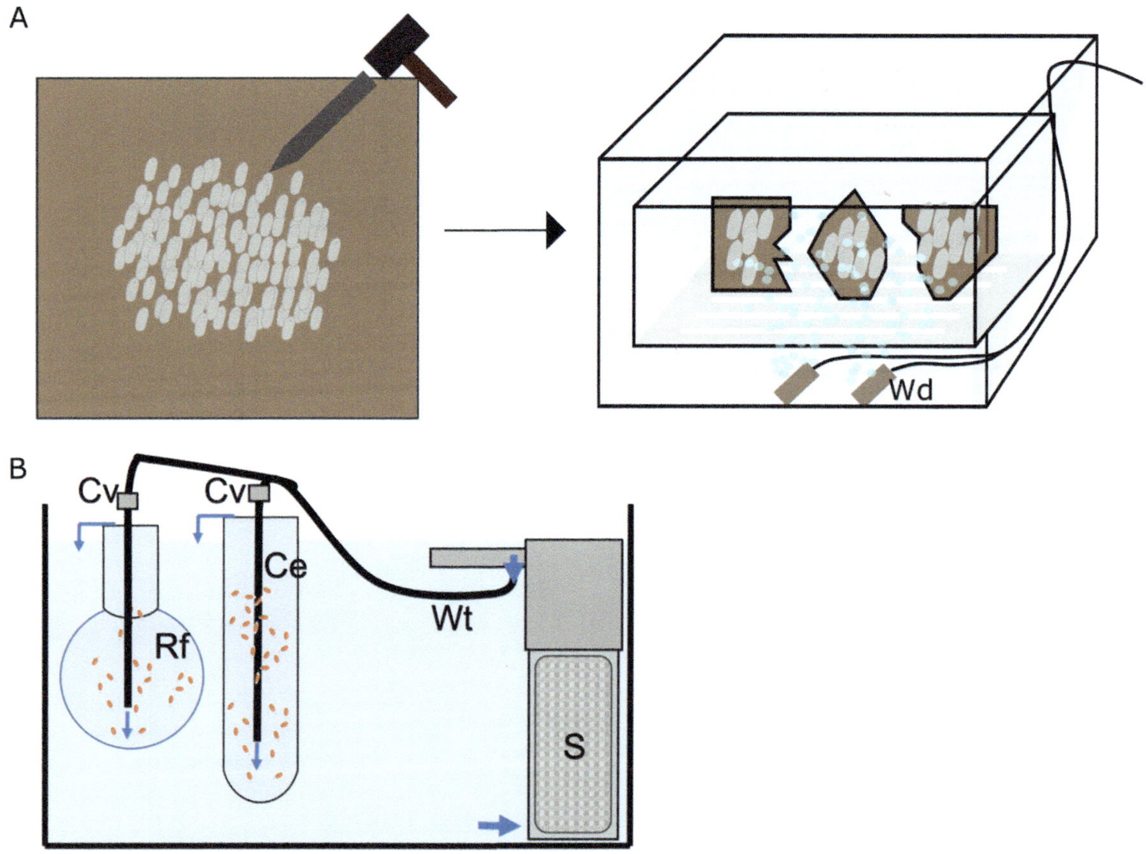

FIGURE 22.4 Schematic diagram of the separated egg rearing systems. A) Illustrates the method developed by Mitchell et al. (2021) consisting of breaking the egg support into small pieces and placing them in a seawater bath above wooden air diffusers (Wd) to ensure oxygenation and movement of the eggs. B) Illustrates the method developed by Yamanaka et al. (2021) requiring the removal of the eggs from the support and placing them in a round-bottomed flask (Rf) or a centrifuge tube (Ct) connected to a water tube (Wt) and control valve (Cv) that ensure water circulation and movement by being connected to a sponge filter (S). Blue arrow indicates water flow circulation.

If the eggs are separated from the support, some adjustments to incubation are required as described in Yamanaka et al. (2021). Eggs need to be incubated in a way that provides gentle suspension and continual water movement. This method consists of placing 100 eggs in a 100 ml centrifuge tube or a 300 ml round bottom flask filled with seawater, with water movement (Figure 22.4B). It is also possible to use commercial egg tumblers usually used for hatching mouth-breeding fishes such as African cichlids and cardinal fish (Haesler et al. 2011). In egg tumblers, eggs are maintained in a compartment under constant water movement, due to water flow and circulation created by aeration in a second compartment. In all egg incubation methods, daily observations are needed to remove dead eggs and avoid rot propagation. On the day of hatching, as with all other described methods, additional movement of water over the eggs is required to allow embryos to successfully hatch. One option involves placing the loose eggs into a dark cylindrical hatching container filled with seawater, maintained at the same temperature as during incubation, and placed on a laboratory shaker. On the night of hatching, embryos are transferred into the hatching container, placed in darkness shaken at 120 rpm for one hour to create a physical stimulus and promote hatching. After one hour of shaking, hatched larvae are transferred into a larval rearing tank. Unhatched embryos are placed back into darkness and provided with another hour of shaking to give them more time to hatch. Such methods allowed reaching 33.6% (±6.89%) survival rates compared to 38.8% (±3.82%) when eggs were still in the presence of the parents (Yamanaka et al. in press). However, it should be noted that this is a relatively low percentage of survival compared to the bulk of the anemonefish larval rearing literature and further research is needed to improve the protocol. Another method that deserves investigation is the incubation of anemonefish eggs in petri dishes placed in a thermo-regulated dark incubator as it is done in zebrafish. Some trials have been performed at the Observatoire Océanologique de Banyuls-sur-Mer and the Okinawa Institute of Science and Technology showing some success, but additional work is necessary to precisely assess the hatching rate and the survival rates following such incubation.

While there are some challenges to using anemonefish in certain laboratory experiments, they are rapidly emerging as a model. Specifically, the development of rearing and hatching methods without parental care is the major advancement in allowing exciting future research that will place anemonefish in the research realm of model freshwater fish species such as zebrafish or medaka, for which egg development and microinjection are well established and reliable. Future efforts on method development should be focused on the estimation of survival rates and adjustments that enable potential improvements.

22.4 CONCLUSION

In this chapter, we provide an overview of the methods used worldwide to maintain, breed, and rear anemonefishes on the laboratory scale. The robustness of anemonefishes, including their capability to adapt to different system types and maintenance conditions (closed or open systems, filtered or natural seawater, various temperature, and salinity tolerance), is an asset for their use as a model organism. It allows researchers flexibility to choose the methods according to the constraints of their laboratory and research goals, instead of being highly constrained by the needs of the species. Recent advancements now allow the rearing of embryos without parental care and in small volumes. Such methods are of particular interest as they put anemonefishes on the map as a potential model organism enabling the use of molecular tools (such as micro-injection), functional experiments, and ecotoxicological approaches. It is also worth adding that the knowledge to date on anemonefish husbandry, breeding, and rearing can also serve as a basis for future research and developments on the captive breeding and rearing of reef fish species, especially other benthic spawners which represent approximately 20% of coral reef fish species.

REFERENCES

Anil M. K., B. Santhosh, B. Prasad, and R. George. 2012. Broodstock development and breeding of black-finned anemone fish *Amphiprion nigripes* Regan, 1908 under captive conditions. *Indian Journal of Fisheries* 59: 77–82.

Anto J., J. Majoris, and R. G. Turingan. 2009. Prey selection and functional morphology through ontogeny of *Amphiprion clarkii* with a congeneric comparison. *Journal of Fish Biology* 75(3): 575–590.

Avella M. A., I. Olivotto, G. Gioacchini, F. Maradonna, and O. Carnevali. 2007. The role of fatty acids enrichments in the larviculture of false percula clownfish *Amphiprion ocellaris*. *Aquaculture* 273(1): 87–95.

Bell J. G., L. A. McEvoy, A. Estevez, R. J. Shields, and J. R. Sargent. 2003. Optimising lipid nutrition in first-feeding flatfish larvae. *Aquaculture* 227(1–4): 211–220.

Calado R., I. Olivotto, M. P. Oliver, and J. Holt. 2017. *Marine Ornamental Species Aquaculture*. Chichester, West Sussex, UK: Wiley-Blackwell.

Callan C. K. 2007. Assessment of the flame angelfish (*Centropyge loriculus*) as a model species in studies on egg and larval quality in marine fishes. http://digitalcommons.library.umaine.edu/etd/126/ (accessed October 6, 2017).

Chambel J., V. Severiano, T. Baptista, S. Mendes, and R. Pedrosa. 2015. Effect of stocking density and different diets on growth of Percula Clownfish, *Amphiprion percula* (Lacepede, 1802). *SpringerPlus* 4(1): 183.

Copeman L. A., C. C. Parrish, J. A. Brown, and M. Harel. 2002. Effects of docosahexaenoic, eicosapentaenoic, and arachidonic acids on the early growth, survival, lipid composition and pigmentation of yellowtail flounder (*Limanda ferruginea*): A live food enrichment experiment. *Aquaculture* 210(1–4): 285–304.

Coughlin D. J., J. R. Strickler, and B. Sanderson. 1992. Swimming and search behaviour in clownfish, Amphiprion perideraion, larvae. *Animal Behaviour* 44: 427–440.

Dhaneesh K., D. K. Nanthini, T. T. Ajith Kumar, T. Balasubramanian, and K. Tissera. 2012b. Breeding, embryonic development and salinity tolerance of Skunk clownfish *Amphiprion akallopisos*. *Journal of King Saud University - Science* 24(3): 201–209.

Dhaneesh K. V., T. T. Ajith Kumar, G. Swagat, and T. Balasubramanian. 2012a. Breeding and mass scale rearing of clownfish *Amphiprion percula*: Feeding and rearing in brackishwater. *Chinese Journal of Oceanology and Limnology* 30(4): 528–534.

Díaz-Jiménez L., M. P. Hernández-Vergara, C. I. Pérez-Rostro, and M. Á Olvera-Novoa. 2021. The effect of two carotenoid sources, background colour and light spectrum on the body pigmentation of the clownfish *Amphiprion ocellaris*. *Aquaculture Research* 52(7): 3052–3061.

Divya S. P., T. T. A. Kumar, R. Rajasekaran, and T. Balasubramanian. 2011. Larval rearing of clownfish using *Brachionus plicatilis* rotifer as starter food. *Science Asia* 37(3): 179–185.

Dixson D. L., P. L. Munday, and G. P. Jones. 2010. Ocean acidification disrupts the innate ability of fish to detect predator olfactory cues. *Ecology Letters* 13(1): 68–75.

Dixson D. L., P. L. Munday, M. Pratchett, and G. P. Jones. 2011. Ontogenetic changes in responses to settlement cues by Anemonefish. *Coral Reefs* 30(4): 903–910.

Donelson J. M. 2015. Development in a warm future ocean may enhance performance in some species. *Journal of Experimental Marine Biology and Ecology* 472: 119–125.

Edwards T. M., H. D. Miller, and L. J. Guillette. 2005. Water quality influences reproduction in female mosquitofish (*Gambusia holbrooki*) from eight Florida springs. *Environmental Health Perspectives* 114(S-1) Supplement 1: 69–75.

Frakes T., and F. H. Hoff. 1982. Effect of high nitrate-N on the growth and survival of juvenile and larval anemonefish, *Amphiprion ocellaris*. *Aquaculture* 29(1–2): 155–158.

Ghosh S., T. T. A. Kumar, K. Nanthinidevi, and T. Balasubrananian. 2011. Hatchery production of Clark's clownfish, *Amphiprion clarkii* (Bennett, 1830) using brackishwater. In *2nd International Conference on Agricultural and Animal Science*, Singapore. IACSIT Press.

Ghosh S., T. T. A. Kumar, K. Nanthinidevi, and T. Balasubramanian. 2012. Reef fish breeding and hatchery production using brackishwater, a sustainable technology with special reference to Clark's clownfish, *Amphiprion clarkii* (Bennett, 1830). *IJESD* 3: 56–60.

Gopakumar G., R. M. George, and S. Jasmine. 1999. Breeding and larval rearing of the clownfish *Amphiprion chrysogaster*. *Marine Fisheries Information Service, Technical and Extension Series* 161: 8–11. CMFRI, Kochi.

Green, B. S., and M. J. McCormick. 2001. Ontogeny of the Digestive and Feeding Systems in the Anemonefish Amphiprion Melanopus. *Environmental Biology of Fishes* 61: 73–83.

Green B. S. 2004. Embryogenesis and oxygen consumption in benthic egg clutches of a tropical clownfish, *Amphiprion melanopus* (Pomacentridae). *Comparative Biochemistry and Physiology Part A* 138(1): 33–38. doi: 10.1016/j.cbpb.2004.02.014

Haesler M. P., C. M. Lindeyer, O. Otti, D. Bonfils, D. Heg, and M. Taborsky. 2011. Female mouthbrooders in control of pre- and postmating sexual selection. *Behavioral Ecology* 22(5): 1033–1041.

Hess S., B. J. M. Allan, A. S. Hoey, M. D. Jarrold, A. S. Wenger, and J. L. Rummer. 2019. Enhanced fast-start performance and anti-predator behaviour in a coral reef fish in response to suspended sediment exposure. *Coral Reefs* 38(1): 103–108.

Ho A. L. F. C., N. M. Orlando Bertran, and J. Lin. 2013a. Dietary esterified astaxanthin concentration effect on dermal coloration and chromatophore physiology in spinecheek anemonefish, *Premnas biaculeatus*. *Journal of the World Aquaculture Society* 44(1): 76–85.

Ho A. L. F. C., S. K. O'Shea, and H. F. Pomeroy. 2013b. Dietary esterified astaxanthin effects on color, carotenoid concentrations, and compositions of clown anemonefish, *Amphiprion ocellaris*, skin. *Aquaculture International* 21(2): 361–374.

Ignatius B., G. Rathore, I. Jagadis, D. Kan, and A. C. C. Victor. 2001. Spawning and larval rearing technique for tropical clown fish *Amphiprion sebae* under captive condition. *Journal of Aquaculture in the tropics* 16(3): 241–249.

Jarrold M. D., P. L. Munday. 2019. Diel CO2 cycles and parental effects have similar benefits to growth of a coral reef fish under ocean acidification. *Biology Letters* 15(2): 20180724. Royal Society.

Job S. D., and D. R. Bellwood. 2000. Light sensitivity in larval fishes: Implications for vertical zonation in the pelagic zone. *Limnology Oceanography* 45(2): 362–371.

Kitahara T. 1983. Behavior of carotenoids in the chum salmon (*Oncorhynchus keta*) during anadromous migration. *Comparative Biochemistry and Physiology -- Part B* 76(1): 97–101.

Klann M., M. Mercader, P. Salis, M. Reynaud, N. Roux, V. Laudet, and L. Besseau. 2021. Anemonefishes. In: *Handbook of Marine Model Organisms in Experimental Biology*, eds A. Boutetand B. Schierwater pp. 443–464. Boca Raton: CRC Press.

Kumar T. T. A., and T. Balasubramanian. 2009. Broodstock development, spawning and larval rearing of the false clown fish, *Amphiprion ocellaris* in captivity using estuarine water. *Current Science* 97(10): 1483–1486.

Kumar T. T. A., M. Gopi, K. V. Dhaneesh, R. Vinoth, S. Ghosh, T. Balasubramanian, and T. Shunmugaraj. 2012. Hatchery production of the clownfish *Amphiprion nigripes* at Agatti island, Lakshadweep, India. *Journal of Environmental Biology* 33: 623–628.

Kumar T. T. A., S. K. Setu, P. Murugesan, and T. Balasubramanian. 2010. Studies on captive breeding and larval rearing of clown fish [al], *Amphiprion sebae* (Bleeker, 1853) using estuarine water. *Indian Journal of Geo-Marine Sciences* 39(1): 6.

Lim K. C., F. M Yusoff, M. Shariff, M. S. Kamarudin. 2018. Astaxanthin as feed supplement in aquatic animals. *Reviews in Aquaculture* 10(3): 738–773.

Madhu K., and R. Madhu. 2007. Influence of lunar rhythm on spawning of clown anemonefish *Amphiprion percula* under captive condition in Andaman and Nicobar Islands. *Journal of the Marine Biological Association of India* 49(1): 58–64.

Madhu K., R. Madhu, L. Krishnan, C. S. Sasidharan, and K. M. Venugopal. 2006a. Spawning and larval rearing of *Amphiprion ocellaris* under captive condition. *Marine Fisheries Information Service, Technical and Extension Series* 188: 1–5.

Madhu K., R. Madhu, G. Gopakumar, and C. S. Sasidharan. 2006b. Breeding, larval rearing and seed production of maroon clown *Premnas biaculeatus* under captive conditions. *Marine Fisheries Information Service, Technical and Extension Series* 190: 1–5.

McMahon S. J., P. L. Munday, M. Y. L. Wong, and J. M. Donelson. 2019. Elevated CO2 and food ration affect growth but not the size-based hierarchy of a reef fish. *Scientific Reports* 9(1): 19706.

Mitchell L. J., V. Tettamanti, J. S. Rhodes, N. J. Marshall, K. L. Cheney, and F. Cortesi. 2021. CRISPR/Cas9-mediated generation of biallelic F0 anemonefish (*Amphiprion ocellaris*) mutants. *PLoS ONE* 16(12): e0261331.

Munday P. L., D. L. Dixson, J. M. Donelson, G. P. Jones, M. S. Pratchett, G. V. Devitsina, and K. B. Døving. 2009. Ocean acidification impairs olfactory discrimination and homing ability of a marine fish. *Proceedings of the National Academy of Sciences* 106(6): 1848–1852.

Naas K. E., I. Huse, and J. Iglesias. 1996. Illumination in first feeding tanks for marine fish larvae. *Aquacultural Engineering* 15(4): 291–300.

Naas K. E., T. N↿ss, and T. Harboe. 1992. Enhanced first feeding of halibut larvae (*Hippoglossus hippoglossus L.*) in green water. *Aquaculture* 105(2): 143–156.

Olivotto I., F. Capriotti, I. Buttino, A. M. Avella, V. Vitiello, F. Maradonna, and O. Carnevali. 2008. The use of harpacticoid copepods as live prey for *Amphiprion clarkii* larviculture: Effects on larval survival and growth. *Aquaculture* 274(2): 347–352.

Pietoyo A., K. W. Hidayat, S. Nurazizah, I. F. Zainul Arifin, D. G. Prabowo, F. T. Widianto, I. Mustakim. 2020. The effect of stocking density on the growth of ocellaris clownfish (*Amphiprion ocellaris* Cuvier, 1830) with recirculation. *Journal of Aquaculture and Fish Health* 9(3): 189.

Richardson D., P. Harrison, and V. Harriott. 1997. Timing of spawning and fecundity of a tropical and subtropical anemonefish (Pomacentridae:Amphiprion) on a high latitude reef on the east coast of Australia. *Marine Ecology Progress Series* 156: 175–181.

Roux N., V. Logeux, N. Trouillard, R. Pillot, K. Magré, P. Salis, D. Lecchini, et al. 2021. A star is born again: Methods for larval rearing of an emerging model organism, the False clownfish *Amphiprion ocellaris. Journal of Experimental Zoology (Molecular and Developmental Evolution)* 336(4): 376–385.

Roux N., P. Salis, S-H. Lee, L. Besseau, and V. Laudet. 2020. Anemonefish, a model for Eco-Evo-Devo. *EvoDevo* 11(1): 20.

Salis P., N. Roux, D. Huang, A. Marcionetti, P. Mouginot, M. Reynaud, O. Salles, et al. 2021. Thyroid hormones regulate the formation and environmental plasticity of white bars in clownfishes. *Proceedings of the National Academy of Sciences* 118(23): e2101634118.

Sargent J., L. McEvoy, A. Estevez, G. Bell, M. Bell, J. Henderson, and D. Tocher. 1999. Lipid nutrition of marine fish during early development: Current status and future directions. *Aquaculture* 179(1–4): 217–229.

Sargent J. R., L. A. McEvoy, and J. G. Bell. 1997. Requirements, presentation and sources of polyunsaturated fatty acids in marine fish larval feeds. *Aquaculture* 155(1–4): 117–127.

Schiedt K., F. J. Leuenberger, M. Vecchi, and E. Glinz. 1985. Absorption, retention and metabolic transformations of carotenoids in rainbow trout, salmon and chicken. *Pure and Applied Chemistry* 57(5): 685–692.

Tanaka Y., A. Yamamoto, T. Kamata, and K. L. Simpson. 1992. Biochemical study on the carotenoids in the anemonefish, *Amphiprion* spp. *Memoirs of Faculty of Fisheries Kagoshima University Kagoshimadai Suisangakubu Kiyo* 41: 1–8.

Thresher R. E. 1984. *Reproduction in Reef Fishes*. Neptune City: T.F.H. Publications.

Tucker L. R. 1958. An inter-battery method for factor analysis. *Psychometrika* 23: 111–136.

Varghese B. 2009. Dietary influence on the egg production and larval viability in True Sebae Clownfish *Amphiprion sebae* Bleeker 1853. *Asian Fisheries Science* 22(1): 7–20.

Vissio P. G., M. J. Darias, M. P. Di Yorio, D. I. Pérez Sirkin, and T. H. Delgadin. 2021. Fish skin pigmentation in aquaculture: The influence of rearing conditions and its neuroendocrine regulation. *General and Comparative Endocrinology* 301: 113662.

Wabnitz C. 2003. *From Ocean to Aquarium: The Global Trade in Marine Ornamental Species*. Nairobi: UNEP/Earthprint (Vol. 17).

Wittenrich M. L. 2007. *The Complete Illustrated Breeder's Guide to Marine Aquarium Fishes*. Neptune City: T.F.H. Publications.

Yamanaka S., Y. Okada, T. Furuta, and M. Kinoshita. 2021. Establishment of culture and microinjection methods for false clownfish embryos without parental care. *Development, Growth & Differentiation* 63(9): 459–466.

Yasir I., and J. G. Qin. 2007. Embryology and early ontogeny of an anemonefish *Amphiprion ocellaris. Journal of the Marine Biological Association of the United Kingdom* 87(4): 1025–1033.

Yasir I., and J. G. Qin. 2010. Effect of dietary carotenoids on skin color and pigments of false clownfish, *Amphiprion ocellaris*, Cuvier. *Journal of the World Aquaculture Society* 41(3): 308–318.

23 Resilience and Adaptation to Local and Global Environmental Change

Celia Schunter, Jennifer M. Donelson, Philip L. Munday, and Timothy Ravasi

CONTENTS

23.1 INTRODUCTION

It is unequivocal that human activities are the primary cause of rapid environmental change in the modern era (Bongaarts 2019; Masson-Delmotte et al. 2021). With the human population expected to reach over 9.6 billion by 2050 (Gerland et al. 2014), the ever-increasing demand for food, energy, water, and materials will induce further shifts in environmental conditions across the globe (Steffen et al. 2015). The concern for plants and animals is that environmental conditions are shifting too rapidly beyond the historical range, and far beyond what has occurred in the recent past, with potentially broad-ranging effects on their physiology, ecology, and behavior (Pörtner and Farrell 2008; Poloczanska et al. 2013; Wong and Candolin 2015). Environmental change is considered a stressor when it is outside the normal range of variation and has negative biological effects (Vinebrooke et al. 2004). Environmental changes may be broad-reaching and are relevant to large parts of this planet, such as climate change which is categorized as a global stressor, while other anthropogenic activities may only be observed at a more localized scale and are considered to be local stressors (Brown et al. 2013). Combined effects of local and global stressors may also affect individuals and populations, and provoke changes to ecological interactions.

The vast majority of anemonefishes live in shallow tropical to sub-tropical coastal waters where they are susceptible to a variety of human-induced global and local stressors. Due to their mutualistic association with anemones, which are essential habitats for benthic juveniles and adults, anemonefishes are susceptible to both the direct effects of global and local stressors and the indirect effects of stressors on the anemones they inhabit. Indeed, marine fishes that have obligate associations with anthozoans (corals and anemones) are predicted to be especially at risk from climate change due to the sensitivity of their habitat to environmental stressors (Pratchett et al. 2018). Predicting the resilience of anemonefishes to environmental change, therefore, requires an understanding of the direct effects on individual fish, the social organization of the group, effects on the host anemone, and their mutualistic relationship (Halpern et al. 2008). Ultimately, we are interested in understanding the capacity of populations and species to recover, and hence respond and cope with these human-induced changes (i.e., resilience). This question of resilience is a multi-step process, starting with understanding the sensitivity of organisms to changes in their environment, both within and outside the historical range of conditions, and then considering how current projections for human-induced change relate to this sensitivity, and what capacity they have to respond through processes of acclimation and adaptation. Due to the cultural importance, economic contribution through aquarium collection and tourism, and the excellent capacity to breed and complete the life cycle in captivity of anemonefish (Chapter 22), this taxonomic group has become a strong model for understanding the potential impacts of environmental change on marine coastal and coral reef fishes (Figure 23.1).

23.2 GLOBAL STRESSORS

Since the industrial era, the emission of anthropogenic greenhouse gases into the atmosphere has increased dramatically, resulting in carbon dioxide (CO_2) levels now being higher than at any time in the past two million years (Masson-Delmotte et al. 2021). This increase in greenhouse gases causes an enhancement of the natural greenhouse effect, with

DOI: 10.1201/9781003125365-28

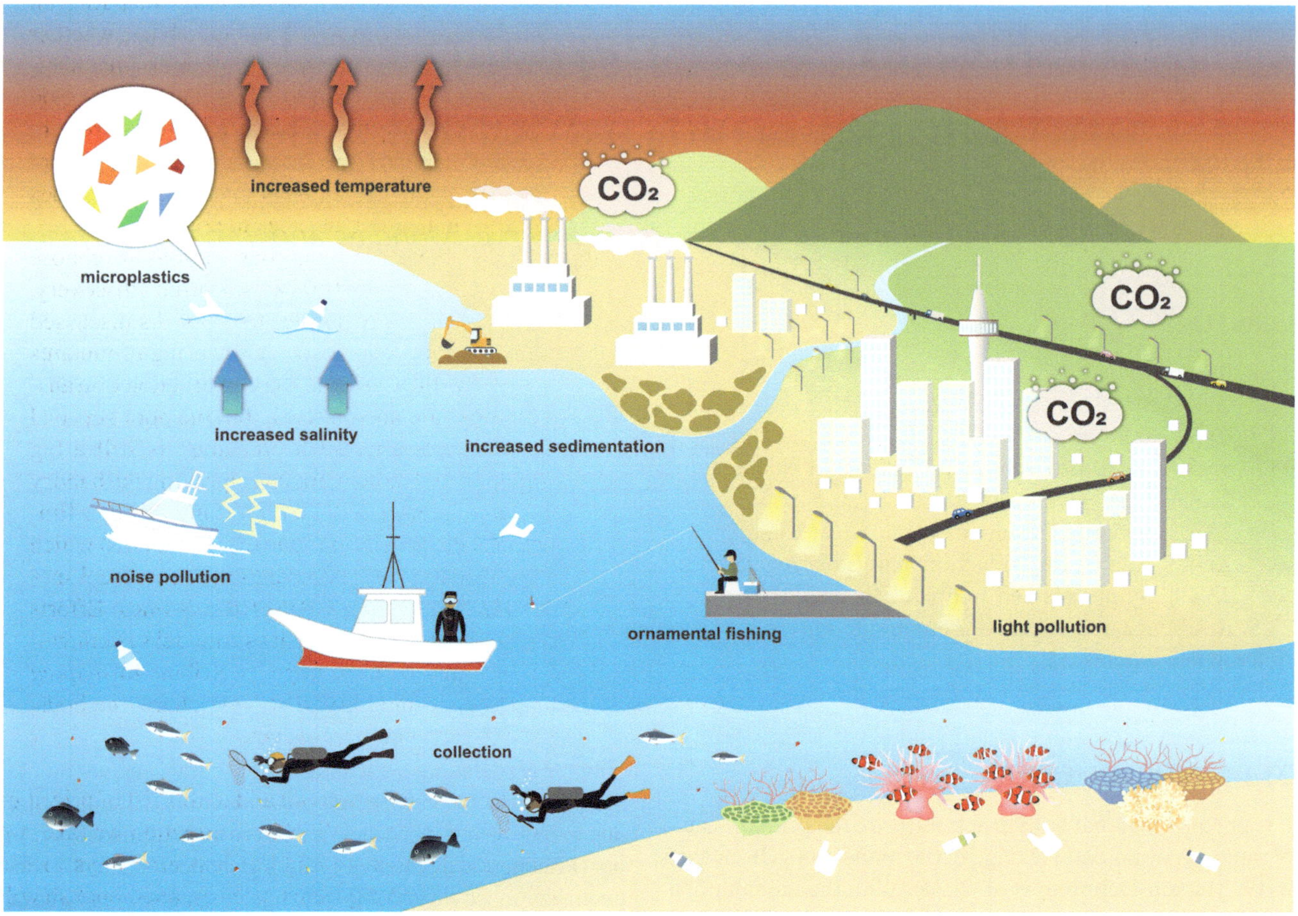

FIGURE 23.1 Global and local environmental stressors due to anthropogenic activities which can affect individuals, populations, and species of anemonefish.

more heat radiation trapped within the atmosphere resulting in global warming. The ocean absorbs much of this excess heat, leading to surface ocean temperatures already warming by 0.3–0.4°C from 1981 to 2019 (Bindoff et al. 2019). Due to the continued burning of fossil fuels, global ocean temperatures are projected to increase up to a further 3°C by 2100 (Pörtner et al. 2019). These changes to global climatic conditions are also leading to an increased frequency and intensity of extreme temperatures, especially extreme warm periods during summer, termed heatwaves (Oliver et al. 2018; Perkins-Kirkpatrick and Lewis 2020). Since 2016, marine heatwaves have increased in frequency and duration resulting in 50–100% more days of marine heatwave annually across the globe (Frölicher et al. 2018; Oliver et al. 2018). As global warming continues, the occurrence of marine heatwave days is expected to further increase, becoming over five times more likely (Oliver et al. 2019; Pörtner et al. 2019). As the vast majority of marine species are ectothermic, marine heatwaves and the increase in average ocean temperature have the potential to negatively affect physiological processes and the fitness of individuals, and ultimately species' resilience in the future (Pörtner and Farrell 2008; Hofmann and Todgham 2010). Moreover, for species such as anemonefishes that have an obligate relationship with another marine organism, the effects of warming and heatwaves on the more sensitive species in the mutualistic partnership may drive the effects on both species.

The anthropogenic production of carbon dioxide (CO_2) is of particular concern for marine species, due to ~30% of all atmospheric production being absorbed by the oceans (Hoegh-Guldberg et al. 2014). This results in a decrease in pH and a change in the relative availability of dissolved carbonate and bicarbonate ions (commonly termed ocean acidification). To date, the average pH of the oceans has reduced by 0.1 units, and depending on future emissions, is predicted to reduce a further 0.2 units (RCP 8.5) by 2081–2100 (Bindoff et al. 2019). When marine organisms are exposed to elevated levels of CO_2 (thus, lower pH), it can result in disturbances to their physiology, life histories, and behavior (Kroeker et al. 2013; Brauner et al. 2019). It is now evident that ecologically important behaviors of marine fish and invertebrates can be impaired by environmental conditions projected to occur in the ocean by 2100 (Munday et al. 2009a; Simpson et al. 2011; Cattano et al. 2018). The alteration of these key ecological processes could have implications for survival and lead to higher mortality and reduced population replenishment (Munday et al. 2009a; Clements and Hunt 2015; Nagelkerken and Munday 2016).

Other physical properties of seawater can also be affected by climate change. For example, over the last 50 years, the surface oceans have also become more saline due to warming climate via altering evaporation rates by 7% with each increased degree of temperature (Durack et al. 2012). Hence, ocean salinities have been increasing and are predicted to further increase in the near future. Stronger and more frequent cyclone and rainfall events that will create periods of low salinity are also predicted for tropical regions (Bindoff et al. 2019). Osmoregulation and associated physiological processes linked to salinity are important to coral reef organisms, including fish. Further global stressors exist, for example, oxygen-depleted zones leading to hypoxia; however, this has not been addressed in anemonefishes, and hence, is not discussed further in this chapter.

23.3 LOCAL STRESSORS

Terrestrial input of sediments to the marine environment is a natural occurrence (Milliman and Meade 1983) that can promote oceanic productivity (Thorrold and McKinnon 1995). However, human activities are greatly affecting the deposition of sediment in marine ecosystems through local actions such as dredging, agriculture, urban development, catchment modification, and deforestation (Kroon et al. 2012). This has dramatically increased the volume and type of sediments, nutrient inputs, and chemical pollution (Chapter 24) that are transported into coastal waters (Kroon et al. 2012). This additional sedimentation is a concern for marine organisms due to the resulting increased turbidity reducing light availability and visual acuity, physiological stress, and physical damage to tissues and organs (Fabricius 2005; Sheridan et al. 2014), leading to mortality and shifts in community composition (Rodgers et al. 2021).

Environmental pollution can come in the form of chemical pollution (see Chapter 24), but also microplastics, light, and noise. In fact, noise pollution has been hailed as one of the most serious forms of pollution by the World Health Organization (World Health Organization 2011) and is now omnipresent (Kunc et al. 2016). In aquatic environments, the concern regarding noise pollution is twofold (Putland et al. 2018). Firstly it is due to the diversity of species and life stages that use sound for orientation and communication. During the larval stage, sound is used for orientation and habitat selection (Vermeij et al. 2010; Kunc et al. 2016; Simpson et al. 2016). Adult fish produce sound to synchronize spawning or in courtship interactions (Mann and Lobel 1997; Amorim et al. 2015), as well as part of territorial defence and antagonistic interactions (Ladich 1997; Mensinger 2013). Secondly, the hearing and vocalization range of fish strongly overlaps with the production range of noise by coastal development, recreational boats, shipping, and their sonar communication (Yan et al. 2002; Hildebrand 2009; Slabbekoorn et al. 2010; Fakan and McCormick 2019). Similarly, light pollution can have effects on the marine environment, with artificial light polluting the coastlines including artificial light at night (Depledge et al. 2010), whereas another aspect of light pollution is increased solar UV radiation (Häder et al. 2007). In fact, most anemonefishes use both sound and light for orientation and communication, and are known to have good hearing and vision (Parmentier et al. 2009; Stieb et al. 2019). Hence, light and noise pollution could affect many aspects of an anemonefish's life, including the location of suitable coral reef habitat and selection of host anemone habitat by larvae, the survivorship of juveniles, and mating of adults. Lastly, microplastics are now widely distributed in marine ecosystems. Ingestion of microplastics has been shown to be an important pollutant in other fish due to its increased presence in the intestinal tract and the limited capacity of fish to expel the plastic pieces (Santana et al. 2021).

The trade of marine ornamentals is a global multimillion-dollar industry that is rapidly expanding (Wabnitz et al. 2003). There are already concerns about overexploitation in some areas of the world due to the fishery being highly selective and harvesting from limited areas, leading to localized depletions or extinctions of target anemones and anemonefish species (Edwards and Shepherd 1992; Gasparini et al. 2005; Shuman et al. 2005; Madduppa et al. 2014). In addition, some fisheries still employ destructive practices like cyanide (Madeira et al. 2020b). The collection of anemones and anemonefishes could affect their capacity to cope with environmental stressors by reducing effective population size, reproductive output, and genetic diversity (Madduppa et al. 2018).

23.4 SENSITIVITY TO ENVIRONMENTAL CHANGE IN ANEMONEFISHES

An overarching pattern of biological response to environmental stressors is that sensitivity can be related to life-stage. For species like anemonefish, their life cycle is such that all life stages other than the larval period are site-attached, with limited potential for migration away from stressful conditions. Early life stages of aquatic species are often found to be the most sensitive to environmental change (Dahlke et al. 2020; Madeira et al. 2020a) and experiences during early life can have lasting implications on later life stages (Metcalfe and Monaghan 2001). This early life sensitivity is generally attributed to incomplete development of physiological systems and organs, relatively high surface area to volume ratios, and limited energy reserves in combination with rapid development, which is energetically demanding (Pankhurst and Munday 2011; Mohammed 2013). Some life stages, including reproduction, are also highly sensitive due to physiological optimization to a narrow range of environmental conditions that provide fitness advantages (Hofmann and Todgham 2010; Pankhurst and Munday 2011).

In the following we outline what is known to date about potential global and local stressors across the different life stages of anemonefishes (Figure 23.2). While we have a relatively solid understanding of how global stressors may impact anemonefish, research on local stressors is more limited (Table 23.1).

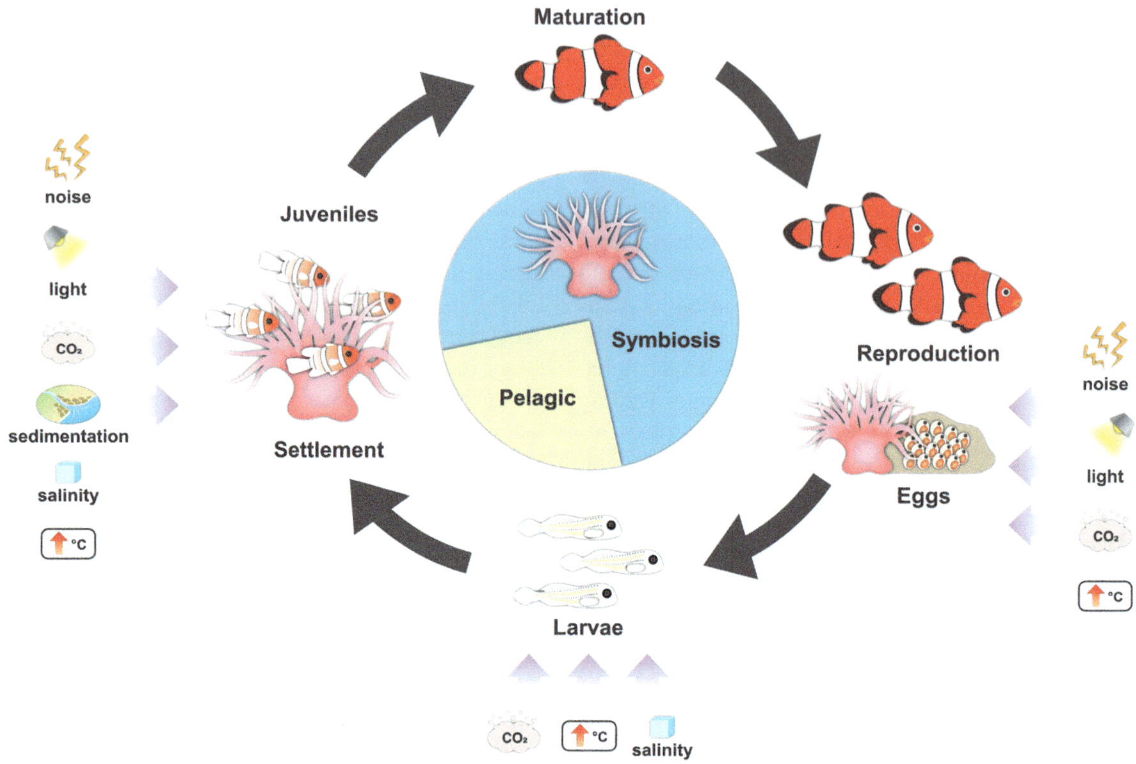

FIGURE 23.2 Life cycle of an anemone fish. The different local and global stressors known to negatively affect each life stage are outlined.

23.4.1 Larval Development

The pelagic larval stage, from hatching to settlement, is a period of high mortality for marine fishes. Consequently, the environmental conditions experienced during this pelagic stage can have substantial implications for which individuals, and ultimately how many, recruit into the benthic population. Global ocean circulation patterns are shifting with the strengthening of poleward boundary currents (Vergés et al. 2014), which has the potential to alter anemonefish larval dispersal as it is already shifting many other topical reef fish beyond their normal distributional range (Booth et al. 2011). Larger and better-conditioned individuals are more likely to survive the selective mortality of early settlement (Searcy and Sponaugle 2001; Hoey and McCormick 2004; Almany and Webster 2006), therefore, growth and condition during larval development can affect recruitment. For anemonefish, water temperature experienced during the pelagic larval period can influence growth rate, swimming speed, metabolic rate, and development rate (Green and Fisher 2004; McLeod et al. 2013), but does not appear to affect critical thermal maximum or enzymatic performance (Illing et al. 2020). Water temperature affects development rate differently depending on whether the temperature is cooler or warmer than the usual summer range. An increase in temperature from 25 to 28°C, shifting within the seasonal range, caused a reduction in development time by approximately one day per degree Celsius (from 12.3 to nine days) in *Amphiprion melanopus* (Green and Fisher 2004). For *Amphiprion percula*, larval development ranged from 10.2 to 15.6 days under both normal summer conditions of 29.2°C and elevated temperatures of 30.7 and 32.2°C. Extended development times were only observed when fish experienced elevated temperature and reduced food availability (McLeod et al. 2013), with larvae that developed at 32.2°C having a ~33% increase in metabolic rate (McLeod et al. 2013). In a similar experiment, *A. percula* larvae exhibited a ~55% increase in metabolic rate when acutely (24 hours) held at 31.5°C in the later part of the development period (McLeod and Clark 2016). In all studies thus far, a general increase in growth rate with rising water temperature is observed (Green and Fisher 2004; McLeod et al. 2013); however, there are limitations when food supply is low at elevated temperatures, resulting in reduced growth (McLeod et al. 2013). Also, while temperature influences the rate and nature of development, when fish are compared at metamorphosis, they are of similar size, condition, and swimming ability (Green and Fisher 2004; McLeod et al. 2013), indicating that larval fish are settling with similar physical and physiological capacity regardless of the temperature experienced in laboratory experiments. The likely effects of ocean warming on larval development in nature may be more accurately predicted by experiments using lower levels of food availability, due to the patchiness or mismatches of food productivity in the future (Hays et al. 2005; Asch et al. 2019).

Near-future predicted acidified ocean conditions reveal variable impacts, and can affect larval growth and survival

TABLE 23.1
Sensitivity of Anemonefish Species to Global and Local Stressors Depending on the Type of Effect Observed (Morphological, Physiological, or Behavioral)

Species	Stressor	Effect: Morphological	Physiological	Behavioral
Amphiprion percula	CO_2	Elevated CO_2 (550, 750, and 1,030 ppm) produced no effect on embryonic duration, egg survival, or size at hatching compared with current-day controls (400 ppm) (Munday et al. 2009b). Settling larvae were 15–18% longer and 47–52% heavier from some CO_2 pairs (550, 750, and 1,030 ppm) compared with controls. Offspring from other pairs were unaffected by CO_2 (Munday et al. 2009b). Elevated CO_2 (1,721 µatm) increased otolith size, but not at 1,050 µatm, compared with control larvae (404 µatm) (Munday et al. 2011). No effects of elevated CO_2 (~1,000 µatm), with or without the presence of a diel CO_2 cycle, on the growth and survival of juveniles (Jarrold and Munday 2018). Parental exposure to elevated CO_2 (~1,000 µatm) increased length but not weight of offspring as juveniles (McMahon et al. 2019).	No effect of elevated CO_2 on maximum swimming speed at settlement (Munday et al. 2009b). Elevated CO_2 (~1,000 µatm) increased reproductive output in terms of number of clutches and number of eggs per clutch in both (Welch and Munday 2016).	Larvae altered olfactory choices, including becoming attracted to predator cues after 4 days exposure to 1,000 ppm CO_2 (Dixson et al. 2010; Munday et al. 2010). Elevated CO_2 (~1,000 µatm) caused a loss of response to predator cues (i.e., reduced feeding) (McMahon et al. 2018). Parental exposure to elevated CO_2 (~1,000 µatm) did not affect the impaired response of juveniles to predator odour (McMahon et al. 2019) Reduced reaction to an auditory stimulus occurred when juveniles developed at 600–900 µatm (Simpson et al. 2011) Behavioral impairments observed with elevated CO_2 were reversed when a Gaba antagonist (gabazine), was administered (Nilsson et al. 2012; Munday et al. 2016). Behavioral impairments were less severe when the elevated CO_2 fluctuated in diel cycles (750 ± 300 µatm) (Jarrold et al. 2017). Altered behavioral response associated with elevated CO_2 is seen in the brain molecular response (Schunter et al. 2021).
	Temperature	Growth rate increased with rising water temperature from 29.2°C (control) to 30.7 and 32.2°C. However, this increase did not occur when food supply was low (McLeod et al. 2013). Larval duration increased with temperature (from 29.2 to 32.2°C), but only when food was low (McLeod et al. 2013). Peak reproduction occurred at 28.74°C (Dhaneesh et al. 2012b) or above 30.0°C (Madhu and Madhu 2007).	Larvae developed at 32.2°C had ~33% increase in metabolic rate compared to 29.2°C controls (McLeod et al. 2013). Larvae exhibited a ~55% increase in metabolic rate when acutely (24 hours) held at 31.5°C in the later part of the development period, compared to 28.5°C control. No effect on standard dynamic action was observed (McLeod and Clark 2016).	
	Local stressors	Exposure to 135 and 180 mgL^{-1} of suspended sediment reduced oxygen diffusion distance on gills but not lamellae length (Hess et al. 2017).	Exposure to 135 and 180 mgL^{-1} of suspended produced no impact on the capacity to uptake oxygen (Hess et al. 2017).	

(Continued)

TABLE 23.1 (CONTINUED)
Sensitivity of Anemonefish Species to Global and Local Stressors Depending on the Type of Effect Observed (Morphological, Physiological, or Behavioral)

Species	Stressor	Effect: Morphological	Physiological	Behavioral
Amphirion ocellaris	**Temperature**	Elevated temperature (32°C) did not affect average growth rate, but induced stronger asymmetrical individual body weight within groups, compared to 28°C (Pham et al. 2021). This pattern was consistent across food levels.	One week exposure to 30°C caused thermal stress in terms of heat shock protein 70 kDa and total ubiquitin increasing in juveniles, compared to 26°C. This stress reduced with one-month exposure (Madeira et al. 2017, 2021). No increased mortality with exposure to 30°C for one month, compared to 26°C. Lipid storage capacity was not affected, fatty acid profiles were unresponsive, lipid metabolic networks were conserved, and fish weight was maintained at elevated temperature (Madeira et al. 2021). Preferred temperature ranged from 27–32°C depending on the holding temperate of the fish (from 20–35°C). Aerobic metabolic performance was maintained across 23–32°C (Velasco-Blanco et al. 2019). Lower temperatures (26–28°C) resulted in lower hatching efficiency compared to 29°C (Soman et al. 2021). Higher temperature (30°C) led to reduced yolk volume compared to 29°C (Soman et al. 2021).	
	Local stressors		~50% of juveniles survived cyanide dose of $50 mgL^{-1}$ and 100% survival at lower dose of $25 mgL^{-1}$ at control temperature 26°C (Madeira et al. 2020b). Additional stress of elevated temperature (29 and 32°C) caused mortality to increase 60% and 20% for the lower dose of $25 mgL^{-1}$ (Madeira et al. 2020b). Artificial light at night (12 h of dim light, mean lx = 26.5) had no impact on the frequency of spawning or fertilization success of breeding adults. But embryos were unable to hatch at all (Forbert et al. 2019).	Increased microplastic correlated with higher activity levels (Nanninga et al. 2020).

(*Continued*)

TABLE 23.1 (CONTINUED)
Sensitivity of Anemonefish Species to Global and Local Stressors Depending on the Type of Effect Observed (Morphological, Physiological, or Behavioral)

Species	Stressor	Effect: Morphological	Effect: Physiological	Effect: Behavioral
Amphiprion melanopus	CO_2	Length, weight, and survival of juveniles were reduced with development at 1,000 µatm (Miller et al. 2012). Parental exposure to elevated CO_2 (1,000 µatm) reversed the negative effects on juvenile growth and survival (Miller et al. 2012). Length and weight reduced in juveniles reared in stable elevated CO_2 compared with control (Jarrold and Munday 2019). Negative effects of elevated CO_2 (1,000 µatm) on length and weight were absent when a diel cycle occurred (1,000 ± 300 µatm) (Jarrold and Munday 2019).	Elevated CO_2 (~1.000 µatm) increased reproductive output, including in terms of number of clutches and number of eggs per clutch (Miller et al. 2013).	Elevated CO_2 (~1,000 µatm) reduced escape performance in juveniles in terms of occurrence, speed, distance, and duration. The effects were partially reversed when parents were exposed to elevated CO_2 (Allan et al. 2014).
	Temperature	Warming from 25 to 28°C increased growth rate of larvae (Green and Fisher 2004). Still metamorphosed at the same size. Warming from 25 to 28°C caused a reduction in development time from 12.3 to 9 days) (Green and Fisher 2004).	At 7 days post-hatching critical swimming speed was faster in larvae from 28°C than 25°C (Green and Fisher 2004). Development of larvae and juveniles at 28.5 or 30°C did not affect critical thermal maximum or enzymatic performance (Illing et al. 2020). Ocean warming caused a decline in the number of clutches, egg size, and survival of eggs at 30°C. Reproduction ceased at 31.5°C. Effects are likely due to negative effects on reproductive hormones (Miller et al. 2015).	Food consumption and foraging activity increased with moderate warming (from 28.5 to 30°C), but then decreased with further temperature increase (31.5°C) in juveniles (Nowicki et al. 2012).
	Salinity		Salinity decreased from 35 psu to 17.5 psu causing oxidative stress through increased plasma cortisol, antioxidant and lip peroxidation levels in juveniles. The addition of prolactin enzyme to low salinity water reduced this stress response (Park et al. 2011). Survival to as low as 10 ppt was possible. Acclimation to low salinity (15 ppt) occurred through the level of prolactin transcripts in the liver (Noh et al. 2013).	
	Local stressors	All levels of suspended sediment (45, 90, 135, and 180 mgL^{-1}) reduced gill lamellae and oxygen diffusion distance (Hess et al. 2017). Boat noise caused increased heart rate but had no effect on yolk volume of embryos (Fakan and McCormick 2019).	All levels of suspended sediment (45, 90, 135, and 180 mgL^{-1}) increased resting metabolic rates and reduced aerobic capacity (Hess et al. 2017).	Exposure to 180 mgL^{-1} caused fish to be more responsive and produce more and enhanced escapes to startle stimuli (Hess et al. 2019).

(Continued)

TABLE 23.1 (CONTINUED)
Sensitivity of Anemonefish Species to Global and Local Stressors Depending on the Type of Effect Observed (Morphological, Physiological, or Behavioral)

		Effect		
Species	**Stressor**	**Morphological**	**Physiological**	**Behavioral**
Amphiprion clarkii	**CO_2**	Elevated CO_2 (800 μatm) increased otolith roughness of larvae (Holmberg et al. 2019).		
	Temperature		Reproduction occurs between 20.5–27°C during spring and summer (Bell 1976; Ochi 1985).	
	Local stressors		UV radiation (0.4 W/m2) increased apoptosis and reactive oxygen species enzymes (Ryu, Choi, Song, and Kil, 2019). High sensitivity to cyanide, no survival after a 60-second dose of 50 mgL^{-1} (Madeira et al. 2020b).	
Amphiprion akallopisos	**Salinity**		Larval survival decreased to 88% at 40 ppt and 76% with lower salinities of 20–25 ppt compared to control 35 ppt (Dhaneesh et al. 2012a). Very high (53–55 ppt) and very low levels of salinity (3–6 ppt) resulted in increased mortality in juveniles (Dhaneesh et al. 2012a).	Very high (53–55 ppt) and very low levels of salinity (3–6 ppt) resulted in decreased appetite and movement in juveniles (Dhaneesh et al. 2012a).
Amphprion chrysopterus	**Local stressors**	Artificial light (4.3 lux) reduced growth in juveniles (Schligler et al. 2021).	Artificial light (4.3 lux) reduced survival of juveniles (Schligler et al. 2021). Boat noise caused increased cortisol levels and increased testosterone levels in adult males (Mills et al. 2020).	Boat noise caused fish to spend more time hiding, reduce distance travelled from the anemone, and increase aggression levels (Mills et al. 2020).
Amphirion frenatus	**Local stressors**			High sensitivity to cyanide, no survival after a 60-second dose of 50 mgL^{-1} (Madeira et al. 2020b).
Amphirion akindynos	**Temperature**		Peak reproduction occurred during summer from 24–25°C (Richardson et al. 1997).	

(Continued)

TABLE 23.1 (CONTINUED)
Sensitivity of Anemonefish Species to Global and Local Stressors Depending on the Type of Effect Observed (Morphological, Physiological, or Behavioral)

		Effect		
Species	**Stressor**	**Morphological**	**Physiological**	**Behavioral**
Amphirion latezonatus	**Temperature**		Maximum specific growth rate occurred between 22–23°C (Rushworth et al. 2011). Peak reproduction occurred during summer from 24–25°C (Richardson et al. 1997).	
Amphiprion ephippium	**Local stressors**		Embryonic heart rate increased when exposed to noise (400–700 Hz) (Simpson et al. 2005).	
Amphiprion rubrocintus	**Local stressors**		Embryonic heart rate increased when exposed to noise (400–700 Hz) (Simpson et al. 2005).	
Premnas biaculeatus	**Temperature**	Fish that developed in either +1.5 or +3°C above control temperatures until one year had enhanced growth (Donelson 2015).	Fish that developed in either +1.5 or +3°C above control temperatures until one year had enhanced aerobic metabolic capacity (Donelson 2015).	

Note: local stressors include sediment, light, noise, microplastics, and chemicals from aquarium collection.

of some marine fishes, but not others (Munday et al. 2019). When the eggs and larvae of *A. percula* were reared in elevated CO_2 (550, 750, and 1,030 ppm), there was no effect on embryonic duration, egg survival, or size at hatching compared with current-day controls (400 ppm) (Munday et al. 2009b). In contrast, elevated CO_2 tended to increase the growth rate of larvae, but this differed between clutches. At settlement, larvae from some parental pairs were 15–18% longer and 47–52% heavier in elevated CO_2 compared with controls, while larvae from other parents were unaffected (Munday et al. 2009b), indicating that responses may differ depending on genetic background. While larger larvae tended to swim faster, there was no direct effect of elevated CO_2 on maximum swimming speed at settlement (Munday et al. 2009b). Hence, in terms of larval growth, effects of elevated CO_2 are variable, but negative effects have not been observed. Elevated CO_2 during larval development has also been reported to affect otolith size, shape, and structure in some anemonefishes. Otoliths are a core part of the fish auditory system. Increased otolith roughness was observed in larval *Amphiprion clarkii* reared at 800 µatm (Holmberg et al. 2019), and increased otolith size in *A. percula* at 1,721 µatm, but not at 1,050 µatm, compared with control larvae reared at 404 µatm (Munday et al. 2011). Whether these changes in otolith size and structure have any ecological consequences is currently unknown, although impaired hearing of juvenile snapper reared in elevated CO_2 has recently been linked to increased fluctuating asymmetry in their otoliths (Radford et al. 2021).

More notable effects of elevated CO_2 have been seen on behavioral phenotypes of anemonefish larvae. *A. percula* was tested for its behavioral response, in particular to predator cues, in various studies mimicking future elevated CO_2 conditions (Dixson et al. 2010; Munday et al. 2010). Larvae became attracted to predator cues after four days of exposure to 1,000 ppm CO_2, whereas under normal conditions they would exhibit a strong avoidance. In elevated CO_2 conditions, fish need to regulate their acid-base balance to avoid acidosis. However, this may cause a Cl- and HCO3- imbalance, which in turn affects the main inhibitory neurotransmitter in the brain, the GABA-ergic pathway, and causes behavioral abnormalities (Nilsson et al. 2012). These behavioral impairments observed in *A. percula* with elevated CO_2 were reversed when gabazine, a GABA antagonist, was administered (Nilsson et al. 2012; Munday et al. 2016). Due to the relatively stable CO_2 levels of the pelagic environment, the stable CO_2 conditions used in these studies are relevant for anemonefish larvae in the future, and could potentially decrease survival in settling larvae.

Far less research has been conducted to understand how stressors other than elevated temperature and CO_2 may affect larval development in anemonefish. The only study evaluating changes in salinity on *Amphiprion akallopisos* found that larval survival decreased from 100% at normal 35 ppt to 88% at higher salinity of 40 ppt and to 76% with lower salinities of 20–25 ppt (Dhaneesh et al. 2012a). Considering the critical nature of environmental conditions during the larval stage for population sustainability, there is a surprising lack of research into how local stressors could impact this crucial life stage.

23.4.2 Juvenile Development

Juvenile development seems to be more robust to elevations in water temperature, compared to larval development. For example, *Amphirion ocellaris* exhibited limited physiological effects to a one-month exposure to 30°C compared with a control of 26°C (Madeira et al. 2017, 2021). After one week, thermal stress was observed in the muscle, with both heat shock protein 70 kDa and total ubiquitin increasing significantly (Madeira et al. 2017). However, with longer exposure, total ubiquitin levels decreased, indicating that fish were able to physiologically respond by maintaining high levels of Hsp70 and through an effective protein turnover, such that there was no increased mortality. In addition, analyses of body condition and lipid content of muscle and liver tissue of juvenile *A. ocellaris* showed that lipid storage capacity was not affected, fatty acid profiles were unresponsive, lipid metabolic networks were conserved, and fish weight was maintained at elevated temperature (Madeira et al. 2021). When combining the effects of elevated water temperature (control: 28°C, elevated: 32°C) with food availability (fed once or three times a day) for 52 days, again, *A. ocellaris* was observed to maintain growth at high temperature regardless of food level (Pham et al. 2021). However, elevated temperature did induce asymmetry in the weight of individual fish within the group tanks. Gene expression of genes involved in hormone activity and appetite, *pomca* and *agrpl,* correlated to patterns of higher food intake at elevated temperature (Pham et al. 2021). Interestingly, in *Amphirion melanopus*, juvenile food consumption and foraging activity increased with moderate warming (28.5–30°C), but then decreased with further temperature increase (31.5°C) (Nowicki et al. 2012). The spinecheek anemonefish, *Premnas biaculeatus*, has also been observed to be robust to elevated water temperature above the usual current-day (Donelson 2015). Fish that developed in either +1.5 or +3°C above control temperatures from settlement to one year old had enhanced aerobic metabolic capacity and growth.

Due to the prevalence of anemonefish in aquaculture, a number of studies have attempted to define the optimal thermal range for particular species. The sub-tropical wide-band anemonefish, *Amphirion latezonatus*, exhibited a maximum specific growth rate between 22–23°C, which is consistent with the sub-tropical distribution of this species (Rushworth et al. 2011). For *A. ocellaris*, the preferred temperature ranged from approximately 27–32°C depending on the holding temperate of the fish (from 20–35°C), while aerobic metabolic performance was maintained across 23–32°C (Velasco-Blanco et al. 2019), consistent with the tropical distribution of this species. Overall, this work suggests that juvenile anemonefishes may be relatively robust to ocean warming, with no performance declines up to 32°C. However, these estimates of thermal sensitivity are far lower than estimates that would be extrapolated from critical maximum limits, with the critical maximum for *A. ocellaris* 3.7–8.1°C warmer (35.7–40.1°C; Paschke et al. 2018; Velasco-Blanco et al. 2019).

In terms of ocean acidification, when elevated CO_2 levels are kept stable during juvenile development, growth rates (Jarrold and Munday 2019) and survival (Miller et al. 2012) have been observed to decline in some anemonefishes. However, these effects are mitigated by parental exposure to elevated CO_2 or the presence of diel CO_2 cycles in the elevated CO_2 treatments. Length and weight were lower in juvenile *A. melanopus* reared in stable elevated CO_2 (1,000 μatm) compared with controls (Jarrold and Munday 2019), as was observed in an earlier study (Miller et al. 2012). However, parental exposure to elevated CO_2 reversed the negative effects of elevated CO_2 on juvenile growth (Miller et al. 2012; Jarrold and Munday 2019) and survival (Miller et al. 2012), demonstrating the importance of multigenerational experiments in assessing the impacts of environmental stressors on fish populations. Furthermore, the negative effects of elevated CO_2 (1,000 μatm) on length and weight in juvenile *A. melanopus* were absent when there was a diel cycle in the elevated CO_2 treatment (1,000 ± 300 μatm). In contrast to *A. melanopus*, there was no detrimental effect of elevated CO_2, with or without the presence of a diel CO_2 cycle, on the growth and survival of juvenile *A. percula* (Jarrold and Munday 2018), indicating that juveniles of this species are broadly tolerant to elevated CO_2, as also observed in larvae of this species. Taken together, these studies suggest that elevated CO_2 is unlikely to substantially impact juvenile growth rates and survival in anemonefish populations under natural conditions, either because the species are tolerant to elevated CO_2, or because negative effects are likely to be buffered by parental exposure and/or daily CO_2 cycles that occur naturally in the coral reef habitats they occupy (Hannan et al. 2020).

As observed in the larval stage, elevated CO_2 can alter ecologically important behaviors in juvenile anemonefishes. In particular, elevated CO_2 caused a loss of response to predator cues (McMahon et al. 2018) and a reduced reaction to an auditory stimulus (Simpson et al. 2011) in juvenile *A. percula*. However, behavioral impairments were less severe when the elevated CO_2 fluctuated in diel cycles (Jarrold et al. 2017). This reduction in the altered behavioral response associated with elevated CO_2 was also seen in the molecular response of the brain in *A. percula*, where few gene expression changes were observed if CO_2 was cycling in a diel pattern, even if average CO_2 levels were higher than current-day levels (Schunter et al. 2021). Hence, while studies show that stable elevated CO_2 can alter the behavior of juvenile *A. percula*, more realistic experiments with future acidified conditions including diel cycles suggest that the effects on juveniles are unlikely to be substantial in natural populations of this species. Negative effects of elevated CO_2 on escape performance in juvenile *A. melanopus* were partially reversed when parents were exposed to elevated CO_2 (Allan et al. 2014), but parental exposure to elevated CO_2 did not have a substantive effect on the impaired response of juveniles to predator odour in elevated CO_2 (McMahon et al. 2019), suggesting that beneficial parental effects on behavioral responses to elevated CO_2 may be limited, as has been observed in other coral reef fishes (Welch et al. 2014).

Hyposalinity exposure has been intensively studied in anemonefishes, as it is used in the treatment of some diseases and can successfully be applied to decrease parasite load (Yokoyama and Shirakashi 2007). In *A. melanopus*, prolactin has been found to be a key enzyme associated with osmoregulation, and in particular salinities down to 15 ppt. When salinity was decreased, prolactin transcription increased in the liver and Na^+/K^+-ATPase expression increased in the gills (Park et al. 2011; Noh et al. 2013). These expression changes levelled off over time, showing an acclimation pattern over a few months to lower salinity levels. Nonetheless, plasma cortisol levels did increase and antioxidant and lip peroxidation increased in juveniles, revealing oxidative stress in the liver, at least after a hyposalinity exposure (Park et al. 2011). The only study evaluating increased salinity in *Amphirion akallopisos* exhibited no stress for juveniles, and only very high (53–55 ppt) and very low levels of salinity (3–6 ppt) resulted in decreased appetite and movement, and eventually mortality in juveniles (Dhaneesh et al. 2012a). Generally, this shows that anemonefish juveniles are tolerant to a range of salinities without substantive impacts on their physiology or survival.

Compared to other life stages, local stressors (with the exception of noise pollution) have been relatively well studied in the juvenile stage. Two studies have investigated the potential effects of suspended sediment on juvenile anemonefish, using relatively short exposures (seven days) to levels regularly occurring on inshore reefs during seasonal flooding events. When exposed to 135 and 180 mgL^{-1} of suspended sediment, the oxygen diffusion distance on gills was reduced in *A. percula*, but there was no impact on the capacity to uptake oxygen or on gill lamellae length (Hess et al. 2017). In contrast, *A. melanopus* were more sensitive to suspended sediment showing shorter gill lamellae, reduced oxygen diffusion distances, increased resting metabolic rates, and reductions in aerobic capacity (45, 90, 135, and 180 mgL^{-1}, Hess et al. 2017). Exposure to high levels of sediment (180 mgL^{-1} for 7 days) also affected escape performance, with fish more responsive, and producing more and enhanced escapes to startle stimuli (Hess et al. 2019). Even when fish were only acutely exposed to suspended sediment, they altered their behavior to be less active and avoid open areas of the testing arena (Hess et al. 2019). Artificial light pollution has also been shown to reduce survival and growth in juveniles of *Amphprion chrysopterus* (~20.5-month exposure, Schligler et al. 2021). In juvenile *A. clarkii*, an increase in UV radiation for up to 14 days increased apoptosis, which was correlated with an increase in reactive oxygen species enzymes (Ryu et al.

2022). Hence, both artificial light and increased solar radiation can have effects on the physiology and survival of juvenile anemone fishes.

Collection methods for the ornamental fish trade have the potential to cause unintended harm to anemonefish. Juvenile *Amphirion frenatus* and *A. clarkii* are highly sensitive to cyanide, with no survival after a 60-second dose of 50 mgL^{-1} (Madeira et al. 2020b). While ~50% of juvenile *A. ocellaris* were able to survive this dose and 100% were able to survive a lower dose of 25 mgL^{-1} at 26°C, the additional stress of elevated temperature (29 and 32°C) caused mortality to increase 60% and 20% for the lower doses (Madeira et al. 2020b). Microplastic pollution was only tested on juvenile *A. ocellaris*, revealing large variation in microplastic intake among individuals within the population, and increased uptake correlating with higher activity levels (Nanninga et al. 2020).

23.4.3 Adults, Reproduction, and Embryogenesis

Considering the broad knowledge of the temperature dependence of reproductive processes in fish (e.g., promoting or inhibiting hormone synthesis, altering hormone structure, and modifying the action of hormones and enzymes within the hypothalamo–pituitary–gonadal [HPG] axis; Pankhurst and Munday 2011), there is surprisingly little known about likely effects of future ocean warming on anemonefish reproduction. Timing of anemonefish reproduction can be affected by both water temperature and lunar period (Richardson et al. 1997; Madhu and Madhu 2007). For *A. latezonatus* and *Amphirion akindynos* from sub-tropical Australia, reproduction peaked when water temperature was warmest from 24 to 25°C (Richardson et al. 1997). For tropical *A. percula*, peak reproduction occurred at either 28.74°C (Dhaneesh et al. 2012b) or above 30.0°C (Madhu and Madhu 2007) depending on location. In contrast, spawning for temperate *A. clarkii* has been observed over a much wider temperature range (between 20.5 and 27°C) during spring and summer (Bell 1976; Ochi 1985). This tendency for reproduction to coincide with the warmest months of the year potentially makes it susceptible to ocean warming and marine heatwaves. Only one study has explored the effects of future projected ocean warming (+1.5: 30°C, +3.0: 31.5°C) on reproduction in *A. melanopus*. (Miller et al. 2015). Elevated temperature caused a decline in reproduction at 30°C, and reproduction ceased at 31.5°C, due to negative effects on plasma 17b-estradiol concentrations. Water temperature also affected the development rate of *A. ocellaris* embryos, with 29°C optimal, but lower temperatures (26–28°C) resulting in lower hatching efficiency, and higher temperature (30°C) leading to reduced yolk volume (Soman et al. 2021).

In contrast to higher temperatures, elevated CO_2 (~1,000 μatm) increased reproductive output in terms of number of clutches and number of eggs per clutch in both *A. melanopus* (Miller et al. 2013) and *A. percula* (Welch and Munday 2016), suggesting that elevated CO_2 could have a stimulatory effect on reproduction in these species. Furthermore, parental exposure to elevated CO_2 can alleviate some negative effects of ocean acidification on their progeny. For example, juvenile *A. melanopus* were smaller when reared in elevated CO_2, but not if their parents had also been exposed to elevated CO_2 (Miller et al. 2012). A similar result was observed by McMahon et al. (2019), who found that parental exposure to elevated CO_2 had a positive effect on length (but not weight) of juvenile *A. percula*.

To our knowledge, studies on local stressor impacts on adults and embryos have predominantly focussed on noise and light pollution. *Amphiprion ephippium* and *A. rubrocintus* embryos increased their heart rate when exposed to noise, and prolonged noise exposure resulted in a lower sensitivity threshold (Simpson et al. 2005). Playback of boat noise had a similar effect on the embryos of another species, *A. melanopus*, causing increased heart rate, although the yolk volume of embryos was not affected (Fakan and McCormick 2019). In adult *A. chrysopterus*, boat noises caused increased cortisol levels and increased testosterone levels in males (Mills et al. 2020). When exposed to boat noise, fish spent more time hiding, decreased the distance travelled from the anemone, and showed increased aggression levels (Mills et al. 2020). Artificial light at night (12 h of dim light, mean lx = 26.5) had no impact on the frequency of spawning or fertilization success of breeding *Amphiprion ocellaris* (Fobert et al. 2019). However, it had dramatic effects on reproductive output with not a single embryo successfully hatching (Fobert et al. 2019).

23.4.4 Symbiosis and Indirect Effects

Anemonefish have an obligate relationship with their anemone habitat that will indirectly affect their resilience to environmental change. Furthermore, this relationship between anemonefish and anemone is mutualistic, which may result in negative feedback when either party is affected (Figure 23.3). When anemones themselves become stressed, their symbiotic relationship with photosynthetic algae (Symbiodiniaceae) can be disrupted, resulting in loss of Symbiodiniaceae, reduced photosynthetic capacity, and bleaching (Weis 2008). Anemones are temperature and irradiance sensitive, with warming of only 1–3°C above normal summer temperatures resulting in bleaching (Hill and Scott 2012; Pryor et al. 2021; Hayashi and Reimer 2020), and elevated light levels (400 μmol photons m^{-2} s^{-1}) exacerbating the effects of temperature (Hill and Scott 2012). All ten species of anemone that host fish have been documented to bleach in nature (Hobbs et al. 2013). Bleaching has been most commonly observed in shallow areas during thermally induced mass bleaching events (Hobbs et al. 2013; Scott and Hoey 2017; Haguenauer et al. 2021), but some bleaching has been attributed to freshwater influx from rain events (Hobbs et al. 2013). Following natural bleaching, anemones can be reduced in both size and abundance (Hobbs et al. 2013) and may become absent from heavily impacted reef areas (Jones et al. 2018). A 34%

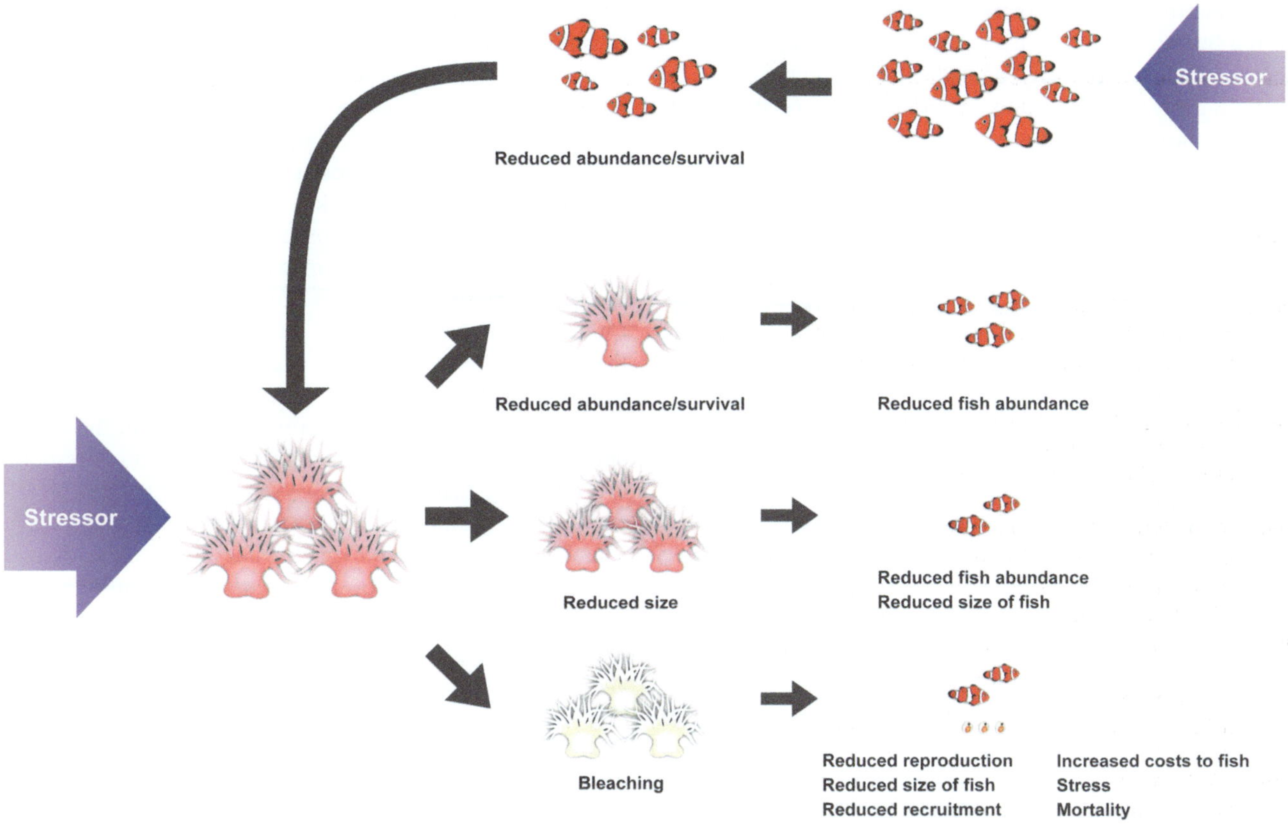

FIGURE 23.3 Pathways of direct and indirect effects of stressors on anemonefish and host anemones.

reduction in anemone size caused by a heatwave event coincided with lower levels of recruitment of the anemone fish *Amphirion polymnus* (Saenz-Agudelo et al. 2011). In contrast, elevated CO_2 has generally been observed to increase photosynthesis in anemones (Suggett et al. 2012; Hoadley et al. 2015) or have no effect in isolation (Pryor et al. 2021). Collection of anemonefish for the aquarium trade could exacerbate the direct effects of environmental stress on anemones. For example, the removal of *A. melanopus* or absence of *Amphiprion bicinctus* from *Entacmaea quadricolor* resulted in negative effects on growth, reproduction, and survival of anemones (Porat and Chadwick-Furman 2004; Frisch et al. 2016). This is likely due to the nutritional benefits fish provide the anemone through excretion and waste products (Holbrook and Schmitt 2004; Roopin and Chadwick 2009), and the protection from predators (Porat and Chadwick-Furman 2004). Effects such as these on the size and health of the host anemones can strongly affect the size and health of the associated anemonefish populations.

Larger anemones host larger groups that contain larger fish (Hattori 2006; Chausson et al. 2018). Following natural bleaching, anemones may reduce in size (~34%), and this can impact the fecundity (~38% reduction) of female *A. polymnus* (Saenz-Agudelo et al. 2011). Similarly, *A. chrysopterus* living on bleached anemones following a marine heatwave spawned less frequently and laid fewer eggs, which suffered higher mortality and produced fewer viable offspring (Beldade et al. 2017). The stress of reproductive adults was observed through elevated cortisol levels and a reduction in reproductive hormone levels (Beldade et al. 2017). A bleached host (*Heteractis crispa*) induced shifts in behavior for *A. akindynos*, with a lack of appropriate anti-predator response (i.e., no longer reducing feeding and sheltering) in the presence of predator stimuli, which resulted in reduced survival in situ during the 72 hours following bleaching (Lönnstedt and Frisch 2014). Interestingly, bleached hosts did not seem to deter *A. polymnus* or *A. latezonatus* juveniles from recruiting (Saenz-Agudelo et al. 2011; Scott and Dixson 2016), but *A. larkia*, *A. ocellaris*, *A. percula* and *P. biaculeatus* avoided bleached hosts (Scott and Dixson 2016). Shortly following experimental bleaching, juvenile *A. chrysopterus* had elevated standard metabolic rate (Norin et al. 2018), however, this effect weakened by four weeks post-bleaching, and fish had significantly reduced metabolic rates by eight weeks post-bleaching (Cortese et al. 2021). This measure of standard metabolic rate (measured at night) is unlikely to tell the full story of the cost of living in a bleached anemone, with more time spent outside the anemone, reduced activity, reduced growth, and reduced survival all indicating negative effects of host bleaching (Cortese et al. 2021). Since the relationship is mutualistic, it is perhaps unsurprising that resident fish can provide an advantage to bleached anemones in their recovery post-bleaching (Pryor et al. 2020). Depletion of anemonefish populations may take a substantial period of time for recovery (e.g., 10 years; Frisch et al. 2019), which will be problematic as stressors increase in diversity and frequency in the future. However, if anemones are available, there is the potential for recovery, as fish can preferentially recruit to vacant hosts (Elliott and Mariscal 2001), further highlighting the essential nature of anemones in the response of this group to environmental change.

The majority of studies investigating the effects of climate change and other environmental stressors on anemonefishes are lab-based, which do not account for mediating effects of biological factors such as predator type and abundance, food availability, and species interactions. More studies on wild fish would enable validation of the results from such studies. A single environmental factor is often studied in the laboratory, with the objective to isolate and understand its effects, whereas in the wild there will be a combination of environmental changes determining the fate of anemonefishes. The interaction between environmental elements will be important, but few studies include this complexity. For example, elevated CO_2 did not affect behavior in *A. melanopus*, but there were effects when combined with elevated temperature (Nowicki et al. 2012). More importantly, laboratory experiments rarely consider the effects of environmental stressors on anemonefishes and their host anemones and the mutualistic interaction between these two partners. A much greater focus on the entire mutualistic system is essential to predicting how this group of fish will respond to environmental stress. Field-based studies will likely be a powerful approach to studying the effects of individual and interacting stressors on the anemone–anemonefish mutualism that ultimately determines the fate of both organisms.

23.5 RESILIENCE AND ADAPTIVE POTENTIAL TO FUTURE ENVIRONMENTAL CHANGE

Change over time in nature is a normal and constant process, and important for evolution. However, it is the diversity of environmental modifications and accelerated pace of change by human activity that is currently shifting this natural ecological succession to an increased state of disturbance or even collapse (Cooper et al. 2020). The resilience of anemonefish, defined here as the ability of species to survive and maintain populations, can involve several processes by which populations may be sustained despite changing environmental conditions, including acclimation via phenotypic plasticity, adaptation, or movement (Figure 23.4).

The mutualistic relationship between anemonefish and anemone puts them at high risk of impacts from environmental change, since environmental stressors can impact both the host and the fish (Figure 23.3). While we have not explicitly reviewed the literature on direct effects of stressors on anemones, they appear to be the more sensitive party to thermal change, with marine heatwaves resulting in high mortality. The strong relationship between fish and habitat will likely mean a restricted ability of anemonefish to persist in the absence of habitat (Munday 2004), or as host quality declines (Froehlich et al. 2021), and limited capacity to simply switch hosts. Of the 28 species of anemonefish, nine species are known to associate with only one species of anemone, five species associate with two species of anemone, and the remaining 14 species associate with three or more species (Litsios et al. 2012). This poses a significant risk for those anemonefishes that associate with one or two host anemone species if the host themselves are unable to survive global and local stressors. Furthermore, collection of either anemones or fishes for the ornamental industry has the potential to impact the capacity for resilience of colonies and populations. Removal of fish can create cascading effects on recruitment and productivity of the remaining fish in the colony, depending on the number and size of fish that are removed (Frisch et al. 2016).

While the mutualistic requirements will play a large role in defining the impact on anemonefishes, there is also the possibility for acclimation to some level of environmental change through phenotypic plasticity. Phenotypic plasticity is traditionally defined as the capacity of a given genotype to render alternative phenotypes under different environmental conditions, but can be more broadly considered as environmentally induced phenotypic variation (Pigliucci 2001; West-Eberhard 2003). In relation to human-induced environmental change, plasticity is usually considered in terms of improving performance in altered conditions and is often measured at the individual level in terms of changes in behavior, physiology, or morphology (Angilletta 2009). Anemonefishes exhibit a large array of plasticity, including social group size and hierarchical queues, change in sex and size, a variety of changes to behaviors including parental care, and molecular adjustments to dealing with environmental change (e.g., Barbasch and Buston 2018; Barbasch et al. 2020). Plasticity may allow individuals to improve performance in a changed environment, and if these changes are passed across generations (e.g., inheritance, parental effects, imprinting), this may allow populations to be more resilient over time (Figure 23.4). For example, research on the effects of elevated temperature on juvenile anemonefish shows a high capacity for plasticity (Donelson 2015; Madeira et al. 2017). Ultimately though, adaptation is needed for populations to overcome trade-offs when adjusting to suboptimal environmental conditions. Rapid adaptation may be possible for anemonefishes if high-standing genetic variation is present in current populations, which previously allowed for adaptive speciation events (Litsios et al. 2012), and may indicate that there is potential for local adaptation. However, ornamental fish collection results in a loss of genetic diversity (Madduppa et al. 2018), which reduces the adaptive capacity of a population to deal with environmental changes. A study following a wild anemonefish population across ten years found little adaptive capacity in fitness and elevated susceptibility to local environmental change (Salles et al. 2020).

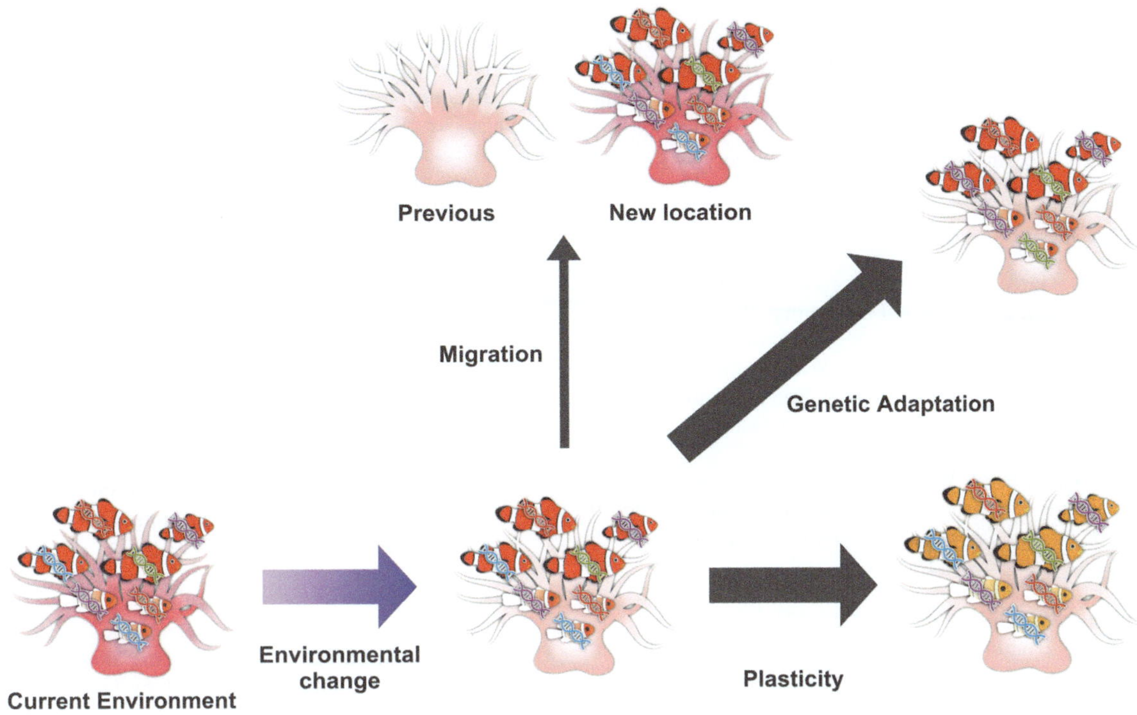

FIGURE 23.4 Different pathways of resilience. Arrow width represents likelihood based on current knowledge. DNA strand symbol and color represent genetic background. Colors of fish represent phenotype.

Finally, species have the potential to shift their location to avoid stressful local conditions. Current species ranges are determined by a diversity of biotic and abiotic factors (Cahill et al. 2014; Lewis et al. 2017; Stuart-Smith et al. 2017), and as such, all species may not readily shift in relation to environmental change. Patterns of range expansion are often related to ecological traits (Auer and King 2014; Bates et al. 2014), and a commonly observed pattern is for generalist species to have a greater capacity to shift (Stuart-Smith et al. 2021). For anemonefish, population establishment in a new area not only needs larval fish to arrive, but given the mutualism, range shifting requires anemones to also arrive (and likely before the fish). In fact, while shifting ranges in anemonefish have been shown to be possible, it is the absence of the preferred anemone species that is hindering this process (Pryor et al. 2022). Investigations into the potential for range shifts along the east coast of Australia revealed a need for vegetated islands and rocky islets to create suitable habitat, and their absences in sub-tropical and temperate regions could limit range shifting (Malcolm and Scott 2017). Additionally, the dispersal potential of anemonefishes may be restricted due to a relatively short pelagic larval phase in comparison to other damselfishes or reef fishes (Thresher et al. 1989; Frédérich and Parmentier 2015) and the high levels of self-recruitment (Jones et al. 2005), which might result in lower levels of population connectivity (Kubisch et al. 2013). Depth as a refuge also may be unlikely as heatwave events have even induced anemone bleaching on mesophotic reefs, albeit to a lesser extent (Haguenauer et al. 2021).

More research is warranted to understand anemonefishes' capacities to acclimate, adapt, or move. The reproductive biology and aquaculture interest in anemonefish (Chapter 22) have invoked their use as model species for aquarium-based research, including the capacity for plasticity and adaptation of reef fish to climate change. For example, anemonefishes are easily sourced from the wild, ample offspring are produced in a clutch, high larval survival is possible during captive rearing, multiple clutches are produced within a breeding season, and they have relatively short generations times, allowing them to be used for cross-generational studies (e.g., Miller et al. 2012; McMahon et al. 2018a). The high site-fidelity of anemones allows populations to be monitored through time and in response to pulse stressors events (Beldade et al. 2017). Multigenerational pedigree studies in nature have also been possible (e.g., Salles et al. 2020). However, some aspects such as benthic spawning, make anemonefish less optimal for experimental breeding designs to study heritability, where multiple males and females need to be cross-bred to produce half-sibling offspring (e.g., Anttila et al. 2014). While anemonefishes are a good study species for examining parental effects, phenotypic plasticity, and genetic diversity within populations, there has been a lack of studies that explore the mutualistic anemone–anemonefish system and its resilience as a whole. Anemonefishes are great models for coral reef fishes with high reliance on a particular habitat which is itself equally or even more sensitive to environmental change, leading to large potential to use anemonefishes and their hosts as small but complex systems to understand the impacts of rapid environmental change.

ACKNOWLEDGEMENTS

All the figures have been generated and credited to Kaori Serakaki from the Communication and Public Relations Division, Media Relations Section of the Okinawa Institute of Science and Technology & Graduate University (OIST).

REFERENCES

Allan, B. J. M., G. M. Miller, K. I. Mccormick, P. Domenici, and P. L. Munday. 2014. Parental effects improve escape performance of juvenile reef fish in a high-CO_2 world. *Proceedings of the Royal Society B: Biological Sciences* 281: 20132179.

Almany, G. R., and M. S. Webster. 2006. The predation gauntlet: Early post-settlement mortality in reef fishes. *Coral Reefs* 25: 19–22.

Amorim, M. C. P., R. O. Vasconcelos, and P. J. Fonseca. 2015. Fish sounds and mate choice. In *Sound Communication in Fishes. Animal Signals and Communication*, vol. 4, ed. F. Ladich, 1–33. Vienna: Springer.

Angilletta, M. J. J. 2009. *Thermal Adaptation: A Theoretical and Empirical Synthesis*. Oxford: Oxford University Press.

Anttila, K., C. S. Couturier, Ø. Øverli, A. Johnsen, G. Marthinsen, G. E. Nilsson, and A. P. Farrell. 2014. Atlantic salmon show capability for cardiac acclimation to warm temperatures. *Nature Communications* 51(5): 1–6.

Asch, R. G., C. A. Stock, and J. L. Sarmiento. 2019. Climate change impacts on mismatches between phytoplankton blooms and fish spawning phenology. *Global Change Biology* 25: 2544–2559.

Auer, S. K., and D. I. King. 2014. Ecological and life-history traits explain recent boundary shifts in elevation and latitude of western North American songbirds. *Global Ecology and Biogeography* 23: 867–875.

Barbasch, T. A., and P. M. Buston. 2018. Plasticity and personality of parental care in the clown anemonefish. *Animal Behaviour* 136: 65–73.

Barbasch, T. A., T. Rueger, M. Srinivasan, M. Y. L. Wong, G. P. Jones, and P. M. Buston. 2020. Substantial plasticity of reproduction and parental care in response to local resource availability in a wild clownfish population. *Oikos* 129: 1844–1855.

Bates, A. E., G. T. Pecl, S. Frusher, A. J. Hobday, T. Wernberg, D. A. Smale, J. M. Sunday, et al. 2014. Defining and observing stages of climate-mediated range shifts in marine systems. *Global Environmental Change* 26: 27–38.

Beldade, R., A. Blandin, R. O'Donnell, and S. C. Mills. 2017. Cascading effects of thermally-induced anemone bleaching on associated anemonefish hormonal stress response and reproduction. *Nature Communications* 8: 716.

Bell, L. 1976. Notes on the nesting success and fecundity of the anemonefish *Amphiprion clarkii* at Miyake-Jima, Japan. *Japanese Journal of Ichthyology* 22: 207–211.

Bindoff, N. W., W. L. Cheung, J. G. Kairo, J. Arístegui, V. A. Guinder, R. Hallberg, R. Hilmi, et al. 2019. Changing ocean, marine ecosystems, and dependent communities. In *IPCC Special Report on the Ocean and Cryosphere in a Changing Climate*, eds H. O. Pörtner et al. https://www.ipcc.ch/srocc/chapter/chapter-5/ (accessed February 10, 2022).

Bongaarts, J. 2019. IPBES, 2019. Summary for policymakers of the global assessment report on biodiversity and ecosystem services of the intergovernmental science-policy platform on biodiversity and ecosystem services. *Population and Development Review* 45(3): 680–681.

Booth, D. J., N. Bond, P. Macreadie, D. J. Booth, N. Bond, and P. Macreadie. 2011. Detecting range shifts among Australian fishes in response to climate change. *Marine and Freshwater Research* 62: 1027–1042.

Brauner, C. J., R. B. Shartau, C. Damsgaard, A. J. Esbaugh, R. W. Wilson, and M. Grosell. 2019. Acid-base physiology and CO2 homeostasis: Regulation and compensation in response to elevated environmental CO2. *Fish Physioloogy* 37: 69–132.

Brown, C. J., M. I. Saunders, H. P. Possingham, and A. J. Richardson. 2013. Managing for interactions between local and global stressors of ecosystems. *PLoS One* 8: e65765.

Cahill, A. E., M. E. Aiello-Lammens, M. C. Fisher-Reid, X. Hua, C. J. Karanewsky, H. Y. Ryu, G. C. Sbeglia, et al. 2014. Causes of warm-edge range limits: Systematic review, proximate factors and implications for climate change. *Journal of Biogeography* 41: 429–442.

Cattano, C., J. Claudet, P. Domenici, and M. Milazzo. 2018. Living in a high CO2 world: A global meta-analysis shows multiple trait-mediated fish responses to ocean acidification. *Ecology Monographs* 88: 320–335.

Chausson, J., M. Srinivasan, and G. P. Jones. 2018. Host anemone size as a determinant of social group size and structure in the orange clownfish (*Amphiprion percula*). *PeerJ* 6: e5841.

Clements, J., and H. Hunt. 2015. Marine animal behaviour in a high CO2 ocean. *Marine Ecology Progress Series* 536: 259–279.

Cooper, G. S., S. Willcock, and J. A. Dearing. 2020. Regime shifts occur disproportionately faster in larger ecosystems. *Nature Communications* 111(11): 1–10.

Cortese, D., T. Norin, R. Beldade, A. Crespel, S. S. Killen, and S. C. Mills. 2021. Physiological and behavioural effects of anemone bleaching on symbiont anemonefish in the wild. *Functional Ecology* 35: 663–674.

Dahlke, F. T., S. Wohlrab, M. Butzin, and H.-O. Pörtner. 2020. Thermal bottlenecks in the life cycle define climate vulnerability of fish. *Science* 369: 65–70.

Depledge, M. H., C. A. J. Godard-Codding, and R. E. Bowen. 2010. Light pollution in the sea. *Marine Pollution Bulletin* 60: 1383–1385.

Dhaneesh, K. V., K. Nanthini Devi, T. T. Ajith Kumar, T. Balasubramanian, and K. Tissera. 2012a. Breeding, embryonic development and salinity tolerance of Skunk clownfish Amphiprion akallopisos. *Journal of King Saud University - Science* 24: 201–209.

Dhaneesh, K. V., T. T. Ajith Kumar, G. Swagat, T. Balasubramanian. 2012b. Breeding and mass scale rearing of clownfish Amphiprion percula: Feeding and rearing in brackishwater. *Chinese Journal of Oceanology and Limnology* 30: 528–534.

Dixson, D. L., P. L. Munday, and G. P. Jones. 2010. Ocean acidification disrupts the innate ability of fish to detect predator olfactory cues. *Ecology Letters* 13: 68–75.

Donelson, J. M. 2015. Development in a warm future ocean may enhance performance in some species. *Journal of Experimental Marine Biology and Ecology* 472: 119–125.

Durack, P. J., S. E. Wijffels, and R. J. Matear. 2012. Ocean salinities reveal strong global water cycle intensification during 1950 to 2000. *Science* 336: 455–458.

Edwards, A. J., and A. D. Shepherd. 1992. Environmental implications of aquarium-fish collection in the Maldives, with proposals for regulation. *Environmental Conservation* 31: 339–348.

Elliott, J. K., and R. N. Mariscal. 2001. Coexistence of nine anemonefish species: Differential host and habitat utilization, size and recruitment. *Marine Biology* 138: 23–36.

Fabricius, K. E. 2005. Effects of terrestrial runoff on the ecology of corals and coral reefs: Review and synthesis. *Marine Pollution Bulletin* 50: 125–146.

Fakan, E. P., and M. I. McCormick. 2019. Boat noise affects the early life history of two damselfishes. *Marine Pollution Bulletin* 141: 493–500.

Fobert, E. K., K. Burke Da Silva, and S. E. Swearer. 2019. Artificial light at night causes reproductive failure in clownfish. *Biology Letters* 15: 20190272.

Frédérich, B., and E. Parmentier. 2015. *Biology of Damselfishes*. Boca Raton: CRC Press/Taylor & Francis.

Frisch, A. J., J. P. A. Hobbs, S. T. Hansen, D. H. Williamson, M. C. Bonin, G. P. Jones, and J. R. Rizzari. 2019. Recovery potential of mutualistic anemone and anemonefish populations. *Fisheries Research* 218: 1–9.

Frisch, A. J., J. R. Rizzari, K. P. Munkres, and J. P. A. Hobbs. 2016. Anemonefish depletion reduces survival, growth, reproduction and fishery productivity of mutualistic anemone–anemonefish colonies. *Coral Reefs* 35: 375–386.

Froehlich, C. Y. M., O. S. Klanten, M. L. Hing, M. Dowton, and M. Y. L. Wong. 2021. Uneven declines between corals and cryptobenthic fish symbionts from multiple disturbances. *Scientific Reports* 11: 1–10.

Frölicher, T. L., E. M. Fischer, and N. Gruber. 2018. Marine heatwaves under global warming. *Nature* 560: 360–364.

Gasparini, J. L., S. R. Floeter, C. E. L. Ferreira, and I. Sazima. 2005. Marine ornamental trade in Brazil. *Biodiversity and Conservation* 14: 2883–2899.

Gerland, P., A. E. Raftery, H. Sevcikova, N. Li, D. Gu, T. Spoorenberg, L. Alkema, et al. 2014. World population stabilization unlikely this century. *Science* 346: 234–237.

Green, B. S., and R. Fisher. 2004. Temperature influences swimming speed, growth and larval duration in coral reef fish larvae. *Journal of Experimental Marine Biology and Ecology* 299: 115–132.

Häder, D., H. Kumar, R. Smith, and R. Worrest. 2007. Effects of solar UV radiation on aquatic ecosystems and interactions with climate change. *Photochemical and Photobiological Sciences* 6: 267–285.

Haguenauer, A., G. Siu, D. Cortese, R. Beldade, and S. C. Mills. 2021. Deep heat : A comparison of water temperature, anemone bleaching, anemonefish density and reproduction between shallow and mesophotic reefs. *Fishes* 6(3): 37.

Halpern, B., S. Walbridge, K. Selkoe, C. Kappel, F. Micheli, C. D'Agrosa, J. Bruno, et al. 2008. A global map of human impact on marine ecosystems. *Science* 319: 948–952.

Hannan, K. D., G. M. Miller, S. A. Watson, J. L. Rummer, K. Fabricius, and P. L. Munday. 2020. Diel pCO_2 variation among coral reefs and microhabitats at Lizard Island, Great Barrier Reef. *Coral Reefs* 39(243): 1–16.

Hattori, A. 2006. Vertical and horizontal distribution patterns of the giant sea anemone Heteractis crispa with symbiotic anemonefish on a fringing coral reef. *Journal of Ethology* 24: 51–57.

Hayashi, K., and J. D. Reimer. 2020. Five-year study on the bleaching of anemonefish-hosting anemones (Cnidaria: Anthozoa: Actiniaria) in subtropical Okinawajima Island. *Regional Studies in Marine Science* 35: 101240.

Hays, G., A. Richardson, and C. Robinson. 2005. Climate change and marine plankton. *Trends in Ecology and Evolution* 20: 337–344.

Hess, S., B. J. M. Allan, A. S. Hoey, M. D. Jarrold, A. S. Wenger, and J. L. Rummer. 2019. Enhanced fast-start performance and anti-predator behaviour in a coral reef fish in response to suspended sediment exposure. *Coral Reefs* 38: 103–108.

Hess, S., L. J. Prescott, A. S. Hoey, S. A. McMahon, A. S. Wenger, and J. L. Rummer. 2017. Species-specific impacts of suspended sediments on gill structure and function in coral reef fishes. *Proceedings of the Royal Society B: Biological Sciences* 284: 20171279.

Hildebrand, J. 2009. Anthropogenic and natural sources of ambient noise in the ocean. *Marine Ecology Progress Series* 395: 5–20.

Hill, R., and A. Scott. 2012. The influence of irradiance on the severity of thermal bleaching in sea anemones that host anemonefish. *Coral Reefs* 31: 273–284.

Hoadley, K. D., D. Rollison, D. T. Pettay, and M. E. Warner. 2015. Differential carbon utilization and asexual reproduction under elevated pCO2 conditions in the model anemone, Exaiptasia pallida, hosting different symbionts. *Limnology and Oceanography* 60: 2108–2120.

Hobbs, J.-P. A., A. J. Frisch, B. M. Ford, M. Thums, P. Saenz-Agudelo, K. A. Furby, and M. L. Berumen. 2013. Taxonomic, spatial and temporal patterns of bleaching in anemones inhabited by anemonefishes. *PLoS One* 8: e70966.

Hoegh-Guldberg, O., C. Rongshuo, E. S. Poloczanska, P. G. Brewer, S. Sundby, K. Hilmi, V. J. Fabry, S. Jung 2014. The ocean. *Climate Change 2014: Impacts, Adaptation, and Vulnerability. Part B: Regional Aspects. Contribution of Working Group II to the Fifth Assessment Report*, Intergovernmental Panel on Climate Change, 1655–1731. Cambridge: Cambridge University Press.

Hoey, A. S., and M. I. McCormick. 2004. Selective predation for low body condition at the larval-juvenile transition of a coral reef fish. *Oecologia* 139: 23–29.

Hofmann, G. E., and A. E. Todgham. 2010. Living in the now: Physiological mechanisms to tolerate a rapidly changing environment. *Annual Review of Physiology* 72: 127–145.

Holbrook, S. J., and R. J. Schmitt. 2004. Growth, reproduction and survival of a tropical sea anemone (Actiniaria): Benefits of hosting anemonefish. *Coral Reefs* 241(24): 67–73.

Holmberg, R. J., E. Wilcox-Freeburg, A. L. Rhyne, M. F. Tlusty, A. Stebbins, S. W. N. Nye Jr., A. Honig, et al. 2019. Ocean acidification alters morphology of all otolith types in Clark's anemonefish (Amphiprion clarkii). *PeerJ* 7: e6152.

Illing, B., A. T. Downie, M. Beghin, and J. L. Rummer. 2020. Critical thermal maxima of early life stages of three tropical fishes: Effects of rearing temperature and experimental heating rate. *Journal of Thermal Biology* 90: 102582.

Jarrold, M. D., C. Humphrey, M. I. McCormick, and P. L. Munday. 2017. Diel CO2 cycles reduce severity of behavioural abnormalities in coral reef fish under ocean acidification. *Scientific Reports* 7: 1–9.

Jarrold, M. D., and P. L. Munday. 2018. Diel CO2 cycles do not modify juvenile growth, survival and otolith development in two coral reef fish under ocean acidification. *Marine Biology* 165: 49.

Jarrold, M. D., and P. L. Munday. 2019. Diel CO_2 cycles and parental effects have similar benefits to growth of a coral reef fish under ocean acidification. *Biology Letters* 15: 20180724.

Jones, A. M., S. Gardner, and W. Sinclair. 2018. Losing 'nemo': Bleaching and collection impacts inshore populations of anemonefish and their hosts the sea anemones. *Fish Biology* 73(3): 753–761.

Jones, G. P., S. Planes, and S. R. Thorrold. 2005. Coral reef fish larvae settle close to home. *Current Biology* 15: 1314–1318.

Kroeker, K. J., R. L. Kordas, R. Crim, I. E. Hendriks, L. Ramajo, G. S. Singh, C. M. Duarte, et al. 2013. Impacts of ocean acidification on marine organisms: Quantifying sensitivities and interaction with warming. *Global Change Biology* 19: 1884–1896.

Kroon, F. J., P. M. Kuhnert, B. L. Henderson, S. N. Wilkinson, A. Kinsey-Henderson, B. Abbott, J. E. Brodie, and R. D. R. Turner. 2012. River loads of suspended solids, nitrogen, phosphorus and herbicides delivered to the Great Barrier Reef lagoon. *Marine Pollution Bulletin* 65: 167–181.

Kubisch, A., T. Degen, T. Hovestadt, and H. J. Poethke. 2013. Predicting range shifts under global change: The balance between local adaptation and dispersal. *Ecography* 36: 873–882.

Kunc, H. P., K. E. McLaughlin, and R. Schmidt. 2016. Aquatic noise pollution: Implications for individuals, populations, and ecosystems. *Proceedings of the Royal Society B: Biological Sciences* 283: 20160839.

Ladich, F. 1997. Agonistic behaviour and significance of sounds in vocalizing fish. *Marine and Freshwater Behaviour and Physiology* 29: 87–108.

Lewis, J. S., M. L. Farnsworth, C. L. Burdett, D. M. Theobald, M. Gray, and R. S. Miller. 2017. Biotic and abiotic factors predicting the global distribution and population density of an invasive large mammal. *Scientific Reports* 7: 44152.

Litsios, G., C. A. Sims, R. O. Wüest, P. B. Pearman, N. E. Zimmermann, and N. Salamin. 2012. Mutualism with sea anemones triggered the adaptive radiation of clownfishes. *BMC Evolutionary Biology* 12: 212.

Lönnstedt, O., and A. Frisch. 2014. Habitat bleaching disrupts threat responses and persistence in anemonefish. *Marine Ecology Progress Series* 517: 265–270.

Madduppa, H. H., J. Timm, and M. Kochzius. 2018. Reduced genetic diversity in the clown anemonefish Amphiprion ocellaris in exploited reefs of Spermonde Archipelago, Indonesia. *Frontiers in Marine Science* 5: 80.

Madduppa, H. H., K. von Juterzenka, M. Syakir, and M. Kochzius. 2014. Socio-economy of marine ornamental fishery and its impact on the population structure of the clown anemonefish Amphiprion ocellaris and its host anemones in Spermonde Archipelago, Indonesia. *Ocean and Coastal Management* 100: 41–50.

Madeira, C., D. Madeira, M. S. Diniz, H. N. Cabral, and C. Vinagre. 2017. Comparing biomarker responses during thermal acclimation: A lethal vs non-lethal approach in a tropical reef clownfish. *Comparative Biochemistry and Physiology Part A: Molecular & Integrative Physiology* 204: 104–112.

Madeira, C., D. Madeira, N. Ladd, C. J. Schubert, M. S. Diniz, C. Vinagre, and M. C. Leal. 2021. Conserved fatty acid profiles and lipid metabolic pathways in a tropical reef fish exposed to ocean warming – An adaptation mechanism of tolerant species? *Science of the Total Environment* 782: 146738.

Madeira, D., C. Madeira, P. M. Costa, C. Vinagre, H.-O. Pörtner, and M. S. Diniz. 2020a. Different sensitivity to heatwaves across the life cycle of fish reflects phenotypic adaptation to environmental niche. *Marine Environmental Research* 162: 105192.

Madeira, D. S., J. Andrade, M. C. Leal, V. Ferreira, R. J. M. Rocha, R. Rosa, and R. Calado. 2020b. Synergistic effects of ocean warming and cyanide poisoning in an ornamental tropical reef fish. *Frontiers in Marine Science* 7: 1–11.

Madhu, K., and R. Madhu. 2007. Influence of lunar rhythm on spawning of clown anemonefish Amphiprion percula under captive condition in Andaman and Nicobar islands. *Journal of Marine Biology Association of India* 49: 58–64.

Malcolm, H., and A. Scott. 2017. Range extensions in anemonefishes and host sea anemones in eastern Australia: Potential constraints to tropicalisation. *Marine and Freshwater Research* 68: 1224–1232.

Mann, D. A., and P. S. Lobel. 1997. Propagation of damselfish (Pomacentridae) courtship sounds. *The Journal of the Acoustical Society of America* 101: 3783–3791.

Masson-Delmotte, V., P. Zhai, A. Pirani, S. L. Connors, C. Péan, S. Berger, N. Caud, et al. 2021. *IPCC, 2021: Summary for Policymakers.* Cambridge: Cambridge University Press.

McLeod, I. M., and T. D. Clark. 2016. Limited capacity for faster digestion in larval coral reef fish at an elevated temperature. *PLoS One* 11: 1–13.

McLeod, I. M., J. L. Rummer, T. D. Clark, G. P. Jones, M. I. McCormick, A. S. Wenger, and P. L. Munday. 2013. Climate change and the performance of larval coral reef fishes: The interaction between temperature and food availability. *Conservation Physiology* 1: 1–12.

McMahon, S. J., J. M. Donelson, and P. L Munday. 2018. Food ration does not influence the effect of elevated CO2 on antipredator behaviour of a reef fish. *Marine Ecology Progress Series* 586: 155–165.

McMahon, S. J., P. L. Munday, M. Y. L. Wong, and J. M. Donelson. 2019. Elevated CO2 and food ration affect growth but not the size-based hierarchy of a reef fish. *Scientific Reports* 9: 19706.

Mensinger, A. F. 2013. Disruptive communication: Stealth signaling in the toadfish. *Journal of Experimental Biology* 217(3): 344–350.

Metcalfe, N. B., and P. Monaghan. 2001. Compensation for a bad start: Grow now, pay later? *Trends in Ecology and Evolution* 16: 254–260.

Miller, G., S. Watson, M. McCormick, and P. Munday. 2013. Increased CO2 stimulates reproduction in a coral reef fish. *Global Change Biology* 19: 3037–3045.

Miller, G. M., F. J. Kroon, S. Metcalfe, and P. L. Munday. 2015. Temperature is the evil twin: Effects of increased temperature and ocean acidification on reproduction in a reef fish. *Ecological Applications* 25: 603–620.

Miller, G. M., S.-A. Watson, J. M. Donelson, M. I. McCormick, and P. L. Munday. 2012. Parental environment mediates impacts of increased carbon dioxide on a coral reef fish. *Nature Climate Change* 2: 858–861.

Milliman, J. D., and R. H. Meade. 1983. World-wide delivery of river sediment to the oceans. *Journal of Geology* 91: 1–21.

Mills, S. C., R. Beldade, L. Henry, D. Laverty, S. L. Nedelec, S. D. Simpson, and A. N. Radford. 2020. Hormonal and behavioural effects of motorboat noise on wild coral reef fish. *Environmental Pollution* 262: 114250.

Mohammed, A. 2013. Why are early life stages of aquatic organisms more sensitive to toxicants than adults? In *New Insights into Toxicity and Drug Testing,* ed. S. J. T. Gowder, pp. 49–62 Fudan: InTech,.

Munday, P. L. 2004. Habitat loss, resource specialization, and extinction on coral reefs. *Global Change Biology* 10: 1642–1647.

Munday, P. L., D. L. Dixson, J. M. Donelson, G. P. Jones, M. S. Pratchett, G. V. Devitsina, and K. B. Døving. 2009a. Ocean acidification impairs olfactory discrimination and homing ability of a marine fish. *Proceedings of theNational Academy of Sciences of the United States of America* 106: 1848–1852.

Munday, P. L., D. L. Dixson, M. I. McCormick, M. Meekan, M. C. O. Ferrari, and D. P. Chivers. 2010. Replenishment of fish populations is threatened by ocean acidification. *Proceedings of the National Academy of Sciences of the United States of America* 107: 12930–12934.

Munday, P. L., J. M. Donelson, D. L. Dixson, and G. G. K. Endo. 2009b. Effects of ocean acidification on the early life history of a tropical marine fish. *Proceedings of the National Academy of Sciences of the United States of America* 276: 3275–3283.

Munday, P. L., V. Hernaman, D. L. Dixson, and S. R.Thorrold. 2011. Effect of ocean acidification on otolith development in larvae of a tropical marine fish. *Biogeosciences* 8: 1631–1641.

Munday, P. L., M. D. Jarrold, and I. Nagelkerken. 2019. Ecological effects of elevated CO2 on marine and freshwater fishes: From individual to community effects. *Fish Physiology* 37: 323–368.

Munday, P. L., M. J. Welch, B. J. M. Allan, S.-A. Watson, S. J. McMahon, and M. I. McCormick. 2016. Effects of elevated CO2 on predator avoidance behaviour by reef fishes is not altered by experimental test water. *PeerJ* 4: e2501.

Nagelkerken, I., and P. L. Munday. 2016. Animal behaviour shapes the ecological effects of ocean acidification and warming: Moving from individual to community-level responses. *Global Change Biology* 22: 974–989.

Nanninga, G. B., A. Scott, and A. Manica. 2020. Microplastic ingestion rates are phenotype-dependent in juvenile anemonefish. *Environmental Pollution* 259: 113855.

Nilsson, G. E., D. L. Dixson, P. Domenici, M. I. McCormick, C. Sørensen, S. A. Watson, and P. L. Munday. 2012. Near-future carbon dioxide levels alter fish behaviour by interfering with neurotransmitter function. *Nature Climate Change* 2: 201–204.

Noh, G. E., S. Rho, Y. J. Chang, B. H. Min, and J.-M. Kim. 2013. Gene encoding prolactin in cinnamon clownfish Amphiprion melanopus and its expression upon acclimation to low salinities. *Aquatic Biosystems* 9: 1.

Norin, T., S. C. Mills, A. Crespel, D. Cortese, S. S. Killen, and R. Beldade. 2018. Anemone bleaching increases the metabolic demands of symbiont anemonefish. *Proceedings of the Royal Society B: Biological Sciences* 285: 20180282.

Nowicki, J. P., G. M. Miller, and P. L. Munday. 2012. Interactive effects of elevated temperature and CO2 on foraging behavior of juvenile coral reef fish. *Journal of Experimental Marine Biology and Ecology* 412: 46–51.

Ochi, H. 1985. Temporal patterns of breeding and larval settlement in a temperate population of the tropical anemonefish, Amphiprion clarkii m ost coral fishes have a pelagic larval stage lasting scores of days after spawning or hatching, after which larval settlement. *Journal of Ichthyology* 32: 248–257.

Oliver, E. C. J., M. T. Burrows, M. G. Donat, A. Sen Gupta, L. V. Alexander, S. E. Perkins-Kirkpatrick, J. A. Benthuysen, et al. 2019. Projected marine heatwaves in the 21st century and the potential for ecological impact. *Frontiers in Marine Science* 6: 734.

Oliver, E. C. J., M. G. Donat, M. T. Burrows, P. J. Moore, D. A. Smale, L. V. Alexander, J. A. Benthuysen, et al. 2018. Longer and more frequent marine heatwaves over the past century. *Nature Communications* 9: 1324.

Pankhurst, N. W., and P. L. Munday. 2011. Effects of climate change on fish reproduction and early life history stages. *Marine and Freshwater Research* 62: 1015–1026.

Park, M. S., H. S. Shin, C. Y. Choi, N. N. Kim, D.-W. Park, G.-S. Kil, and J. Lee. 2011. Effect of hypoosmotic and thermal stress on gene expression and the activity of antioxidant enzymes in the cinnamon clownfish, *Amphiprion melanopus. Animal Cells and Systems* 15(3): 219–225.

Parmentier, E., O. Colleye, and D. Mann. 2009. Hearing ability in three clownfish species. *Journal of Experimental Biology* 212: 2023–2026.

Paschke, K., J. Agüero, P. Gebauer, F. Díaz, and M. Mascaró. 2018. Comparison of aerobic scope for metabolic activity in aquatic ectotherms with temperature related metabolic stimulation : A novel approach for aerobic power budget. *Frontiers in Physiology* 9: 1438.

Perkins-Kirkpatrick, S. E., and S. C. Lewis. 2020. Increasing trends in regional heatwaves. *Nature Communications* 11: 3357.

Pham, L. P., A. E. O. Jordal, M. V. Nguyen, and I. Rønnestad. 2021. Food intake, growth, and expression of neuropeptides regulating appetite in clown anemonefish (Amphiprion ocellaris) exposed to predicted climate changes. *General and Comparative Endocrinology* 304: 11371

Pigliucci, M. 2001. *Phenotypic Plasticity: Beyond Nature and Nurture*. Baltimore: John Hopkins University Press.

Poloczanska, E. S., C. J. Brown, W. J. Sydeman, W. Kiessling, D. S. Schoeman, P. J. Moore, K. Brander, et al. 2013. Global imprint of climate change on marine life. *Nature Climate Change* 3: 919–925.

Porat, D., and N. E. Chadwick-Furman. 2004. Effects of anemonefish on giant sea anemones: Expansion behavior, growth, and survival. In *Coelenterate Biology 2003*, eds D. G. Fautin, et al., 513–520. Dordrecht: Springer Netherlands.

Pörtner, H.-O., D. C. Roberts, V. Masson-Delmotte, P. Zhai, M. Tignor, E. Poloczanska, K. Mintenbeck, et al. 2019. *IPCC, 2019: Summary for Policymakers.* In *Climate Change and Land: An IPCC Special Report on Climate Change, Desertification, Land Degradation, Sustainable Land Management, Food Security, and Greenhouse Gas Fluxes in Terrestrial Ecosystems.* Geneva: IPCC Geneva.

Pörtner, H. O., and A. Farrell. 2008. Physiology and climate change. *Science* 322(5902): 690–692.

Pratchett, M. S., C. A. Thompson, A. S. Hoey, P. F. Cowman, and S. K. Wilson. 2018. Effects of coral bleaching and coral loss on the structure and function of reef fish assemblages. In *Coral Bleaching. Ecological Studies Pattern, Processes, Causes and Consequences*, vol. 233, eds M. van Oppen and J. Lough, 265–294 New York: Springer.

Pryor, S. H., L. Andrews, B. P. Kelaher, A. Tagliafico, and A. Scott. 2021. Ocean temperature, but not acidification, causes sea anemone bleaching under a near-future climate scenario. *Coral Reefs* 40: 355–364.

Pryor, S. H., R. Hill, D. L. Dixson, N. J. Fraser, B. P. Kelaher, and A. Scott. 2020. Anemonefish facilitate bleaching recovery in a host sea anemone. *Scientific Reports* 10: 18586.

Pryor, S. H., H. A. Malcolm, B. P. Kelaher, K. L. Davis, and A. Scott. 2022. Habitat limits the poleward establishment of anemonefishes in a climate change hotspot. *Estuarine, Coastal and Shelf Science* 264: 107662.

Putland, R. L., N. D. Merchant, A. Farcas, and C. A. Radford. 2018. Vessel noise cuts down communication space for vocalizing fish and marine mammals. *Global Change Biology* 24: 1708–1721.

Radford, C. A., S. P. Collins, P. L. Munday, and D. Parsons. 2021. Ocean acidification effects on fish hearing. *Proceedings of the Royal Society B: Biological Sciences* 288: 20202754.

Richardson, D. L., P. L. Harrison, and V. J. Harriott. 1997. Timing of spawning and fecundity of a tropical and subtropical anemonefish (*Pomacentridae: Amphiprion*) on a high latitude reef on the east coast of Australia *Marine Ecology Progress Series* 156: 175–181.

Rodgers, K. S., M. P. Stefanak, A. O. Tsang, J. J. Han, A. T. Graham, and Y. O. Stender. 2021. Impact to coral reef populations at Hā'ena and Pila'a, Kaua'i, following a record 2018 freshwater flood event. *Diversity* 13: 1–27.

Roopin, M., and N. E. Chadwick. 2009. Benefits to host sea anemones from ammonia contributions of resident anemonefish. *Journal of Experimental Marine Biology and Ecology* 370: 27–34.

Rushworth, K. J. W., S. D. A. Smith, K. L. Cowden, and S. W. Purcell. 2011. Optimal temperature for growth and condition of an endemic subtropical anemonefish. *Aquaculture* 318: 479–482.

Ryu, T., M. Herrera, B. Moore, M. Izumiyama, E. Kawai, V. Laude, and T. Ravasi. 2022. A chromosome-scale genome assembly of the false clownfish, *Amphiprion ocellaris. G3 Genes|Genomes|Genetics* 12(5): jkac074.

Saenz-Agudelo, P., G. P. Jones, S. R. Thorrold, and S. Planes. 2011. Detrimental effects of host anemone bleaching on anemonefish populations. *Coral Reefs* 302(30): 497–506.

Salles, O. C., G. R. Almany, M. L. Berumen, G. P. Jones, P. Saenz-Agudelo, M. Srinivasan, S. R. Thorrold, et al. 2020. Strong habitat and weak genetic effects shape the lifetime reproductive success in a wild clownfish population. *Ecology Letters* 23: 265–273.

Santana, M. F. M., A. L. Dawson, C. A. Motti, L. van Herwerden, C. Lefevre, and F. J. Kroon. 2021. Ingestion and depuration of microplastics by a planktivorous coral reef fish, *Pomacentrus amboinensis. Frontiers in Environmental Science* 9: 641135.

Schligler, J., D. Cortese, R. Beldade, S. E. Swearer, and S. C. Mills. 2021. Long-term exposure to artificial light at night in the wild decreases survival and growth of a coral reef fish. *Proceedings of the Royal Society B: Biological Sciences* 288: 20210454.

Schunter, C., M. D. Jarrold, P. L. Munday, and T. Ravasi. 2021. Diel pCO_2 fluctuations alter the molecular response of coral reef fishes to ocean acidification conditions. *Molecular Ecology* 30(20): 5105–5118.

Scott, A., and D. L. Dixson. 2016. Reef fishes can recognize bleached habitat during settlement: Sea anemone bleaching alters anemonefish host selection. *Proceedings of the Royal Society B: Biological Sciences* 283: 20152694.

Scott, A., and A. S. Hoey. 2017. Severe consequences for anemonefishes and their host sea anemones during the 2016 bleaching event at Lizard Island, Great Barrier Reef. *Coral Reefs* 36: 873–873.

Searcy, S. P., and S. Sponaugle. 2001. Selective mortality during the larval-juvenile transition in two coral reef fishes. *Ecology* 82: 2452–2470.

Sheridan, C., P. Grosjean, J. Leblud, C. V. Palmer, A. Kushmaro, and I. Eeckhaut. 2014. Sedimentation rapidly induces an immune response and depletes energy stores in a hard coral. *Coral Reefs* 33: 1067–1076.

Shuman, C. S., G. Hodgson, and R. F. Ambrose. 2005. Population impacts of collecting sea anemones and anemonefish for the marine aquarium trade in the Philippines. *Coral Reefs* 24, 564–573.

Simpson, S. D., P. L. Munday, M. L. Wittenrich, R. Manassa, D. L. Dixson, M. Gagliano, and H. Y. Yan. 2011. Ocean acidification erodes crucial auditory behaviour in a marine fish. *Biology Letters* 7: 917–920.

Simpson, S. D., A. N. Radford, S. L. Nedelec, M. C. O. Ferrari, D. P. Chivers, M. I. McCormick, and M. G. Meekan. 2016. Anthropogenic noise increases fish mortality by predation. *Nature Communications* 7: 10544.

Simpson, S. D., H. Y. Yan, M. L. Wittenrich, and M. G. Meekan. 2005. Response of embryonic coral reef fishes (Pomacentridae: Amphiprion spp.) to noise. *Marine Ecology Progress Series* 287: 201–208.

Slabbekoorn, H., N. Bouton, I. van Opzeeland, A. Coers, C. ten Cate, and A. N. Popper. 2010. A noisy spring: The impact of globally rising underwater sound levels on fish. *Trends in Ecology & Evolution* 25: 419–427.

Soman, M., N. K. Chadha, K. Madhu, R. Madhu, P. B. Sawant, and B. Francis. 2021. Optimization of temperature improves embryonic development and hatching efficiency of false clown fish, Amphiprion ocellaris Cuvier, 1830 under captive condition. *Aquaculture* 536: 736417.

Steffen, W., W. Broadgate, L. Deutsch, O. Gaffney, and C. Ludwig. 2015. The trajectory of the Anthropocene: The great acceleration. *The Anthropocene Review* 2: 81–98.

Stieb, S. M., F. de Busserolles, K. L. Carleton, F. Cortesi, W.-S. Chung, B. E. Dalton, L. A. Hammond, et al. 2019. A detailed investigation of the visual system and visual ecology of the Barrier Reef anemonefish, Amphiprion akindynos. *Scientific Reports* 9: 1–14.

Stuart-Smith, R. D., G. J. Edgar, and A. E. Bates. 2017. Thermal limits to the geographic distributions of shallow-water marine species. *Nature Ecology and Evolution* 1: 1846–1852.

Stuart-Smith, R. D., C. Mellin, A. E. Bates, and G. J. Edgar. 2021. Habitat loss and range shifts contribute to ecological generalization among reef fishes. *Nature Ecology and Evolution* 5: 656–662.

Suggett, D. J., J. M. Hall-Spencer, R. Rodolfo-Metalpa, T. G. Boatman, R. Payton, D. T. Pettay, V. R. Johnson, et al. 2012. Sea anemones may thrive in a high CO 2 world. *Global Change Biology* 18: 3015–3025.

Thorrold, S. R., and A. D. McKinnon. 1995. Response of larval fish assemblages to a riverine plume in coastal waters of the central Great Barrier Reef lagoon. *Limnology and Oceanography* 40: 181–189.

Thresher, R. E., P. L. Colin, and L. J. Bell. 1989. Planktonic duration, distribution and population structure of western and central pacific damselfishes (Pomacentridae). *Copeia* 1989: 420.

Velasco-Blanco, G., A. D. Re, F. Díaz, L. Ibarra-Castro, M. I. A. de la Parra, L. E. Rodríguez-Ibarra, and C. Rosas. 2019. Thermal preference, tolerance, and thermal aerobic scope in clownfish Amphiprion ocellaris (Cuvier, 1830) predict its aquaculture potential across tropical regions. *International Aquatic Research* 11: 187–197.

Vergés, A., P. D. Steinberg, M. E. Hay, A. G. B. Poore, A. H. Campbell, E. Ballesteros, K. L. Heck, et al. 2014. The tropicalization of temperate marine ecosystems: Climate-mediated changes in herbivory and community phase shifts. *Proceedings of the Royal Society B: Biological Sciences* 281: 20140846.

Vermeij, M. J. A., K. L. Marhaver, C. M. Huijbers, I. Nagelkerken, and S. D. Simpson. 2010. Coral larvae move toward reef sounds. *PLoS One* 5:e10660.

Vinebrooke, R. D., K. L. Cottingham, M. S. Jon Norberg, S. I. Dodson, S. C. Maberly, and U. Sommer. 2004. Impacts of multiple stressors on biodiversity and ecosystem functioning: The role of species co-tolerance. *Oikos* 104: 451–457.

Wabnitz, C., M. Taylor, E. Green, and T. Razak. 2003. From ocean to aquarium: The global trade in marine ornamental species. UNEP-WCMC Biodiversity Series 17. https://www.unep-wcmc.org/resources-and-data/from-ocean-to-aquarium--the-global-trade-in-marine-ornamental-species (accessed February 1, 2022).

Weis, V. M. 2008. Cellular mechanisms of Cnidarian bleaching: Stress causes the collapse of symbiosis. *Journal of Experimental Biology* 211: 3059–3066.

Welch, M. J., and P. L. Munday. 2016. Contrasting effects of ocean acidification on reproduction in reef fishes. *Coral Reefs* 35: 485–493.

Welch, M. J., S.-A. Watson, J. Q. Welsh, M. I. McCormick, and P. L. Munday. 2014. Effects of elevated CO2 on fish behaviour undiminished by transgenerational acclimation. *Nature Climate Change* 4: 1086–1089.

West-Eberhard, M. J. 2003. *Developmental Plasticity and Evolution*. Oxford: Oxford University Press.

Wong, B. B. M., and U. Candolin. 2015. Behavioral responses to changing environments. *Behavioral Ecology* 26: 665–673.

World Health Organization. 2011. *Burden of Disease from Environmental Noise: Quantification of Healthy Life Years Lost in Europe*. Geneva: World Health Organization.

Yan, H., S. Simpson, and M. Wittenrich. 2002. Ontogenetic development of mechanosensory ability of a pomacentrid coral fish, the red saddleback clownfish (*Amphiprion ephippium*). *The Journal of the Acoustical Society of America* 112: 2204–2204.

Yokohama, H., and S. Shirakashi. 2007. Evaluation of hyposalinity treatment on infection with Enteromyxum leei (Myxozoa) using anemonefish Amphiprion spp. as experimental host. *Bulletin of the European Association of Fish Pathologists* 27: 74–78.

24 Anemonefishes as Models in Ecotoxicology

Simon Pouil, Marc Besson, and Marc Metian

CONTENTS

24.1 INTRODUCTION

The in-depth study of the effects of biological and chemical agents on the biology of organisms strongly relies on suitable model species, for which husbandry methods and various resources such as genomes, proteomes, and transcriptomes are available. These resources allow for a comprehensive examination of the compound's ecotoxicity (e.g., at multiple scales of biological organization from the effects on genes to populations, and at multiple life stages) (Segner and Baumann 2016).

In fish, the zebrafish (*Danio rerio*), medaka (*Oryzias latipes*), fathead minnow (*Pimephales promelas*), and three-spined stickleback (*Gasterosteus aculeatus*) are the most common model species used to investigate the ecotoxicity of hazardous biological and chemical agents (Norrgren 2012). Although responses to toxic agents can be evolutionary conserved across fish species (Gunnarsson et al. 2008; Villeneuve et al. 2014), the fact that these model fishes are all freshwater or euryhaline (i.e., able to tolerate a wide range of salinity) species most often prevent the knowledge gained from them to be applied on marine species (Hsu et al. 2014; Kong et al. 2008). Indeed, such species not only live in ecosystems with distinct abiotic and biotic characteristics but also present different life-history strategies and metabolisms, requiring system-specific model species.

This lack of diversity in fish model species in ecotoxicology is particularly striking with respect to coral reef fishes, for which there is no such model species while being one of the most diverse groups of vertebrate species, on which more than 500 million people depend on for subsistence (Hoegh-Guldberg et al. 2019). Coral reef fishes play a key role in fuelling the exceptional productivity of coral reefs (Brandl et al. 2019), but very little is known regarding the pollutant bioaccumulation and biomagnification processes within the coral reef fish food web (Briand et al. 2018), from which various trophic levels are traditionally consumed by human populations. Moreover, most coral reef fishes have a bipartite life cycle with larvae developing in the ocean before returning to settlement habitats such as seagrass beds and mangroves and eventually recruiting into coral reefs (Sale 2004), which makes them potential vector of pollutants between all these marine ecosystems. Altogether, these examples demonstrate the urgent need to draw research avenues on coral reef fish ecotoxicology, not only to investigate the effects of pollutants on these populations but also to better understand their impacts on ecosystem functioning and human health.

DOI: 10.1201/9781003125365-29

Among coral reef fishes, anemonefishes or Amphiprioninae, which belong to the Pomacentridae family and live with anemones, a well-known example of mutualism (Feeney et al. 2019), have been extensively studied in the past decades as evidenced in the previous chapters of this book. Briefly, these species have been used as models in myriad ecology and evolution studies investigating, for example, the adaptation of the fish population to climate change (e.g., Jarrold et al. 2017; Lehmann et al. 2019; McLeod et al. 2013), the dispersal and connectivity patterns of fish population in coral reefs (e.g., Planes et al. 2009), the larval recruitment and habitat selection processes in coastal marine fishes (e.g., Scott and Dixson 2016), and the social organization and sex changes in fish hierarchical groups (e.g., Buston 2003; Wong et al. 2016).

In this chapter, we highlight how anemonefishes can also serve as relevant fish models to examine the ecotoxicology of hazardous biological and chemical agents in the marine environment. We then review the ecotoxicological effects of various hazardous compounds on anemonefishes and present how future research using them as models will promote our knowledge of fish ecotoxicology.

24.2 THE EMERGENCE OF ANEMONEFISHES AS MODEL ORGANISMS IN ECOTOXICOLOGY

The anemonefish *Amphiprion ocellaris* (Cuvier 1830), or false clownfish, is the most widely used coral reef fish in experimental studies because it is one of the rare coral reef fish species that can be entirely and easily reared in aquaria, and which is readily largely available in the aquaculture market (Pouil et al. 2020). *A. ocellaris* is not the only anemonefish in this case, since the life cycle of, at least, 25 anemonefish species has so far been controlled in aquaria (Pouil et al. 2020). This extensive expertise on the biology and life cycle of multiple anemonefish species makes them relevant coral reef fish models for eco-evo-devo sciences (Roux et al. 2020). With a relatively short embryonic and larval development, anemonefishes are relevant model species to assess the impact of contaminants on the postembryonic development at molecular and endocrinological levels (Roux et al. 2020). Rearing techniques for anemonefishes were relatively well documented in the literature (e.g., Calado et al. 2017; Divya et al. 2011; Kumar et al. 2012; Madhu et al. 2006). Recently, a husbandry detailed protocol has been published providing a detailed description of the anemonefish husbandry system as well as live prey culturing protocols (Roux et al. 2021). The authors built the anemonefish husbandry system from the ones developed for zebrafish with some adaptations. Briefly, the recirculating system of artificial seawater was composed of 30 63-L rearing tanks placed on three shelves while the treatment of the outlet water was done in an 800-L technical sump tank below the rearing tanks. The sump technical tank was equipped with filtering foams, a phosphate reactor filled with resin, a skimmer, a UV sterilizer, 100 kg of live rocks for biological filtration, and a lift pump allowing the return of filtered seawater to the rearing tanks. One of the advantages of such a system is to be flexible with the possibility to change the number of the tanks and/or their volumes. This first step makes it possible to envisage a standardized breeding protocol necessary for using anemonefish as model species. As rearing volume can be an issue for several days of ecotoxicology experiments, especially when the compound is either rare, toxic, and/or expensive, Roux et al. (2021) developed a protocol for larval rearing protocol in small glass beakers (less than 1 L), useful for toxicity assessment where chemicals can be rare, expensive, and/or very toxic. While mass breeding reproduction protocols exist in private companies, they are not disclosed outside them. Thus, Roux et al.'s open-access technical paper makes breeding techniques of *A. ocellaris* available to the whole scientific community.

The commercial availability of numerous, genetically and phenotypically diverse anemonefish strains, especially for *A. ocellaris*, is an asset for ecotoxicological studies where responses from laboratory tests are inferred for those of the more diverse wild populations. Furthermore, the fact that most anemonefish species have a comparable life history allows for the use of similar housing materials and husbandry protocols, which could facilitate experiments involving multiple species. Other coral reef fish species, such as benthic spawners from the Apogonidae, Gobiidae, and Pseudochromidae families (Calado et al. 2017), can also be successfully bred in captivity while having shorter life cycles than anemonefish. However, the availability of these species is much more limited (Pouil et al. 2020) and the limited knowledge and molecular tools available for these species are limiting factors for using them as biological models. The relative proximity of anemonefish with established fish models such as zebrafish and medaka, in which extensive chemical screenings have already been performed, is another asset for the use of anemonefishes in ecotoxicology in comparison to other coral reef fish species (Roux et al. 2020). For example, an inhibitor of tyrosine kinase receptors known to decrease iridophore number in zebrafish (Fadeev et al. 2016) has been successfully used to show that white bars in *A. ocellaris* are formed by iridophores (Salis et al. 2019).

24.3 CURRENT KNOWLEDGE OF ANEMONEFISHES' ECOTOXICOLOGY

24.3.1 Data Survey

The biological and ecological characteristics of anemonefishes make them promising fish model organisms for ecotoxicological research (see the previous section for details), which could promote our ability to evaluate the effects of pollutants on coral reefs as well as marine ecosystems more generally. However, ecotoxicological research on anemonefishes is still limited. A systematic review has been performed using Web of Science (WOS) covering > 12,000 scholarly journals and providing a satisfactory representation of international mainstream scientific research (Moed 2006). Only ten peer-review articles involving bioaccumulation or toxicity assays through exposure to chemicals in anemonefish were found (Figure 24.1).

Our finding highlights the current need to get a more comprehensive dataset regarding the sensitivity of anemonefish to different classes of chemicals. Such data are required to consider them as proper complementary fish models. In the following sections, we summarized the current knowledge available about toxicity for some classes of chemicals on anemonefish. For the sake of clarity, the information is presented in chronological order. The overall observed effects are summarized in Figure 24.2.

24.3.2 Nitrogen Compounds

Several studies investigated the toxicity of ammonia (NH_3), nitrites (NO_2^-), and nitrates (NO_3^-) in anemonefish (Frakes and Hoff 1982; Medeiros et al. 2016; Rodrigues et al. 2014). The purpose of the aforementioned studies is mainly for ornamental fish production because nitrogen compounds are metabolic by-products excreted by fish and elevated concentrations of these products can have deleterious effects on fish production (Calado et al. 2017). Inventories of the major forms of nitrogen in the ocean revealed that the mean concentrations in the euphotic zone were 0.1, 0.3, and 7 mmol m^{-3} for NO_2^-, ammonium (NH_4^+) and NO_3^-, respectively

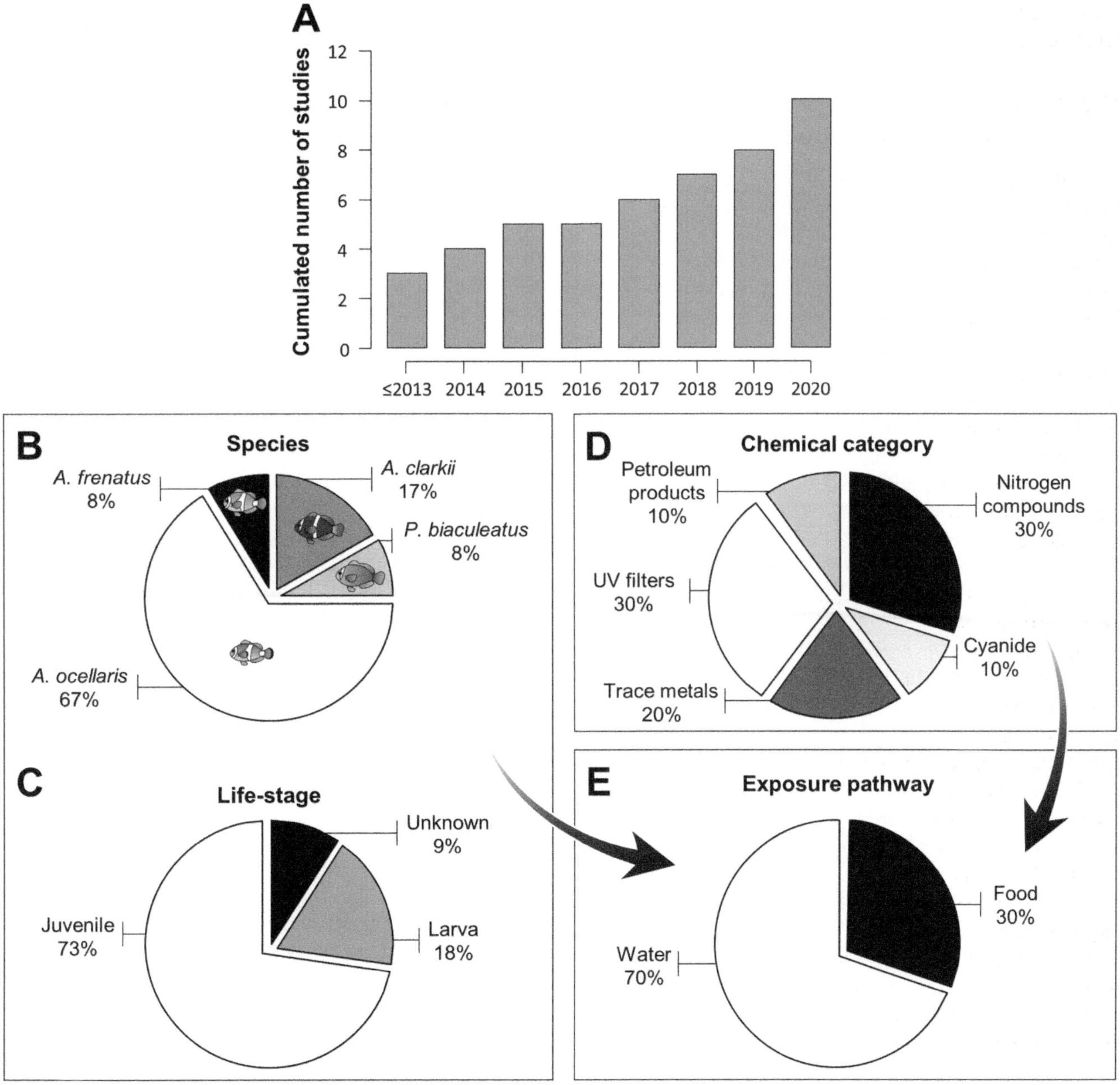

FIGURE 24.1 General trends in ecotoxicology research using anemonefish. (A) Cumulated number of studies published that experimentally examined the effects of chemicals on anemonefish fish species. Respective proportion of (B) anemonefish species, (C) life stages, (D) chemical category, and (E) exposure pathways.

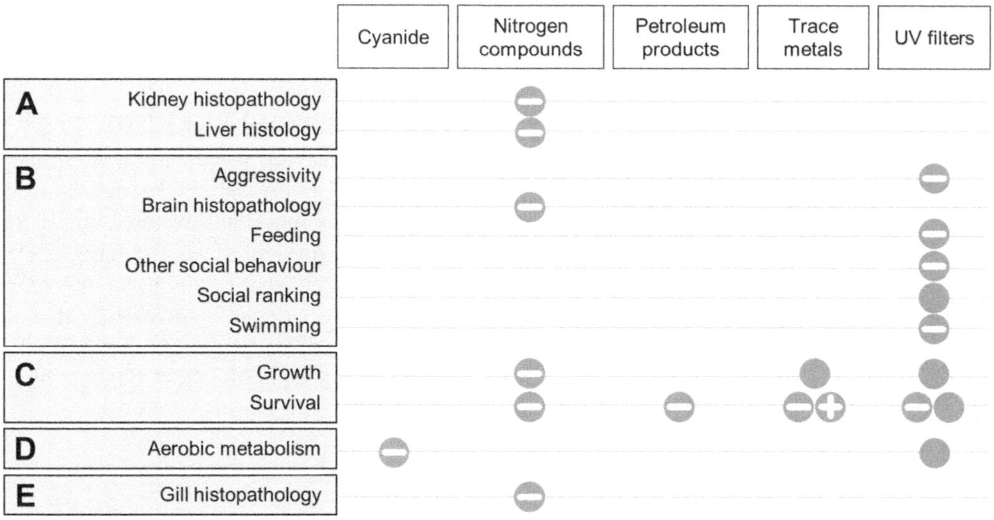

FIGURE 24.2 Ecotoxicological effects of different chemicals on anemonefish biological functions and systems: (A) excretory functions, (B) behavioral and sensory functions, (C) fitness indicators, (D) metabolism indicators, and (E) respiration functions. Solid circles indicate no effect while the symbols "–" and "+" indicate negative and positive effects, respectively.

(Gruber 2008). These compounds can also be found in the coastal tropical environment due to human activities (de Barros Marangoni et al. 2020). Frakes and Hoff (1982) were among the first to investigate the effects of nitrogenous compounds in anemonefish. They observed reduced growth in juvenile *A. ocellaris* exposed to 100 mg L^{-1} of NO_3^- while survival of larvae was three times lower when they reared at this NO_3^- concentration. More recently, Rodrigues et al. (2014) evaluated the median lethal concentration values (LC_{50}) and the histopathological effects of NH_3 on juvenile maroon clownfish *Premnas biaculeatus* exposed for 96 h to six concentrations (0.39–1.93 mg L^{-1} of NH_3-N). The 24 and 96 h LC_{50} values of NH_3-N were 1.68 and 0.89 mg L^{-1} (Table 24.1) respectively while fish exposed to different ammonia concentrations displayed histopathological alterations in the gills, kidney, liver, and brain. Such results have been confirmed in *A. ocellaris* juveniles by Medeiros et al. (2016) who exposed fish to six concentrations of NH_3 (0.23–1.63 mg L^{-1} of NH_3-N) and eight concentrations of NO_{2-} (26.3–202.2 mg L^{-1} of NO_2^--N). Authors found 24 h and 96 h LC_{50} of 1.06 and 0.75 mg L^{-1}, respectively, for NH_3-N, and 188.3 and 108.8 mg L^{-1}, respectively, for NO_2^--N. These results indicate that maroon clownfish are relatively sensitive to NH_3 and NO_2^- compared to other marine finfish (Medeiros et al. 2016; Rodrigues et al. 2014). Histological analysis showed that both nitrogenous compounds induced tissue lesions (Figure 24.2).

24.3.3 Petroleum Products

Exploitation of commercial quantities of oil and gas can impact coral reef ecosystems with the discovery of fields in shallow tropical seas (Neff et al. 2000). When crude or refined oil is accidentally spilt, these compounds are subject to several biological, chemical, and physical processes (i.e., weathering) that change the properties of oils. Neff et al. (2000) investigated the effects of weathering on the toxicity of three crude oils and a diesel fuel on marine organisms. Several tropical/subtropical and temperature model organisms including invertebrates (echinoderms and shrimps) and fish (silverside *Menidia beryllina* and anemonefish *A. clarkii*) were used. The water-accommodated fraction (WAF) of the four oils and their weathered fractions were prepared at a concentration of 28 g L^{-1} of oil in seawater. All the 96-h static-acute toxicity tests were performed with serial dilutions of 0, 8, 16, 64, or 75, and 100% of the WAF. The 96 h LC_{50} were ranging from 35% to > 100% of the WAF depending on the three crude oils and their weathered fractions (Table 24.1). Results were comparable between silverside and anemonefish fish showing similar sensitivity to the three oils. The 96 h LC_{50} values for diesel were ranging from 54 to 79% depending on the weathered fractions in silverside while 96 h LC_{50} was always > 100% in anemonefish suggesting a lesser sensitivity.

24.3.4 Trace Metals

Toxicity of trace metals has been poorly investigated in anemonefish. Interestingly, the only information on bioaccumulation and toxicity of metals is from two studies focused on essential elements. Furuta et al. (2005) examined the effect of Cu addition to rearing water on the survival *A. ocellaris*. The survival rates at 80 and 160 µg Cu L^{-1} were 65 and 80%, respectively while the survival rate was only 30% in the control conditions control in newly hatched larvae reared for 14 d. The positive effect of copper addition on the survival rate was confirmed with fish from seven different spawning events. The reason why the Cu supply in the rearing water improves the survival of larval anemonefish remains unclear (Furuta et al. 2005). Jacob et al. (2017)

TABLE 24.1
Median Lethal Doses of Different Chemicals in Anemonefish Species

Chemical category	Chemical	Species	LC_{50}	Exposure (h)	Reference
Nitrogen compounds	NH_3-N	*A. ocellaris*	1.68	24	Rodrigues et al. (2014)
		A. ocellaris	0.89	96	Rodrigues et al. (2014)
		A. ocellaris	1.06	24	Medeiros et al. (2016)
		A. ocellaris	0.83	48	Medeiros et al. (2016)
		A. ocellaris	0.75	72	Medeiros et al. (2016)
		A. ocellaris	0.75	96	Medeiros et al. (2016)
	NO_{2^-}-N	*A. ocellaris*	188.3	24	Medeiros et al. (2016)
		A. ocellaris	151	48	Medeiros et al. (2016)
		A. ocellaris	124.1	72	Medeiros et al. (2016)
		A. ocellaris	108.8	96	Medeiros et al. (2016)
Petroleum products	Wonnich crude	*A. clarkii*	35[a]	96	Neff et al. (2000)
	Wonnich 150°C+	*A. clarkii*	69[a]	96	Neff et al. (2000)
	Wonnich 200°C+	*A. clarkii*	> 100[a]	96	Neff et al. (2000)
	Campbell condensate	*A. clarkii*	39[a]	96	Neff et al. (2000)
	Campbell 150°C+	*A. clarkii*	81[a]	96	Neff et al. (2000)
	Campbell 200°C+	*A. clarkii*	> 100[a]	96	Neff et al. (2000)
	Agincourt crude	*A. clarkii*	> 100[a]	96	Neff et al. (2000)
	Agincourt 150°C+	*A. clarkii*	> 100[a]	96	Neff et al. (2000)
	Agincourt 200°C+	*A. clarkii*	> 100[a]	96	Neff et al. (2000)
	Agincourt 250°C+	*A. clarkii*	> 100[a]	96	Neff et al. (2000)
	Australian diesel	*A. clarkii*	> 100[a]	96	Neff et al. (2000)
	Diesel 200°C+	*A. clarkii*	> 100[a]	96	Neff et al. (2000)
	Diesel 250°C+	*A. clarkii*	> 100[a]	96	Neff et al. (2000)
Other	Cyanide	*A. ocellaris*[b]	28.45	96	Madeira et al. (2020)
		A. ocellaris[c]	50	96	Madeira et al. (2020)

Note: LC_{50}, expressed as mg L^{-1} excepted otherwise mentioned.
[a] LC_{50} expressed as per cent water-accommodated fraction (WAF).
[b] Small: 25.00 ± 0.03 mm, 0.30 ± 0.09 g.
[c] Medium: 38.00 ± 0.02 mm, 1.12 ± 0.21 g.

assessed the trophic transfer of three essential elements (Co, Mn, and Zn) in juvenile *A. ocellaris* exposed to two pH values simulating present-day conditions vs. acidified conditions (Δ pH = 0.5). Assimilation efficiencies (AEs) of three essential elements as well as other kinetic parameters, determined over a 20-d period following a single-feeding with radiolabelled pellets, were not affected by this experimental pH change although AEs were element dependent ($AE_{Zn} > AE_{Mn} > AE_{Co}$).

24.3.5 UV Filters

UV filters are recent anthropogenic pollutants encountered in the marine environment. Toxicity studies involving fish species have been mainly focused on UV filters used in sunscreens because of their potential ecological risk and due to their occurrence and persistence in aquatic ecosystems (Tovar-Sanchez et al. 2020). Anemonefish inhabiting shallow waters frequented by swimmers are relevant models for such assessments. We found three recent peer-reviewed studies assessing the effects of UV filters on anemonefish. Barone et al. (2019) investigated acute toxicity of TiO_2-based vs. oxybenzone-based sunscreens on *A. ocellaris*. Mortality, swimming, and feeding behavior were compared in juvenile fish exposed for 97 h to concentrations ranging from 0 to 100 mg L^{-1} of sunscreens mixed in seawater. They found that at the highest exposure concentration, oxybenzone-containing sunscreen had a negative impact on mortality, leading to 25% death over the 97-h exposure period. This concentration also impaired behavior with exposed fish showing abnormal swimming during the test while all of them stop to eat over the first 49 h. Mortality rate induced by TiO_2-based sunscreen exposure was much lower (< 7%) while normal swimming and feeding have been recovered at the end of the 97-h exposure period. Effects of another UV filter, the octocrylene (OC), on the physiology of *A. ocellaris* exposed through their food were assessed by Lucas et al. (2020). Juvenile fish were exposed for two months by feeding them using artificial dry food spikes with OC at a concentration of 10 μg g^{-1} dry weight and aerobic metabolic scope (AS) has been assessed as an indicator of the physiological state. The authors concluded that dietary exposure to OC at the tested concentration did not influence the aerobic metabolism of

A. ocellaris. Effects of chronic exposure to UV filter benzophenone-3 (BP-3) on social behaviors were investigated in juvenile *A. ocellaris* dietarily exposed to BP-3 (from 0 and 1,000 ng g^{-1}) over 90 d (Chen et al. 2018). Survival and growth were not affected by the BP-3 exposure except for a higher weight for the dominant fish while social rankings and intra-colonial social behaviors were not significantly affected by the BP-3 exposure (Figure 24.2).

24.3.6 Cyanide

Illegal cyanide fishing, mostly for marine ornamental trade, is one of the major anthropogenic threats to Indo-Pacific coral reefs, targeting a multitude of coral fish species (Calado et al. 2017; Madeira et al. 2020). In a recent study, Madeira et al. (2020) assessed the toxicity effects of cyanide in eight species of Pomacentridae including three *Amphiprion* species (*A. clarkii*, *A. frenatus*, and *A. ocellaris*). Juveniles of each species were exposed for 60 s to 50 mg L^{-1} of cyanide at 26°C. Only two species survived after 96 h with 50% survival for *A. ocellaris* and 20% for *Chromis cyanea*. In a second experiment, juveniles of *A. ocellaris* were exposed to different concentrations (0, 6.25, 12.5, 25, 50, and 100 mg L^{-1}) of cyanide for 96-h LC_{50} determination. 96-h LC_{50} were ranging from 20 to 53 mg L^{-1} depending on the fish size (Table 24.1).

24.4 HOW ANEMONEFISH MAY FUEL ADVANCEMENTS IN ECOTOXICOLOGY

24.4.1 Full Life-Cycle Fish Tests

Most of the ecotoxicological fish tests are primarily focused on acute or short-term exposure to chemicals providing toxicity values useful for regulatory decision making. Nevertheless, the long-term impacts of chemicals are still largely unknown while such information is important especially for persistent organic and inorganic pollutants. Some standardized chronic full life-cycle and multi-generational exposures have been implemented for fish such as the Fish Sexual Development Test (OECD Test No. 234) and the Extended One-Generation Reproduction Test (OECD Test No. 240) based on medaka and fathead minnow (OECD 2015). The implementation of such tests requires being able to carry out the complete life cycle of the model species in captivity, which remains complex today for most of the coral fish species. Anemonefish are among the only reef species whose rearing and reproduction can be relatively easily undertaken in captivity (Pouil et al. 2020). In contrast to many coral reef fishes that spawn in the open ocean, anemonefishes are benthic spawners and produce relatively big eggs (between 1.5 to 3.5 mm in length and 0.8 to 1.85 mm depending on species; Calado et al. 2017) glued to a support making them quite easy to handle. Most of the anemonefish embryos hatch, at 28°C, seven days post-fertilization (Calado et al. 2017). The spawning interval is short with reproduction events observed every two weeks in *A. ocellaris* in laboratory (Roux et al. 2021). Larvae can be easily fed on conventional zooplankton and show very fast growth rates (larval phase $\leqslant$ 15 d; Calado et al. 2017). Under optimal rearing conditions, juveniles up to 2.5 cm in length can be produced in five months in *A. percula* (Johnston et al. 2003). This reasonable rearing time is compatible to perform chronic exposure tests from eggs, or maternal transfer as suggested by Lucas et al. (2020), to juveniles. Nevertheless, sexual maturity is achieved late, often $>$ 1 year and spawning usually starts one to three months after the pair is established but sometimes it takes up to one year (Calado et al. 2017). In addition, mass rearing of anemonefish in laboratory facilities can be used to develop specific and reproducible strains well characterized both genetically and phenotypically, as it was done for zebrafish (Meyer et al. 2013) making a strong advantage to investigating bioaccumulation, organotropism, and the fate of contaminants. Nevertheless, managing genetic diversity in laboratory strains should be taken into consideration from the perspective of extrapolating the results of ecotoxicological laboratory tests to wild populations as we explained in section 24.2 of this chapter.

24.4.2 Single vs. Multi-Stressor Experiments in Laboratory

Past experimental studies demonstrated that climate change can affect many aspects of the biology and ecology of anemonefishes (see Chapter 23). For example, ocean acidification may lead to disruption of multiple sensory abilities in several species of anemonefishes (Dixson et al. 2010; Munday et al. 2009a; Simpson et al., 2011). While the veracity of these results is currently being debated (Clark et al. 2020), other studies also demonstrated that ocean acidification can affect anemonefishes' early life history (Munday et al. 2009b; Munday et al. 2011) and reproduction success (Kannan et al. 2020). Thus, anemonefishes appear to be particularly relevant models to investigate how the bioaccumulation and toxicity of a given chemical are affected by projected environmental conditions, and, conversely, whether the sensitivity of anemonefishes to climate change can be increased when co-exposed to a pollutant. However, very few studies have examined such multi-stressor scenarios combining ecotoxicology and environmental change in anemonefishes. As described in Section 24.3.4, Jacob et al. (2017) assessed the trophic transfer of three metals in juvenile *A. ocellaris* exposed to projected future pCO_2 levels over the next two centuries (pH 7.5) as well as present-day conditions (pH 8.0) and found no effect of acidification neither on the assimilation efficiency of the metals in anemonefish nor on the stomach pH. Interestingly, Madeira et al. (2020) found that cyanide exposure at increased temperatures (i.e., +3 and +6°C above present-day scenarios) led to higher mortality rates in juveniles *A. ocellaris*, when compared to the same exposure at control temperature (26°C) highlighting the potential synergistic effects of ocean warming and toxicity in anemonefish.

24.4.3 Laboratory and *In Situ* Models

Some aspects of clownfish ecology not only make them relevant as laboratory models as we have shown earlier but also open new horizons for their potential use in the field. In situ experiments, over the past decade, have received increased attention and acceptance as ways to complement traditional laboratory experiments by improving the connection between pollutant exposure (i.e., external bioavailable concentration) and the intrinsic sensitivity of the organisms (i.e., bioaccumulation, effects) under realistic conditions. Nevertheless, such an approach remains challenging to implement in fish. One of the most important considerations in performing in situ ecotoxicology studies is to examine the technique-related artefacts which can significantly influence test outcomes and the difficulties in establishing adequate controls to accurately interpret organism responses (Ferrari et al. 2013). In the saddleback clownfish *A. polymnus*, Jones et al. (2005), using parentage by DNA genotyping, found that one-third of settled juveniles had returned to a two-hectare natal area, with many settling< 100 m from their birth site, although another study found that self-recruitment can be highly variable (Nanninga et al. 2015). This represents the smallest scale of dispersal known for any marine fish species with a pelagic larval phase (Jones et al. 2005). This means that individuals can be tagged and tracked over time without altering their habitat and their life as well. Non-lethal recordings, samplings, and biometry can be performed over time on the same individuals depending on the objective of the study. Using anemonefish, in situ ecotoxicological studies could be performed with laboratory-reared organisms with a known life history, as well as indigenous organisms. In addition, Thorrold et al. (2006) described a new technique for transgenerational marking of embryonic otoliths in *A. melanopus*. The approach is based on the maternal transfer of ^{137}Ba from spawning females, exposed to the isotope, to egg material that is ultimately incorporated into the otoliths of embryos. The authors found that female *A. melanopus* continued to produce marked larvae over multiple clutches and for at least 90 days after a single injection. This technique can be extended by using different combinations of stable Ba isotopes, allowing marking fish from multiple populations and thus performing inter-population surveys over extended periods.

24.4.4 Availability of Genetic and Transcriptomic Data

Because of the increasing use of anemonefish species as fish models in a wide range of biological studies their genetic, physiological, and ecological backgrounds are relatively well characterized and multiple useful tools are available. These include the genomes of around a dozen anemonefish species (Marcionetti et al. 2019) as well as life-stage and tissue-specific transcriptomes in *A. ocellaris* and *A. percula* (Maytin et al. 2018; Salis et al. 2019, 2021b). A detailed description of the larval neuro-anatomy of *A. ocellaris* is also available (Jacob et al. 2016) as well as high-resolution time-lapse videos and descriptions of embryonic development in the same species (Salis et al. 2021a). The availability of such tools makes anemonefish relevant for assessing responses to contaminants at different scales: from molecular to individual endpoints and may help to cross the bridge from the individual to population levels.

24.5 REMAINING CHALLENGES AND FUTURE PERSPECTIVES

24.5.1 The Need for Standardized Husbandry

Although some recent advances have been made (see Roux et al. 2021), to date, the development of standardized husbandry protocols for anemonefish failed to keep pace with the increasing use of these species in laboratories for ecotoxicology studies. Some studies are still performed using purchased individuals from commercial suppliers without their genetic origin and their life-history traits being known (Chen et al. 2018; Jacob et al. 2017). A variety of husbandry methods are currently used based on artificial (e.g., Roux et al. 2021) or filtered natural seawater (e.g., Kumar et al. 2012) mostly in recirculating systems but also in flow-through systems. Such differences contribute to extensive variations in fish density, water chemistry, temperature and light conditions while feeding protocols are highly variable among the studies involving anemonefish. In addition, even in recent studies on husbandry protocols for anemonefish, growth, and survival performances, especially during the early stages of life are not always recorded (e.g., Roux et al. 2021) although this is an important aspect in ecotoxicology experiments requiring production of a sufficient number of healthy individuals on a regular basis. Since several species of clownfish have been produced for many years in private aquaculture farms for the ornamental fish market, it would be interesting to promote synergistic activities between academic research institutes and the private sector to optimize anemonefish husbandry protocols (Pouil et al. 2020).

24.5.2 From Individual to Population-Level Responses

A major challenge in ecotoxicology is to link responses highlighted at the individual level in the laboratory to population- and ecosystem-level responses in the field (Vighi and Villa 2013). Responses can be variable between natural populations of a given species. There is substantial evidence that genetic variation, at both the level of the individual and population, has a significant effect on behavior, fitness, and response to chemicals in fish. Coe et al. (2009) showed that the wild zebrafish were significantly more variable than the laboratory strains for several measures of genetic variability, including allelic richness and expected heterozygosity. While, to our knowledge, the genetic diversity of captive-bred anemonefish strains was never assessed, Madduppa et al. (2018) demonstrated that ornamental fishery contributes to a reduction in population size and genetic diversity in *A. ocellaris* populations in the Spermonde Archipelago, one of the main collection sites for ornamental fish in Indonesia. The potential lack of genetic variation in captive-bred anemonefish should be given due consideration for any study which attempts to extrapolate the results of ecotoxicological laboratory tests to wild populations. Therefore, the degree to which captive-bred strains are representative of wild anemonefish populations and congeneric species should be validated.

24.6 CONCLUSION

Although coral reefs are among the most sensitive aquatic ecosystems to anthropogenic threats, there is currently no clearly established model species of coral fish species used in ecotoxicology. Current standard fish-based tests in ecotoxicology are mostly limited to freshwater model species limiting knowledge gained to marine species. Due to their relatively easy husbandry, anemonefishes have been widely used in research examining the biology and ecology of coral reefs and coral reef fishes, leading to the development of various molecular tools and husbandry methods for these species. While our knowledge of the ecotoxicity of biotic and abiotic agents is still limited for this fish family, anemonefishes present many assets that make them promising species for examining fish ecotoxicology in future marine environments and coral reef ecosystems. However, key challenges remain, such as the standardization of husbandry protocols and the difficulty to extrapolate individual responses at population and community levels, which is true for any other fish model used in experimental work. We highlighted that anemonefish, especially *A. ocellaris*, have the potential to be a model in ecotoxicology. While they have barely been used in the past, they are now more and more considered for investigating the effects of some substances (e.g., sunscreens) and their use could be definitely extended as highlighted in this chapter. It is now clear that it is time to use an adequate fish model in order to properly assess the risk coral reefs are facing and will face with the increased number of contaminants or stressors in this high-diversity ecosystem. Therefore, the information provided here constitutes the first foundation to optimize ecotoxicology studies based on coral fish species using the promising anemonefishes as models.

REFERENCES

Barone, A. N., C. E. Hayes, J. J. Kerr, R. C. Lee, and D. B. Flaherty. 2019. Acute toxicity testing of TiO_2-based vs. oxybenzone-based sunscreens on clownfish (*Amphiprion ocellaris*). *Environmental Science and Pollution Research* 26(14): 14513–20.

Brandl, S. J., L. Tornabene, C. H. Goatley, J. M. Casey, R. A. Morais, I. M. Côté, C. C. Baldwin, et al. 2019. Demographic dynamics of the smallest marine vertebrates fuel coral reef ecosystem functioning. *Science* 364(6446): 1189–92.

Briand, M. J., P. Bustamante, X. Bonnet, C. Churlaud, and Y. Letourneur. 2018. Tracking trace elements into complex coral reef trophic networks. *Science of the Total Environment* 612: 1091–104.

Buston, P. M. 2003. Mortality is associated with social rank in the clown anemonefish (*Amphiprion percula*). *Marine Biology* 143(4): 811–15.

Calado, R., I. Olivotto, M. P. Oliver, and G. J. Holt. 2017. *Marine Ornamental Species Aquaculture*. Oxford: Wiley Blackwell.

Chen, T. H., C. Y. Hsieh, F. C. Ko, and J. O. Cheng. 2018. Effect of the UV-filter benzophenone-3 on intra-colonial social behaviors of the false clown anemonefish (*Amphiprion ocellaris*). *Science of the Total Environment* 644: 1625–29.

Clark, T. D., G. D. Raby, D. G. Roche, S. A. Binning, B. Speers-Roesch, F. Jutfelt, and J. Sundin. 2020. Ocean acidification does not impair the behaviour of coral reef fishes. *Nature* 577: 370–75.

Coe, T. S., P. B. Hamilton, A. M. Griffiths, D. J. Hodgson, M. A. Wahab, and C. R. Tyler. 2009. Genetic variation in strains of zebrafish (*Danio rerio*) and the implications for ecotoxicology studies. *Ecotoxicology* 18(1): 144–50.

de Barros Marangoni, L. F., C. Ferrier-Pagès, C. Rottier, A. Bianchini, and R. Grover. 2020. Unravelling the different causes of nitrate and ammonium effects on coral bleaching. *Scientific Reports* 10(1): 1–14.

Divya, S. P., T. T. A. Kumar, R. Rajasekaran, and T. Balasubramanian. 2011. Larval rearing of clownfish using *Brachionus plicatilis* rotifer as starter food. *Science Asia* 37(3): 179–85.

Dixson, D. L., P. L. Munday, and G. P. Jones. 2010. Ocean acidification disrupts the innate ability of fish to detect predator olfactory cues. *Ecology Letters* 13(1): 68–75.

Fadeev, A., J. Krauss, A. P. Singh, and C. Nüsslein-Volhard. 2016. Zebrafish leucocyte tyrosine kinase controls iridophore establishment, proliferation and survival. *Pigment Cell & Melanoma Research* 29(3): 284–96.

Feeney, W. E., R. M. Brooker, L. N. Johnston, J. D. J. Gilbert, M. Besson, D. Lecchini, D. L. Dixson, et al. 2019. Predation drives recurrent convergence of an interspecies mutualism. *Ecology Letters* 22: 256–64.

Ferrari, J. D., O. Geffard, and A. Chaumot. 2013. In situ bioassays in ecotoxicology. In *Encyclopedia of Aquatic Ecotoxicology*, eds J.-F. Férard, and C. Blaise, 623–41. London: Springer Science & Business Media.

Frakes, T., and F. H. Hoff Jr. 1982. Effect of high nitrate-N on the growth and survival of juvenile and larval anemonefish, *Amphiprion ocellaris*. *Aquaculture* 29(1–2): 155–58.

Furuta, T., N. Iwata, K. Kikuchi, and K. Namba. 2005. Effects of copper on survival and growth of larval false clown anemonefish *Amphiprion ocellaris*. *Fisheries Science* 71(4): 884–88.

Gunnarsson, L., A. Jauhiainen, E. Kristiansson, O. Nerman, and D. G. Joakim Larsson. 2008. Evolutionary conservation of human drug targets in organisms used for environmental risk assessments. *Environmental Science & Technology* 42: 5807–13.

Gruber, N. 2008. The marine nitrogen cycle: Overview and challenges. In *Nitrogen in the Marine Environment*, eds D. G. Capone, D. A. Bronk, M. R. Mulholland, and E. J. Carpenter, 1–43. Amsterdam: Elsevier.

Hoegh-Guldberg, O., L. Pendleton, and A. Kaup. 2019. People and the changing nature of coral reefs. *Regional Studies in Marine Science* 30: 100699.

Hsu, H. H., L. Y. Lin, Y. C. Tseng, J. L. Horng, and P. P. Hwang. 2014. A new model for fish ion regulation: Identification of ionocytes in freshwater-and seawater-acclimated medaka (*Oryzias latipes*). *Cell and Tissue Research* 357(1): 225–43.

Jacob, H., M. Metian, R. M. Brooker, E. Duran, N. Nakamura, N. Roux, P. Masanet, et al. 2016. First description of the neuroanatomy of a larval coral reef fish *Amphiprion ocellaris*. *Journal of Fish Biology* 89(3): 1583–91.

Jacob, H., S. Pouil, D. Lecchini, F. Oberhänsli, P. Swarzenski, and M. Metian. 2017. Trophic transfer of essential elements in the clownfish *Amphiprion ocellaris* in the context of ocean acidification. *Plos one* 12(4): e0174344.

Jarrold, M. D., C. Humphrey, M. I. McCormick, and P. L. Munday. 2017. Diel CO_2 cycles reduce severity of behavioural abnormalities in coral reef fish under ocean acidification. *Scientific Reports* 7(1): 10153.

Johnston, G., H. Kaiser, T. Hecht, and L. Oellermann. 2003. Effect of ration size and feeding frequency on growth, size distribution and survival of juvenile clownfish, *Amphiprion percula*. *Journal of Applied Ichthyology* 19(1): 40–43.

Jones, G. P., S. Planes, and S. R. Thorrold. 2005. Coral reef fish larvae settle close to home. *Current Biology* 15(14): 1314–18.

Kannan, G., S. Ayyappan and Y. Mariasingarayan. 2020. Ocean acidification impacts on hatching success and reproductive tissue damage in anemonefish. *Indian Journal of Natural Sciences* 10(58): 17992–8000.

Kong, R. Y. C., J. P. Giesy, R. S. S. Wu, E. X. H. Chen, M. W. L Chiang, P. L. Lim, B. B. H. Yuen, et al. 2008. Development of a marine fish model for studying in vivo molecular responses in ecotoxicology. *Aquatic Toxicology* 86(2): 131–41.

Kumar, T. T. A., M. Gopi, K. V. Dhaneesh, R. Vinoth, S. Ghosh, T. Balasubramanian, and T. Shunmugaraj. 2012. Hatchery production of the clownfish *Amphiprion nigripes* at Agatti island, Lakshadweep, India. *Journal of Environmental Biology* 33(3): 623–28.

Lehmann, R., D. J. Lightfoot, C. Schunter, C. T. Michell, H. Ohyanagi, K. Mineta, S. Foret, et al. 2019. Finding Nemo's genes: A chromosome-scale reference assembly of the genome of the orange clownfish *Amphiprion percula*. *Molecular Ecology Resources* 19(3): 570–85.

Lucas, J., V. Logeux, A. M. Rodrigues, D. Stien, and P. Lebaron. 2020. Trophic contamination by octocrylene does not affect aerobic metabolic scope in juveniles clownfish. *Annals of Environmental Science and Toxicology* 4(1): 50–54.

Madduppa, H. H., J. Timm, and M. Kochzius. 2018. Reduced genetic diversity in the clown anemonefish *Amphiprion ocellaris* in exploited reefs of Spermonde Archipelago, Indonesia. *Frontiers in Marine Science* 5: 80.

Madeira, D., J. Andrade, M. C. Leal, V. Ferreira, R. J. M. Rocha, R. Rosa, and R. Calado. 2020. Synergistic effects of ocean warming and cyanide poisoning in an ornamental tropical reef fish. *Frontiers in Marine Science* 7: 246.

Madhu, K., R. Madhu, L. Krishnan, C. S. Sasidharan, and K. M. Venugopal. 2006. Spawning and larval rearing of *Amphiprion ocellaris* under captive condition. *Marine Fisheries Information Service. Technical and Extension Series* 188: 1–5.

Marcionetti, A., V. Rossier, N. Roux, P. Salis, V. Laudet, and N Salamin. 2019. Insights into the genomics of clownfish adaptive radiation: Genetic basis of the mutualism with sea anemones. *Genome Biology and Evolution* 11(3): 869–82.

Maytin, A. K., S. W. Davies, G. E. Smith, S. P. Mullen, and P. M. Buston. 2018. *De novo* transcriptome assembly of the clown anemonefish (*Amphiprion percula*): A new resource to study the evolution of fish color. *Frontiers in Marine Science* 5: 284.

McLeod, I. M., J. L. Rummer, T. D. Clark, G. P. Jones, M. I. McCormick, A. S. Wenger, and P. L. Munday. 2013. Climate change and the performance of larval coral reef fishes: The interaction between temperature and food availability. *Conservation Physiology* 1(1): cot024.

Medeiros, R. S., B. A. Lopez, L. A. Sampaio, L. A. Romano, and R. V. Rodrigues. 2016. Ammonia and nitrite toxicity to false clownfish *Amphiprion ocellaris*. *Aquaculture International* 24(4): 985–93.

Meyer, B. M., J. M. Froehlich, N. J. Galt, and P. R. Biga. 2013. Inbred strains of zebrafish exhibit variation in growth performance and myostatin expression following fasting. *Comparative Biochemistry and Physiology Part A: Molecular & Integrative Physiology* 164(1): 1–9.

Moed, H. F. 2006. *Citation Analysis in Research Evaluation (Vol. 9)*. London: Springer Science & Business Media.

Munday, P. L., D. L. Dixson, J. M. Donelson, G. P. Jones, M. S. Pratchett, G. V. Devitsina, and K. B. Døving. 2009a. Ocean acidification impairs olfactory discrimination and homing ability of a marine fish. *Proceedings of the National Academy of Sciences of the United States of America* 106(6): 1848–52.

Munday, P. L., J. M. Donelson, D. L. Dixson, and G. G. Endo. 2009b. Effects of ocean acidification on the early life history of a tropical marine fish. *Proceedings of the Royal Society B: Biological Sciences* 276(1671): 3275–83.

Munday, P. L., V. Hernaman, D. L. Dixson, and S. R. Thorrold. 2011. Effect of ocean acidification on otolith development in larvae of a tropical marine fish. *Biogeosciences* 8(6): 1631–41.

Nanninga, G. B., P. Saenz-Agudelo, P. Zhan, I. Hoteit, and M. L. Berumen. 2015. Not finding Nemo: Limited reef-scale retention in a coral reef fish. *Coral Reefs* 34(2): 383–92.

Neff, J. M., S. Ostazeski, W. Gardiner, and I. Stejskal. 2000. Effects of weathering on the toxicity of three offshore Australian crude oils and a diesel fuel to marine animals. *Environmental Toxicology and Chemistry: An International Journal* 19(7): 1809–21.

Norrgren, L. 2012. Fish models for ecotoxicology. *Acta Veterinaria Scandinavica* 54: S14.

OECD. 2015. *OECD Guidelines for the Testing of Ahemicals*. Paris: OECD Publishing.

Planes, S., G. P. Jones, and S. R. Thorrold. 2009. Larval dispersal connects fish populations in a network of marine protected areas. *Proceedings of the National Academy of Sciences of the United States of America* 106(14): 5693–97.

Pouil, S., M. F. Tlusty, A. L. Rhyne, and M. Metian. 2020. Aquaculture of marine ornamental fish: Overview of the production trends and the role of academia in research progress. *Reviews in Aquaculture* 12(2): 1217–30.

Rodrigues, R. V., L. A. Romano, M. H. Schwarz, B. Delbos, and L. A. Sampaio. 2014. Acute tolerance and histopathological effects of ammonia on juvenile maroon clownfish *Premnas biaculeatus* (Block 1790). *Aquaculture Research* 45(7): 1133–39.

Roux, N., V. Logeux, N. Trouillard, R. Pillot, K. Magré, P. Salis, D. Lecchini, et al. 2021. A star is born again: Methods for larval rearing of an emerging model organism, the false clownfish *Amphiprion ocellaris*. *Journal of Experimental Zoology Part B: Molecular and Developmental Evolution* 336(4): 376–85.

Roux, N., P. Salis, S. H. Lee, L. Besseau, and V. Laudet. 2020. Anemonefish, a model for Eco-Evo-Devo. *EvoDevo* 11(1): 1–11.

Sale, P. F. 2004. Connectivity, recruitment variation, and the structure of reef fish communities. *Integrative and Comparative Biology* 44(5): 390–99.

Salis, P., S. H. Lee, N. Roux, D. Lecchini, and V. Laudet. 2021a. The real Nemo Movie: Description of embryonic development in *Amphiprion ocellaris* from first division to hatching. *Developmental Dynamics* 250(11): 1651–1667.

Salis, P., N. Roux, D. Huang, A. Marcionetti, P. Mouginot, M. Reynaud, O. Salles, et al. 2021b. Thyroid hormones regulate the formation and environmental plasticity of white bars in clownfishes. *Proceedings of the National Academy of Sciences* 118(23): e2101634118.

Salis, P., T. Lorin, V. Lewis, C. Rey, A. Marcionetti, M-L. Escande, N. Roux, et al. 2019. Developmental and comparative transcriptomic identification of iridophore contribution to white barring in clownfish. *Pigment Cell & Melanoma Research* 32(3): 391–402.

Scott, A., and D. L. Dixson. 2016. Reef fishes can recognize bleached habitat during settlement: Sea anemone bleaching alters anemonefish host selection. *Proceedings of the Royal Society B: Biological Sciences* 283(1831): 20152694.

Segner, H., and L. Baumann. 2016. What constitutes a model organism in ecotoxicology? *Integrated Environmental Assessment and Management* 12(1): 195–205.

Simpson, S. D., P. L. Munday, M. L. Wittenrich, R. Manassa, D. L. Dixson, M. Gagliano, and H. Y. Yan. 2011. Ocean acidification erodes crucial auditory behaviour in a marine fish. *Biology letters* 7(6): 917–20.

Thorrold, S. R., G. P. Jones, S. Planes, and J. A. Hare. 2006. Transgenerational marking of embryonic otoliths in marine fishes using barium stable isotopes. *Canadian Journal of Fisheries and Aquatic Sciences* 63(6): 1193–97.

Tovar-Sanchez, A., D. Sánchez-Quiles, and J. Blasco. 2020. *Sunscreens in Coastal Ecosystems*. London: Springer.

Vighi, M., and S. Villa. 2013. Ecotoxicology: The challenges for the 21st century. *Toxics* 1(1): 18–35.

Villeneuve, D. L., D. Crump, N. Garcia-Reyero, M. Hecker, T. H. Hutchinson, C. A. LaLone, B. Landesmann, et al. 2014. Adverse outcome pathway (AOP) development I: Strategies and principles. *Toxicological Sciences* 142: 312e320.

Wong, M., C. Uppaluri, A. Medina, J. Seymour, and P. M. Buston. 2016. The four elements of within-group conflict in animal societies: An experimental test using the clown anemonefish, *Amphiprion percula*. *Behavioral Ecology and Sociobiology* 70(9): 1467–75.

25 Saving Nemo

Extinction Risk, Conservation Status, and Effective Management Strategies for Anemonefishes

Geoffrey P. Jones, Maya Srinivasan, Gemma F. Galbraith, Michael L. Berumen, and Serge Planes

CONTENTS

25.1 INTRODUCTION

Interest in the extinction risk and conservation of anemonefishes might not have begun in 2003, but the Disney movie *Finding Nemo* sparked an interest in the conservation status of the very small coral reef fishes that continues today. The movie is directly quoted in the titles of numerous papers, including "Finding Nemo" (Ollerton et al. 2007), "Losing Nemo" (Jones et al. 2008), "Not Finding Nemo" (Nanninga et al. 2015), and "Trying to Find Nemo" (Scott and Baird 2015). There is a website dedicated to saving Nemo (www.savingnemo.org) and a documentary film initially titled *Saving Nemo* (Sharkbay Films, 2011), although it was later forced to change its name to *Filmstar Fish: The Struggle for Survival.* The popularity of the movie has no doubt contributed to a huge demand for anemonefish in the aquarium trade and concerns over impacts on wild populations (Jones et al. 2008; Burke da Silva and Nedoskyo 2016). However, at the same time, we have seen an enormous interest in the life history and ecology of anemonefishes, with research focusing on human impacts, extinction risks, and effective management strategies. Almost all of the key papers on threats to anemonefish and their conservation have been published from 2003 onwards, with accelerating interest in recent years (Figure 25.1). So, what does this research tell us about the threat of extinction for anemonefishes? How are we impacting them and what makes them threatened? And

DOI: 10.1201/9781003125365-30

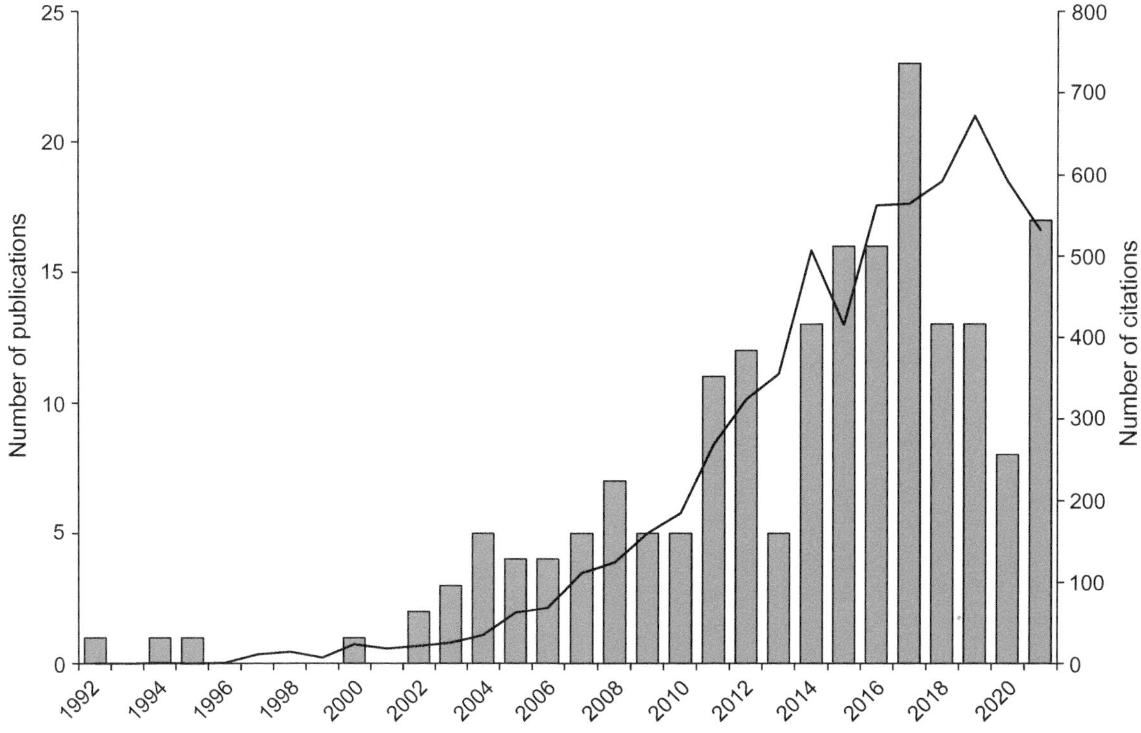

FIGURE 25.1 Results of a Web of Science search showing the number of publications (histograms) and number of citations (line graph) of papers on human impacts on and conservation of anemonefishes, showing the rapid increase since 2003.

for those that are critically endangered, what can we do about it? That is, how do we go about saving Nemo?

By 2012, a high proportion of species in families represented in the movie *Finding Nemo* had been assessed by the IUCN (International Union for the Conservation of Nature) and 16% of species had been classified as threatened (McClenachan et al. 2012). However, poor young Nemo himself did not make the cut. As of 2021, no anemonefish species has been placed in any of the endangered species categories on the IUCN Redlist. The United States National Marine Fisheries Service (NMFS) and the National Oceanic and Atmospheric Administration (NOAA) responded to a petition to have *Amphiprion percula* (aka Nemo) classified as endangered under the US Endangered Species Act (Rauch 2015; Maison and Graham 2016). It was concluded, based on population estimates ranging from 13 and 18 million, that it is not in any danger of extinction either now or in the near future. While this is clearly consistent with the current IUCN criteria, such population estimates are rare and unreliable for marine species. Even if correct, it is questionable whether or not these numbers reflect the actual likelihood of extinction. There is a growing list of suspected and confirmed neo-extinctions of marine fishes (Roberts and Hawkins 1999; Hawkins et al 2000; Dulvy et al 2004), many of which appear to have occurred following catastrophic declines from numbers that were historically much higher than the IUCN thresholds for endangered or vulnerable species.

While there are 28 currently recognized species, the highest biodiversity of anemonefish, like most coral reef fishes, is concentrated in Southeast Asia and the Western Pacific, where up to 12 species may co-occur in the same area (Figure 25.2). The location of the biodiversity hotspot is unfortunate in the sense that it can be largely overlaid with a high degree of coral reef habitat degradation, a high dependence on marine resources, and numerous problems associated with implementing effective management strategies. Fortunately, many of these species have broad semi-overlapping Indo-Pacific distributions that include places where the impacts are likely to be less severe. Many of the smaller range species lie at the outer limits of the *Amphiprion/Premnas* distribution, where the level of threat is likely to vary from species to species (McClanahan et al. 2021). Our understanding of anemonefish biodiversity is likely a work in progress, with two new species recently recognized for the Pacific (*A. barberi*, Allen et al. 2008; and *A. pacificus*, Allen et al. 2010). Based on the spatial and morphological variation in the region, Drew et al. (2008) suggested that there may be even more species in the Pacific. There are also likely to be more hidden species in the coral triangle (Timm et al. 2008), which will be a conservation concern. One species, *A. leucokranos*, no longer exists, but only because it was determined to be a hybrid formed when a female *A. chrysopterus* mates with a male *A. sandaracinos* (Gainsford et al. 2020). The threat to anemonefish biodiversity requires an assessment, not just of the likelihood of losing one or more of the 28 species, but the extent to which we are losing local populations or seeing a decline in genetic diversity that will affect a species ability to adapt to environmental change.

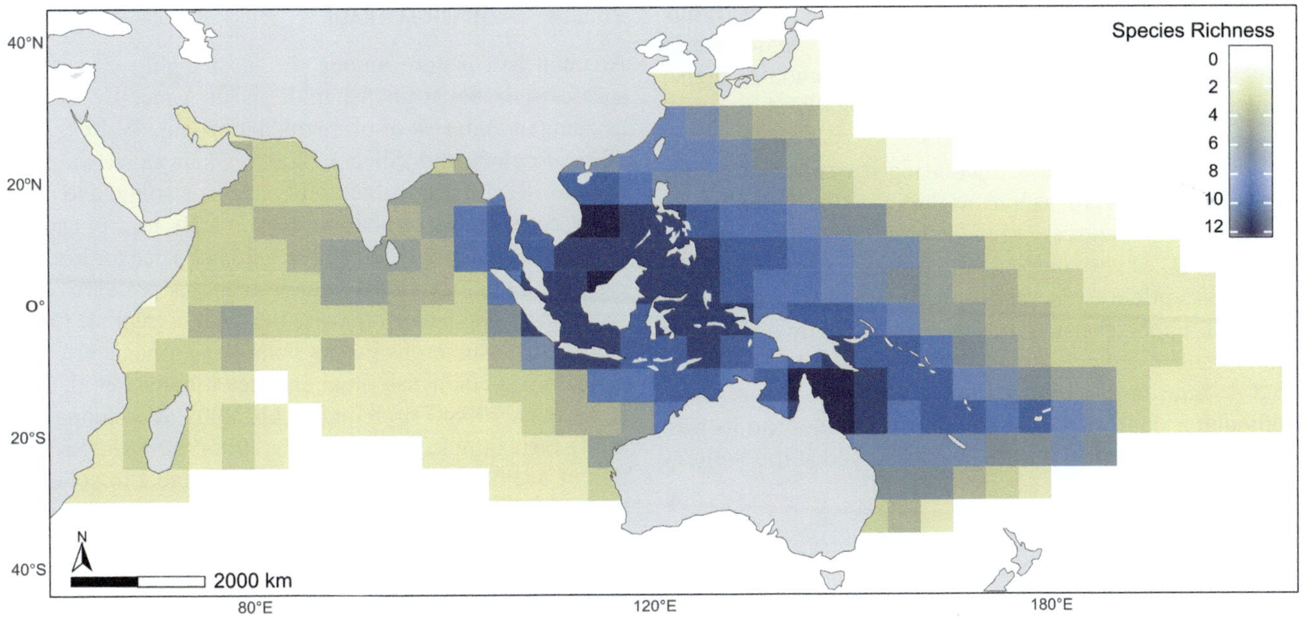

FIGURE 25.2 Biodiversity heat map for 28 anemonefish species calculated from overlapping extent of occurrence (EOO km^2). EOO was estimated as minimum convex hulls: the smallest polygon in which no internal angle exceeds 180 degrees, and which contains all the sites of occurrence (IUCN 2001, 2012, 2019). Hulls were constructed from known occurrence records obtained fishbase.org (Froese and Pauly 2021) and overlaid with a global grid of 5° × 5° cells in R 4.1.1 (R Core Team, 2021). Species richness was calculated as the total number of overlapping hulls per cell.

The aims of this chapter are: (1) to assess the main threats to regional and local anemonefish biodiversity as a result of local population declines and possible extinction; (2) to evaluate the life history and ecological characteristics of anemonefishes (and their anemones) that exacerbate the risk of extinction; (3) to quantify relationships among different rarity traits (specialization, geographic range, and depth range) that may expose anemonefish species to double or triple jeopardy; (4) to calculate the actual area of occurrence of anemonefishes within the geographic ranges and assess these in relation to the IUCN endangered species categories; and (5) to consider pre-emptive and effective management strategies that can aid population recovery, protect wild breeding populations into the future, and reduce the likelihood of local or global extinction.

25.2 THREATENING PROCESSES

25.2.1 Anemone Bleaching

The likelihood of extinction needs to be assessed in the context of the multiple factors having a negative impact on anemonefishes and/or their critical habitats, all of which can be traced back to human activities. Of these, ocean warming and associated bleaching of anemones and their surrounding coral habitat are likely to be the most extreme and largest-scale threats. Almost all anemone symbionts bleach and bleaching can occur across a broad geographic range (Hobbs et al. 2013; Burke da Silva and Nedoskyo 2016) and from shallow to mesophotic reefs (Haguenauer et al. 2021). In some locations, bleaching is now being observed regularly over multiple years (Hobbs et al. 2013; Hayashi and Reimer 2020). While anemones can recover and fish may survive, numerous papers have observed fewer fish where anemones have bleached (Hattori 2002; Jones et al. 2008). Hattori (2002) found that *A. perideraion* went locally extinct following bleaching mortality of *H. crispa* in the 1998 El Nino, while the generalist *A. clarkii* survived in less preferred hosts. Even if anemonefish survive in bleached anemones, it is known to impair reproductive activity and reduce fecundity (Saenz Agudelo et al. 2011; Beldade et al. 2017), increase metabolic stress (Norin et al. 2018), reduce metabolic rate (Cortese et al. 2021), and disrupt antipredator behavior (Lönnstedt and Frisch 2014), all of which may affect the future persistence of populations. Interestingly, bleaching does not appear to affect nematocysts, so anemones can still defend themselves, which may aid in their recovery (Hoepner et al. 2019).

25.2.2 Ocean Warming

Ocean warming can have direct negative effects on anemonefish, independent of habitat bleaching, due to thermal sensitivity during the larval and juvenile stages. McLeod et al. (2013) showed that larval *A. percula* take longer to develop and settle at higher temperatures, which may reduce numbers recruiting to the adult population. Nowicki et al. (2012) demonstrated that juvenile *A. melanopus* needed to forage and consume more at higher temperatures, perhaps because of lower energy efficiency. Experiments on *A. melanopus* also show temperature has a strong negative effect on reproduction at 1.5 degrees above current ambient conditions (Miller et al. 2015). On the other hand, future warming may enhance performance in *P. biaculeatus*, which has a higher aerobic scope and potentially higher growth and condition at temperatures slightly above present-day conditions (Donelson 2015).

Fishes may respond to increasing ocean temperatures by shifting poleward, as long as they can find suitable habitat and food resources (Munday et al. 2008). Subtropical anemonefishes with narrow latitudinal ranges are particularly susceptible to increasing temperatures as they may be near upper thermal tolerances. Malcolm and Scott (2017) suggested the Australian endemics *A. akindynos* and *A. latezonatus* have a limited ability to move south, although a small range shift in *A. latezonatus* and its host anemone *E. quadricolor* has been observed at the Solitary Islands, NSW.

25.2.3 Ocean Acidification

Experimental studies on larval and juvenile fishes have indicated numerous effects of elevated pCO_2 on growth, survival, physiological condition, otolith morphology, and sensory behavior (Munday et al. 2019; Munday et al. 2020). This has raised long-term concerns over ocean acidification or the lower predicted pH by mid-to-late this century as a result of increasing CO_2 in the atmosphere. Although controversial, ocean acidification appears to have particularly strong effects on the olfactory and auditory sensory mechanisms of larval anemonefishes (Munday et al. 2009, 2010; Dixson et al. 2010; Simpson et al. 2011; Nilsson et al. 2012). Given that they rely so heavily on olfaction and sound for finding reefs, finding anemones, and avoiding predators, acidification is likely to have a strong negative effect on their settlement and survival. At low pH, *A. percula* larvae completely lose their ability to discriminate the smell of their host anemone (Munday et al. 2009) and their innate ability to detect predator olfactory cues is impaired, with some larvae actually becoming attracted to predators (Dixson et al. 2010). Anemonefish reproduction also appears to be affected by acidification. Miller et al. (2013) showed that *A. melanopus* reproduction was stimulated at low pH and females had a higher fecundity, although larvae tended to have smaller yolks. Recently, Holmberg et al. (2018) showed that acidification alters otolith morphology in *A. clarkii*, with a dramatic negative effect on settlement competency.

The future effects of acidification will occur in combination with rising temperatures, and impacts will depend on how these drivers interact. Nowicki et al. (2012) found temperature had a greater effect on the foraging behavior of *A. melanopus* than low pH, and the interaction between the two caused a reduction in food consumption. Similarly, Miller et al. (2015) showed that the effects of temperature on *A. melanopus* reproduction were much stronger than the effects of acidification, but the negative effect of pCO_2 on offspring quality was more pronounced at higher temperatures.

25.2.4 Aquarium Trade

Anemonefish feature among the most highly sought-after species in the aquarium fish trade and are generally regarded as being at high risk of overexploitation (e.g., Wabnitz et al. 2003; Roelofs and Silcock 2008; Okemwa et al. 2016; Biondo 2018). Their vulnerability can be attributed to popularity, accessibility on shallow coastal reefs, ease of capture, and market value. There is often a higher price for rare species or rare color morphs, which increases pressure on the species and populations that are the most susceptible (Militz et al. 2018). Numerous papers point to significant effects of collecting anemonefish on abundance (Shuman et al. 2005; Jones et al. 2008; Frisch and Hobbs 2009; Madduppa et al. 2014). For example, Shuman et al. (2005) show a large depletion of anemonefish numbers in areas subject to collecting in the Philippines, compared with unfished areas. Similarly, the collecting of *A. ocellaris* has a huge negative impact on abundance at Spermonde Archipelago (Madduppa et al. 2014). Here there is also a reduced genetic diversity in *A. ocellaris* that can be attributed to aquarium fish collecting (Madduppa et al. 2018). While depletion of anemonefish numbers can be quite rapid as an industry develops, recovery can be extremely slow, even if there is a complete moratorium on collection (Frisch et al. 2019).

25.2.5 Coastal Development

Direct loss of habitat due to coastal development, including increasing sedimentation and nutrient enrichment associated with deforestation, agricultural activities, and marine dredging, represents a major potential threat to coral reef fishes (Wenger et al. 2015, 2017). However, clear evidence for impacts on anemonefishes is hard to find, either for the fish themselves or the anemone habitat. Hayashi et al. (2019a, 2019b) documented the low abundance and diversity of clownfish species directly on the coastline near urban developments in Okinawa. Long-term declines in *A. bicinctus* and the host anemone *Entacmaea* in the Gulf of Eilat have been linked to pollution and coastal development (Howell et al. 2016). Anemones may be more resilient to sedimentation than corals (Liu et al. 2015), which may explain why anemonefish-associated anemones can be found in turbid waters. However, direct effects of sedimentation on anemonefishes have been linked to prolonged larval development (Wenger et al. 2014), adverse effects on gill function (Hess et al. 2015, 2017), and altered anti-predator behavior (Hess et al. 2019), all of which may negatively impact on population size.

25.3 RISK FACTORS: LIFE HISTORY AND ECOLOGICAL TRAITS

25.3.1 Habitat Specialization

The multiple human impacts listed here and acting together may pose a risk of extinction for any fish species, but most anemonefishes share life history and ecological traits that exacerbate these threats. The most important of these is their high degree of specialization and obligate dependence on a small range of species of a single habitat-forming organism, the anemones (Allen 1975; Fautin and Allen 1992; Burke da Silva and Nedoskyo 2016). Only ten anemone species are colonized by the 28 anemonefish species and there are eight anemonefish species associated with a single anemone species (Burke da Silva and Nedoskyo 2016). Two species are associated with a particular morph of a single anemone species, with *P. biaculatus* associated with the solitary morph of *Entacmaea* and *A. melanopus* associated with the colonial morph (Srinivasan et al. 1999). The survival of the vast majority of the anemonefish species is dependent on just four to five preferred anemone species. Anemonefishes, perhaps more than any other reef fishes, share the fate of all highly specialized animals. Their distribution and abundance are completely linked to their hosts, and so if the hosts disappear, the fish will disappear too.

25.3.2 Mutual Dependence

Another major problem for anemonefishes is that not only are they dependent upon their anemone hosts, but the anemones are just as dependent on them (Fautin and Allen 1992; Burke da Silva and Nedoskyo 2016). Experiments show that when all fish are collected, anemones are often eaten by their predators such as butterflyfish and angelfish, resulting in their death (Bradshaw 1994; Frisch et al. 2016). This is a real problem for the aquarium fish collecting industry, as it is not sustainable unless no anemones are left vacant (Frisch et al. 2016). Anemonefish also appear to be necessary to aid recovery from bleaching (Pryor et al. 2020) but have the ability to avoid bleached anemones if they have the choice (Scott and Dixson 2016). So, anemonefish have a susceptibility that they share with all other obligate mutualists – if one partner goes extinct, the other will go extinct at the same time.

25.3.3 Low Density

For whatever reason, the anemones that host anemonefishes are never particularly abundant on coral reefs. Hence, they almost always have low population densities resulting from naturally low densities of hosts (Srinivasan et al. 1997; Scott and Baird 2015; Steinberg et al. 2020; Hayashi et al. 2019b). Highly specialized species are even more likely to exhibit low abundance when their preferred habitats are rare (Jones et al. 2002; Munday 2004).

25.3.4 Low Connectivity

The emerging evidence is that dispersal distances in anemonefishes can be limited, with a high degree of self-recruitment within populations on isolated reefs (see Chapter 20). Low connectivity may explain slow recovery when local populations are severely depleted or become locally extinct (Bonin et al. 2016; Frisch et al. 2019). This may be a particular problem for endemics that occupy relatively few isolated reefs, with subpopulations being completely dependent on self-recruitment (Steinberg et al. 2016; van der Meer et al. 2012). In this case, local extinction may be a stepping stone to global extinction. However, self-recruitment has its benefits in a stable environment, in terms of promoting local population persistence and local adaptation (Jones et al. 2009; Jones 2015).

25.3.5 Small Geographic Range

A large proportion of anemonefishes have broad Indo-Pacific distributions, and these will only be exposed to global threats such as increasing temperatures or ocean acidification. However, there are also numerous small range species, either endemic to isolated island groups (e.g., *A. chagosensis*, *A. chrysogaster*) or with small latitudinal ranges on mainland coasts (e.g., *A. omanensis*, *A. latezonatus*) usually near the periphery of the global range of anemonefishes (Figure 25.3). The vast majority of neo-extinctions in the animal kingdom have been species with small ranges or island endemics that have been exposed to habitat loss, exotic pests, and diseases. The few recorded extinctions of marine fishes in recent times have all been small-range species (Roberts and Hawkins 1999; Hawkins et al. 2000; Dulvy et al. 2004). Small-range anemonefishes are clearly susceptible to environmental disturbances that impact the scale of their distribution. They may also be more sensitive than large-range species to global change, especially subtropical species that are likely to be adapted to cooler water environments.

25.3.6 Depth Range

A final risk factor is the narrow and shallow water depth distributions of the majority of anemonefish species (Fautin and Allen 1992). This exposes them to any human impacts that tend to be more severe in shallow water, such as warming water, bleaching, and coastal sedimentation. Some species, such as *A. percula* are most abundant in water less than 3–4 m in depth. This makes a large proportion of the population completely accessible to aquarium fish collectors, without the need for underwater breathing apparatus. Deep water surveys on the GBR have shown that some species like *A. akindynos* and *A. perideraion* can be abundant on mesophotic reefs, suggesting they may have a *depth-refuge* from shallow water disturbances (Bridge et al. 2012). However, this may not apply to the majority of species which may be much less abundant at depth and appear to be susceptible to deep water warming and bleaching (e.g., *A. chrysopterus*, Haguenauer et al. 2021).

25.4 RARITY TRAITS: DOUBLE AND TRIPLE JEOPARDY

A species with any one of the aforementioned traits would attract conservation attention in an environment that is showing signs of increasing and proliferating threats. The

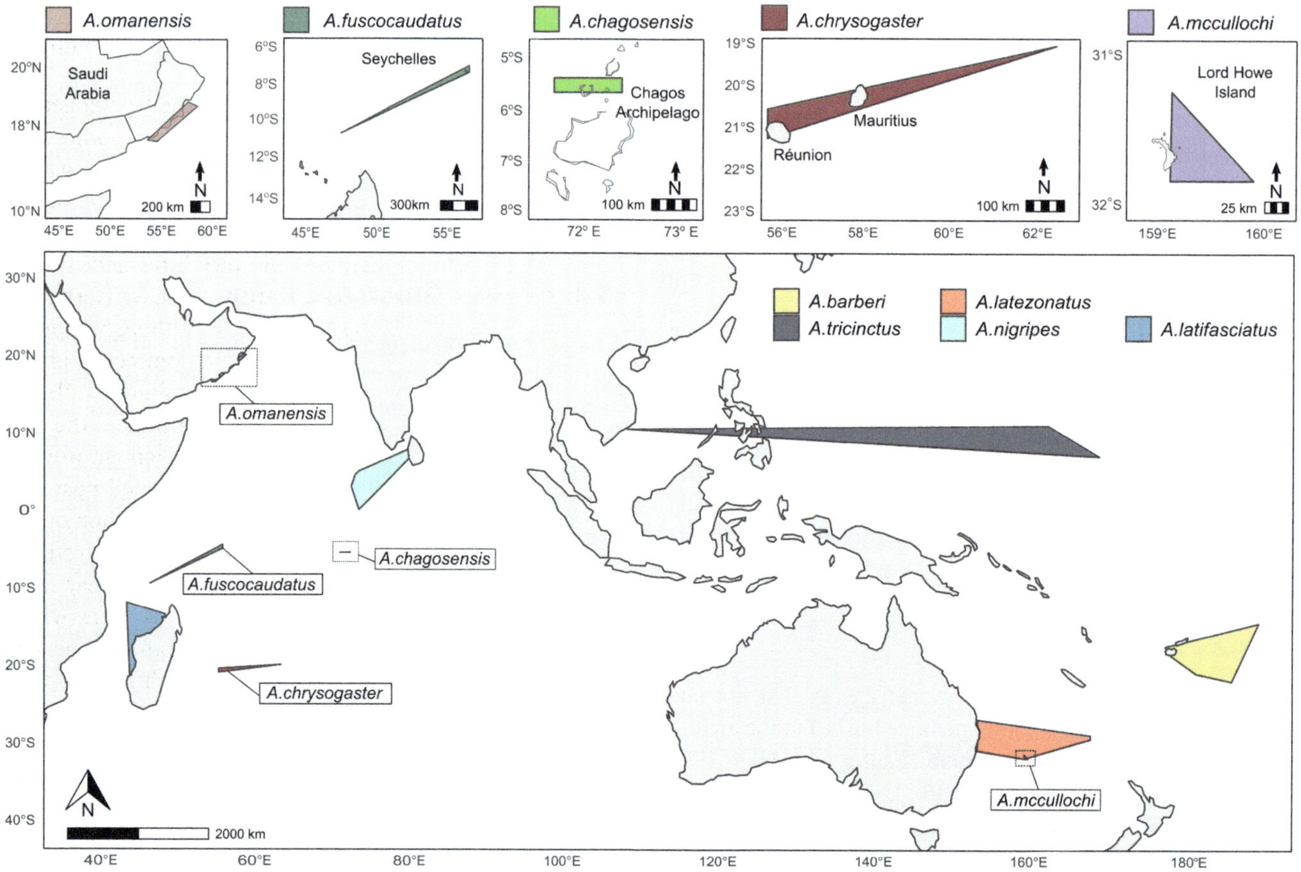

FIGURE 25.3 Extent of occurrence (EOO) for the ten anemone fish species with smallest geographical ranges. EOO is plotted as minimum convex hulls; the smallest polygon in which no internal angle exceeds 180 degrees, and which contains all the sites of occurrence (IUCN 2001, 2012, 2019). Hulls were constructed using known occurrence records from fishbase.org (Froese and Pauly 2021) in R 4.1.1 (R Core Team 2021). Occurrence records for *A. chagosensis* were limited to three points on the same axis and therefore a 0.1 decimal degree buffer was applied to calculate an approximate EOO using the package ConR (Dauby et al. 2017, 2020).

problem anemonefish face is that they can have a combination of life history and ecological traits that multiply the risk of extinction. This can be especially true when different aspects of rarity, including small geographic range, low abundance, and high specialization are linked (Rabinowitz 1981; Jones et al. 2002). A species with a combination of any two of these traits is considered to have *double jeopardy* of extinction, and for species exhibiting all three traits, it is *triple jeopardy* (Jones et al. 2002; Munday 2004). Our analyses show that there are strong relationships among these risk factors in anemonefishes. Using data on geographic range sizes from Fishbase and an index of specialization based on the number of anemones occupied (Fautin 1991; Burke da Silva and Nedoskyo 2016), we show that range size is positively related to decreased specialization (Figure 25.4a). Hence small-range species are exposed to the double risk associated with human impacts on the area in which they live and on the anemone on which they depend. Similarly, depth range declines with increasing specialization on host anemones, so the most specialized species are the most restricted to shallow water (Figure 25.4b). Clearly, some anemonefishes have triple jeopardy. These are the species specialized on a single host anemone which have a small geographic range and also have a narrow depth range. This exposes them to a much greater range of threats than they would have had if they possessed only one of these traits.

Small geographic range and high specialization are also likely to be associated with a low total population size, which would also constitute triple jeopardy for anemonefishes. What little information we have suggests that anemonefish breeding populations are limited by the number of their preferred hosts (see Chapter 18) and they are generally found at low densities. However, the triple jeopardy would only hold if population densities were not related to geographic range. Some evidence suggests that for marine fishes, small-range endemics tend to have higher population densities than their widespread counterparts at the same locations (Hobbs et al. 2010, 2011). This has not been evaluated for anemonefishes, although McClanahan et al. (2021) showed that the endemic *A. chrysogaster* at Mauritius is moderately abundant, is broadly distributed around the island, and has a large depth range. A complete understanding of extinction risks for island endemics will require a greater effort in estimating population densities and total population size.

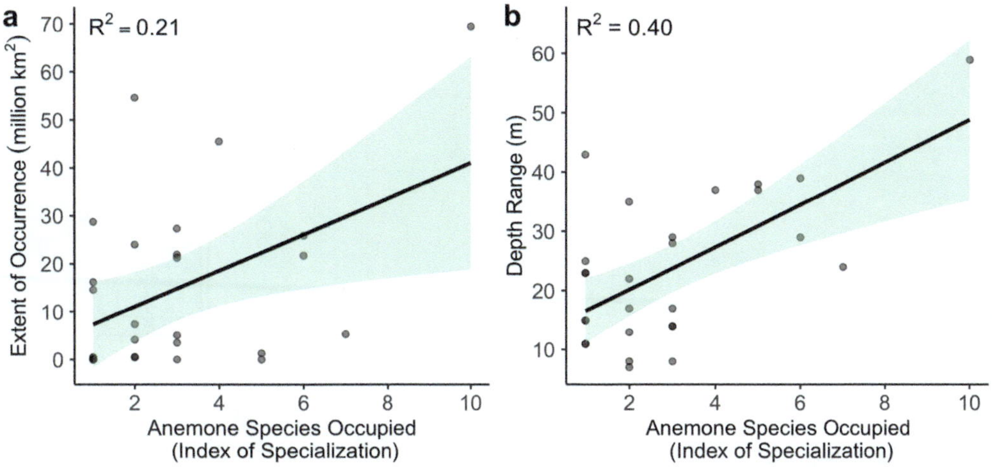

FIGURE 25.4 Relationships between the number of anemone species occupied (Index of Specialization) and a) geographic range as extent of occurrence (EOO, million km^2) (glm, $p = 0.017$, $t = 2.56$) and b) known depth range (m) (glm, $p < 0.001$, $t = 4.06$) for 27 anemonefish species. Depth range and known occurrence records were obtained from fishbase.org (Froese and Pauly 2021). EOO was calculated as minimum convex hulls constructed from occurrence records. All analyses were performed in R 4.1.1 (R Core Team 2021) using the packages ConR (Dauby 2020), rCAT (Moat 2020), rsq (Zahng 2021), and glmmTMB (Brooks et al. 2017). Generalized linear models were fit with Gaussian error family and identity link. *A. pacificus* is not included in either model as habitat use for this species is not known.

25.5 CONSERVATION STATUS

To date, only 15 of the 28 species have been assessed by the IUCN and all have been classified globally as "Least Concern". However, at this stage, this assessment does not include most of the species with the smallest ranges (see Figure 25.3). In terms of regional assessments, the Redlist website lists *A. clarkii* as endangered in the Red Sea, but no data on this assessment is available. Few anemonefish species have been listed as endangered by any country through their national endangered species legislation. As stated earlier, the United States has assessed *A. percula* for its Pacific territories under its Endangered Species Act and it has been considered at no risk of extinction now or in the near future (Rauch 2015; Maison and Graham 2016).

The IUCN criterion that seems most applicable to the real threat of extinction for anemonefishes concerns not just geographic range or *extent of occurrence* (EOO), but their *area of occupancy* (AOO) within their geographic range. AOO reflects the fact that a taxon will not usually occur throughout the full area of its EOO, which may contain unsuitable or unoccupied habitats (IUCN 2001, 2012). That is, over what actual area have they been observed. The IUCN considers species with an AOO of less than 10 km^2 as Critically Endangered, < 500 km^2 as Endangered, and $< 2{,}000$ km^2 as Vulnerable (Criterion B2). Using data from Fishbase (Froese and Pauly 2021) on geographic range and confirmed locations, we show the AOOs for the endemic anemonefishes can be extremely small (Figure 25.5). On this basis, three species approach the threshold to be classified as Critically Endangered and as many as 23 species would be classified as Endangered (Figure 25.5). In combination with a small extent of occurrence, low numbers of locations, and estimates of the reef area within these ranges (Allan Coral Atlas 2020), we suggest that these species urgently require an evaluation by the IUCN and by the government agencies of the countries where they are endemic. We acknowledge that accurate estimates of AOO require extensive known occurrence records which is currently problematic for data deficient species. The Fishbase online database we used was the only source of confirmed occurrence data available for all 28 species, but at this stage, these records are not complete. It is noteworthy that the two species with the smallest geographic ranges, *A. chagosensis* (EOO = 4,056 km^2) and *A. mccullochi* (EOO = 1,317 km^2), qualify as Endangered on the basis that their EOOs are less than 5,000 km^2.

25.6 EFFECTIVE MANAGEMENT STRATEGIES

There are numerous options for protecting anemonefishes that will vary in their effectiveness depending on the species, the location and the most significant threats. The ability to implement effective management will depend on the political will and socioeconomic circumstances that prevail. Here, we will just highlight a few management options that should work, based on the literature or the biology of the species.

25.6.1 Marine Reserves

Marine reserves or no-take marine protected areas established to protect biodiversity are known to protect anemonefishes where they can be well-managed. Several studies have shown higher numbers of anemonefishes in marine reserves compared to adjacent areas subject to collecting (Shuman et al. 2005; Jones et al. 2008; Madduppa et al. 2014). Scott et al. (2011) showed a long-term increase in

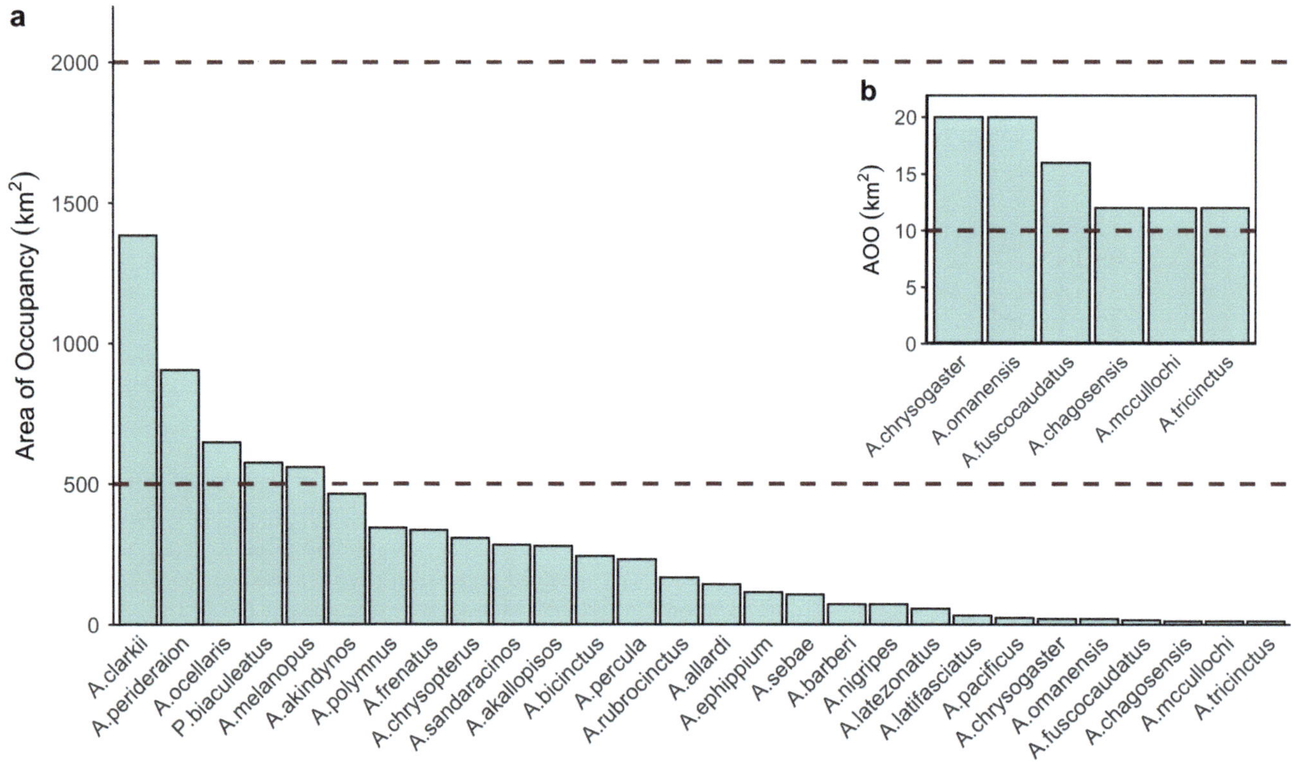

FIGURE 25.5 Histogram of area of occupancy (AOO, km^2) for 28 anemonefish species. IUCN Red List threat categories are indicated by dashed lines following Criterion B2 (IUCN 2019) and show: a) five species as Vulnerable (AOO $<$ 2,000 km^2) and 23 species as Endangered (AOO $<$ 500 km^2). Inset b) six species with smallest AOO and dashed line to show threshold for IUCN Red List category Critically Endangered (AOO $<$ 10 km^2). Three species are within 2 km^2 of this threshold. AOO estimates were made using known occurrence records from fishbase.org (Froese and Pauly 2021) at the recommended reference scale of 4 km^2 (2 × 2 km) for occupied cells to assess Red List criterion B2. AOO is taken as the total area of occupied cells in a uniform grid within a given extent of occupancy (EOO) (IUCN 2001, 2012, 2019). Analysis was performed in R 4.1.1 (R Core Team 2021) with the packages rCAT (Moat 2020) and ConR (Dauby 2020).

A. akindynos abundance in marine reserves at the Solitary Islands. Bonin et al. (2016) found higher numbers of *A. melanopus* in protected areas at the Keppel Islands, following a long period of historic collecting of anemonefishes. Genetic analyses show that despite protection, the effective population size was extremely small (~750 breeders), so there are questions about how big reserves need to be to protect a population large enough to avoid local extinction. It is important to monitor the success of reserves in protecting anemonefishes, as there can be unexpected outcomes. For example, McClanahan (1994) showed that *A. allardi* thrives in fished areas where it has become associated with high numbers of sea urchins that are thriving due to overexploitation of a triggerfish predator. Where marine reserves were established, the anemonefish went locally extinct.

Studies on self-recruitment in anemonefishes show the benefit of marine reserves as local sanctuaries for species (Almany et al. 2007; Jones et al. 2009). A high proportion of juveniles return to the natal population, ensuring protection that carries through to the next generation (Salles et al. 2016, 2020). This has been recorded for five generations of *A. percula* in a small island reserve in Papua New Guinea; however, this population is largely protected by its remoteness, rather than any effective management actions. Marine reserves are likely to be an effective first line of defence for endemic species where levels of self-recruitment are expected to be extremely high (van der Meer et al. 2012; Steinberg et al. 2016). However, recovery from past or widespread impacts is likely to be slow (Frisch et al. 2019). Sato et al. (2017) make the point that reserves may be of limited value from the point of view of supporting anemonefish collecting in fished areas through larval dispersal.

Marine reserves will no doubt be more effective for anemonefishes if specific information on their distribution and abundance is taken into account when selecting sites for reserves. There should be a high priority for *anemonefish hotspots* or places where local species diversity is high or a species of concern is unusually abundant. The Solitary Islands marine park is a good example of this, where locations with high anemonefish densities are well protected (Scott et al. 2011). In Kimbe Bay, Papua New Guinea, *A. percula* is unusually abundant on fringing reefs surrounding small offshore islands, compared to emergent reefs with no islands (Dixson et al. 2011). A marine park planning exercise that prioritized the protection of these reefs because of their habitat diversity could not have been better designed for protecting this iconic species (Green et al. 2009).

25.6.2 Catch Regulations

In our view, given the likely endangered species status of species with very small areas of occurrence, all collecting should be banned and trade deemed illegal. For species clearly overfished in an area, moratoriums on collecting need to be in place, such as has occurred at the Keppel Islands. Such moratoriums may need to remain in place for the long term for sufficient recovery to occur (Frisch and Hobbs 2009; Frisch et al. 2016), and future catch levels would need to be tightly controlled. In circumstances where catches can be reliably controlled, a ban on catching adult fishes and controlled levels of harvesting juveniles would be extremely effective. Juveniles would be of higher value for the aquarium fish market and limited harvest would not impact the size of the breeding population. However, this could only work for anemonefish species that live in large social groups and not for species that only occur in pairs.

25.6.3 Protecting Anemones

Marine reserves and other measures that focus on anemonefishes will not protect anemones from extrinsic disturbances such as global warming, sedimentation, and pollution. There is no silver bullet for protecting anemones from these disturbances, and the increasing levels of anemone bleaching are a huge concern. Banning the collecting of anemones that support anemonefishes should be the number one management priority. Studies on how to reduce the impacts of warming water and pollution on anemones should be the number one research priority. The evidence suggests that the reproductive success of anemonefishes is critically linked, not just to anemones, but to anemones at particularly high-quality locations (Salles et al. 2020). There should be a premium on identifying and targeting the protection of these important sources of future generations.

25.6.4 Captive Breeding

There is a long tradition of captive breeding and release in terrestrial conservation and it seems a very attractive option for enhancing depleted anemonefish populations. Most species have now been bred in captivity (Olivotto and Geffroy 2017), and juveniles can be readily released in the wild where vacant individuals of preferred anemones can be found. Direct supply of aquarium-reared juveniles has the potential to take the pressure off collecting from wild populations (Burke da Silva and Nedoskyo 2016). This topic has been discussed elsewhere in this book (Chapter 22), but there are many reasons why captive breeding and release of anemonefish should be a last resort. In marine systems, captive breeding does not have a good record in reducing wild catches and there is a lot of potential for conflict when it comes to competing sources of income. For restoring anemonefish populations, the emphasis should really be on breeding and out-planting anemones to restore numbers to historic levels. The methods for propagating anemones exist and there is clearly a huge market for anemones that host anemonefish for the aquarium trade (Fraser et al. 2021). Borrowing from this technology for conservation purposes seems like the best way forward, especially if there is scope for artificial selection for bleaching-resistant strains.

25.7 CONCLUSIONS

If we set out to design a fish species that would have a high risk of extinction in an era of rapid environmental change, it would probably look and be like an anemonefish! It would, of course, be so cute that everybody would like to have one, even though it is designed to be rare, so there can never be enough to go around. We would keep coming up with new ways to negatively impact its population size or degrade its environment. We would give it every life-history trait we could think of that would reduce its ability to withstand all of these changes. We would design it to live in a single habitat that we know is highly sensitive to warming water and then crank up the temperature. Perhaps we would put it out on some remote island and hope it can sustain itself there, away from as many human impacts as possible.

The only saving grace to saving Nemo is that we have not seen a species of anemonefish go extinct – yet. That in itself seems like a miracle. We still have 28 species and maybe there are even a few more hidden away for safe keeping. Perhaps this means that they do have a secret for survival that we do not fully understand. Our recent research into larval dispersal shows an incredible ability to navigate their way to suitable habitat, so long as there is some suitable habitat left to find. Perhaps it means that the management actions we have taken for at least some species have been effective. If so, we just need to find ways to expand and tailor these management efforts to all species, whether that is going to be full protection, marine reserves, or sustainable harvest strategies. Or perhaps we have just been lucky that we still have 28 species. One thing the movie *Finding Nemo* teaches us is that survival always depends on a bit of luck.

ACKNOWLEDGEMENTS

We would like to acknowledge the College of Science and Engineering and the ARC Centre of Excellence for Coral Reef Studies at James Cook University for logistic support. We also thank the Mahonia Na Dari Research and Conservation Centre and the Walindi Plantation Resort for supporting our research in Papua New Guinea. Nick Murray, Gilles Dauby, Luis Verde Arregoitia, and Justin Moat all provided valuable insights and guidance to our spatial analysis.

REFERENCES

Allen Coral Atlas. 2020. Imagery, maps and monitoring of the world's tropical coral reefs. https://allencoralatlas.org/ (accessed June 28, 2022).

Allen, G. R. 1975. *Anemonefishes*. Neptune City, NJ: TFH Publications.

Allen, G. R., J. Drew, and S. Fenner. 2010. *Amphiprion pacificus*, a new species of anemonefish (Pomcaentridae) from Fiji, Tonga, Samoa and Wallis Island. *Aquaculture* 16:129–138.

Allen, G. R., J. Drew, and L. Kaufman. 2008. *Amphiprion barberi*, a new species of anemonefish (Pomacentridae) from Fiji, Tonga and Samoa. *Aqua International Journal of Ichthyology* 14:105–114.

Almany, G. R., M. L. Berumen, S. R. Thorrold, S. Planes, and G. P. Jones. 2007. Local replenishment of coral reef fish populations in a marine reserve. *Science* 316:742–744.

Beldade, R., A. Blandin, R. O'Donnell, and S. C. Mills. 2017. Cascading effects of thermally-induced anemone bleaching on associated anemonefish hormonal stress response and reproduction. *Nature Communications* 8:716.

Biondo, M. V. 2018. Importation of marine ornamental fishes to Switzerland. *Golbal Ecology and Conservation* 15: e00418.

Bonin, M. C., H. B. Harrison, D. H. Williamson, A. J. Frisch, P. Saenz-Agudelo, M. L. Beruman, and G. P. Jones. 2016. The role of marine reserves in the replenishment of a locally-impacted population of anemonefish on the Great Barrier Reef. *Molecular Ecology* 25:487–499.

Bradshaw, E. 1994. *Dynamic interactions of the mutualistic association between the spine-cheeked anemonefish Premnas biaculeatus and the host actinian Entacmaea quadricolor.* BSc Honours thesis, James Cook University.

Bridge, T., A. Scott, and D. Steinberg. 2012. Abundance and diversity of anemonefishes and their host sea anemones at two mesophotic sites on the Great Barrier Reef, Australia. *Coral Reefs* 31:1057–1062.

Brooks, M. E., K. Kristensen, K. j. van Benthem, A. Magnusson, C. W. Berg, A. Nielsen, H. Skaug, M. Maechler, and B. M. Bolker. 2017. glmmTMB balances speed and flexibility among packages for zero-inflated generalized linear mixed modeling. *The R Journal* 9(2):378–400.

Burke da Silva, K., and A. Nedoskyo. 2016. Sea anemones and anemonefishes: A match made in heaven. In *The Cnidaria, Past, Present and Future*, eds S. Goffredo and Z. Dubinsky, 425–438. New York: Springer.

Cortese, D., T. Norin, R. Beldade, A. Crespel, S. S. Killen, and S. C. Mills. 2021. Physiological and behavioural effects of anemone bleaching on symbiont anemonefish in the wild. *Functional Ecology* 35:663–674.

Dauby, G. 2020. ConR: Computation of parameters used in preliminary assessment of conservation, status. R package version 1.3.0. https://CRAN.R-project.org/package=ConR (accessed February 2, 2022).

Dauby, G., T. Stévart, V. Droissart, A. Cosiaux, V. Deblauwe, M. Simo-Droissart, M. S. M. Sosef, et al. 2017. ConR: An R package to assist large-scale multispecies preliminary conservation assessments using distribution data. *Ecology and Evolution* 7:11292.

Dixson, D. L., G. P. Jones, P. L. Munday, M. S. Pratchett, M. Srinivasan, S. Planes, and S. R. Thorrold. 2011. Terrestrial chemical cues help coral reef fish larvae locate settlement habitat surrounding islands. *Ecology and Evolution* 1:586–595.

Dixson, D. L., P. L. Munday, and G. P. Jones. 2010. Ocean acidification impairs innate ability of fish to avoid predators. *Ecology Letters* 13:68–75.

Donelson, J. M. 2015. Development in a warm future ocean may enhance performance in some species. *Journal of Experimental Marine Biology and Ecology* 472:119–125.

Drew, J., G. R. Allen, L. Kaufman, and P. H. Barber. 2008. Endemism and regional color and genetic differences in five putatively cosmopolitan reef fishes. *Conservation Biology* 22:965–975.

Dulvy, N. K., J. R. Ellis, N. B. Goodwin, A. Grant, J. D. Reynolds, and S. Jennings. 2004. Methods of assessing extinction risk in marine fishes. *Fish and Fisheries* 5:255–276.

Fautin, D. G. 1991. The anemonefish symbiosis. What is known and what is not. *Symbiosis* 10(1–3): 23–46.

Fautin, D. G., and G. R. Allen. 1992. *Anemone Fishes and Their Host Sea Anemones.* Perth, WA: Western Australian Museum.

Fraser, N., S. Mangubhai, K. Hall, and A. Scott. 2021. Sea anemones in the marine aquarium trade: Market preferences indicate opportunities for mariculture and conservation Aquatic Conservation. *Marine and Freshwater Ecosystems* 31(12): 3594–3606.

Frisch, A. J., and J. P. A. Hobbs. 2009. *Rapid Assessment of Anemone and Anemonefish Populations at the Keppel Islands. A Report to the Great Barrier Reef Marine Park Authority.* Townsville, QLD: Great Barrier Reef Marine Park Authority.

Frisch, A. J., J. P. A. Hobbs, S. T. Hansen, D. H. Williamson, M. C.Bonin, G. P. Jones, and J. R. Rizzari. 2019. Recovery potential of mutualistic anemone and anemonefish populations. *Fisheries Research* 218:1–9.

Frisch, A. L., J. R. Rizzari, K. P. Munkres, and J. P. A. Hobbs. 2016. Anemonefish depletion reduces survival, growth, reproduction and fishery productivity of mutualistic anemone-anemonefish colonies. *Coral Reefs* 35:375–386.

Froese, R., and Pauly, D. (eds). 2021. FishBase.World, Wide Web electronic publication. http://www.fishbase.se (accessed February 2, 2022).

Gainsford, A., G. P. Jones, J. P. A. Hobbs, F. M. Heindler, and L. van Herwerden. 2020. Species integrity, introgression, and genetic variation across a coral reef fish hybrid zone. *Ecology and Evolution* 10:11998–12014.

Green, A., S. E. Smith, G. Lipsett-Moore, C. Groves, N. Peterson, S. Sheppard, P. Lokani, et al. 2009. Designing a resilient network of marine protected areas for Kimbe Bay, Papua New Guinea. *Oryx* 43:488–498.

Haguenauer, A., F. Zuberer, G. Siu, D. Cortese, R. Beldade, and S. C. Mills. 2021. Deep heat: A comparison of water temperature, anemone bleaching, anemonefish density and reproduction between shallow and mesophotic reefs. *Fishes* 6:37.

Hattori, A. 2002. Small and large anemonefishes can coexist using same patchy resources on a coral reef, before habitat destruction. *Journal of Animal Ecology* 71:824–831.

Hawkins, J. P., C. M. Roberts, and V. Clark. 2000. The threatened status of restricted-range coral reef fish species. *Animal Conservation* 3:81–88.

Hayashi, K., and J. D. Reimer. 2020. Five-year study on the bleaching of anemonefish-hosting anemones (Cnidaria: Anthozoa: Actiniaria) in subtropical Okinawajima Island. *Regional Studies in Marine Science* 35:101240.

Hayashi, K., K. Tachihara, and J. D. Reimer. 2019a. Loss of natural coastline influences species diversity of anemonefish and host anemones in the Ryuku Archipelago. *Aquatic Conservation: Marine and Freshwater Ecosystems* 31:15–27.

Hayashi, K., K. Tachihara, and J. D. Reimer. 2019b. Low density populations of anemonefish with low replenishment rates on a reef edge with anthropogenic impacts. *Environmental Biology of Fishes* 102:41–54.

Hess, S., B. J. M. Allan, A. S. Hoey, M. D. Jarrold, A. S. Wenger, and J. L. Rummer. 2019. Enhanced fast-start performance and anti-predator behaviour in a coral reef fish in response to suspended sediment exposure. *Coral Reefs* 38:103–108.

Hess, S., L. J. Prescott, A. S. Hoey, S. A. McMahon, A. S. Wenger, and J. L. Rummer 2017. Species-specific impacts of suspended sediments on gill structure and function in coral reef fishes. *Proceedings of the Royal Society B: Biological Sciences* 284:20171279.

Hess, S., A. S. Wenger, T. D. Ainsworth, and J. L. Rummer. 2015. Exposure of clownfish larvae to suspended sediment levels found on the Great Barrier Reef: Impacts on gill structure and microbiome. *Scientific Reports* 5:10561.

Hobbs, J. P. A., A. J. Frisch, B. M. Ford, M. Thums, P. Saenz-Agudelo, K. A. Furby, and M. L. Beruman. 2013. Taxonomic, spatial and temporal patterns of bleaching in anemones inhabited by anemonefishes. *PLOS ONE* 8:e70966.

Hobbs, J. P. A., G. P. Jones, and P. L. Munday. 2010. Rarity and extinction risk in coral reef angelfishes on isolated islands: Interrelationships among abundance, geographic range size and specialisation. *Coral Reefs* 29:1–11.

Hobbs, J. P. A., G. P. Jones, and P. L. Munday. 2011. Extinction risk in endemic marine fishes. *Conservation Biology* 25:1053–1055.

Hoepner, C. M., C. A. Abott, and K. Burke da Silva. 2019. The ecological importance of toxicity: Sea anemones maintain toxic defence when bleached. *Toxins* 11:266.

Holmberg, R. J., E. Wilcox-Freeburg, A. L. Rhyne, M. F. Tlusty, A. Stebbins, S. W. Nye Jr, A. Honig, et al. 2018. Ocean acidification alters morphology of all otolith types in Clark's anemonefish (*Amphiprion clarkii*). *PeerJ* 7:e6152.

Howell, J., T. L. Goulet, and D. Goulet. 2016. Anemonefish musical chairs and the plight of the two-band anemonefish, *Amphiprion bicinctus*. *Environmental Biology of Fishes* 99:873–886.

IUCN. 2001. *IUCN Red List Categories and Criteria: Version 3.1.* IUCN Species Survival Commission. IUCN, Gland, Switzerland and Cambridge.

IUCN. 2012. *IUCN Red List Categories and Criteria: Version 3.1.* Second edition. Gland, Switzerland and Cambridge.

IUCN Standards and Petitions Committee. 2019. *Guidelines for Using the IUCN Red List Categories and Criteria. Version 14.* Prepared by the Standards and Petitions Committee. http://www.iucnredlist.org/documents/RedListGuidelines.pdf (accessed February 2, 2022).

Jones, A. M., S. Gardner, and W. Sinclair. 2008. Losing 'Nemo': Bleaching and collection appear to reduce inshore populations of anemonefishes. *Journal of Fish Biology* 73:753–761.

Jones, G. P. 2015. Mission impossible: Unlocking the secrets of dispersal in coral reef fishes. In *Ecology of Fishes on Coral Reefs: The Functioning of an Ecosystem in a Changing World*, ed. C. Mora, 16–27. Cambridge: Cambridge University Press.

Jones, G. P., G. R. Almany, G. R. Russ, P. F. Sale, R. S. Steneck, M. J. H. van Oppen, and B. L. Willis. 2009. Larval retention and connectivity among populations of corals and reef fishes: History, advances and challenges. *Coral Reefs* 28:307–325.

Jones, G. P., M. J. Caley, and P. L. Munday. 2002. Rarity in coral reef fish communities. In *Coral Reef Fishes. Dynamics and Diversity in a Complex Ecosystem*, ed. P. F. Sale, 81–101. San Diego, CA: Academic Press.

Liu, P. J., M. C. Hsin, Y. H. Huang, T-Y. Fan, P-J. Meng, C-C. Lu, and H-J. Lin.2015. Nutrient enrichment coupled with sedimentation favors sea anemones over corals. *PLOS ONE* 10(4): e0125175.

Lönnstedt, O. M., and A. J. Frisch. 2014. Habitat bleaching disrupts threat responses and persistence in anemonefishes. *Marine Ecology Progress Series* 517:265–270.

Madduppa, H. H., J. Timm, and M. Kochzius. 2018. Reduced genetic diversity in the clown anemonefish Amphiprion ocellaris in exploited reefs of Spermonde Archipelago, Indonesia. *Frontiers in Marine Science* 5:80.

Madduppa, H. H., K. von Juterzenka, M. Syakir, and M. Kochzius. 2014. Socio-economy of marine ornamental fishery and its impact on the population structure of the clown anemonefish *Amphiprion ocellaris* and its host anemones in Spermonde Archipelago, Indonesia. *Ocean and Coastal Management* 100:41–50.

Maison, K. A., and K. S. Graham. 2016. *Status Review Report: Orange Clownfish (Amphiprion percula).* NOAA Technical Memorandum NMFS-PIFSC-52.

Malcolm, H. A. C., and A. Scott. 2017. Range extensions in anemonefishes and host sea anemones in eastern Australia: Potential constraints to tropicalisation. *Marine and Freshwater Research* 68:1224–1232.

McClanahan, T. R. 1994. Endemic survives adversity. *Coral Reefs* 13:104.

McClanahan, T., V. Munbodhe, J. Naggea, N. Muthiga, and R. Bhagooli. 2021. Rare coral and reef fish species status, possible extinctions, and associated environmental perceptions in Mauritius. *Conservation Science and Practice* 2021:e527.

McClenachan, L., A. B. Cooper, K. E. Carpenter, and N. K. Dulvy. 2012. Extinction risk and bottlenecks in the conservation of charismatic marine species. *Conservation Letters* 5:73–80.

McLeod, I. M., J. L. Rummer, T. D. Clark, G. P. Jones, M. I. McCormick, A. S. Wenger, P. L. Munday. 2013. Climate change and the performance of larval coral reef fishes: The interaction between temperature and food availability. *Conservation Physiology* 1:cot024.

Militz, T. A., S. Foale, J. Kinch, and P. C. Southgate. 2018. Natural rarity places clownfish colour morphs at risk of targeted and opportunistic exploitation in a marine aquarium fishery. *Aquatic Living Resources* 31:18.

Miller, G. M., F. J. Kroon, S. Metcalfe, and P. L. Munday. 2015. Temperature is the evil twin: Effects if increased temperature and ocean acidification on reproduction in a reef fish. *Ecological Applications* 25:603–620.

Miller, G. M., S. A. Watson, M. I. McCormick, and P. L. Munday. 2013. Increased CO_2 stimulates reproduction in a coral reef fish. *Global Change Biology* 19:3037–3045.

Moat, J. 2020. *rCAT: Conservation Assessment Tools.* R package version 0.1.6. https://CRAN.R-project.org/package=rCAT (accessed February 2, 2022).

Munday, P. L. 2004. Habitat loss, resource specialization, and extinction on coral reefs. *Global Change Biology* 10:1642–1647.

Munday, P. L., D. L. Dixson, J. M. Donelson, G. P. Jones, M. S. Pratchett, G. V. Devitsina, and K. B. Døving. 2009. Ocean acidification impairs olfactory discrimination and homing ability of a marine fish. *Proceedings of the National Academy of Science* 106:1848–1852.

Munday, P. L., D. L. Dixson, M. I. McCormick, M. Meekan, M. C. O. Ferrari, and D. P. Chivers. 2010. Replenishment of fish populations is threatened by ocean acidification. *Proceedings of the National Academy of Sciences* 107(29):12930–12934.

Munday, P. L., D. L. Dixson, M. J. Welch, D. P. Chivers, P. Domenici, M. Grosell, R. M. Heuer, et al. 2020. Methods matter in repeating ocean acidification studies. *Nature* 586:E20–E24.

Munday, P. L., M. D. Jarrold, and I. Nagelkerken. 2019. Ecological effects of elevated CO2 on marine and freshwater fishes: From individual to community effects. In *Fish Physiology, Vol. 37*, eds Grosell, M., Munday, P. L., Farrell, A. P., and C. J. Brauner, 323–368. Amsterdam: Elsevier Inc.

Munday, P. L., G. P. Jones, M. S. Pratchett, and A. J. Williams. 2008. Climate change and the future for coral reef fishes. *Fish and Fisheries* 9:261–285.

Nanninga, G. B., P. Saenz-Agudelo, P. Zhan, L. Hoteit, and M. L. Berumen. 2015. Not finding Nemo: Limited reef-scale retention in a coral reef fish. *Coral Reefs* 34:383–392.

Nilsson, G. E., D. L. Dixson, P. Domenici, M. I. McCormick, C. Sørensen, S-A. Watson, and P. L. Munday. 2012. Near-future carbon dioxide levels alter fish behaviour by interfering with neurotransmitter function. *Nature Climate Change* 2(3):201–204.

Norin, T., S. C. Mills, A. Crespel, D. Cortese, S. S. Killen, and R. Beldade. 2018. Anemone bleaching increases the metabolic demands of symbiont anemonefish. *Proceedings of the Royal Society B* 285:20180282.

Nowicki, J. P., G. M. Miller, and P. L. Munday. 2012. Interactive effects of elevated temperature and CO_2 on foraging behavior of juvenile coral reef fish. *Journal of Experimental Marine Biology and Ecology* 412:46–51.

Okemwaa, G. M., B. Kaunda-Arara, E. N. Kimani, and B. Ogutu. 2016. Catch composition and sustainability of the marine aquarium fishery in Kenya. *Fisheries Research* 183:19–31.

Olivotto, I., and B. Geffroy. 2017. Clownfish. In *Marine Ornamental Species Aquaculture, First Edition*, eds R. Calado, I. Olivotto, M. P. Oliver and G. J. Holt, 177–199. Hoboken, NJ: John Wiley & Sons Ltd.

Ollerton, J., D. McCollin, D. G. Fautin and G. R. Allen. 2007. Finding NEMO: Nestedness engendered by mutualistic organization in anemonefish and their hosts. *Proceedings of the Royal Society B* 274:591–598.

Pryor, S. H., R. Hill, D. L. Dixson, N. J. Fraser, B. P. Kelaher, and A. Scott. 2020. Anemonefish facilitate bleaching recovery in a host sea anemone. *Scientific Reports* 10:18586.

R Core Team. 2021. *R: A Language and Environment for Statistical Computing.* R Foundation for Statistical Computing, Vienna, Austria. https://www.R-project.org/ (accessed February 2, 2022).

Rabinowitz, D. 1981. Seven forms of rarity. In *The Biological Aspects of Rare Plant Conservation*, ed. H. Synge, 205–217. New York: John Wiley & Sons Ltd.

Rauch, S. D. III. 2015. Endangered and threatened wildlife and plants; Notice of 12-month finding on a petition to list the orange clownfish as threatened or endangered under the Endangered Species Act. *Federal Register* 80:51235–51247.

Roberts, C. M., and J. P. Hawkins. 1999. Extinction risk in the sea. *Trends in Ecology and Evolution* 14:241–246.

Roelofs, A., and R. Silcock. 2008. *A Sustainability Assessment of Marine Fish Species Collected in the Queensland Marine Aquarium Trade.* Brisbane, QLD: Department of Primary Industries and Fisheries.

Saenz-Agudelo, P., G. P. Jones, S. R. Thorrold, and S. Planes. 2011. Detrimental effects of host anemone bleaching on anemonefish populations. *Coral Reefs* 30:497–506.

Salles, O. C., G. R Almany, M. L. Berumen, G. P. Jones, P. Saenz-Agudelo, M. Srinivasan, S. R. Thorrold, et al. 2020. Strong habitat and weak genetic effects shape the lifetime reproductive success in a wild clownfish population. *Ecology Letters* 23:265–273.

Salles, O. C., B. Pujol, J. A. Maynard, G. R. Almany, M. L.Beruman, G. P. Jones, P. Saenz-Agudelo, et al. 2016. First genealogy for a wild marine fish population reveals multi-generational philopatry. *Proceedings of the National Academy of Science* 113:13245–13250.

Sato, M., K. Honda, W. H. Uy, D. I. Baslot, T. G. Genovia, Y. Nakamura, L. P. C. Bernardo, et al. 2017. Marine protected area restricts demographic connectivity: Dissimilarity in a marine environment can function as an ecological barrier. *Ecology and Evolution* 7:7859–7817.

Scott, A., and A. H. Baird. 2015. Trying to find Nemo: Low abundance of sea anemones and anemonefishes on central and southern mid-shelf reefs in the Great Barrier Reef. *Marine Biodiversity* 45:327–331.

Scott, A., H. A. Malcolm, C. Damiano, and D. L. Richardson. 2011. Long-term increases in abundance of anemonefish and their host anemones in an Australian marine protected area. *Marine and Freshwater Research* 62:187–196.

Scott, A., K. J. W. Rushworth, S. J. Dalton, and S. D. A. Smith. 2016. Subtropical anemonefish *Amphiprion latezonatus* recorded in two additional host sea anemone species. *Marine Biodiversity* 46:327–328.

Shuman, C. S., G. Hodgson, and R. F. Ambrose. 2005. Population impacts of collecting sea anemones and anemonefish for the marine aquarium trade in the Philippines. *Coral Reefs* 24:564–573.

Simpson, S. D., P. L. Munday, M. L. Wittenrich, R. Manassa, D. L. Dixson, M. Gagliano, and H. Y. Yan. 2011. Ocean acidification erodes crucial auditory behaviour in a marine fish. *Biology Letters* 7:917–920.

Srinivasan, M. 1997. *Ecology and Life-Histories of Anemonefishes: A Multi-Scale Study of Patterns and Processes.* BSc Honours thesis, James Cook University.

Srinivasan, M., G. P. Jones, and M. J. Caley. 1999. Experimental evaluation of the roles of habitat selection and interspecific competition in determining patterns of host use in two anemonefishes. *Marine Ecology Progress Series* 186:283–292.

Steinberg, R., M. van der Meer, E. Walker, M. L. Berumen, J.-P. A. Hobbs, and L. van Herwerden. 2016. Genetic connectivity and self-replenishment of inshore and offshore populations of the endemic anemonefish, *Amphiprion latezonatus. Coral Reefs* 35: 959–970.

Steinberg, R. K., M. H. van der Meer, M. S. Pratchett, L. van Herwerden, and J. P. A. Hobbs. 2020. Keep your friends close and your anemones closer – ecology of endemic wideband anemonefish, *Amphiprion latezonatus. Environmental Biology of Fishes* 103:1513–1526.

Timm, J., M. Figiel, and M. Kochzius. 2008. Contrasting patterns in species boundaries and evolution of anemonefishes (Amphiprioninae, Pomacentridae) in the centre of marine biodiversity. *Molecular Phylogenetics and Evolution* 49:268–276.

van der Meer, M. H., J. P. A. Hobbs, G. P. Jones, and L. van Herwerden. 2012. Genetic connectivity among and self-replenishment within island populations of a restricted range subtropical reef fish. *PLOS One* 7: e49660.

Wabnitz, C., M. Taylor, E. Green, and T. Razak. 2003. *From Ocean to Aquarium: The Global Trade in Marine Ornamental Species.* Cambridge: UNEP-WCMC.

Wenger, A. S., K. E. Fabricius, G. P. Jones, and J. E. Brodie. 2015. Sedimentation, eutrophication and pollution: Effects on coral reef fishes. In *Ecology of Fishes on Coral Reefs: The Functioning of an Ecosystem in a Changing World*, ed. Mora, C., 145–153. Cambridge: Cambridge University Press.

Wenger, A. S., M. I. McCormick, G. G. K. Endo, I. M. McLeod, F. Kroon, and G. P. Jones. 2014. Suspended sediment alters larval development in a coral reef fish. *Journal of Experimental Biology* 217:1122–1128.

Wenger, A. S., C. A. Rawson, S. Wilson, S. J. Newman, M. J. Travers, S. Atkinson, N. Browne, et al. 2017. Management strategies to minimize the dredging impacts of coastal development on fish and fisheries. *Fish and Fisheries* 18:967–985.

Zhang, D. 2021. *rsq: R-Squared and Related Measures*. R package version 2.2. https://CRAN.R-project.org/package=rsq (accessed February 2, 2022).

Conclusion

Anemonefish Revisited – Future Questions for a Model System

Timothy Ravasi, Geoffrey P. Jones, and Vincent Laudet

At the end of this book, we think it is important to highlight some themes that connect the different chapters and offer promising and exciting future research paths, using anemonefish as model organisms. By model organisms we do not mean that anemonefishes are somehow typical of all fishes, but that their characteristics make them perfect for expanding the boundaries of certain fields of research, to work out what fishes can do, and how they can do it. Our perspectives are undoubtedly biased by our respective experiences and interests in science, but we would like to highlight ten key priorities for the next phase of anemonefish research.

1. Even though a large amount of work has been done to understand the life history and relationships among individuals within an anemonefish colony, much remains to be done. First, our knowledge is biased toward the most studied species, namely the clownfish *Amphiprion percula* and *A. ocellaris*. It seems clear that the structure of the colonies, with a dominant female, a subdominant male, and juveniles ranked by order, and a sex change system allowing a queue for access to reproduction, is existing in all species. But the respective strength of the female, the male, and the juvenile, the influence of the size and species of the sea anemone, and the precise rules governing these colonies will vary from one species to another and this diversity remains poorly studied. Just by scrutinizing the six species present in Okinawa waters (*A. ocellaris*, *A. clarkii*, *A. frenatus*, *A. polymnus*, *A. perideraion*, and *A. sandaracinos*) we realized that there are many aspects that we still do not fully understand and that will require further studies. In addition, many basic ecological questions are not solved: what are the predators of anemonefishes when they are settled in their sea anemone? Is predation pressure an important aspect explaining their behavior or pigmentation? What are they eating? What are the relative roles of competition for food and living space? What is their level of infection by parasites? Is this changing from one environment to another? All these questions remain to be solved and will require thorough studies.
2. Despite decades of work on the symbiotic relationship between anemonefishes and anemones, many aspects of the relationship remain a mystery. We do not fully understand the mechanisms by which the anemonefish avoids being stung by sea anemone tentacles. There is obviously a specificity of interactions with some species being generalists (that is living in any sea anemone) while other species associate with few sea anemones only, but the respective importance of ecological or physiological processes in explaining these specificity rules remain to be fully understood. Also, the symbiosis is in fact a "ménage à trois" since the sea anemone themselves are symbiotic with dinoflagellates of the family Symbiodiniaceae. However, if and how the presence of anemonefish influences Symbiodiniaceae and vice versa is still poorly understood. Moreover, anemonefish colonies are even more complex because many other organisms like fish, shrimps, or crabs are often present. Also, some anemonefish species are willing to share their sea anemone hosts with other anemonefish species (the ultimate example being the case of *A. chrysopterus* and *A sandaracinos* that can reproduce and give rise to the hybrid *A. leucokranos*), whereas others never do this, and we still do not know why there are such differences. We have started to realize that there are precise rules governing some associations, but here again, we are far from understanding the complex mechanisms at work.
3. The pigmentation patterns of anemonefish coupled with their bold behavior make them iconic and popular species, but what is the precise function of these brilliant colors? Are they used for camouflage? For aposematism or the warning of putative predators of the toxicity of their host? Or for social interactions? Opinions diverge quite strongly in the field, and we can anticipate that future work will soon be done favouring one or several of these hypotheses. Furthermore, it is important to investigate "how" these color patterns emerge. As discussed in the relevant chapter, there is a strong

antero-posterior system that controls the formation of this pigment pattern and at first glance, the rules governing it are quite different from the rules deciphered using other model fish species, such as zebrafish. Are genes controlling pigment pattern formation in anemonefishes different from those of zebrafish? Is there some sort of tinkering mechanism that is using the same genes but in a different way to give rise to these different patterns? This is still unknown. Lastly, there is an incredible diversity in the pigment patterns in anemonefishes with significant interspecific and intraspecific variation as well as many direct environmental effects giving rise to phenotypic plasticity. What are the rules governing these effects? How is the environment acting on the patterning system? Is there any connection between the pigmentation and the vision as one can imagine if there is a social function?

4. Many aspects of the whole life history of anemonefish are still mysterious. The larval stage is probably the most vexing enigma in this area. It is very difficult to imagine how the larvae can survive in the pelagic environment when they are very sensitive to condition changes in laboratory husbandries. We simply ignore almost everything about larvae in the ocean. What do they eat? Are they actively swimming? Are they living in groups? How do they know where they are and in which direction to swim? We are convinced they are not simply passive propagules carried by the currents, but then how to study this? The recent observation that many fish larvae are in surface slicks that act as nurseries suggests that in the pelagic realm there is a hidden complexity that largely escaped our detection so far, and this also very much applies to anemonefish larvae. They are captured very rarely, but this is not surprising given that anemonefishes are not very numerous and they lay a lot fewer eggs than many other fish species (for example, a couple laying clutch twice a month with a maximum of 1,000 eggs each time and the breeding season lasting seven months represents only 14,000 eggs per breeding pair per year in Okinawa). Other aspects of the life cycle that require more studies include: how are the young juveniles accepted into colonies? How is the sex change triggered in physiological terms and how do the sex hormones act on the behavior of the fishes? And last but not the least, what are the biological mechanisms that allow anemonefishes to live so long?
5. On the topic of larvae, anemonefishes have been an outstanding model group for providing the first empirical estimates of marine larval fish dispersal and population connectivity in marine fishes. For the first time, they have enabled us to understand the scale at which some marine populations work and how they should be managed. The relevant chapter shows that we now have data on self-recruitment and connectivity for half of the described anemonefish species. It is going to be exciting to see this expanded to the others, given the incredible diversity that has already been described. Genetic parentage analysis is going to be supplemented by a host of new molecular techniques to detect short and long-distance dispersal events. By finding the connections between adults and offspring in field populations, we can now realistically model populations over multiple generations, to predict how they will respond to, and evolve in response to anthropogenic pressures. We will also be able to quantify the inter-generational conservation benefits of management actions such as marine reserves, for which we are still mostly reliant on short-term data.
6. We have seen that clownfish colonies are complex microsocieties. We know that visual communication is important in regulating the relations between those fishes. However, has been known for a long time now that acoustic communication is also of great importance. What is the respective importance of vision and hearing? Is there also chemical communication between the individuals? What is the impact of the personality of individual fishes in the communication within these microsocieties? This brings us to the more general question of the molecular underpinnings of behavior. We do think that anemonefishes are powerful models to better understand the molecular and neurobiological basis of complex behaviors. The advantage of these models is that this question can be tackled both in the lab and in the wild and that the repertoire of possible behaviors is large.
7. The relatively easy rearing protocols available for anemonefishes make them a great model for performing aquarium-based manipulative experiments and studying the potential of coral reef fishes to acclimate, adapt, or move in response to anthropogenic stressors such as climate change. Multiple clutches are produced within a breeding season, and they are therefore also a good model for transgenerational studies. Their mutualistic association with the anemone and the fact that they tend to spend their entire life and reproductive cycle in the same sea anemone open up the possibility to perform long-term monitoring studies in-situ and make them a good model to study the effect of climate stressors in the wild. The recent advances in molecular and genomic technologies for non-model species (discussed in the following) also dramatically increased our knowledge of how anemonefishes may acclimate, adapt, and evolve under the pressure of anthropogenic stressors at an unprecedented molecular resolution. Because of

these, we strongly think anemonefishes are great systems to understand the impacts of rapid environmental change and they will help us to answer some of the unsolved questions still remaining, for example, what is the parental contributions to the observed acclimation? And is there an interplay between epigenetics mechanisms that facilitate the rapid transgenerational adaptation to future climate?

8. In the past ten years, several advances in genomics technologies such as chromosome-scale genomes assemblies, transcriptomics, and proteomics applied to non-model species substantially transformed the role of anemonefish as a group in the understanding of evolution, ecology, and genetics of coral reef fishes. The availability of several chromosomes-scale genomes for anemonefish species, for example for *A. percula*, *A. ocellaris*, and *A. clarkii* among others, allowed, for the first time, to define an accurate phylogeny of this group of fishes and highlighted interesting aspects of their mutualistic lifestyle with anemones, their unique color patterns, and their development. Transcriptomics and proteomics measurements for different developmental stages or under climate stressors increased our knowledge of the molecular mechanisms controlling anemonefishes responses to future climate conditions, sex change, social structure, and development. The currently available and future genome-wide datasets might have a huge impact on anemonefish research and will further establish these fishes as important model organisms to study coral reef fishes. However, it remains to be seen if these genomic resources will be useful to establish methods commonly use in model organisms, such as CRISPR/Cas9, to perform functional studies. And also, does the mobilization of transposable elements (TEs) have any important underlying role in the adaption of coral reef fishes to a new environment?
9. Anemonefishes have the potential to provide the first reliable understanding of the extinction risks of marine fishes and the effectiveness of actions taken to minimize these risks. They allow the measuring of all the critical parameters for successfully applying the IUCN endangered species criteria, including absolute breeding population size, geographic range and area of occupancy, and the number and size of small subpopulations. We can undertake realistic quantitative population viability analyses to model populations under different scenarios designed to prioritize conservation measures. While anemonefishes have many characteristics that we know exacerbate the risk of extinction, such as extreme habitat specialization and small area of occupancy, we also know that species with small, geographically restricted populations can and do persist. Anemonefishes have many adaptations to survive, including an incredible ability to find suitable habitat, whether near home or on distant reefs, as long as that habitat exists. We will need to adapt our conservation practices to find effective ways to protect or restore anemone habitat if we want to ensure we do not lose any of the 28 precious anemonefish species over the challenging decades to come.
10. Although there are numerous reasons to choose anemonefish to fast-track scientific discovery, there are still many hurdles to cross. As discussed in the relevant chapter the engineering of mutants using CRISPR/Cas9 and the construction of transgenic lines are now in reach, but the poor survival rate of larvae after such procedure is still a big limitation and more work on improving husbandry and larval rearing will be important. Another limitation is the generation time (ca. 18 months) which is problematic when compared to other model species for constructing CRISPR/Cas9 lines. Efforts to improve the anemonefish as a model system will continue and will, we hope, contribute to shaping an active community that will share methods, reagents, and new approaches.

No doubt other researchers could add to this list of priorities and interesting questions. Whichever direction we take, we must continuously improve and develop new ways to use anemonefishes as model systems. Clearly, anemonefish will never be a biological model as efficient as the zebrafish or the medaka. However, we are convinced they have great potential as a marine fish model to allow us to answer a wide range of questions that could not currently be tackled for most other groups. By focusing on anemonefishes, we will not only learn more about the ecology, biology, and evolution of these unique fishes but also develop new models that will ultimately be tested on other marine fishes. No doubt, the next book on anemonefishes will answer many of the questions posed here. We will eagerly await to see what the new questions will be.

Index

For Product Safety Concerns and Information please contact our EU representative GPSR@taylorandfrancis.com Taylor & Francis Verlag GmbH, Kaufingerstraße 24, 80331 München, Germany

Batch number: 10731517

Printed by Printforce, the Netherlands